T0178161

Deep Learning for NLP and Speech Recognition

Uday Kamath • John Liu • James Whitaker

Deep Learning for NLP and Speech Recognition

 Springer

Uday Kamath
Digital Reasoning Systems Inc.
McLean
VA, USA

John Liu
Intelluron Corporation
Nashville
TN, USA

James Whitaker
Digital Reasoning Systems Inc.
McLean
VA, USA

ISBN 978-3-030-14598-9   ISBN 978-3-030-14596-5   (eBook)
https://doi.org/10.1007/978-3-030-14596-5

This Springer imprint is published by the registered company Springer Nature Switzerland AG.
The registered company address is: Gewerbestrasse 11, 6330 Cham, Switzerland

To my parents Krishna and Bharathi, my wife Pratibha, the kids Aaroh and Brandy, my family and friends for their support.
—Uday Kamath

To Catherine, Gabrielle Kaili-May, Eugene and Tina for inspiring me always.
—John Liu

To my mother Nancy for her constant support, my family, and my friends who have blessed my life with love.
—James Whitaker

Foreword

The publication of this book is a perfect timing. Existing books on deep learning either focus on theoretical aspects or are largely manuals for tools. But this book presents an unprecedented analysis and comparison of deep learning techniques for natural language and speech processing, closing the substantial gap between theory and practice. Each chapter discusses the theory underpinning the topics, and an exceptional collection of 13 case studies in different application areas is presented. They include classification via distributed representation, summarization, machine translation, sentiment analysis, transfer learning, multitask NLP, end-to-end speech, and question answering. Each case study includes the implementation and comparison of state-of-the-art techniques, and the accompanying website provides source code and data. This is extraordinarily valuable for practitioners, who can experiment firsthand with the methods and can deepen their understanding of the methods by applying them to real-world scenarios.

This book offers a comprehensive coverage of deep learning, from its foundations to advanced and recent topics, including word embedding, convolutional neural networks, recurrent neural networks, attention mechanisms, memory-augmented networks, multitask learning, domain adaptation, and reinforcement learning. The book is a great resource for practitioners and researchers both in industry and academia, and the discussed case studies and associated material can serve as inspiration for a variety of projects and hands-on assignments in a classroom setting.

Associate Professor at GMU   Carlotta Domeniconi, PhD
Fairfax, VA, USA
February 2019

Natural language and speech processing applications such as virtual assistants and smart speakers play an important and ever-growing role in our lives. At the same time, amid an increasing number of publications, it is becoming harder to identify the most promising approaches. As the Chief Analytics Officer at Digital Reasoning and with a PhD in Big Data Machine Learning, Uday has access to both the practical and research aspects of this rapidly growing field. Having authored

Mastering Java Machine Learning, he is uniquely suited to break down both practical and cutting-edge approaches. This book combines both theoretical and practical aspects of machine learning in a rare blend. It consists of an introduction that makes it accessible to people starting in the field, an overview of state-of-the-art methods that should be interesting even to people working in research, and a selection of hands-on examples that ground the material in real-world applications and demonstrate its usefulness to industry practitioners.

Research Scientist at DeepMind   Sebastian Ruder, PhD
London, UK
February 2019

A few years ago, I picked up a few text-books to study topics related to artificial intelligence—such as natural language processing and computer vision. My memory of reading these text-books largely consisted of staring helplessly out of the window. Whenever I attempted to implement the described concepts and math, I wouldn't know where to start. This is fairly common in books written for academic purposes; they mockingly leave the actual implementation "as an exercise to the reader." There are a few exceptional books that try to bridge this gap, written by people who know the importance of going beyond the math all the way to a working system. This book is one of those exceptions—with it's discussions, case studies, code snippets, and comprehensive references, it delightfully bridges the gap between learning and doing.

I especially like the use of Python and open-source tools out there. It's an opinionated take on implementing machine learning systems—one might ask the following question: "Why not X," where X could be Java, C++, or Matlab? However, I find solace in the fact that it's the most popular opinion, which gives the readers an immense support structure as they implement their own ideas. In the modern Internet-connected world, joining a popular ecosystem is equivalent to having thousands of humans connecting together to help each other—from Stack Overflow posts solving an error message to GitHub repositories implementing high-quality systems. To give you perspective, I've seen the other side, supporting a niche community of enthusiasts in machine learning using the programming language Lua for several years. It was a daily struggle to do new things—even basic things such as making a bar chart—precisely because our community of people was a few orders of magnitude smaller than Python's.

Overall, I hope the reader enjoys a modern, practical take on deep learning systems, leveraging open-source machine learning systems heavily, and being taught a lot of "tricks of the trade" by the incredibly talented authors, one of whom I've known for years and have seen build robust speech recognition systems.

Research Engineer at Facebook AI Research (FAIR)   Soumith Chintala, PhD
New York, NY, USA
February 2019

Preface

Why This Book?

With the widespread adoption of deep learning, natural language processing (NLP), and speech applications in various domains such as finance, healthcare, and government and across our daily lives, there is a growing need for one comprehensive resource that maps deep learning techniques to NLP and speech and provides insights into using the tools and libraries for real-world applications. Many books focus on deep learning theory or deep learning for NLP-specific tasks, while others are cookbooks for tools and libraries. But, the constant flux of new algorithms, tools, frameworks, and libraries in a rapidly evolving landscape means that there are few available texts that contain explanations of the recent deep learning methods and state-of-the-art approaches applicable to NLP and speech, as well as real-world case studies with code to provide hands-on experience. As an example, you would find it difficult to find a single source that explains the impact of neural attention techniques applied to a real-world NLP task such as machine translation across a range of approaches, from the basic to the state-of-the-art. Likewise, it would be difficult to find a source that includes accompanying code based on well-known libraries with comparisons and analysis across these techniques.

This book provides the following all in one place:

- A comprehensive resource that builds up from elementary deep learning, text, and speech principles to advanced state-of-the-art neural architectures

- A ready reference for deep learning techniques applicable to common NLP and speech recognition applications

- A useful resource on successful architectures and algorithms with essential mathematical insights explained in detail

- An in-depth reference and comparison of the latest end-to-end neural speech processing approaches

- A panoramic resource on leading-edge transfer learning, domain adaptation, and deep reinforcement learning architectures for text and speech

- Practical aspects of using these techniques with tips and tricks essential for real-world applications

- A hands-on approach in using Python-based libraries such as Keras, TensorFlow, and PyTorch to apply these techniques in the context of real-world case studies

In short, the primary purpose of this book is to provide a single source that addresses the gap between theory and practice using case studies with code, experiments, and supporting analysis.

Who Is This Book for?

This book is intended to introduce the foundations of deep learning, natural language processing, and speech, with an emphasis on application and practical experience. It is aimed at NLP practitioners, graduate students in Engineering and Computer Science, advanced undergraduates, and anyone with the appropriate mathematical background who is interested in an in-depth introduction to the recent deep learning approaches in NLP and speech. Mathematically, we expect that the reader is familiar with multivariate calculus, probability, linear algebra, and Python programming.

Python is becoming the *lingua franca* of data scientists and researchers for performing experiments in deep learning. There are many libraries with Python-enabled bindings for deep learning, NLP, and speech that have sprung up in the last few years. Therefore, we use both the Python language and its accompanying libraries for all case studies in this book. As it is unfeasible to fully cover every topic in a single book, we present what we believe are the key concepts with regard to NLP and speech that will translate into application. In particular, we focus on the intersection of those areas, wherein we can leverage different frameworks and libraries to explore modern research and related applications.

What Does This Book Cover?

The book is organized into three parts, aligning to different groups of readers and their expertise. The three parts are:

- **Machine Learning, NLP, and Speech Introduction**. The first part has three chapters that introduce readers to the fields of NLP, speech recognition, deep learning, and machine learning with basic hands-on case studies using Python-based tools and libraries.

- **Deep Learning Basics**. The five chapters in the second part introduce deep learning and various topics that are crucial for speech and text processing, including word embeddings, convolutional neural networks, recurrent neural networks, and speech recognition basics.
- **Advanced Deep Learning Techniques for Text and Speech**. The third part has five chapters that discuss the latest research in the areas of deep learning that intersect with NLP and speech. Topics including attention mechanisms, memory-augmented networks, transfer learning, multitask learning, domain adaptation, reinforcement learning, and end-to-end deep learning for speech recognition are covered using case studies.

Next, we summarize the topics covered in each chapter.

- In the **Introduction**, we introduce the readers to the fields of deep learning, NLP, and speech with a brief history. We present the different areas of machine learning and detail different resources ranging from books to datasets to aid readers in their practical journey.

- The **Basics of Machine Learning** chapter provides a refresher of basic theory and important practical concepts. Topics covered include the learning process, supervised learning, data sampling, validation techniques, overfitting and underfitting of the models, linear and nonlinear machine learning algorithms, and sequence data modeling. The chapter ends with a detailed case study using structured data to build predictive models and analyze results using Python tools and libraries.

- In the **Text and Speech Basics** chapter, we introduce the fundamentals of computational linguistics and NLP to the reader, including lexical, syntactic, semantic, and discourse representations. We introduce language modeling and discuss applications such as text classification, clustering, machine translation, question answering, automatic summarization, and automated speech recognition, concluding with a case study on text clustering and topic modeling.

- The **Basics of Deep Learning** chapter builds upon the machine learning foundation by introducing deep learning. The chapter begins with a fundamental analysis of the components of deep learning in the multilayer perceptron (MLP), followed by variations on the basic MLP architecture and techniques for training deep neural networks. As the chapter progresses, it introduces various architectures for both supervised and unsupervised learning, such as multiclass MLPs, autoencoders, and generative adversarial networks (GANs). Finally, the material is combined into the case study, analyzing both supervised and unsupervised neural network architectures on a spoken digit dataset.

- For the **Distributed Representations** chapter, we investigate distributional semantics and word representations based on vector space models such as word2vec and GloVe. We detail the limitations of word embeddings including antonymy and polysemy and the approaches that can overcome them. We also investigate extensions of embedding models, including subword, sentence, concept, Gaussian, and hyperbolic embeddings. We finish the chapter with a case study that dives into how embedding models are trained and their applicability to document clustering and word sense disambiguation.

- The **Convolutional Neural Networks** chapter walks through the basics of convolutional neural networks and their applications to NLP. The main strand of discourse in the chapter introduces the topic by starting from fundamental mathematical operations that form the building blocks, explores the architecture in increasing detail, and ultimately lays bare the exact mapping of convolutional neural networks to text data in its various forms. Several topics such as classic frameworks from the past, their modern adaptations, applications to different NLP tasks, and some fast algorithms are also discussed in the chapter. The chapter ends with a detailed case study using sentiment classification that explores most of the algorithms mentioned in the chapter with practical insights.

- The **Recurrent Neural Networks** chapter presents recurrent neural networks (RNNs), allowing the incorporation of sequence-based information into deep learning. The chapter begins with an in-depth analysis of the recurrent connections in deep learning and their limitations. Next, we describe basic approaches and advanced techniques to improve performance and quality in recurrent models. We then look at some applications of these architectures and their application in NLP and speech. Finally, we conclude with a case study applying and comparing RNN-based architectures on a neural machine translation task, analyzing the effects of the network types (RNN, GRU, LSTM, and Transformer) and configurations (bidirectional, number of layers, and learning rate).

- The **Automatic Speech Recognition** chapter describes the fundamental approaches to automatic speech recognition (ASR). The beginning of the chapter focuses on the metrics and features commonly used to train and validate ASR systems. We then move toward describing the statistical approach to speech recognition, including the base components of an acoustic, lexicon, and language model. The case study focuses on training and comparing two common ASR frameworks, CMUSphinx and Kaldi, on a medium-sized English transcription dataset.

- The **Attention and Memory-Augmented Networks** chapter introduces the reader to the attention mechanisms that have played a significant role in neural techniques in the last few years. Next, we introduce the related topic of memory-augmented networks. We discuss most of the neural-based memory networks, ranging from memory networks to the recurrent entity networks in enough detail for the user to understand the working of each technique. This chapter is unique as it has two case studies, the first one for exploring the attention mechanism and the second for memory networks. The first case study extends the machine translation case study started in Chap. 7 to examine the impact of different attention mechanisms discussed in this chapter. The second case study explores and analyzes different memory networks on the question-answering NLP task.

- The **Transfer Learning: Scenarios, Self-Taught Learning, and Multitask Learning** chapter introduces the concept of transfer learning and covers multitask learning techniques extensively. This case study explores multitask learning techniques for NLP tasks such as part-of-speech tagging, chunking, and named entity recognition and analysis. Readers should expect to gain insights into real, practical aspects of applying the multitask learning techniques introduced here.

- The **Transfer Learning: Domain Adaptation** chapter probes into the area of transfer learning where the models are subjected to constraints such as having fewer data to train on, or situations when data on which to predict is different from data it has trained on. Techniques for domain adaptation, few-shot learning, one-shot learning, and zero-shot learning are covered in this chapter. A detailed case study is presented using Amazon product reviews across different domains where many of the techniques discussed are applied.

- The **End-to-End Speech Recognition** chapter combines the ASR concepts in Chap. 8 with the deep learning techniques for end-to-end recognition. This chapter introduces mechanisms for training end-to-end sequence-based architectures with CTC and attention, as well as explores decoding techniques to improve quality further. The case study extends the one presented in Chap. 8 by using the same dataset to compare two end-to-end techniques, Deep Speech 2 and ESPnet (CTC-Attention hybrid training).

- In the **Deep Reinforcement for Text and Speech** chapter, we review the fundamentals of reinforcement learning and discuss their adaptation to deep sequence-to-sequence models, including deep policy gradient, deep Q-learning, double DQN, and DAAC algorithms. We investigate deep reinforcement learning approaches to NLP tasks including information extraction, text summarization, machine translation, and automatic speech recognition. We conclude with a case study on the application of deep policy gradient and deep Q-learning algorithms to text summarization.

Acknowledgments

The construction of this book would not have been possible without the tremendous efforts of many people. Firstly, we want to thank Springer, especially our editor Paul Drougas, for working very closely with us and seeing this to fruition. We want to thank Digital Reasoning for giving us the opportunity to work on many real-world NLP and speech problems that have had a significant impact on our work here. We would specifically like to thank Maciek Makowski and Gabrielle Liu for reviewing and editing the content of this book, as well as those that have provided support in engineering expertise, performing experiments, content feedback, and suggestions (in alphabetical order): Mona Barteau, Tim Blass, Brandon Carl, Krishna Choppella, Wael Emara, Last Feremenga, Christi French, Josh Gieringer, Bruce Glassford, Kenneth Graham, Ramsey Kant, Sean Narenthiran, Curtis Ogle, Joseph Porter, Drew Robertson, Sebastian Ruder, Amarda Shehu, Sarah Sorensen, Samantha Terker, Michael Urda.

McLean, VA, USA   Uday Kamath
Nashville, TN, USA   John Liu
McLean, VA, USA   James Whitaker

Contents

Part I Machine Learning, NLP, and Speech Introduction

1 **Introduction** 3
   1.1   Machine Learning 5
        1.1.1   Supervised Learning 5
        1.1.2   Unsupervised Learning 6
        1.1.3   Semi-Supervised Learning and Active Learning 7
        1.1.4   Transfer Learning and Multitask Learning 7
        1.1.5   Reinforcement Learning 7
   1.2   History 7
        1.2.1   Deep Learning: A Brief History 8
        1.2.2   Natural Language Processing: A Brief History 11
        1.2.3   Automatic Speech Recognition: A Brief History 15
   1.3   Tools, Libraries, Datasets, and Resources for the Practitioners 18
        1.3.1   Deep Learning 18
        1.3.2   Natural Language Processing 19
        1.3.3   Speech Recognition 20
        1.3.4   Books 21
        1.3.5   Online Courses and Resources 21
        1.3.6   Datasets 22
   1.4   Case Studies and Implementation Details 25
   References 27

2 **Basics of Machine Learning** 39
   2.1   Introduction 39
   2.2   Supervised Learning: Framework and Formal Definitions 40
        2.2.1   Input Space and Samples 40
        2.2.2   Target Function and Labels 41
        2.2.3   Training and Prediction 41
   2.3   The Learning Process 42
   2.4   Machine Learning Theory 43

        2.4.1   Generalization–Approximation Trade-Off via the
Vapnik–Chervonenkis Analysis 43
        2.4.2   Generalization–Approximation Trade-off via the
Bias–Variance Analysis 46
        2.4.3   Model Performance and Evaluation Metrics 47
        2.4.4   Model Validation 50
        2.4.5   Model Estimation and Comparisons..................... 53
        2.4.6   Practical Tips for Machine Learning 54
  2.5   Linear Algorithms 55
        2.5.1   Linear Regression 55
        2.5.2   Perceptron 58
        2.5.3   Regularization..................................... 59
        2.5.4   Logistic Regression 61
        2.5.5   Generative Classifiers 64
        2.5.6   Practical Tips for Linear Algorithms 66
  2.6   Non-linear Algorithms 67
        2.6.1   Support Vector Machines............................. 68
        2.6.2   Other Non-linear Algorithms 69
  2.7   Feature Transformation, Selection, and Dimensionality Reduction . . 69
        2.7.1   Feature Transformation 70
        2.7.2   Feature Selection and Reduction 71
  2.8   Sequence Data and Modeling................................ 72
        2.8.1   Discrete Time Markov Chains 72
        2.8.2   Discriminative Approach: Hidden Markov Models 73
        2.8.3   Generative Approach: Conditional Random Fields 75
  2.9   Case Study 78
        2.9.1   Software Tools and Libraries 78
        2.9.2   Exploratory Data Analysis (EDA) 78
        2.9.3   Model Training and Hyperparameter Search.............. 79
        2.9.4   Final Training and Testing Models 83
        2.9.5   Exercises for Readers and Practitioners 85
  References 85

3   Text and Speech Basics 87
  3.1   Introduction 87
        3.1.1   Computational Linguistics............................ 87
        3.1.2   Natural Language 88
        3.1.3   Model of Language 89
  3.2   Morphological Analysis 90
        3.2.1   Stemming 91
        3.2.2   Lemmatization 92

3.3   Lexical Representations 92
       3.3.1   Tokens 92
       3.3.2   Stop Words 93
       3.3.3   N-Grams 93
       3.3.4   Documents 94
3.4   Syntactic Representations................................... 96
       3.4.1   Part-of-Speech 97
       3.4.2   Dependency Parsing.................................. 99
3.5   Semantic Representations.................................... 101
       3.5.1   Named Entity Recognition 102
       3.5.2   Relation Extraction................................. 103
       3.5.3   Event Extraction 104
       3.5.4   Semantic Role Labeling.............................. 104
3.6   Discourse Representations 105
       3.6.1   Cohesion 105
       3.6.2   Coherence 105
       3.6.3   Anaphora/Cataphora 105
       3.6.4   Local and Global Coreference 106
3.7   Language Models 106
       3.7.1   N-Gram Model 107
       3.7.2   Laplace Smoothing................................... 107
       3.7.3   Out-of-Vocabulary 108
       3.7.4   Perplexity 108
3.8   Text Classification 109
       3.8.1   Machine Learning Approach........................... 109
       3.8.2   Sentiment Analysis.................................. 110
       3.8.3   Entailment 112
3.9   Text Clustering 113
       3.9.1   Lexical Chains 114
       3.9.2   Topic Modeling...................................... 114
3.10   Machine Translation 115
       3.10.1   Dictionary Based 115
       3.10.2   Statistical Translation............................. 116
3.11   Question Answering 116
       3.11.1   Information Retrieval Based 117
       3.11.2   Knowledge-Based QA 118
       3.11.3   Automated Reasoning 118
3.12   Automatic Summarization 119
       3.12.1   Extraction Based.................................... 119
       3.12.2   Abstraction Based 120
3.13   Automated Speech Recognition................................ 120
       3.13.1   Acoustic Model 120
3.14   Case Study 122
       3.14.1   Software Tools and Libraries 123
       3.14.2   EDA 123

        3.14.3 Text Clustering 126
        3.14.4 Topic Modeling....................................... 129
        3.14.5 Text Classification 131
        3.14.6 Exercises for Readers and Practitioners 133
    References 134

Part II Deep Learning Basics

4   Basics of Deep Learning 141
    4.1   Introduction 141
    4.2   Perceptron Algorithm Explained 143
        4.2.1 Bias 143
        4.2.2 Linear and Non-linear Separability 146
    4.3   Multilayer Perceptron (Neural Networks) 146
        4.3.1 Training an MLP 147
        4.3.2 Forward Propagation 148
        4.3.3 Error Computation 149
        4.3.4 Backpropagation...................................... 150
        4.3.5 Parameter Update 152
        4.3.6 Universal Approximation Theorem 153
    4.4   Deep Learning 154
        4.4.1 Activation Functions 155
        4.4.2 Loss Functions 161
        4.4.3 Optimization Methods 162
    4.5   Model Training 165
        4.5.1 Early Stopping 165
        4.5.2 Vanishing/Exploding Gradients 166
        4.5.3 Full-Batch and Mini-Batch Gradient Decent 167
        4.5.4 Regularization....................................... 167
        4.5.5 Hyperparameter Selection 171
        4.5.6 Data Availability and Quality 172
        4.5.7 Discussion 174
    4.6   Unsupervised Deep Learning 175
        4.6.1 Energy-Based Models 175
        4.6.2 Restricted Boltzmann Machines 176
        4.6.3 Deep Belief Networks 178
        4.6.4 Autoencoders 178
        4.6.5 Sparse Coding....................................... 182
        4.6.6 Generative Adversarial Networks 182
    4.7   Framework Considerations................................... 183
        4.7.1 Layer Abstraction 184
        4.7.2 Computational Graphs 185
        4.7.3 Reverse-Mode Automatic Differentiation 186
        4.7.4 Static Computational Graphs 186
        4.7.5 Dynamic Computational Graphs........................ 187

	4.8		Case Study 187
			4.8.1	Software Tools and Libraries 187
			4.8.2	Exploratory Data Analysis (EDA) 188
			4.8.3	Supervised Learning 189
			4.8.4	Unsupervised Learning 193
			4.8.5	Classifying with Unsupervised Features 196
			4.8.6	Results 198
			4.8.7	Exercises for Readers and Practitioners 198
		References 199

5	Distributed Representations 203
	5.1		Introduction 203
	5.2		Distributional Semantics................................... 203
			5.2.1	Vector Space Model 203
			5.2.2	Word Representations 205
			5.2.3	Neural Language Models............................. 206
			5.2.4	word2vec 208
			5.2.5	GloVe 219
			5.2.6	Spectral Word Embeddings 221
			5.2.7	Multilingual Word Embeddings 222
	5.3		Limitations of Word Embeddings 222
			5.3.1	Out of Vocabulary 222
			5.3.2	Antonymy 223
			5.3.3	Polysemy 224
			5.3.4	Biased Embeddings 227
			5.3.5	Other Limitations 227
	5.4		Beyond Word Embeddings................................... 227
			5.4.1	Subword Embeddings 228
			5.4.2	Word Vector Quantization 228
			5.4.3	Sentence Embeddings 230
			5.4.4	Concept Embeddings 232
			5.4.5	Retrofitting with Semantic Lexicons 233
			5.4.6	Gaussian Embeddings 234
			5.4.7	Hyperbolic Embeddings 236
	5.5		Applications 238
			5.5.1	Classification...................................... 239
			5.5.2	Document Clustering 239
			5.5.3	Language Modeling 240
			5.5.4	Text Anomaly Detection 241
			5.5.5	Contextualized Embeddings 242
	5.6		Case Study 243
			5.6.1	Software Tools and Libraries 243
			5.6.2	Exploratory Data Analysis 243
			5.6.3	Learning Word Embeddings 244
			5.6.4	Document Clustering 256

        5.6.5   Word Sense Disambiguation 257
        5.6.6   Exercises for Readers and Practitioners 259
    References 259

6 Convolutional Neural Networks 263
    6.1   Introduction 263
    6.2   Basic Building Blocks of CNN 264
        6.2.1   Convolution and Correlation in Linear Time-Invariant
                Systems 264
        6.2.2   Local Connectivity or Sparse Interactions 265
        6.2.3   Parameter Sharing 266
        6.2.4   Spatial Arrangement 266
        6.2.5   Detector Using Nonlinearity 270
        6.2.6   Pooling and Subsampling 271
    6.3   Forward and Backpropagation in CNN 273
        6.3.1   Gradient with Respect to Weights $\frac{\partial E}{\partial \mathbf{W}}$ 274
        6.3.2   Gradient with Respect to the Inputs $\frac{\partial E}{\partial \mathbf{X}}$ 275
        6.3.3   Max Pooling Layer 276
    6.4   Text Inputs and CNNs 276
        6.4.1   Word Embeddings and CNN 277
        6.4.2   Character-Based Representation and CNN 280
    6.5   Classic CNN Architectures 281
        6.5.1   LeNet-5 282
        6.5.2   AlexNet 283
        6.5.3   VGG-16 285
    6.6   Modern CNN Architectures 285
        6.6.1   Stacked or Hierarchical CNN 286
        6.6.2   Dilated CNN 287
        6.6.3   Inception Networks 288
        6.6.4   Other CNN Structures 289
    6.7   Applications of CNN in NLP 292
        6.7.1   Text Classification and Categorization 293
        6.7.2   Text Clustering and Topic Mining 294
        6.7.3   Syntactic Parsing 294
        6.7.4   Information Extraction 294
        6.7.5   Machine Translation 295
        6.7.6   Summarizations 296
        6.7.7   Question and Answers 296
    6.8   Fast Algorithms for Convolutions 297
        6.8.1   Convolution Theorem and Fast Fourier Transform 297
        6.8.2   Fast Filtering Algorithm 297
    6.9   Case Study 300
        6.9.1   Software Tools and Libraries 300
        6.9.2   Exploratory Data Analysis 301
        6.9.3   Data Preprocessing and Data Splits 301

       6.9.4   CNN Model Experiments 303
       6.9.5   Understanding and Improving the Models.............. 307
       6.9.6   Exercises for Readers and Practitioners 309
  6.10 Discussion 310
  References 310

7   Recurrent Neural Networks 315
  7.1   Introduction 315
  7.2   Basic Building Blocks of RNNs 316
       7.2.1   Recurrence and Memory 316
       7.2.2   PyTorch Example 317
  7.3   RNNs and Properties.. 318
       7.3.1   Forward and Backpropagation in RNNs 318
       7.3.2   Vanishing Gradient Problem and Regularization 323
  7.4   Deep RNN Architectures 327
       7.4.1   Deep RNNs... 327
       7.4.2   Residual LSTM 328
       7.4.3   Recurrent Highway Networks 329
       7.4.4   Bidirectional RNNs 329
       7.4.5   SRU and Quasi-RNN 331
       7.4.6   Recursive Neural Networks 331
  7.5   Extensions of Recurrent Networks 333
       7.5.1   Sequence-to-Sequence 334
       7.5.2   Attention 335
       7.5.3   Pointer Networks 336
       7.5.4   Transformer Networks 337
  7.6   Applications of RNNs in NLP 339
       7.6.1   Text Classification 339
       7.6.2   Part-of-Speech Tagging and Named Entity
             Recognition 340
       7.6.3   Dependency Parsing................................. 340
       7.6.4   Topic Modeling and Summarization 340
       7.6.5   Question Answering................................. 341
       7.6.6   Multi-Modal 341
       7.6.7   Language Models 341
       7.6.8   Neural Machine Translation 343
       7.6.9   Prediction/Sampling Output 346
  7.7   Case Study 348
       7.7.1   Software Tools and Libraries 349
       7.7.2   Exploratory Data Analysis 349
       7.7.3   Model Training..................................... 355
       7.7.4   Results 362
       7.7.5   Exercises for Readers and Practitioners 363

    7.8   Discussion 364
        7.8.1 Memorization or Generalization 364
        7.8.2 Future of RNNs 365
    References 365

8 Automatic Speech Recognition 369
    8.1   Introduction 369
    8.2   Acoustic Features 370
        8.2.1 Speech Production 370
        8.2.2 Raw Waveform 371
        8.2.3 MFCC 372
        8.2.4 Other Feature Types 376
    8.3   Phones 377
    8.4   Statistical Speech Recognition 379
        8.4.1 Acoustic Model: $P(X|W)$ 381
        8.4.2 *LanguageModel* : $P(W)$ 385
        8.4.3 HMM Decoding 386
    8.5   Error Metrics 387
    8.6   DNN/HMM Hybrid Model 388
    8.7   Case Study 391
        8.7.1 Dataset: Common Voice 392
        8.7.2 Software Tools and Libraries 392
        8.7.3 Sphinx 392
        8.7.4 Kaldi 396
        8.7.5 Results 401
        8.7.6 Exercises for Readers and Practitioners 402
    References 403

Part III Advanced Deep Learning Techniques for Text and Speech

9 Attention and Memory Augmented Networks 407
    9.1   Introduction 407
    9.2   Attention Mechanism 408
        9.2.1 The Need for Attention Mechanism 409
        9.2.2 Soft Attention 410
        9.2.3 Scores-Based Attention 411
        9.2.4 Soft vs. Hard Attention 412
        9.2.5 Local vs. Global Attention 412
        9.2.6 Self-Attention 413
        9.2.7 Key-Value Attention 414
        9.2.8 Multi-Head Self-Attention 415
        9.2.9 Hierarchical Attention 416
        9.2.10 Applications of Attention Mechanism in Text and Speech .. 418
    9.3   Memory Augmented Networks 419
        9.3.1 Memory Networks 419

9.3.2   End-to-End Memory Networks......................... 422
9.3.3   Neural Turing Machines 424
9.3.4   Differentiable Neural Computer 428
9.3.5   Dynamic Memory Networks 431
9.3.6   Neural Stack, Queues, and Deques.................... 434
9.3.7   Recurrent Entity Networks 437
9.3.8   Applications of Memory Augmented Networks in Text
         and Speech 440
9.4   Case Study 440
9.4.1   Attention-Based NMT 440
9.4.2   Exploratory Data Analysis 441
9.4.3   Question and Answering 450
9.4.4   Dynamic Memory Network............................ 455
9.4.5   Exercises for Readers and Practitioners 459
References 460

10   Transfer Learning: Scenarios, Self-Taught Learning, and Multitask
      Learning 463
10.1   Introduction 463
10.2   Transfer Learning: Definition, Scenarios, and Categorization 464
10.2.1   Definition 465
10.2.2   Transfer Learning Scenarios 466
10.2.3   Transfer Learning Categories 466
10.3   Self-Taught Learning 467
10.3.1   Techniques 468
10.3.2   Theory 469
10.3.3   Applications in NLP................................. 470
10.3.4   Applications in Speech 470
10.4   Multitask Learning 471
10.4.1   Techniques 471
10.4.2   Theory 480
10.4.3   Applications in NLP................................. 480
10.4.4   Applications in Speech Recognition.................... 482
10.5   Case Study 482
10.5.1   Software Tools and Libraries 482
10.5.2   Exploratory Data Analysis 483
10.5.3   Multitask Learning Experiments and Analysis 484
10.5.4   Exercises for Readers and Practitioners 489
References 489

11   Transfer Learning: Domain Adaptation 495
11.1   Introduction 495
11.1.1   Techniques 496
11.1.2   Theory 513

11.1.3 Applications in NLP..................................... 515
11.1.4 Applications in Speech Recognition..................... 516
11.2 Zero-Shot, One-Shot, and Few-Shot Learning................... 517
11.2.1 Zero-Shot Learning 517
11.2.2 One-Shot Learning 520
11.2.3 Few-Shot Learning 521
11.2.4 Theory 522
11.2.5 Applications in NLP and Speech Recognition 522
11.3 Case Study 523
11.3.1 Software Tools and Libraries 524
11.3.2 Exploratory Data Analysis 524
11.3.3 Domain Adaptation Experiments 525
11.3.4 Exercises for Readers and Practitioners 530
References 531

12 End-to-End Speech Recognition 537
12.1 Introduction 537
12.2 Connectionist Temporal Classification (CTC) 538
12.2.1 End-to-End Phoneme Recognition 541
12.2.2 Deep Speech 541
12.2.3 Deep Speech 2 543
12.2.4 Wav2Letter 544
12.2.5 Extensions of CTC 545
12.3 Seq-to-Seq 546
12.3.1 Early Seq-to-Seq ASR 548
12.3.2 Listen, Attend, and Spell (LAS) 548
12.4 Multitask Learning 549
12.5 End-to-End Decoding 551
12.5.1 Language Models for ASR 551
12.5.2 CTC Decoding 552
12.5.3 Attention Decoding 555
12.5.4 Combined Language Model Training.................... 556
12.5.5 Combined CTC–Attention Decoding 557
12.5.6 One-Pass Decoding 558
12.6 Speech Embeddings and Unsupervised Speech Recognition 559
12.6.1 Speech Embeddings 559
12.6.2 Unspeech... 560
12.6.3 Audio Word2Vec 560
12.7 Case Study 561
12.7.1 Software Tools and Libraries 561
12.7.2 Deep Speech 2 562
12.7.3 Language Model Training 564
12.7.4 ESPnet 566

　　　　12.7.5　Results 570
　　　　12.7.6　Exercises for Readers and Practitioners 571
　　References 571

13 Deep Reinforcement Learning for Text and Speech 575
　13.1　Introduction 575
　13.2　RL Fundamentals 575
　　　　13.2.1　Markov Decision Processes 576
　　　　13.2.2　Value, Q, and Advantage Functions 577
　　　　13.2.3　Bellman Equations 578
　　　　13.2.4　Optimality 579
　　　　13.2.5　Dynamic Programming Methods 580
　　　　13.2.6　Monte Carlo 582
　　　　13.2.7　Temporal Difference Learning 583
　　　　13.2.8　Policy Gradient 586
　　　　13.2.9　Q-Learning 587
　　　　13.2.10 Actor-Critic 588
　13.3　Deep Reinforcement Learning Algorithms 590
　　　　13.3.1　Why RL for Seq2seq 590
　　　　13.3.2　Deep Policy Gradient 591
　　　　13.3.3　Deep Q-Learning 592
　　　　13.3.4　Deep Advantage Actor-Critic 596
　13.4　DRL for Text 597
　　　　13.4.1　Information Extraction 597
　　　　13.4.2　Text Classification 601
　　　　13.4.3　Dialogue Systems 602
　　　　13.4.4　Text Summarization 603
　　　　13.4.5　Machine Translation 605
　13.5　DRL for Speech 605
　　　　13.5.1　Automatic Speech Recognition 606
　　　　13.5.2　Speech Enhancement and Noise Suppression 606
　13.6　Case Study 607
　　　　13.6.1　Software Tools and Libraries 607
　　　　13.6.2　Text Summarization 608
　　　　13.6.3　Exploratory Data Analysis 608
　　　　13.6.4　Exercises for Readers and Practitioners 612
　　References 612

Future Outlook 615
　End-to-End Architecture Prevalence 615
　Transition to AI-Centric 615
　Specialized Hardware 616
　Transition Away from Supervised Learning 616
　Explainable AI 616
　Model Development and Deployment Process 617

Democratization of AI 617
NLP Trends 617
Speech Trends 618
Closing Remarks 618

Index 619

Notation

Calculus

$\approx$	Approximately equal to
$|\mathbf{A}|$	L_1 norm of matrix $\mathbf{A}$
$\|\mathbf{A}\|$	L_2 norm of matrix $\mathbf{A}$
$\frac{da}{db}$	Derivative of a with respect to b
$\frac{\partial a}{\partial b}$	Partial derivative of a with respect to b
$\nabla_x Y$	Gradient of Y with respect to x
$\nabla_{\mathbf{X}} Y$	Matrix of derivatives of Y with respect to $\mathbf{X}$

Datasets

$\mathcal{D}$	Dataset, a set of examples and corresponding targets, $\{(\mathbf{x}_1, y_1), (\mathbf{x}_2, y_2), \ldots, (\mathbf{x}_n, y_n)\}$
$\mathcal{X}$	Space of all possible inputs
$\mathcal{Y}$	Space of all possible outputs
y_i	Target label for example i
$\hat{y}_i$	Predicted label for example i
$\mathcal{L}$	Log-likelihood loss
Ω	Learned parameters

Functions

$f : \mathbb{A} \to \mathbb{B}$	A function f that maps a value in the set $\mathbb{A}$ to set $\mathbb{B}$
$f(\mathbf{x}; \theta)$	A function of $\mathbf{x}$ parameterized by θ. This is frequently reduced to $f(\mathbf{x})$ for notational clarity.
$\log x$	Natural log of x
$\sigma(a)$	Logistic sigmoid, $\frac{1}{1+\exp -a}$
$[\![a \neq b]\!]$	A function that yields a 1 if the condition contained is true, otherwise it yields 0
$\arg\min_x f(x)$	Set of arguments that minimize $f(x)$, $\arg\min_x f(x) = \{x \mid f(x) = \min_{x'} f(x')\}$
$\arg\max_x f(x)$	Set of arguments that maximize $f(x)$, $\arg\max_x f(x) = \{x \mid f(x) = \max_{x'} f(x')\}$

Linear Algebra

a Scalar value (integer or real)

$\begin{bmatrix} a_1 \\ \vdots \\ a_n \end{bmatrix}$ Vector containing elements a_1 to a_n

$\begin{bmatrix} a_{1,1} & \cdots & a_{1,n} \\ \vdots & \ddots & \vdots \\ a_{m,1} & \cdots & a_{m,n} \end{bmatrix}$ A matrix with m rows and n columns

$A_{i,j}$ Value of matrix $\mathbf{A}$ at row i and column j

$\mathbf{a}$ Vector (dimensions implied by context)

$\mathbf{A}$ Matrix (dimensions implied by context)

$\mathbf{A}^\mathsf{T}$ Transpose of matrix $\mathbf{A}$

$\mathbf{A}^{-1}$ Inverse of matrix $\mathbf{A}$

$\mathbf{I}$ Identity matrix (dimensionality implied by context)

$\mathbf{A} \cdot \mathbf{B}$ Dot product of matrices $\mathbf{A}$ and $\mathbf{B}$

$\mathbf{A} \times \mathbf{B}$ Cross product of matrices $\mathbf{A}$ and $\mathbf{B}$

$\mathbf{A} \circ \mathbf{B}$ Element-wise (Hadamard) product

$\mathbf{A} \otimes \mathbf{B}$ Kronecker product of matrices $\mathbf{A}$ and $\mathbf{B}$

$\mathbf{a} ; \mathbf{b}$ Concatenation of vectors $\mathbf{a}$ and $\mathbf{b}$

Probability

$\mathbb{E}$ Expected value

$P(A)$ Probability of event A

$P(A|B)$ Probability of event A given event B

$X \sim \mathcal{N}(\mu, \sigma^2)$ Random variable X sampled from a Gaussian (Normal) distribution with μ mean and σ^2 variance

Sets

$\mathbb{A}$ A set

$\mathbb{R}$ Set of real numbers

$\mathbb{C}$ Set of complex numbers

$\emptyset$ Empty set

$\{a, b\}$ Set containing the elements a and b

$\{1, 2, \ldots n\}$ Set containing all integers from 1 to n

$\{a_1, a_2, \ldots a_n\}$ Set containing n elements

$a \in \mathbb{A}$ Value a is a member of the set $\mathbb{A}$

$[a, b]$ Set of real values from a to b, including a and b

$[a, b)$ Set of real values from a to b, including a but excluding b

$a_{1:m}$ Set of elements $\{a_1, a_2, \ldots, a_m\}$ (used for notational convenience)

Most of the chapters, unless and otherwise specified, assume the notations given above.

Chapter 1
Introduction

In recent years, advances in machine learning have led to significant and widespread improvements in how we interact with our world. One of the most portentous of these advances is the field of deep learning. Based on artificial neural networks that resemble those in the human brain, deep learning is a set of methods that permits computers to learn from data without human supervision and intervention. Furthermore, these methods can adapt to changing environments and provide continuous improvement to learned abilities. Today, deep learning is prevalent in our everyday life in the form of Google's search, Apple's Siri, and Amazon's and Netflix's recommendation engines to name but a few examples. When we interact with our email systems, online chatbots, and voice or image recognition systems deployed at businesses ranging from healthcare to financial services, we see robust applications of deep learning in action.

Human communication is at the core of developments in many of these areas, and the complexities of language make computational approaches increasingly difficult. With the advent of deep learning, however, the burden shifts from producing rule-based approaches to learning directly from the data. These deep learning techniques open new fronts in our ability to model human communication and interaction and improve human–computer interaction.

Deep learning saw explosive growth, attention, and availability of tools following its success in computer vision in the early 2010s. Natural language processing soon experienced many of these same benefits from computer vision. Speech recognition, traditionally a field dominated by feature engineering and model tuning techniques, incorporated deep learning into its feature extraction methods resulting in strong gains in quality. Figure 1.1 shows the popularity of these fields in recent years.

The age of big data is another contributing factor to the performance gains with deep learning. Unlike many traditional learning algorithms, deep learning models continue to improve with the amount of data provided, as illustrated in Fig. 1.2.

Perhaps one of the largest contributors to the success of deep learning is the active community that has developed around it. The overlap and collaboration between academic institutions and industry in the open source has led to a virtual cornucopia

© Springer Nature Switzerland AG 2019
U. Kamath et al., *Deep Learning for NLP and Speech Recognition*,
https://doi.org/10.1007/978-3-030-14596-5_1

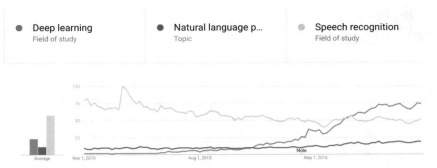

Fig. 1.1: Google trends for deep learning, natural language processing, and speech recognition in the last decade

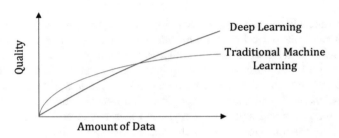

Fig. 1.2: Deep learning benefits heavily from large datasets

of tools and libraries for deep learning. This overlap and influence of the academic world and the consumer marketplace has also led to a shift in the popularity of programming languages, as illustrated in Fig. 1.3, specifically towards Python.

Python has become the go-to language for many analytics applications, due to its simplicity, cleanliness of syntax, multiple data science libraries, and extensibility (specifically with C++). This simplicity and extensibility have led to most top deep learning frameworks to be built on Python or adopt Python interfaces that wrap high-performance C++ and GPU-optimized extensions.

This book seeks to provide the reader an in-depth overview of deep learning techniques in the fields of text and speech processing. Our hope is for the reader to walk away with a thorough understanding of natural language processing and leading-edge deep learning techniques that will provide a basis for all text and speech processing advancements in the future. Since "practice makes for a wonderful companion," each chapter in this book is accompanied with a case study that walks through a practical application of the methods introduced in the chapter.

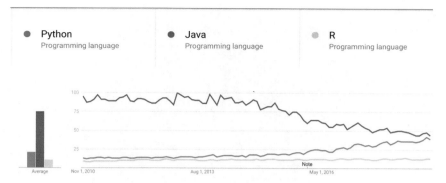

Fig. 1.3: Google trends for programming languages such as Java, Python, and R which are used in data science and deep learning in the last decade

1.1 Machine Learning

Machine learning is quickly becoming commonplace in many of the applications we use daily. It can make us more productive, help us make decisions, provide a personalized experience, and gain insights about the world by leveraging data. The field of AI is broad, encompassing search algorithms, planning and scheduling, computer vision, and many other areas. Machine learning, a subcategory of AI, is composed of three areas: supervised learning, unsupervised learning, and reinforcement learning. Deep learning is a collection of learning algorithms that has been applied to each of these three areas, as shown in Fig. 1.4. Before we go further, we explain how exactly deep learning applies.

Each of these areas will be explored thoroughly in the chapters of this book.

1.1.1 Supervised Learning

Supervised learning relies on learning from a dataset with labels for each of the examples. For example, if we are trying to learn movie sentiment, the dataset may be a set of movie reviews and the labels are the 0–5 star rating.

There are two types of supervised learning: classification and regression (Fig. 1.5).

Classification maps an input into a fixed set of categories, for example, classifying an image as either a cat or dog.

Regression problems, on the other hand, map an input to a real number value. An example of this is trying to predicting the cost of your utility bill or the stock market price.

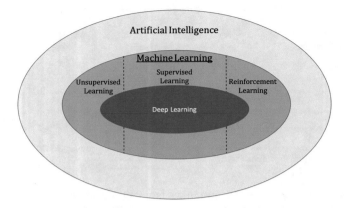

Fig. 1.4: The area of deep learning covers multiple areas of machine learning, while machine learning is a subset of the broader AI category

Movie Review	Label
I really enjoyed it! I though it was clever...	Positive
I have seen better films.	Negative
I can't believe I wasted my time on that...	Negative
Honestly, I was pleasantly surprised...	Positive

New example: It was spectacular! ⟨ ? → Positive / → Negative

(a)

Movie Review	Score
I really enjoyed it! I though it was clever...	5
I have seen better films.	2
I can't believe I wasted my time on that...	1
Honestly, I was pleasantly surprised...	4

New example: It was spectacular! → $1 \leq score \leq 5$

(b)

Fig. 1.5: Supervised learning uses a labeled dataset to predict an output. In a classification problem, (**a**) the output will be labeled as a category (e.g., positive or negative), while in a regression problem, (**b**) the output will be a value

1.1.2 Unsupervised Learning

Unsupervised learning determines categories from data where there are no labels present. These tasks can take the form of clustering, grouping similar items together, or similarity, defining how closely a pair of items is related. For example, imagine we wanted to recommend a movie based on a person's viewing habits. We could cluster users based on what they have watched and enjoyed, and evaluate whose viewing habits most match the person to whom we are recommending the movie.

1.1.3 Semi-Supervised Learning and Active Learning

In many situations when it is not possible to label or annotate the entire dataset due to either cost or lack of expertise or other constraints, learning jointly from the labeled and unlabeled data is called semi-supervised learning. Instead of expert labeling of data, if the machine provides insight into which data should be labeled, the process is called active learning.

1.1.4 Transfer Learning and Multitask Learning

The basic idea behind "transfer learning" is to help the model adapt to situations it has not previously encountered. This form of learning relies on tuning a general model to a new domain. Learning from many tasks to jointly improve the performance across all the tasks is called multitask learning. These techniques are becoming the focus in both deep learning and NLP/speech.

1.1.5 Reinforcement Learning

Reinforcement learning focuses on maximizing a reward given an action or set of actions taken. The algorithms are trained to encourage certain behavior and discourage others. Reinforcement learning tends to work well on games like chess or go, where the reward may be winning the game. In this case, a number of actions must be taken before the reward is reached.

1.2 History

> You don't know where you're going until you know where you've been.—James Baldwin

It is impossible to separate the current approaches to natural language processing and speech from the extensive histories that accompany them. Many of the advancements discussed in this book are relatively new in comparison to those presented elsewhere, and, because of their novelty, it is important to understand how these ideas developed over time to put the current innovations into proper context. Here, we present a brief history of deep learning, natural language processing, and speech recognition.

1.2.1 Deep Learning: A Brief History

There has been much research in both the academic and industrial fields that has led to the current state of deep learning and its recent popularity. The goal of this section is to give a brief timeline of research that has influenced deep learning, although we might not have captured all the details (Fig. 1.6). Schmidhuber [Sch15] has comprehensively captured the entire history of neural networks and various research that led to today's deep learning. In the early 1940s, S. McCulloch and W. Pitts modeled how the brain works using a simple electrical circuit called the *threshold logic unit* that could simulate intelligent behavior [MP88]. They modeled the first neuron with inputs and outputs that generated 0 when the "weighted sum" was below a threshold and 1 otherwise. The weights were not learned but adjusted. They coined the term *connectionism* to describe their model. Donald Hebb in his book "The Organization of Behaviour (1949)" took the idea further by proposing how neural pathways can have multiple neurons firing and strengthening over time with usage, thus laying the foundation of complex processing [Heb49].

According to many, Alan Turing in his seminal paper "Computing Machinery and Intelligence" laid the foundations of artificial intelligence with several criteria to validate the "intelligence" of machines known as the "Turing test" [Tur95]. In 1959, the discovery of simple cells and complex cells that constitute the primary visual cortex by Nobel Laureates Hubel and Wiesel had a wide-ranging influence in many fields including the design of neural networks. Frank Rosenblatt extended the McCulloch–Pitts neuron using the term *Mark I Perceptron* which took inputs, generated outputs, and had linear thresholding logic [Ros58]. The weights in the perceptron were "learned" by successively passing the inputs and reducing the difference between the generated output and the desired output. Bernard Widrow and Marcian Hoff took the idea of perceptrons further to develop Multiple ADAptive LINear Elements (MADALINE) which were used to eliminate noise in phone lines [WH60].

Marvin Minsky and Seymour Papert published the book *Perceptrons* which showed the limitations of perceptrons in learning the simple exclusive-or function (XOR) [MP69]. Because of a large number of iterations required to generate the output and the limitations imposed by compute time they conclusively proved that multilayer networks could not use perceptrons. Years of funding dried because of this and effectively limited research in the neural networks, appropriately called the "The First AI Winter."

In 1986, David Rumelhart, Geoff Hinton, and Ronald Williams published the seminal work "Learning representations by back-propagating errors" which showed how a multi-layered neural network could not only be trained effectively using a relatively simple procedure but how "hidden" layers can be used to overcome the weakness of perceptrons in learning complex patterns [RHW88]. Though there was much research in the past in the form of various theses and research, the works of Linnainmaa, S., P. Werbos, Fukushima, David Parker, Yann Le Cun, and Rumelhart et al. have considerably broadened the popularity of neural networks [Lin70, Wer74, Fuk79, Par85, LeC85].

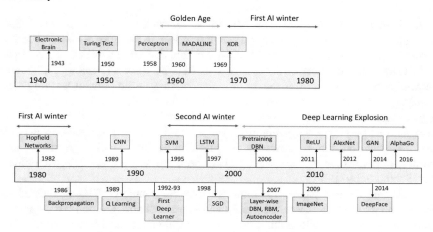

Fig. 1.6: Highlights in deep learning research

LeCun et al. with their research and implementation led to the first widespread application of neural networks to the recognition of handwritten digits used by the U.S. Postal Service [LeC+89]. This work was an important milestone in deep learning history, as it showed how convolution operations and weight sharing could be effective for learning features in modern convolutional neural networks (CNNs). George Cybenko showed how the feed-forward networks with finite neurons, a single hidden layer, and non-linear sigmoid activation function could approximate most complex functions with mild assumptions [Cyb89]. Cybenko's research along with Kurt Hornik's work led to the further rise of neural networks and their application as "universal approximator functions" [Hor91]. The seminal work of Yann Le Cun et al. resulted in widespread practical applications of CNNs such as the reading bank checks [LB94, LBB97].

Dimensionality reduction and learning using unsupervised techniques were demonstrated in Kohen's work titled "Self-Organized Formation of Topologically Correct Feature Maps" [Koh82]. John Hopfield with his Hopfield Networks created one of the first recurrent neural networks (RNNs) that served as a content-addressable memory system [Hop82]. Ackley et al. in their research showed how Boltzmann machines modeled as neural networks could capture probability distributions using the concepts of particle energy and thermodynamic temperature applied to the networks [AHS88]. Hinton and Zemel in their work presented various topics of unsupervised techniques to approximate probability distributions using neural networks [HZ94]. Redford Neal's work on the "belief net," similar to Boltzmann machines, showed how it could be used to perform unsupervised learning using much faster algorithms [Nea95].

Christopher Watkins' thesis introduced "Q Learning" and laid the foundations for reinforcement learning [Wat89]. Dean Pomerleau in his work at CMU's NavLab showed how neural networks could be used in robotics using supervised techniques and sensor data from various sources such as steering wheels [Pom89]. Lin's thesis showed how robots could be taught effectively using reinforcement learning techniques [Lin92]. One of the most significant landmarks in neural networks history is when a neural network was shown to outperform humans in a relatively complex task such as playing Backgammon [Tes95]. The first very deep learning network that used the concepts of unsupervised pre-training for a stack of recurrent neural networks to solve the credit assignment problem was presented by Schmidhuber [Sch92, Sch93].

Sebastian Thrun's paper "Learning To Play the Game of Chess" showed the shortcomings of reinforcement learning and neural networks in playing a complex game like Chess [Thr94]. Schraudolph et al. in their research further highlighted the issues of neural networks in playing the game Go [SDS93]. Backpropagation, which led to the resurgence of neural networks, was soon considered a problem due to issues such as vanishing gradients, exploding gradients, and the inability to learn long-term information, to name a few [Hoc98, BSF94]. Similar to how CNN architectures improved neural networks with convolution and weight sharing, the "long short-term memory (LSTM)" architecture introduced by Hochreiter and Schmidhuber overcame issues with long-term dependencies during backpropagation [HS97]. At the same time, statistical learning theory and particularly support vector machines (SVM) were fastly becoming a very popular algorithm on a wide variety of problems [CV95]. These changes contributed to "The Second Winter of AI."

Many in the deep learning community normally credit the Canadian Institute for Advanced Research (CIFAR) for playing a key role in advancing what we know as deep learning today. Hinton et al. published a breakthrough paper in 2006 titled "A Fast Learning Algorithm for Deep Belief Nets" which led to the resurgence of deep learning [HOT06a]. The paper not only presented the name *deep learning* for the first time but showed the effectiveness of layer-by-layer training using unsupervised methods followed by supervised "fine-tuning" in achieving the state-of-the-art results on the MNIST character recognition dataset. Bengio et al. published another seminal work following this, which gave insights into why deep learning networks with multiple layers can hierarchically learn features as compared to shallow neural networks or support vector machines [Ben+06]. The paper gave insights into why pre-training with unsupervised methods using DBNs, RBMs, and autoencoders not only initialized the weights to achieve optimal solutions but also provided good representations of data that can be learned. Bengio and LeCun's paper "Scaling Algorithms Towards AI" reiterated the advantages of deep learning through architectures such as CNN, RBM, DBN, and techniques such as unsupervised pre-training/fine-tuning inspiring the next wave of deep learning [BL07]. Using non-linear activation functions such as rectified linear units overcame many of the issues with the backpropagation algorithm [NH10, GBB11]. Fei-Fei Li, head of artificial intelligence lab at Stanford University, along with other researchers launched ImageNet, which collected a large number of images and showed the usefulness of data in important tasks such as object recognition, classification, and clustering [Den+09].

At the same time, following Moore's law, computers were getting faster, and graphic processor units (GPUs) overcame many of the previous limitations of CPUs. Mohamed et al. showed a huge improvement in the performance of a complex task such as speech recognition using deep learning techniques and achieved huge speed increases on large datasets with GPUs [Moh+11]. Using the previous networks such as CNNs and combining them with a ReLU activation, regularization techniques, such as dropout, and the speed of the GPU, Krizhevsky et al. attained the smallest error rates on the ImageNet classification task [KSH12]. Winning the ILSVRC-2012 competition by a huge difference between the CNN-based deep learning error rate of 15.3% and the second best at 26.2% put the attention of both academics and industry onto deep learning. Goodfellow et al. proposed a generative network using adversarial methods that addressed many issues of learning in an unsupervised manner and is considered a path-breaking research with wide applications [Goo+14].

Many companies such as Google, Facebook, and Microsoft started replacing their traditional algorithms with deep learning using GPU-based architectures for speed. Facebook's DeepFace uses deep networks with more than 120 million parameters and achieves the accuracy of 97.35% on a Labeled Faces in the Wild (LFW) dataset, approaching human-level accuracy by improving the previous results by an unprecedented 27% [Tai+14]. Google Brain, a collaboration between Andrew Ng and Jeff Dean, resulted in large-scale deep unsupervised learning from YouTube videos for tasks such as object identification using 16,000 CPU cores and close to one billion weights! DeepMind's AlphGo's beat Lee Sedol of Korea, an internationally top-ranked Go player, highlighting an important milestone in overall AI and deep learning.

1.2.2 Natural Language Processing: A Brief History

Natural language processing (NLP) is an exciting field of computer science that deals with human communication. It encompasses approaches to help machines understand, interpret, and generate human language. These are sometimes delineated as natural language understanding (NLU) and natural language generation (NLG) methods. The richness and complexity of human language cannot be underestimated. At the same time, the need for algorithms that can comprehend language is ever growing, and natural language processing exists to fill this gap. Traditional NLP methods take a linguistics-based approach, building up from base semantic and syntactic elements of a language, such as part-of-speech. Modern deep learning approaches can sidestep the need for intermediate elements and may learn its own hierarchical representations for generalized tasks.

As with deep learning, in this section we will try to summarize some important events that have shaped natural language processing as we know it today.

We will give a brief overview of important events that impacted the field up until 2000 (Fig. 1.7). For a very comprehensive summary, we refer the reader to a well-documented outline in Karen Jones's survey [Jon94]. Since neural architectures and deep learning have, in general, had much impact in this area and are the focus of the book, we will cover these topics in more detail.

Though there were traces of interesting experiments in the 1940s, the IBM-Georgetown experiment of 1954 that showcased machine translation of around 60 sentences from Russian to English can be considered an important milestone [HDG55]. Though constrained with computing resources in the form of software and hardware, some of the challenges of syntactic, semantic, and linguistic variety were discovered, and an attempt was made to address them. Similar to how AI was experiencing the golden age, many developments took place between 1954–1966, such as the establishment of conferences including the Dartmouth Conference in 1956, the Washington International Conference in 1958, and the Teddington International Conference on Machine Translation of Languages and Applied Language Analysis in 1961. In the Dartmouth Conference of 1956, John McCarthy coined the term "artificial intelligence." In 1957, Noam Chomsky published his book *Syntactic Structures*, which highlighted the importance of sentence syntax in language understanding [Cho57]. The invention of the phrase-structure grammar also played an important role in that era. Most notably, the attempts at the Turing test by software such as LISP by John McCarthy in 1958 and ELIZA (the first chatbot) had a great influence not only in NLP but in the entire field of AI.

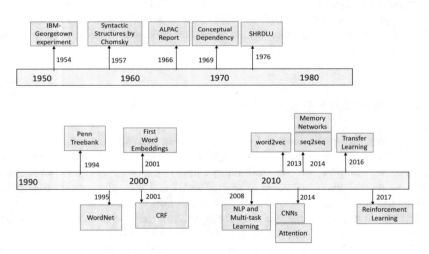

Fig. 1.7: Highlights in natural language processing research

In 1964, the United States National Research Council (NRC) set up a group known as the Automatic Language Processing Advisory Committee (ALPAC) to evaluate the progress of NLP research. The ALPAC report of 1966 highlighted the difficulties surrounding machine translation from the process itself to the cost of implementation and was influential in reducing funding, nearly putting a halt to NLP research [PC66]. This phase of the 1960s–1970s was a period in the study of world knowledge that emphasized semantics over syntactical structures. Grammars such as case grammar, which explored relations between nouns and verbs, played an interesting role in this era. Augmented transition networks was another search algorithm approach for solving problems like the best syntax for a phrase. Schank's conceptual dependency, which expressed language in terms of semantic primitives without syntactical processing, was also a significant development [ST69]. SHRDLU was a simple system that could understand basic questions and answer in natural language using syntax, semantics, and reasoning. LUNAR by Woods et al. was the first of its kind: a question-answering system that combined natural language understanding with a logic-based system. Semantic networks, which capture knowledge as a graph, became an increasingly common theme highlighted in the work of Silvio Ceccato, Margaret Masterman, Quillian, Bobrow and Collins, and Findler, to name a few [Cec61, Mas61, Qui63, BC75, Fin79]. In the early 1980s, the grammatico-logical phase began where the linguists developed different grammar structures and started associating meaning in phrases concerning users' intention. Many tools and software such as Alvey natural language tools, SYSTRAN, METEO, etc. became popular for parsing, translation, and information retrieval [Bri+87, HS92].

The 1990s were an era of statistical language processing where many new ideas of gathering data such as using the corpus for linguistic processing or understanding the words based on its occurrence and co-occurrence using probabilistic-based approaches were used in most NLP-based systems [MMS99]. A large amount of data available through the World Wide Web across different languages created a high demand for research in areas such as information retrieval, machine translation, summarization, topic modeling, and classification [Man99]. An increase in memory and processing speed in computers made it possible for many real-world applications to start using text and speech processing systems. Linguistic resources, including annotated collections such as the Penn Treebank, British National Corpus, Prague Dependency Treebank, and WordNet, were beneficial for academic research and commercial applications [Mar+94, HKKS99, Mil95]. Classical approaches such as n-grams and a bag-of-words representation with machine learning algorithms such as multinomial logistic regression, support vector machines, Bayesian networks, or expectation–maximization were common supervised and unsupervised techniques for many NLP tasks [Bro+92, MMS99]. Baker et al. introduced the FrameNet project which looked at "frames" to capture semantics such as entities and relationships and this led to semantic role labeling, which is an active research topic today [BFL98].

In the early 2000s, the Conference on Natural Language Learning (CoNLL) shared-tasks resulted in much interesting NLP research in areas such as chunking, named entity recognition, and dependency parsing to name a few [TKSB00, TKSDM03a, BM06]. Lafferty et al. proposed conditional random fields (CRF), which have become a core part of most state-of-the-art frameworks in sequence labeling where there are interdependencies between the labels [LMP01].

Bengio et al. in the early 2000s proposed the first neural language model which used mapping of n previous words using a lookup table feeding into a feed-forward network as a hidden layer and generating an output that is smoothed into a softmax layer to predict the word [BDV00]. Bengio's research was the first usage of the "dense vector representation" instead of the "one-hot vector" or bag-of-words model in NLP history. Many language models based on recurrent neural networks and long short-term memory, which were proposed later, have become the state of the art [Mik+10b, Gra13]. Papineni et al. proposed the bilingual evaluation understudy (BLEU) metric which is used even today as a standard metric for machine translations [Pap+02]. Pang et al. introduced sentiment classification, which is now one of the most popular and widely studied NLP tasks [PLV02]. Hovy et al. introduced OntoNotes, a large multilingual corpus with multiple annotations used in a wide variety of tasks such as dependency parsing and coreference resolution [Hov+06a]. The distant supervision technique, by which existing knowledge is used to generate patterns that can be used to extract examples from large corpora, was proposed by Mintz et al. and is used in a variety of tasks such as relation extraction, information extraction, and sentiment analysis [Min+09].

The research paper by Collobert and Weston was instrumental not only in highlighting ideas such as pre-trained word embeddings and convolutional neural networks for text but also in sharing the lookup table or the embedding matrix for multitask learning [CW08]. Multitask learning can learn multiple tasks at the same time and has recently become one of the more recent core research areas in NLP. Mikolov et al. improved the efficiency of training the word embeddings proposed by Bengio et al. by removing the hidden layer and having an approximate objective for learning that gave rise to "word2vec," an efficient large-scale implementation of the embeddings [Mik+13a, Mik+13b]. Word2vec has two implementations: (a) continuous bag-of-words (CBOW), which predicts the center word given the nearby words, and (b) skip-gram, which does the opposite and predicts the nearby words. The efficiency gained from learning on a large corpus of data enabled these dense representations to capture various semantics and relationships. Word embeddings used as representations and pre-training of these embeddings on a large corpus for any neural-based architecture are standard practice today. Recently many extensions to word embeddings, such as projecting word embeddings from different languages into the same space and thus enabling "transfer learning" in an unsupervised manner for various tasks such as machine translation, have gained lots of interest [Con+17].

Sutskever's Ph.D. thesis which introduced the Hessian-free optimizer to train recurrent neural networks efficiently on long-term dependencies was a milestone in reviving the usage of RNNs especially in NLP [Sut13]. Usage of convolutional neural networks on text surged greatly after advances made by Kalchbrenner et al. and Kim et al. [KGB14, Kim14]. CNNs are now widely used across many NLP tasks because of their dependency on the local context through convolution operation, making it highly parallelizable. Recursive neural networks, which provide a recursive hierarchical structure to the sentences and are inspired by linguistic approach, became another important neural architecture in the neural-based NLP world [LSM13].

Sutskever et al. proposed sequence-to-sequence learning as a general neural framework composed of an encoder neural network processing inputs as a sequence and a decoder neural network predicting the outputs based on the input sequence states and the current output states [SVL14]. This framework has found a wide range of applications such as constituency parsing, named entity recognition, machine translation, question-answering, and summarization. Google started replacing its monolithic phrase-based machine translation models with sequence-to-sequence neural MT models [Wu+16]. Character-based rather than word-based representations overcome many issues such as out of vocabulary and have been part of research in deep learning based systems for various NLP tasks [Lam+16, PSG16]. The attention mechanism by Bahdanau et al. is another innovation that has been widely popular in different neural architectures for NLP and speech [BCB14b]. Memory augmented networks with various variants such as memory networks, neural Turing machines, end-to-end memory networks, dynamic memory networks, differentiable neural computers, and recurrent entity networks have become very popular in the last few years for complex natural language understanding and language modeling tasks [WCB14, Suk+15, GWD14, Gra+16, Kum+16, Gre+15, Hen+16]. Adversarial learning and using adversarial examples have recently become common for distribution understanding, testing the robustness of models, and transfer learning [JL17, Gan+16]. Reinforcement learning is another emerging field in deep learning and has applications in NLP, specifically in the areas where there is a temporal dependencies and non-differentiable optimization zones where gradient-based methods fail. Modeling dialog systems, machine translation, text summarization, and visual storytelling among others have seen the benefits of reinforcement techniques [Liu+18, Ran+15, Wan+18, PXS17].

1.2.3 Automatic Speech Recognition: A Brief History

Automatic speech recognition (ASR) is quickly becoming a mainstay in human–computer interaction. Most of the tools used today have an option for speech recognition for various types of dictation tasks, whether it is composing a text message, playing music through a home-connected device, or even text-to-speech applications with virtual assistants. Although many of the techniques have recently gained popularity, research and development of ASR began in the middle of the twentieth century (Fig. 1.8).

The earliest research in ASR began in the 1950s. In 1952, Bell Laboratories created a system to recognize the pronunciation of isolated digits from a single speaker, using formant frequencies (frequencies that correlate to human speech for certain sounds) from the speech power spectrum. Many research universities built systems to recognize specific syllables and vowels for a single talker [JR05b].

In the 1960s, small vocabulary and acoustic phonetic-based tasks became prime research areas, leading to many techniques centered around dynamic programming and frequency analysis. IBM's Shoebox was able to recognize not only digits but also words such as "sum" and "total" and use these in the arithmetic computations to give results. The researchers in University College in England could analyze phonemes for recognition of vowels and consonants [JR05a].

In the 1970s, research moved towards medium-sized vocabulary tasks and continuous speech. The dominant techniques were various types of pattern recognition and clustering algorithms. Dynamic time warping (DTW) was introduced to handle time variability, aligning an input sequence of features to an output sequence of classes. "Harpy," a speech recognizer from Carnegie Mellon University, was capable of recognizing speech with a vocabulary of 1011 words. One of the main achievements of this work was the introduction of the graph search to "decode" lexical representations of words with a set of rules and a finite state network [LR90]. The methods that would optimize this capability, however, were not introduced until the 1990s. A recognition system called Tangora [JBM75] was created by IBM to provide a "voice-activated typewriter." This effort introduced a focus on large vocabularies and the sequence of words for grammars, which led to the introduction of language models for speech. During this era, AT&T also played a significant role in ASR, focusing heavily on speaker-independent systems. Their work, therefore, focused more heavily on what is called the acoustic model, dealing with the analysis of the speech patterns across speakers. By the mid-late 1970s, hidden Markov models were used to model spectral variations for discrete speech.

In the 1980s, the fundamental approach of ASR shifted to a statistical foundation, specifically HMM methods for modeling transitions between states. By the mid-1980s, HMMs had become the dominant technique for ASR (and remains one of the most prominent today). This shift to HMMs allowed many other advancements such as speech decoding frameworks with FSTs. The 1980s saw the introduction of neural networks for speech recognition. Their ability to approximate any function made them an exciting candidate for predicting the state transitions while still relying on the HMM to handle the temporal nature of continuous speech. Various toolkits were created during this period to support ASR, such as Sphinx [Lee88] and DECIPHER [Mur+89] from SRI.

In the 1990s, many advancements in machine learning were incorporated into ASR, which led to improved accuracy. Many of these were software which became available commercially, such as Dragon which had a dictionary of 80,000 words and the ability to train the software to the user's voice. Many toolkits were created to support ASR in the late 1980s and 1990s such as HTK [GW01] from Cambridge, a hidden Markov model toolkit.

The time delay neural network (TDNN) [Wai+90] was one of the earliest applications of deep learning to speech recognition. It utilized stacked 2D convolutional layers to perform phone classification. The benefits of this approach were that it was shift-invariant (not requiring a segmentation); however, the width of the network limits the context window. The TDNN approach was comparable to early HMM-based approaches; however, it did not integrate with HMMs and was difficult to use in large vocabulary settings [YD14].

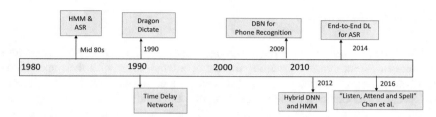

Fig. 1.8: Highlights in ASR

In the 2000s, the focus on machine learning advancements continued. In [MDH09] deep belief networks were applied to phone recognition, achieving state-of-the-art performance on the TIMIT corpus.[1] These networks learn unsupervised features for better acoustic robustness. In [Dah+12] a hybrid DNN and context-dependent (CD) hidden Markov model was introduced that extended the advancements of the DNN and achieved substantial improvements for large vocabulary speech recognition. Deep neural networks continued to advance the state-of-the-art during the 2000s, and the DNN/HMM hybrid model became the dominant approach.

Since 2012, deep learning has been applied to the sequence portion of the ASR task, replacing the HMM for many of the techniques, moving towards end-to-end models for speech recognition. With their introduction, many of the modern methods have been making their way into ASR, such as attention [Cho+15] [KHW17], and RNN transducers [MPR08]. The incorporation of sequence-to-sequence architectures with larger datasets allows the models to learn the acoustic and linguistic dependencies directly from the data, leading to higher quality.

End-to-end research has continued to develop in recent years, focusing on improving some of the difficulties that arise from end-to-end models; however, hybrid architectures tend to remain more popular in production, due to the usefulness of lexicon models in decoding. For a more detailed survey of the history of ASR, [TG01] is recommended.

---

[1] http://www.ldc.upenn.edu/Catalog/CatalogEntry.jsp?catalogId=LDC93S1.

1.3 Tools, Libraries, Datasets, and Resources for the Practitioners

There are a myriad of open-source resources available to the reader interested in building NLP, deep learning, or speech analytics. In the section below, we provide a list of the more popular libraries and datasets. This is in no way an exhaustive list, as our goal is to familiarize the reader with the wide range of available frameworks and resources.

1.3.1 Deep Learning

As with the techniques in NLP, deep learning frameworks have also seen tremendous progress in recent years. Numerous frameworks exist, each with their specialization. The most popular deep learning frameworks are: TensorFlow, PyTorch, Keras, MXNet, CNTK, Chainer, Caffe2, PaddlePaddle, and Matlab.[2] The main component of a modern deep learning framework is efficiency in linear algebra capabilities, as this applies to deep learning, with support CPU and GPU computation (dedicate hardware like TPUs [Jou16] are becoming increasingly popular as well). All relevant Python frameworks encompass both CPU and GPU support. The implementation differences tend to focus on the trade-offs between the intended end user (researcher vs. engineer vs. data scientist).

We look at the top deep learning frameworks from the ones mentioned previously and give a brief description of them. In order to do this, we compare the Google Trends for each of the frameworks and focus on the top 3.[3] As shown in Fig. 1.9, the top framework (worldwide) is Keras, followed by TensorFlow, and then PyTorch. Additional information about the frameworks used in the case studies will be given throughout the book.

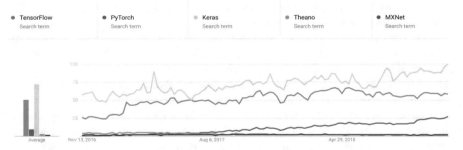

Fig. 1.9: Google trends for deep learning frameworks worldwide

---

[2] Theano exists as another popular framework; however, major development has discontinued given the popularity of more recent frameworks. It is therefore not included in this book.

[3] Although this is a single, statistically insignificant, data point, the Google Trends mechanism is useful and roughly correlates with other evaluations such as number of contributors, GitHub popularity, number of articles written, and books written for the various frameworks.

The following are popular open-source frameworks for building neural networks:

- **TensorFlow**: TensorFlow is a computational library based on data flow graphs. These graphs have nodes representing mathematical operations and edges representing tensors that flow between them. Written in Python, TensorFlow was developed by the Google Brain team.

- **Keras**: Keras is a simple, high-level, Python library developed to enable fast prototyping and experimentation. It can run on top of TensorFlow and CNTK, and is now part of the TensorFlow core library. Keras contains implementations of common neural network components and numerous architecture examples.

- **PyTorch**: PyTorch is a Python package for rapid prototyping of neural networks. It is based on Torch, an extremely fast computational framework, and provides dynamic graph computation at runtime. PyTorch was developed by the Facebook AI research team.

- **Caffe**: Caffe is a high-performance C++ framework for building deep learning architectures that can natively support distributed and multi-GPU execution. The current version, Caffe2, is the backend used by Facebook in production.

- **CNTK**: renamed the Microsoft Cognitive Toolkit, CNTK is a computational framework based on directed graphs. It supports the Python, C#, or C++ languages, and was developed by Microsoft Research.

- **MXNet**: MXNet is a high-performance computational framework written in C++ with native GPU support incubated by the Apache Project.

- **Chainer**: Chainer is a pure Python-based framework with dynamic computational graph capability defined at runtime.

1.3.2 Natural Language Processing

The following resources are some of the more popular open-source toolkits for natural language processing:

- **Stanford CoreNLP**: A Java-based toolkit of linguistic analysis tools for processing natural language text. CoreNLP was developed by Stanford University.

- **NLTK**: The Natural Language Toolkit, or NLTK for short, is an open-source suite of libraries for symbolic and statistical natural language processing for English. It was developed by the University of Pennsylvania.

- **Gensim**: A Python-based, open-source toolkit that focuses on vector space and topic modeling of text documents.

- **spaCy**: A high-performance Python-based toolkit for advanced natural language processing. SpaCy is open-source and supported by Explosion AI.

- **OpenNLP**: An open-source machine learning toolkit for processing text in natural language. OpenNLP is a sponsored by the Apache Project.

- **AllenNLP**: An NLP research library built in PyTorch.

1.3.3 Speech Recognition

The following resources are some of the more popular open-source toolkits for speech recognition.[4]

1.3.3.1 Frameworks

- **Sphinx**: An ASR toolkit developed by Carnegie Mellon University, focused on production and application development.

- **Kaldi**: An open-source C++ ASR framework, built for research-focused speech processing as well as for professional use.

- **ESPnet**: An end-to-end deep learning-based, ASR framework, inspired by Kaldi and written with PyTorch and Chainer backends.

1.3.3.2 Audio Processing

- **SoX**: An audio manipulation toolkit and library. It implements many file formats and is useful for playing, converting, and manipulating audio files [NB18].

- **LibROSA**: A Python package for audio analysis, commonly used for feature extraction and digital signals processing (DSP) [McF+15].

1.3.3.3 Additional Tools and Libraries

- **KenLM**: A high-performance, n-gram language modeling toolkit commonly integrated with ASR frameworks.

- **LIME**: Local interpretable model-agnostic explanation (LIME), a local and model-agnostic explainer for machine learning and deep learning models.

---

[4] Code repositories containing specific implementations that do not provide a full framework may be used, but are not included on this list.

1.3.4 Books

The fields of NLP and machine learning are very broad and cannot all be contained in a single resource. Here we recognize various books that provide deeper explanations, for supplementary information. The Elements of Statistical Learning by Hastie et al. gives a good base to machine learning and statistical techniques [HTF01]. Learning From Data by Abu-Mostafa et al. also provides insights into theories in machine learning in a more simplified and understandable manner [AMMIL12]. Deep Learning (Adaptive Computation and Machine Learning series) [GBC16] presents an advanced theory-focused book. This is widely considered the foundational book for deep learning, moving from the fundamentals for deep learning (Linear Algebra, Probability Theory, and Numerical computation) to an exploration of many architecture implementations and approaches. Foundations of Statistical Natural Language Processing [MS99] is a comprehensive resource on statistical models for natural language processing, providing an in-depth mathematical foundations for implementing NLP tools. Speech and Language Processing [Jur00] provides an introduction to NLP and speech, with both breadth and depth in many areas of statistical NLP. More recent editions explore neural network applications.

Neural Network Methods in Natural Language Processing (Synthesis Lectures on Human Language Technologies)[Gol17] presents neural network applications on language data, moving from the introduction of machine learning and neural networks to specialized neural architectures for NLP applications. Automatic Speech Recognition: A Deep Learning Approach by Yu et al. gives a thorough introduction to ASR and deep learning techniques [YD15].

1.3.5 Online Courses and Resources

Below we list some of the online courses where the experts in the field teach topics related to deep learning, NLP, and speech and are extremely beneficial.

- **Natural Language Processing with Deep Learning**
  http://web.stanford.edu/class/cs224n/
- **Deep Learning for Natural Language Processing**
  http://www.cs.ox.ac.uk/teaching/courses/2016-2017/dl
- **Neural Networks for NLP**
  http://phontron.com/class/nn4nlp2017/schedule.html
- **Deep Learning Specialization**
  https://www.deeplearning.ai/deep-learning-specialization/
- **Deep Learning Summer School**
  https://vectorinstitute.ai/2018/11/07/vector-institute-deep-learning-and-reinforcement-learning-2018-summer-school/
- **Convolutional Neural Networks for Visual Recognition**
  http://cs231n.stanford.edu/

- **Neural Networks for Machine Learning**
  http://www.cs.toronto.edu/~hinton/coursera_lectures.html
- **Neural Networks**
  http://info.usherbrooke.ca/hlarochelle/neural_networks/content.html
- **Practical Deep Learning For Coders**
  https://course.fast.ai/
- **Intro to Deep Learning with PyTorch**
  https://www.udacity.com/course/deep-learning-pytorch--ud188

1.3.6 Datasets

Any end-to-end deep learning application is reliant on data. While most organizations rely on data collection as a fundamental part of their strategy, there are many publicly available datasets for researchers, hobbyists, and practitioners.[5]

Kaggle[6] is one of the most popular sources for datasets and competitions for machine learning and data science. With thousands of datasets and competitions available, an active community of over one million registered users, and funded competitions through its platform, Kaggle serves as a strong starting point for not only datasets but also techniques for many tasks.

The Linguistic Data Consortium[7] combines and sells datasets from a variety of universities, corporations, and research labs. It primarily focuses on linguistic data and language resources, with close to 1000 datasets.

Stanford NLP group also releases a number of datasets for NLP, specifically for the training of their CoreNLP models.[8]

- **Text Similarity**:

Dataset	Description
SentEval [CK18]	Evaluation library for sentence embeddings, comparing their effectiveness on 17 downstream tasks
Quora Question Pairs [ZCZ]	Collection of 400,000 question pairs of potentially duplicate questions from the Quora website with the goal of identifying duplicates

---

[5] Additional text tasks and datasets are captured with associated papers at https://nlpprogress.com/.

[6] https://www.kaggle.com.

[7] https://www.ldc.upenn.edu/.

[8] https://nlp.stanford.edu/data/.

- **Text Clustering and Classification**:

Dataset	Description
Reuters-21,578 [Zdr+18]	A collection of 21,578 articles that appeared on Reuters newswire in 1987
Open ANC [MIG02]	Approximately 15 million spoken and written words of American English from a variety of sources annotated for syntactic structure
MASC [Ide+10]	A subset of approximately 500,000 words of contemporary American English drawn from the Open American National Corpus annotated for syntactic, semantic, and discourse structure

- **Dependency Parsing**:

Dataset	Description
Penn Treebank [MMS93]	A corpus of part-of-speech annotated American English, consisting of 4.5 million words

- **Entity Extraction**:

Dataset	Description
CoNLL 2003 [TKSDM03b]	Newswire text tagged with various types of entities, such as location, organization, person, etc.
WNUT2017 [Der+17]	A dataset composed of annotated tweets, YouTube comments, and other internet sources, tagged with different entities
OntoNotes [Hov+06b]	A multilingual corpus with many labels, such as part-of-speech, parse trees, and entities (two million tokens in version 5)

- **Relation Extraction**:

Dataset	Description
NYT Corpus [RYM10]	New York Times corpus with relational labels for distantly related entities
SemEval-2010 (Task 8) [Hen+09]	A semantic relationship dataset, containing types of relationships between, such as entity-origin and cause–effect

- **Semantic Role Labeling**:

Dataset	Description
OntoNotes [Pra+13]	A dataset of 1.7 million words focused on modeling the role of an entity in text

- **Machine Translation**:

Dataset	Description
Tatoeba	A collection of multilingual sentence pairs from the Tatoeba website https://tatoeba.org
WMT 2014 [Sch18]	An English–French (and additional English–German) dataset consisting of sentences taken from many sources, such as common crawl, UN corpus, and new commentary
Multi30k [Ell+16]	Crowdsourced descriptions of images in multiple languages

- **Text Summarization**:

Dataset	Description
CNN/Daily Mail [Nal+16]	Approximately 300,000 news stories from CNN and Daily Mail websites with corresponding summary bullet points
Cornell Newsroom [GNA18]	Over 1.3 million news articles and summaries from 38 major publications between 1998 and 2017
Google Dataset [FA13]	A sentence-compression task with 200,000 examples that focuses on deleting words to produce a compressed structure of the original, longer sentence
DUC	A smaller sentence summarization task with newswire and document data, containing 500 documents
Webis-TLDR-17 Corpus [Sye+18]	A dataset containing Reddit posts with the "too long; didn't read (TL;DR)" summaries for over three million posts

- **Question Answer**:

Dataset	Description
bAbI [Wes+15]	A collection of tasks and associated datasets to evaluate NLP models, specifically geared towards Q&A
NewsQA [Tri+16]	A set of 100,000 challenging question–answer pairs for news articles from CNN
SearchQA [Dun+17]	A general question-answering set for search QA

- **Speech Recognition**:

Dataset	Description
AN4 [Ace90]	A small, alphanumeric dataset containing randomly generated words, numbers, and letters
WSJ [PB92]	A general purpose, large vocabulary, speech recognition dataset, consisting of 400 h of spoken audio and transcripts
LibriSpeech [Pan+15]	1000 h Dataset containing read speech from audiobooks from the LibriVox audiobook project
Switchboard [GHM92]	Conversational dataset, consisting of more than 240 h of audio. The combined test with CallHome English set is referred to as Hub5'00
TED-LIUM [RDE12]	452 h of TED talks with corresponding transcripts and alignment information. The most recent version (3) [Her+18] has twice as much data as the previous versions
CHiME [Vin+16]	Is a speech challenge that has consisted of various tasks and datasets throughout the years. Some of the tasks include speech separation, recognition, speech processing in noisy environments, and multi-channel recognition
TIMIT [Gar+93]	An ASR dataset for phonetic studies with 630 speakers reading phonetically rich sentences

1.4 Case Studies and Implementation Details

In this book, the goal is not only to inform, but also to enable the reader to practice what is learned. In each of the following chapters a case study explores chapter concepts in detail and provides a chance for hands-on practice. The case studies and supplementary code are written in Python, utilizing a various deep learning frameworks. In most cases, deep learning depends on high-performance C++ or CUDA libraries to perform computation. In our experience, the installation process can become an extremely tedious process, especially when getting started with deep learning. We attempt to limit this difficulty by providing docker [Mer14] images and a GitHub repository for each case study. Docker is a simple yet powerful tool that provides a virtual machine-like environment for high-level (Python) and low-level (C++ and CUDA) libraries to operate in isolation from the operating system.

Directions for accessing and running the code are provided in the repository. (https://github.com/SpringerNLP).

The case studies are as follows:

- **Chapter 2**: An introduction to machine learning classification of the Higgs Boson Challenge. This introduces basic concepts of machine learning along with elements of data science.

- **Chapter 3**: A text clustering, topic modeling, and text classification based on the Reuters-21,578 dataset is used to show some fundamental approaches to NLP.

- **Chapter 4**: The fundamentals of supervised and unsupervised deep learning are introduced on spoken digit recognition using the FSDD dataset.

- **Chapter 5**: Embedding methods are introduced with a focus on text based representations on the Open American National Corpus.

- **Chapter 6**: Text classification is explored with a variety of convolutional neural network-based methods on Twitter airlines dataset.

- **Chapter 7**: Various recurrent neural networks architectures are compared for neural machine translation to perform English-to-French translation.

- **Chapter 8**: Traditional HMM-based speech recognition is explored on the Common Voice dataset, using Kaldi and CMUSphinx.

- **Chapter 9**: This chapter has two distinct case studies. In the first, neural machine translation from Chap. 7 is extended, exploring various attention-based architectures. In the second, memory augmented networks are compared for question-answering tasks based on the bAbI dataset.

- **Chapter 10**: Understanding multitask learning with different architectures on NLP tasks such as part-of-speech tagging, chunking, and named entity recognition is the focus of the case study in this chapter.

- **Chapter 11**: Transfer learning and specifically domain adaptation using various techniques on the Amazon Review dataset is performed.

- **Chapter 12**: Continuing the ASR case study from Chap. 8, end-to-end techniques are applied to speech recognition leveraging CTC and attention on the Common Voice dataset.

- **Chapter 13**: Two popular reinforcement learning algorithms are applied to the task of abstractive text summarization using the Cornell Newsroom dataset.

References

[AMMIL12] Yaser S. Abu-Mostafa, Malik Magdon-Ismail, and Hsuan-Tien Lin. *Learning From Data*. AMLBook, 2012. ISBN: 1600490069, 9781600490064.

[Ace90] Alejandro Acero. "Acoustical and environmental robustness in automatic speech recognition". In: *Proc. of ICASSP*. 1990.

[AHS88] David H. Ackley, Geoffrey E. Hinton, and Terrence J. Sejnowski. "Neurocomputing: Foundations of Research". In: ed. by James A. Anderson and Edward Rosenfeld. MIT Press, 1988. Chap. A Learning Algorithm for Boltzmann Machines, pp. 635–649.

[BCB14b] Dzmitry Bahdanau, Kyunghyun Cho, and Yoshua Bengio. "Neural Machine Translation by Jointly Learning to Align and Translate". In: *CoRR* abs/1409.0473 (2014).

[BFL98] Collin F. Baker, Charles J. Fillmore, and John B. Lowe. "The Berkeley FrameNet Project". In: *Proceedings of the 17th International Conference on Computational Linguistics - Volume 1*. COLING '98. Association for Computational Linguistics, 1998, pp. 86–90.

[BSF94] Y. Bengio, P. Simard, and P. Frasconi. "Learning Long-term Dependencies with Gradient Descent is Difficult". In: *Trans. Neur. Netw.* 5.2 (Mar. 1994), pp. 157–166.

[BDV00] Yoshua Bengio, Réjean Ducharme, and Pascal Vincent. "A Neural Probabilistic Language Model". In: *Proceedings of the 13th International Conference on Neural Information Processing Systems*. Denver, CO: MIT Press, 2000, pp. 893–899.

[BL07] Yoshua Bengio and Yann Lecun. "Scaling learning algorithms towards AI". In: *Large-scale kernel machines*. Ed. by L. Bottou et al. MIT Press, 2007.

[Ben+06] Yoshua Bengio et al. "Greedy Layer-wise Training of Deep Networks". In: *Proceedings of the 19th International Conference on Neural Information Processing Systems*. NIPS'06. Canada: MIT Press, 2006, pp. 153–160.

[BC75] Daniel G. Bobrow and Allan Collins, eds. *Representation and Understanding: Studies in Cognitive Science*. Academic Press, Inc., 1975.

[Bri+87] Ted Briscoe et al. "A Formalism and Environment for the Development of a Large Grammar of English". In: *Proceedings of the 10th International Joint Conference on Artificial Intelligence - Volume 2*. Morgan Kaufmann Publishers Inc., 1987, pp. 703–708.

[Bro+92] Peter F. Brown et al. "Class-based N-gram Models of Natural Language". In: *Comput. Linguist.* 18.4 (Dec. 1992), pp. 467–479.

[BM06] Sabine Buchholz and Erwin Marsi. "CoNLL-X Shared Task on Multilingual Dependency Parsing". In: *Proceedings of the Tenth*

	Conference on Computational Natural Language Learning. Association for Computational Linguistics, 2006, pp. 149–164.

[Cec61]   S. Ceccato. "Linguistic Analysis and Programming for Mechanical Translation". In: Gordon and Breach Science, 1961.

[Cho57]   Noam Chomsky. *Syntactic Structures*. Mouton and Co., 1957.

[Cho+15]   Jan K Chorowski et al. "Attention-based models for speech recognition". In: *Advances in neural information processing systems*. 2015, pp. 577–585.

[CW08]   Ronan Collobert and Jason Weston. "A Unified Architecture for Natural Language Processing: Deep Neural Networks with Multi-task Learning". In: *Proceedings of the 25th International Conference on Machine Learning*. ACM, 2008, pp. 160–167.

[CK18]   Alexis Conneau and Douwe Kiela. "SentEval: An Evaluation Toolkit for Universal Sentence Representations". In: *arXiv preprint arXiv:1803.05449* (2018).

[Con+17]   Alexis Conneau et al. "Supervised Learning of Universal Sentence Representations from Natural Language Inference Data". In: *EMNLP*. Association for Computational Linguistics, 2017, pp. 670–680.

[CV95]   Corinna Cortes and Vladimir Vapnik. "Support-Vector Networks". In: *Mach. Learn.* 20.3 (Sept. 1995), pp. 273–297.

[Cyb89]   G. Cybenko. "Approximation by superpositions of a sigmoidal function". In: *Mathematics of Control, Signals, and Systems (MCSS)* 2 (1989). URL: http://dx.doi.org/10.1007/BF02551274.

[Dah+12]   George E Dahl et al. "Context-dependent pre-trained deep neural networks for large-vocabulary speech recognition". In: *IEEE Transactions on audio, speech, and language processing* 20.1 (2012), pp. 30–42.

[Den+09]   J. Deng et al. "ImageNet: A Large-Scale Hierarchical Image Database". In: *CVPR09*. 2009.

[Der+17]   Leon Derczynski et al. "Results of the WNUT2017 shared task on novel and emerging entity recognition". In: *Proceedings of the 3rd Workshop on Noisy User-generated Text*. 2017, pp. 140–147.

[Koh82]   Bhuwan Dhingra, Kathryn Mazaitis, and William W Cohen. "Quasar: Datasets for Question Answering by Search and Reading". In: *arXiv preprint arXiv:1707.03904* (2017).

[Dun+17]   Matthew Dunn et al. "SearchQA: A new Q&A dataset augmented with context from a search engine". In: *arXiv preprint arXiv:1704.05179* (2017).

[Ell+16]   Desmond Elliott et al. "Multi30k: Multilingual English-German image descriptions". In: *arXiv preprint arXiv:1605.00459* (2016).

[FA13]   Katja Filippova and Yasemin Altun. "Overcoming the lack of parallel data in sentence compression". In: *Proceedings of the 2013 Conference on Empirical Methods in Natural Language Processing*. 2013, pp. 1481–1491.

[Fin79]   Nicholas V. Findler, ed. *Associative Networks: The Representa-tion and Use of Knowledge by Computers*. Academic Press, Inc., 1979.ISBN: 0122563808.

[Fuk79]   K. Fukushima. "Neural network model for a mechanism of pat-tern recognition unaffected by shift in position - Neocognitron". In: *Trans. IECE* J62-A(10) (1979), pp. 658–665.

[Gan+16]   Yaroslav Ganin et al. "Domain-adversarial Training of Neural Net-works". In: *J. Mach. Learn. Res.* 17.1 (Jan. 2016), pp. 2096–2030.

[Gar+93]   John S Garofolo et al. "DARPA TIMIT acoustic-phonetic contin-uous speech corpus CD-ROM. NIST speech disc 1-1.1". In: *NASA STI/Recon technical report n* 93 (1993).

[GW01]   James Glass and Eugene Weinstein. "SPEECHBUILDER: Facil-itating spoken dialogue system development". In: *Seventh Eu-ropean Conference on Speech Communication and Technology*. 2001.

[GBB11]   Xavier Glorot, Antoine Bordes, and Yoshua Bengio. "Deep Sparse Rectifier Neural Networks." In: *AISTATS*. Vol. 15. JMLR.org, 2011, pp. 315–323.

[GHM92]   John J Godfrey, Edward C Holliman, and Jane McDaniel. "SWITCHBOARD: Telephone speech corpus for research and de-velopment". In: *Acoustics, Speech, and Signal Processing, 1992. ICASSP-92., 1992 IEEE International Conference on*. Vol. 1. 1992, pp. 517–520.

[Gol17]   Yoav Goldberg. "Neural network methods for natural language processing". In: *Synthesis Lectures on Human Language Technolo-gies* 10.1 (2017), pp. 1–309.

[GBC16]   Ian Goodfellow, Yoshua Bengio, and Aaron Courville. "Deep learning (adaptive computation and machine learning series)". In: *Adaptive Computation and Machine Learning series* (2016), p. 800.

[Goo+14]   Ian J. Goodfellow et al. "Generative Adversarial Nets". In: *Pro-ceedings of the 27th International Conference on Neural Informa-tion Processing Systems - Volume 2*. NIPS'14. MIT Press, 2014, pp. 2672–2680.

[Gra13]   Alex Graves. "Generating Sequences With Recurrent Neural Net-works." In: *CoRR* abs/1308.0850 (2013).

[GWD14]   Alex Graves, Greg Wayne, and Ivo Danihelka. "Neural Turing Ma-chines". In: *CoRR* abs/1410.5401 (2014).

[Gra+16]   Alex Graves et al. "Hybrid computing using a neural network with dynamic external memory". In: *Nature* 538.7626 (Oct. 2016), pp. 471–476. ISSN: 00280836.

[Gre+15]   Edward Grefenstette et al. "Learning to Transduce with Un-bounded Memory". In: *Advances in Neural Information Process-ing Systems 28: Annual Conference on Neural Information Pro-*

cessing Systems 2015, December 7–12, 2015, Montreal, Quebec, Canada. 2015, pp. 1828–1836.

[GNA18]   Max Grusky, Mor Naaman, and Yoav Artzi. "NEWSROOM: A Dataset of 1.3 Million Summaries with Diverse Extractive Strategies". In: *Proceedings of the 2018 Conference of the North American Chapter of the Association for Computational Linguistics: Human Language Technologies.* Association for Computational Linguistics, 2018, pp. 708–719.

[HKKS99]   Eva Hajicová, Ivana Kruijff-Korbayová, and Petr Sgall. "Prague Dependency Treebank: Restoration of Deletions". In: *Proceedings of the Second International Workshop on Text, Speech and Dialogue.* Springer-Verlag, 1999, pp. 44–49.

[HTF01]   Trevor Hastie, Robert Tibshirani, and Jerome Friedman. *The Elements of Statistical Learning.* Springer Series in Statistics. Springer New York Inc., 2001.

[Heb49]   Donald O. Hebb. *The organization of behavior: A neuropsychological theory.* Wiley, 1949.

[Hen+16]   Mikael Henaff et al. "Tracking the World State with Recurrent Entity Networks". In: *CoRR* abs/1612.03969 (2016).

[Hen+09]   Iris Hendrickx et al. "Semeval-2010 task 8: Multi-way classification of semantic relations between pairs of nominals". In: *Proceedings of the Workshop on Semantic Evaluations: Recent Achievements and Future Directions.* Association for Computational Linguistics. 2009, pp. 94–99.

[Her+18]   François Hernandez et al. "TED-LIUM 3: twice as much data and corpus repartition for experiments on speaker adaptation". In: *arXiv preprint arXiv:1805.04699* (2018).

[HZ94]   G. E. Hinton and R. S. Zemel. "Autoencoders, Minimum Description Length and Helmholtz Free Energy". In: *Advances in Neural Information Processing Systems (NIPS) 6.* Ed. by J. D. Cowan, G. Tesauro, and J. Alspector. Morgan Kaufmann, 1994, pp. 3–10.

[HOT06a]   Geoffrey E. Hinton, Simon Osindero, and Yee-Whye Teh. "A Fast Learning Algorithm for Deep Belief Nets". In: *Neural Comput.* 18.7 (July 2006), pp. 1527–1554.

[Hoc98]   Sepp Hochreiter. "The Vanishing Gradient Problem During Learning Recurrent Neural Nets and Problem Solutions". In: *Int. J. Uncertain. Fuzziness Knowl.-Based Syst.* 6.2 (Apr. 1998), pp. 107–116.

[HS97]   Sepp Hochreiter and Jürgen Schmidhuber. "Long Short-Term Memory". In: *Neural Comput.* 9.8 (Nov. 1997), pp. 1735–1780.

[Hop82]   J. J. Hopfield. "Neural networks and physical systems with emergent collective computational abilities". In: *Proceedings of the National Academy of Sciences of the United States of America* 79.8 (Apr. 1982), pp. 2554–2558.

[Hor91]   Kurt Hornik. "Approximation Capabilities of Multilayer Feedforward Networks". In: *Neural Netw.* 4.2 (Mar. 1991), pp. 251–257.

[Hov+06a]   Eduard Hovy et al. "OntoNotes: The 90% Solution". In: *Proceedings of the Human Language Technology Conference of the NAACL, Companion Volume: Short Papers.* NAACL-Short '06. New York, New York: Association for Computational Linguistics, 2006, pp. 57–60.

[Hov+06b]   Eduard Hovy et al. "OntoNotes: the 90% solution". In: *Proceedings of the human language technology conference of the NAACL, Companion Volume: Short Papers.* Association for Computational Linguistics. 2006, pp. 57–60.

[HDG55]   W. John Hutchins, Leon Dostert, and Paul Garvin. "The Georgetown- I.B.M. experiment". In: *In.* John Wiley And Sons, 1955, pp. 124–135.

[HS92]   William J. Hutchins and Harold L. Somers. *An introduction to machine translation.* Academic Press, 1992.

[Ide+10]   Nancy Ide et al. "MASC: the Manually Annotated Sub-Corpus of American English." In: *LREC.* European Language Resources Association, June 4, 2010.

[JBM75]   Frederick Jelinek, Lalit Bahl, and Robert Mercer. "Design of a linguistic statistical decoder for the recognition of continuous speech". In: *IEEE Transactions on Information Theory* 21.3 (1975), pp. 250–256.

[JL17]   Robin Jia and Percy Liang. "Adversarial Examples for Evaluating Reading Comprehension Systems". In: *Proceedings of the 2017 Conference on Empirical Methods in Natural Language Processing.* Association for Computational Linguistics, 2017, pp. 2021–2031.

[Jon94]   Karen Sparck Jones. "Natural Language Processing: A Historical Review". In: *Current Issues in Computational Linguistics: In Honour of Don Walker.* Springer Netherlands, 1994, pp. 3–16.

[Jou16]   Norm Jouppi. "Google supercharges machine learning tasks with TPU custom chip". In: *Google Blog, May* 18 (2016).

[JR05a]   B. H. Juang and L. R. Rabiner. "Automatic speech recognition - A brief history of the technology development". In: *Elsevier Encyclopedia of Language and Linguistics* (2005).

[JR05b]   Biing-Hwang Juang and Lawrence R Rabiner. "Automatic speech recognition-a brief history of the technology development". In: *Georgia Institute of Technology. Atlanta Rutgers University and the University of California. Santa Barbara* 1 (2005), p. 67.

[Jur00]   Daniel Jurafsky. "Speech and language processing: An introduction to natural language processing". In: *Computational linguistics, and speech recognition* (2000).

[KGB14]   Nal Kalchbrenner, Edward Grefenstette, and Phil Blunsom. "A Convolutional Neural Network for Modelling Sentences". In: Association for Computational Linguistics, 2014, pp. 655–665.

[KHW17]   Suyoun Kim, Takaaki Hori, and Shinji Watanabe. "Joint CTC attention based end-to-end speech recognition using multi-task learning". In: *Acoustics, Speech and Signal Processing (ICASSP), 2017 IEEE International Conference on*. IEEE. 2017, pp. 4835–4839.

[Kim14]   Yoon Kim. "Convolutional Neural Networks for Sentence Classification". In: 2014, pp. 1746–1751.

[Koh82]   T. Kohonen. "Self-Organized Formation of Topologically Correct Feature Maps". In: *Biological Cybernetics* 43.1 (1982), pp. 59–69.

[KSH12]   Alex Krizhevsky, Ilya Sutskever, and Geoffrey E. Hinton. "ImageNet Classification with Deep Convolutional Neural Networks". In: *Proceedings of the 25th International Conference on Neural Information Processing Systems - Volume 1*. Curran Associates Inc., 2012, pp. 1097–1105.

[Kum+16]   Ankit Kumar et al. "Ask Me Anything: Dynamic Memory Networks for Natural Language Processing". In: *Proceedings of the 33nd International Conference on Machine Learning, ICML 2016, New York City, NY, USA, June 19–24, 2016*. 2016, pp. 1378–1387.

[LMP01]   John D. Lafferty, Andrew McCallum, and Fernando C. N. Pereira. "Conditional Random Fields: Probabilistic Models for Segmenting and Labeling Sequence Data". In: *Proceedings of the Eighteenth International Conference on Machine Learning*. Morgan Kaufmann Publishers Inc., 2001, pp. 282–289.

[Lam+16]   Guillaume Lample et al. "Neural Architectures for Named Entity Recognition." In: *HLT-NAACL*. The Association for Computational Linguistics, 2016, pp. 260–270.

[LeC85]   Y. LeCun. "Une procédure d'apprentissage pour réseau a seuil asymmetrique (a Learning Scheme for Asymmetric Threshold Networks)". In: *Proceedings of Cognitiva 85*. 1985, pp. 599–604.

[LeC+89]   Y. LeCun et al. "Backpropagation Applied to Handwritten Zip Code Recognition". In: *Neural Computation* 1.4 (1989), pp. 541–551.

[LB94]   Yann LeCun and Yoshua Bengio. "Word-level training of a handwritten word recognizer based on convolutional neural networks". In: *12th IAPR International Conference on Pattern Recognition, Conference B: Pattern Recognition and Neural Networks, ICPR 1994, Jerusalem, Israel, 9–13 October, 1994, Volume 2*. 1994, pp. 88–92.

[LBB97]   Yann LeCun, Léon Bottou, and Yoshua Bengio. "Reading checks with multilayer graph transformer networks". In: *1997 IEEE International Conference on Acoustics, Speech, and Signal Processing,*

ICASSP '97, Munich, Germany, April 21–24, 1997. 1997, pp. 151–154.

[Lee88] Kai-Fu Lee. "On large-vocabulary speaker-independent continuous speech recognition". In: *Speech communication* 7.4 (1988), pp. 375–379.

[Lin92] Long-Ji Lin. "Reinforcement Learning for Robots Using Neural Networks". UMI Order No. GAX93-22750. PhD thesis. Pittsburgh, PA, USA, 1992.

[Lin70] S. Linnainmaa. "The representation of the cumulative rounding error of an algorithm as a Taylor expansion of the local rounding errors". MA thesis. Univ. Helsinki, 1970.

[Liu+18] Bing Liu et al. "Dialogue Learning with Human Teaching and Feedback in End-to-End Trainable Task-Oriented Dialogue Systems". In: *Proceedings of the 2018 Conference of the North American Chapter of the Association for Computational Linguistics: Human Language Technologies, Volume 1 (Long Papers)*. Association for Computational Linguistics, 2018, pp. 2060–2069.

[LR90] Bruce Lowerre and Raj Reddy. "The HARPY speech understanding system". In: *Readings in speech recognition*. Elsevier, 1990, pp. 576–586.

[LSM13] Minh-Thang Luong, Richard Socher, and Christopher D Manning. "Better Word Representations with Recursive Neural Networks for Morphology". In: *CoNLL-2013* (2013), p. 104.

[MIG02] C. Macleod, N. Ide, and R. Grishman. "The American National Corpus: Standardized Resources for American English". In: *Proceedings of 2nd Language Resources and Evaluation Conference (LREC)*. 2002, pp. 831–836.

[Man99] Inderjeet Mani. *Advances in Automatic Text Summarization*. Ed. by Mark T. Maybury. MIT Press, 1999.

[MMS99] Christopher D Manning, Christopher D Manning, and Hinrich Schütze. *Foundations of statistical natural language processing*. MIT press, 1999.

[MS99] Christopher D. Manning and Hinrich Schütze. *Foundations of Statistical Natural Language Processing*. MIT Press, 1999.

[Mar+94] Mitchell Marcus et al. "The Penn Treebank: Annotating Predicate Argument Structure". In: *Proceedings of the Workshop on Human Language Technology*. Association for Computational Linguistics, 1994, pp. 114–119.

[MMS93] Mitchell P Marcus, Mary Ann Marcinkiewicz, and Beatrice Santorini. "Building a large annotated corpus of English: The Penn Treebank". In: *Computational linguistics* 19.2 (1993), pp. 313–330.

[Mas61] Margaret Masterman. "Semantic message detection for machine translation using an interlingua". In: *Proceedings of the Interna-*

tional Conference on Machine Translation. Her Majesty's Stationery Office, 1961, pp. 438–475.

[MP88]   Warren S. McCulloch and Walter Pitts. "Neurocomputing: Foundations of Research". In: MIT Press, 1988. Chap. A Logical Calculus of the Ideas Immanent in Nervous Activity, pp. 15–27.

[McF+15]   Brian McFee et al. "librosa: Audio and music signal analysis in python". In: *Proceedings of the 14th python in science conference.* 2015, pp. 18–25.

[Mer14]   Dirk Merkel. "Docker: lightweight Linux containers for consistent development and deployment". In: *Linux Journal* 2014.239 (2014), p. 2.

[Mik+10b]   Tomas Mikolov et al. "Recurrent neural network based language model." In: *INTERSPEECH.* Ed. by Takao Kobayashi, Keikichi Hirose, and Satoshi Nakamura. ISCA, 2010, pp. 1045–1048.

[Mik+13a]   Tomas Mikolov et al. "Distributed Representations of Words and Phrases and their Compositionality". In: *Advances in Neural Information Processing Systems 26.* Ed. by C. J. C. Burges et al. Curran Associates, Inc., 2013, pp. 3111–3119.

[Mik+13b]   Tomas Mikolov et al. "Efficient Estimation of Word Representations in Vector Space". In: *CoRR* abs/1301.3781 (2013).

[Mil95]   George A. Miller. "WordNet: A Lexical Database for English". In: *Commun. ACM* 38.11 (Nov. 1995), pp. 39–41.

[MP69]   Marvin Minsky and Seymour Papert. Perceptrons: *An Introduction to Computational Geometry.* Cambridge, MA, USA: MIT Press, 1969.

[Min+09]   Mike Mintz et al. "Distant Supervision for Relation Extraction Without Labeled Data". In: *Proceedings of the Joint Conference of the 47th Annual Meeting of the ACL and the 4th International Joint Conference on Natural Language Processing of the AFNLP: Volume 2 - Volume 2.* ACL '09. Association for Computational Linguistics, 2009, pp. 1003–1011.

[MDH09]   Abdel-rahman Mohamed, George Dahl, and Geoffrey Hinton. "Deep belief networks for phone recognition". In: *Nips workshop on deep learning for speech recognition and related applications.* Vol. 1. 9. Vancouver, Canada. 2009, p. 39.

[Moh+11]   Abdel-rahman Mohamed et al. "Deep Belief Networks using discriminative features for phone recognition". In: *ICASSP.* IEEE, 2011, pp. 5060–5063.

[MPR08]   Mehryar Mohri, Fernando Pereira, and Michael Riley. "Speech recognition with weighted finite-state transducers". In: *Springer Handbook of Speech Processing.* Springer, 2008, pp. 559–584.

[Mur+89]   Hy Murveit et al. "SRI's DECIPHER system". In: *Proceedings of the workshop on Speech and Natural Language.* Association for Computational Linguistics. 1989, pp. 238–242.

[NH10]   Vinod Nair and Geoffrey E. Hinton. "Rectified Linear Units Improve Restricted Boltzmann Machines". In: *Proceedings of the 27th International Conference on International Conference on Machine Learning*. ICML'10. Omnipress, 2010, pp. 807–814.

[Nal+16]   Ramesh Nallapati et al. "Abstractive text summarization using sequence-to-sequence RNNs and beyond". In: *arXiv preprint arXiv:1602.06023* (2016).

[Nea95]   Radford M Neal. "Bayesian learning for neural networks". PhD thesis. University of Toronto, 1995.

[NB18]   Lance Norskog and Chris Bagwell. "Sox-Sound eXchange". In: (2018).

[Pan+15]   Vassil Panayotov et al. "LibriSpeech: an ASR corpus based on public domain audio books". In: *Acoustics, Speech and Signal Processing (ICASSP), 2015 IEEE International Conference on*. 2015, pp. 5206–5210.

[PLV02]   Bo Pang, Lillian Lee, and Shivakumar Vaithyanathan. "Thumbs Up?: Sentiment Classification Using Machine Learning Techniques". In: *Proceedings of the ACL-02 Conference on Empirical Methods in Natural Language Processing - Volume 10*. Association for Computational Linguistics, 2002, pp. 79–86.

[Pap+02]   Kishore Papineni et al. "BLEU: A Method for Automatic Evaluation of Machine Translation". In: *Proceedings of the 40th Annual Meeting on Association for Computational Linguistics*. Association for Computational Linguistics, 2002, pp. 311–318.

[Par85]   D. B. Parker. *Learning-Logic*. Tech. rep. TR-47. Center for Comp. Research in Economics and Management Sci., MIT, 1985.

[PB92]   Douglas B Paul and Janet M Baker. "The design for the Wall Street Journal-based CSR corpus". In: *Proceedings of the workshop on Speech and Natural Language*. 1992, pp. 357–362.

[PXS17]   Romain Paulus, Caiming Xiong, and Richard Socher. "A Deep Reinforced Model for Abstractive Summarization". In: *CoRR* abs/1705.04304 (2017).

[PC66]   John R. Pierce and John B. Carroll. *Language and Machines: Computers in Translation and Linguistics*. Washington, DC, USA: National Academy of Sciences/National Research Council, 1966.

[PSG16]   Barbara Plank, Anders Søgaard, and Yoav Goldberg. "Multilingual Part-of-Speech Tagging with Bidirectional Long Short-Term Memory Models and Auxiliary Loss". In: *Proceedings of the 54th Annual Meeting of the Association for Computational Linguistics (Volume 2: Short Papers)*. Association for Computational Linguistics, 2016, pp. 412–418.

[Pom89]   Dean A. Pomerleau. "Advances in Neural Information Processing Systems 1". In: Morgan Kaufmann Publishers Inc., 1989. Chap. ALVINN: An Autonomous Land Vehicle in a Neural Network, pp. 305–313.

[Pra+13]   Sameer Pradhan et al. "Towards robust linguistic analysis using OntoNotes". In: *Proceedings of the Seventeenth Conference on Computational Natural Language Learning*. 2013, pp. 143–152.

[Qui63]   R Quillian. *A notation for representing conceptual information: an application to semantics and mechanical English paraphrasing.* 1963.

[Ran+15]   Marc'Aurelio Ranzato et al. "Sequence Level Training with Recurrent Neural Networks". In: *CoRR* abs/1511.06732 (2015).

[RYM10]   Sebastian Riedel, Limin Yao, and Andrew McCallum. "Modeling relations and their mentions without labeled text". In: *Joint European Conference on Machine Learning and Knowledge Discovery in Databases*. Springer. 2010, pp. 148–163.

[Ros58]   F. Rosenblatt. "The Perceptron: A Probabilistic Model for Information Storage and Organization in The Brain". In: *Psychological Review* (1958), pp. 65–386.

[RDE12]   Anthony Rousseau, Paul Deléglise, and Yannick Esteve. "TEDLIUM: an Automatic Speech Recognition dedicated corpus." In: *LREC*. 2012, pp. 125–129.

[RHW88]   David E. Rumelhart, Geoffrey E. Hinton, and Ronald J. Williams. "Neurocomputing: Foundations of Research". In: ed. by James A. Anderson and Edward Rosenfeld. MIT Press, 1988. Chap. Learning Representations by Back-propagating Errors, pp. 696–699.

[ST69]   Roger C. Schank and Larry Tesler. "A Conceptual Dependency Parser for Natural Language". In: *Proceedings of the 1969 Conference on Computational Linguistics*. COLING '69. Association for Computational Linguistics, 1969, pp. 1–3.

[Sch92]   J. Schmidhuber. "Learning Complex, Extended Sequences Using the Principle of History Compression". In: *Neural Computation* 4.2 (1992), pp. 234–242.

[Sch93]   J. Schmidhuber. *Habilitation thesis*. 1993.

[Sch15]   J. Schmidhuber. "Deep Learning in Neural Networks: An Overview". In: *Neural Networks* 61 (2015), pp. 85–117.

[SDS93]   Nicol N. Schraudolph, Peter Dayan, and Terrence J. Sejnowski. "Temporal Difference Learning of Position Evaluation in the Game of Go". In: *Advances in Neural Information Processing Systems 6, [7th NIPS Conference, Denver, Colorado, USA, 1993]*. 1993, pp. 817–824.

[Sch18]   H. Schwenk. "WMT 2014 EN-FR". In: (2018).

[Suk+15]   Sainbayar Sukhbaatar et al. "End-To-End Memory Networks". In: *Advances in Neural Information Processing Systems 28: Annual Conference on Neural Information Processing Systems 2015, December 7–12, 2015, Montreal, Quebec, Canada*. 2015, pp. 2440–2448.

[Sut13]   Ilya Sutskever. "Training recurrent neural networks". In: *Ph.D. Thesis from University of Toronto, Toronto, Ont., Canada* (2013).

[SVL14]   Ilya Sutskever, Oriol Vinyals, and Quoc V. Le. "Sequence to Sequence Learning with Neural Networks". In: *Proceedings of the 27th International Conference on Neural Information Processing Systems - Volume 2*. MIT Press, 2014, pp. 3104–3112.

[Sye+18]   Shahbaz Syed et al. *Dataset for generating TL;DR*. Feb. 2018.

[Tai+14]   Yaniv Taigman et al. "DeepFace: Closing the Gap to Human-Level Performance in Face Verification". In: *CVPR*. IEEE Computer Society, 2014, pp. 1701–1708.

[Tes95]   Gerald Tesauro. "Temporal Difference Learning and TD-Gammon". In: *Commun. ACM* 38.3 (Mar. 1995), pp. 58–68.

[Thr94]   Sebastian Thrun. "Learning to Play the Game of Chess". In: *Advances in Neural Information Processing Systems 7, [NIPS Conference, Denver, Colorado, USA, 1994]*. 1994, pp. 1069–1076.

[TKSB00]   Erik F. Tjong Kim Sang and Sabine Buchholz. "Introduction to the CoNLL-2000 Shared Task: Chunking". In: *Proceedings of the 2Nd Workshop on Learning Language in Logic and the 4th Conference on Computational Natural Language Learning - Volume 7*. ConLL'00. Association for Computational Linguistics, 2000, pp. 127–132.

[TKSDM03a]   Erik F. Tjong Kim Sang and Fien De Meulder. "Introduction to the CoNLL-2003 Shared Task: Language-independent Named Entity Recognition". In: *Proceedings of the Seventh Conference on Natural Language Learning at HLT-NAACL 2003 - Volume 4*. Association for Computational Linguistics, 2003, pp. 142–147.

[TKSDM03b]   Erik F Tjong Kim Sang and Fien De Meulder. "Introduction to the CoNLL-2003 shared task: Language-independent named entity recognition". In: *Proceedings of the seventh conference on Natural language learning at HLT-NAACL 2003-Volume 4*. Association for Computational Linguistics. 2003, pp. 142–147.

[TG01]   Edmondo Trentin and Marco Gori. "A survey of hybrid ANN/HMM models for automatic speech recognition". In: *Neurocomputing* 37.1–4 (2001), pp. 91–126.

[Tri+16]   Adam Trischler et al. "NewsQA: A machine comprehension dataset". In: *arXiv preprint arXiv:1611.09830* (2016).

[Tur95]   A. M. Turing. "Computers &Amp; Thought". In: MIT Press, 1995. Chap. Computing Machinery and Intelligence, pp. 11–35.

[Vin+16]   Emmanuel Vincent et al. "The 4th CHiME speech separation and recognition challenge". In: (2016).

[Wai+90]   Alexander Waibel et al. "Phoneme recognition using time-delay neural networks". In: *Readings in speech recognition*. Elsevier, 1990, pp. 393–404.

[Wan+18]   Xin Wang et al. "No Metrics Are Perfect: Adversarial Reward Learning for Visual Storytelling". In: *Proceedings of the 56th Annual Meeting of the Association for Computational Linguistics*

(Volume 1: Long Papers). Association for Computational Linguistics, 2018, pp. 899–909.

[Wat89]   Christopher John Cornish Hellaby Watkins. "Learning from Delayed Rewards". PhD thesis. Cambridge, UK: King's College, 1989.

[Wer74]   P. J. Werbos. "Beyond Regression: New Tools for Prediction and Analysis in the Behavioral Sciences". PhD thesis. Harvard University, 1974.

[WCB14]   Jason Weston, Sumit Chopra, and Antoine Bordes. "Memory Networks". In: *CoRR* abs/1410.3916 (2014).

[Wes+15]   Jason Weston et al. "Towards AI-Complete Question Answering: A Set of Prerequisite Toy Tasks". In: *CoRR* abs/1502.05698 (2015).

[WH60]   Bernard Widrow and Marcian E. Hoff. "Adaptive Switching Circuits". In: *1960 IRE WESCON Convention Record, Part 4*. IRE, 1960, pp. 96–104.

[Wu+16]   Yonghui Wu et al. "Google's neural machine translation system: Bridging the gap between human and machine translation". In: *arXiv preprint arXiv:1609.08144* (2016).

[YD14]   Dong Yu and Li Deng. *Automatic Speech Recognition - A Deep Learning Approach*. Springer, 2014.

[YD15]   Dong Yu and Li Deng. *Automatic Speech Recognition: A Deep Learning Approach*. Springer, 2015.

[ZCZ]   X. Zhang Z. Chen H. Zhang and L. Zhao. *Quora question pairs*.

[Zdr+18]   Anna Zdrojewska et al. "Comparison of the Novel Classification Methods on the Reuters-21578 Corpus." In: *MISSI*. Vol. 833. Springer, 2018, pp. 290–299.

Chapter 2
Basics of Machine Learning

2.1 Introduction

The goal of this chapter is to review basic concepts in machine learning that are applicable or relate to deep learning. As it is not possible to cover every aspect of machine learning in this chapter, we refer readers who wish to get a more in-depth overview to textbooks, such as *Learning from Data* [AMMIL12] and *Elements of Statistical Learning Theory* [HTF09].

We begin by giving the basic **learning framework** for supervised machine learning and the general learning process. We then discuss some core concepts of machine learning theory, such as **VC analysis** and **bias–variance trade-off**, and how they relate to **overfitting**. We guide the reader through various model evaluation, performance, and validation metrics. We discuss some basic linear classifiers starting with discriminative ones, such as linear regression, perceptrons, and logistic regression. We then give the general principle of non-linear transformations and highlight support vector machines and other non-linear classifiers. In the treatment of these topics, we introduce core concepts, such as **regularization**, **gradient descent**, and more, and discuss their impact on effective training in machine learning. Generative classifiers, such as naive Bayes and linear discriminant analysis, are introduced next. We then demonstrate how basic non-linearity can be achieved through linear algorithms via transformations. We highlight common feature transformations, such as feature selection and dimensionality reduction techniques. Finally, we introduce the reader to the world of sequence modeling through **Markov chains**. We provide necessary details in two very effective methods of sequence modeling: **hidden Markov models** and **conditional random fields**.

We conclude the chapter with a detailed case study of supervised machine learning using a real-world problem and dataset to carry out a systematic, evidence-based machine learning process that allows putting into practice the concepts related in this chapter.

© Springer Nature Switzerland AG 2019
U. Kamath et al., *Deep Learning for NLP and Speech Recognition*,
https://doi.org/10.1007/978-3-030-14596-5_2

2.2 Supervised Learning: Framework and Formal Definitions

As discussed in Chap. 1, supervised machine learning is the task of learning from answers (*labels*, or the *ground truth*) provided by an *oracle* in a generalized manner. A simple example would be learning to distinguish *apples* from *oranges*. The process of supervised learning is shown schematically in Fig. 2.1, and we will refer to it for most of the chapters in this book. Let us now describe each of the components of the supervised learning process.

2.2.1 Input Space and Samples

The population of all possible data for a particular learning problem (e.g., discriminating *apples* from *oranges*) is represented by an arbitrary set $\mathcal{X}$. Samples can be drawn independently from the population $\mathcal{X}$ with a probability distribution $P(\mathcal{X})$, which is unknown. They can be represented formally as:

$$\mathbf{X} = \mathbf{x_1}, \mathbf{x_2}, \ldots, \mathbf{x_n} \tag{2.1}$$

Note that $\mathbf{X} \subseteq \mathcal{X}$. An individual data point in the set $\mathbf{X}$ drawn from the input space $\mathcal{X}$, also referred to as an *instance* or an *example*, is normally represented in vector form as $\mathbf{x_i}$ of d *dimensions*. The elements of a vector $\mathbf{x_i}$ are also referred to as *features* or *attributes*. For example, *apples* and *oranges* can be defined in terms of $\{shape, size, color\}$, i.e., using $d = 3$ features/attributes. The features can be *categorical* or *nominal*, such as $color = \{red, green, orange, yellow\}$. Alternatively, they can be ordinal. In the latter case, the features can be *discrete* (taking on a finite number of values) or *continuous*; e.g., each feature $i \in [d]$ can be a scalar in $\mathbb{R}$, yielding a feature space of $\mathbb{R}^d$:

$$\mathbf{x_i} = x_{i1}, x_{i2}, \ldots, x_{id} \tag{2.2}$$

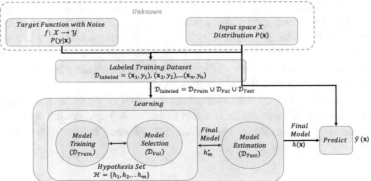

Fig. 2.1: The schematic summarizes the supervised learning process

This set of features can be also seen as d-dimensional vectors $\mathbf{f} = f_1, f_2, \ldots, f_d$, which is useful in various feature transformation and selection processes.

The whole input data and corresponding labels can be viewed in matrix form as:

$$\mathbf{X} = \begin{bmatrix} x_{11} & x_{12} & \cdots & x_{1d} \\ x_{21} & x_{22} & \cdots & x_{2d} \\ \vdots & \vdots & \ddots & \vdots \\ x_{n1} & x_{n2} & \cdots & x_{nd} \end{bmatrix}, \quad \mathbf{Y} = \begin{bmatrix} y_1 \\ y_2 \\ \vdots \\ y_n \end{bmatrix} \tag{2.3}$$

In the above representation, row i of the matrix X stores sample x_i, and its label y_i can be found in row i of the matrix Y.

Alternatively, we can linearly represent the input data and corresponding labels as a labeled dataset $\mathcal{D}_{labeled}$:

$$\mathcal{D}_{labeled} = (\mathbf{x_1}, y_1), (\mathbf{x_2}, y_2), ..(\mathbf{x_n}, y_n) \tag{2.4}$$

2.2.2 Target Function and Labels

Beside the probability distribution $P(X)$, another unknown entity is the *target function* or the *ideal function* that maps the input space $\mathcal{X}$ to the output space $\mathcal{Y}$. This function is formally represented as $[f : \mathcal{X} \to \mathcal{Y}]$. The objective in machine learning is to find an approximation function close to the unknown target function f.

The output space $\mathcal{Y}$ represents all possible values that the target function f can map inputs to. Generally, when the values are *categorical*, finding an approximation to f is known as the *classification* problem, and when the values are *continuous*, the problem is known as *regression*. When the output can only take two values, the problem is known as *binary classification* (e.g., apples and oranges). In regression, $y_i \in \mathbb{R}$.

Sometimes it is advantageous to think not in terms of an exact mapping or a deterministic output for an instance $(\mathbf{x}, y)$, but instead in terms of a target joint probability distribution $P(\mathbf{x}, y) = P(\mathbf{x})P(y|\mathbf{x})$. The latter better accommodates real-world data that contain noise or stochasticity, and we shall have more to say on this later.

2.2.3 Training and Prediction

The entire learning process can now be defined in terms of finding the **approximate** function $h(\mathbf{x})$ from a large hypothesis space $\mathcal{H}$, such that $h(\mathbf{x})$ can effectively *fit* the given data points $\mathcal{D}_{labeled}$ in such a way that $h(\mathbf{x}) \approx f(\mathbf{x})$. The measure of success is normally quantified by *error* (alternatively referred to as *empirical risk*, *loss*, or *cost* function), which measures the discrepancy between $h(\mathbf{x})$ and the unknown target function $f(\mathbf{x})$. Specifically,

$$Error = E(h(\mathbf{x}), f(\mathbf{x})) \approx e(h(\mathbf{x}), y) \tag{2.5}$$

where $E(h(\mathbf{x}), f(\mathbf{x}))$ is the real error from the target function $f(\mathbf{x})$, which is an unknown and approximated by the error obtained through data and labels given by $e(h(\mathbf{x}), y)$.

In the classification domain, the single-point error on a datum $(\mathbf{x}, y)$ can be binary valued (recording a mismatch) and formally written as:

$$E(\mathbf{x}, y) = [\![h(\mathbf{x}) \neq y]\!] \tag{2.6}$$

The $[\![ ]\!]$ represents a function that yields 1 when the values are not equal and 0 otherwise. In the regression domain, the single-point error on a datum $(\mathbf{x}, y)$ can be the squared error:

$$E(\mathbf{x}, y) = (h(\mathbf{x}) - y))^2 \tag{2.7}$$

The *training error* over the entire labeled dataset can be measured using the mean over individual (single-point) errors for classification and regression, as:

$$E_{labeled}(h) = \frac{1}{N} \sum_{n=1}^{N} [\![h(\mathbf{x}_n) \neq y_n]\!] \tag{2.8}$$

$$E_{labeled}(h) = \frac{1}{N} \sum_{n=1}^{N} (h(\mathbf{x}) - y_n)^2 \tag{2.9}$$

The *prediction* or *out-of-sample error* can be computed using the expected value on the unseen datum $(\mathbf{x}, y)$:

$$E_{out}(h) = \mathbb{E}_{\mathbf{x}}[e(h(\mathbf{x}), y)] \tag{2.10}$$

2.3 The Learning Process

Machine learning is the process that seeks to answer three questions:

1. How to train model parameters from labeled data?
2. How to select hyperparameters for a model given labeled data?
3. How to estimate the out-of-sample error from labeled data?

This process is generally done by logically dividing the entire labeled dataset $\mathcal{D}_{Labeled}$ into three components: (a) the training set $\mathcal{D}_{Train}$, (b) the validation set $\mathcal{D}_{Val}$, and (c) the test set $\mathcal{D}_{Test}$, as shown in Fig. 2.2. The training set, $\mathcal{D}_{Train}$, is used to train a given model or hypothesis and learn the model parameters that minimize the training error E_{Train}. The validation set, $\mathcal{D}_{Val}$, is used to select the best parameters or models that minimize the validation error E_{Val}, which serves as a proxy to the out-of-sample error. Finally, the test set, $\mathcal{D}_{Test}$, is used to estimate the *unbiased error* of the model trained with the best parameters over $\mathcal{D}_{Val}$ and with learned parameters over $\mathcal{D}_{Train}$. The *unbiased* error gives a good estimate of the error on unseen future data.

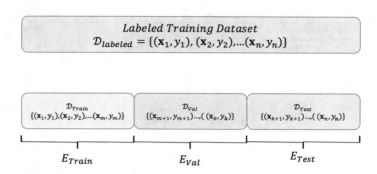

Fig. 2.2: The labeled dataset $\mathcal{D}_{Labeled}$ is split into the training $\mathcal{D}_{Train}$, validation $\mathcal{D}_{Val}$, and testing $\mathcal{D}_{Test}$ datasets

2.4 Machine Learning Theory

In this section we will review basic theory associated with machine learning to address core issues in any learning scenario.

2.4.1 Generalization–Approximation Trade-Off via the Vapnik–Chervonenkis Analysis

The process of fitting a hypothesis function or a model to the labeled dataset can lead to a core problem in machine learning that is known as **overfitting**. The issue here is that we have used all the labeled data points to reduce the error, but this leads to poor **generalization**. Let us consider a simple, one-dimensional dataset generated using $\sin(x)$ as the target function with Gaussian noise added. We can illustrate the issue of overfitting using a hypothesis set of polynomial functions. The different degrees can be treated as parameters via which one can obtain different hypothesis functions with which to fit the labeled data. Figure 2.3 shows how by changing the degree parameter we can reduce the training error significantly by effectively increasing the **complexity** of the model. Figure 2.3 shows how the choice of the hypothesis function can result in **underfitting** or **overfitting**. The hypothesis function that is a polynomial of degree 1 poorly approximates the target function due to its lack of complexity. In contrast, the hypothesis function that is so complex (degree 15 also in Fig. 2.3) has even modeled the noise in the training data, resulting in **overfitting**. Finding the right hypothesis that matches the given resources (training data) in such a way that there is a balance between the approximation and generalization trade-off is the holy grail of machine learning.

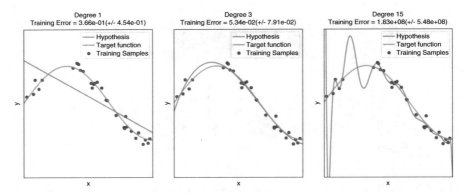

Fig. 2.3: Illustration of underfitting and overfitting in fitting a target function with Gaussian noise with different-degree polynomials

Thus, there are two distinct errors to consider: (a) the **in-sample training error**, given by $E_{train}(h)$, that measures the approximation aspect of the trade-off, and (b) the **out-of sample error**, given by $E_{out}(h)$, that has to be estimated and measures the generalization aspect of the trade-off.

The *probably approximately correct (PAC) learnability* theorem provides the following theoretical bound between the two errors in terms of the probability of the model being *approximately correct*:

$$P[|E_{train}(h) - E_{out}(h)| > \varepsilon] \leq 4m_{\mathcal{H}}(2N)e^{\left(\frac{-\varepsilon^2 N}{8}\right)} \tag{2.11}$$

The equation bounds the probability that the absolute difference between the two errors $E_{train}(h)$ and $E_{out}(h)$ is smaller than ε by something known as the growth function $m_{\mathcal{H}}$ and the number of training data samples N. It has been shown that even with an infinite hypothesis space of the learning algorithm (such as a perceptron, as we will discuss later), the growth function is finite. In particular, the growth function has a tight upper bound measured in terms of the Vapnik–Chervonenkis (VC) dimension d_{VC}, which is the largest N that can be shattered; i.e., $m_{\mathcal{H}}(N) = 2^N$. This makes $m_{\mathcal{H}}(N)$ polynomial in the number of data points [Vap95]. Thus,

$$m_{\mathcal{H}}(N) \leq N^{d_{VC}} + 1 \tag{2.12}$$

$$m_{\mathcal{H}}(N) \sim N^{d_{VC}}$$ (2.13)

Thus, Eq. 2.11 can be rewritten as:

$$E_{out}(h) \leq E_{train}(h) + O\left(\sqrt{\frac{d_{VC}\log(N)}{N}}\right)$$ (2.14)

The VC dimension, given by d_{VC}, correlates with the model complexity, and the above equation can be further rewritten as:

$$E_{out}(h) \leq E_{train}(h) + \underbrace{\Omega(d_{VC})}_{penalty}$$ (2.15)

Figure 2.4 captures the relationship between $E_{out}(h)$ and $E_{train}(h)$ in the above equations. When the model complexity is below an optimum threshold d^*_{VC}, both the training error and the out-of-sample error are decreasing. Choosing any model to represent the data below this optimal threshold will lead to underfitting. When the model complexity is above the threshold, the training error $E_{train}(h)$ still decreases, but the out-of-sample error $E_{out}(h)$ increases, and choosing any model with that complexity will lead to overfitting.

> The PAC analysis in terms of the VC dimension gives an upper bound of the out-of-sample error given the training set and is independent of both the target function $f : \mathcal{X} \to \mathcal{Y}$ and the probability distribution according to which samples are drawn from the population. Recall that both the target function and the probability distribution are unknown.

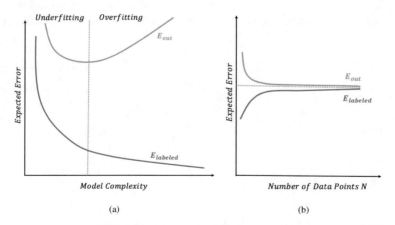

(a)   (b)

Fig. 2.4: Model complexity and learning curves. (**a**) Relationship between training error, out-of-sample error, and model complexity. (**b**) Learning curves for relationship between training error, out-of-sample error, and number of data points

2.4.2 Generalization–Approximation Trade-off via the Bias–Variance Analysis

The bias–variance analysis is another way of measuring or quantifying the generalization–approximation trade-off. The analysis is generally done using regression with mean squared error as the success measure, but it can be modified for classification [HTF09]. The equation for bias–variance trade-off is given by:

$$\mathbb{E}_{\mathbf{x}}[(y - h(\mathbf{x}))^2] = \underbrace{\mathbb{E}_{\mathbf{x}}[(h(\mathbf{x}) - \bar{h}(\mathbf{x}))^2]}_{Variance} + \underbrace{(f(\mathbf{x}) - \bar{h}(\mathbf{x}))^2}_{Bias^2} + \underbrace{\mathbb{E}[(y - f(\mathbf{x}))^2]}_{Noise} \quad (2.16)$$

The idea of the bias–variance trade-off is to decompose the out-of-sample regression error $(y - h(\mathbf{x}))^2$ in terms of three quantities:

Variance: The term $(h(\mathbf{x}) - \bar{h}(\mathbf{x}))^2$ corresponds to the variance of $h(\mathbf{x})$ and is caused by having too many hypotheses in the $\mathcal{H}$ set. The term $\bar{h}(\mathbf{x})$ corresponds to the average hypothesis from the entire set $\mathcal{H}$.

Bias: The term $(f(\mathbf{x}) - \bar{h}(\mathbf{x}))^2$ corresponds to the systematic error caused by not having a good or sufficiently complex hypothesis to approximate the target function $f(\mathbf{x})$.

Noise: The term $(y - f(\mathbf{x}))^2$ corresponds to the inherent noise present in the data.

In general, a simple model suffers from a large bias, whereas a complex model suffers from a large variance. To illustrate the bias–variance trade-off, we use again a one-dimensional $\sin(x)$ as a target function with added Gaussian noise to generate data points. We fit polynomial regression with various degrees, as shown in Fig. 2.5.

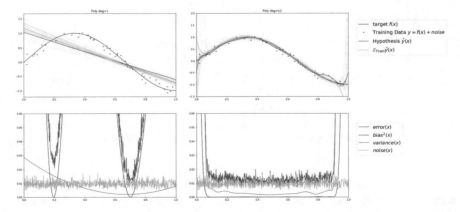

Fig. 2.5: Bias, variance, and noise errors for two hypotheses, i.e., polynomials with degree 1 and degree 12

The bias–variance trade-off is clearly evident in Fig. 2.5 and Table 2.1, which lists the variance, bias, and noise in each case. A simple degree 1 model has higher bias error contributing towards underfitting while a complex degree 12 shows huge variance error contributing towards overfitting.

Table 2.1: Bias, variance, and noise errors for polynomials of degree 1 and 12

Hypothesis	Bias error	Variance error	Noise error	Total error
Degree 1	0.1870	0.0089	0.0098	0.2062
Degree 12	0.0453	2.4698	0.0098	2.5249

2.4.3 Model Performance and Evaluation Metrics

Above, we have evaluated the performance of an algorithm or model using classification or regression error as a metric of success. In general, there are many metrics defined for supervised learning (in both the classification and regression domains) that depend on the size of the data, the distribution of the labels, problem mapping, and more. Below, we describe a few.

2.4.3.1 Classification Evaluation Metrics

We will consider the simple case of binary classification. In the classification domain, the simplest visualization of the success of a model is normally described using the **confusion matrix**, shown in Fig. 2.6. *Accuracy* and *classification error* are informative measures of success when the data is balanced in terms of the classes; that is, the classes have similar sizes. When the data is *imbalanced*, i.e., one class is represented in larger proportion over the other class in the dataset, these measures become biased towards the majority class and give a wrong estimate of success. In such cases, base measures, such as *true positive rate (TPR)*, *false positive rate (FPR)*, *true negative rate (TNR)*, and *false negative rate (FNR)*, become useful. For instance, metrics such as *F1 score* and *Matthews correlation coefficient (MCC)* combine the base measures to give an overall measure of success. Definitions are listed below.

Predicted Class

Actual Class	*Class Positive*	*Class Negative*
Class Positive	*True Positive (TP)*	*False Negative (FN)*
Class Negative	*False Positive (FP)*	*True Negative (TN)*

Fig. 2.6: Confusion metrics for binary classes

1. **True positive rate (TPR) or recall or hit rate or sensitivity**

$$TPR = \frac{TP}{(TP+FN)} \tag{2.17}$$

2. **Precision or positive predictive value**

$$Precision = \frac{TP}{(TP+FP)} \tag{2.18}$$

3. **Specificity**

$$Specificity = \frac{TN}{(TN+FP)} \tag{2.19}$$

4. **Negative prediction value**

$$NPV = \frac{TN}{(TN+FN)} \tag{2.20}$$

5. **Miss rate or false negative rate**

$$FNR = \frac{FN}{(TP+FN)} \tag{2.21}$$

6. **Accuracy**

$$Accuracy = \frac{TN + TP}{(TP + FN + FP + TN)} \tag{2.22}$$

7. **F1 score**

$$F1 = 2 \frac{Precision \times Recall}{(Precision + Recall)} \tag{2.23}$$

8. **Matthews correlation coefficient (MCC)**

$$MCC = 2 \frac{TP \times TN - FP \times FN}{\sqrt{(TP + FP) \times (TP + FN) \times (TN + FP) \times (TN + FN)}} \tag{2.24}$$

Many classification models provide not only a prediction of the class, but also a *confidence value* between 0 and 1 for each data point. The confidence threshold can control the performance of the classifier in terms of TPR and FPR. The curve that plots TPR and FPR for a classifier at various thresholds is known as the *receiver-operating characteristic (ROC)* curve. Similarly, precision and recall can be plotted at different thresholds, giving the *precision-recall curve (PRC)*. The areas under each curve are then respectively known as auROC and auPRC and are popular metrics of performance. In particular, auPRC is generally considered to be an informative metric in the presence of imbalanced classes.

2.4.3.2 Regression Evaluation Metrics

In the regression domain, where the predicted output is a real number that is compared with the actual value (another real number), many variants of squared errors are employed as evaluation metrics. We list a few below.

1. **Average prediction error** is given by:

$$\bar{y} = \frac{\sum_{i=1}^{n}(y_i - \hat{y}_i)}{n} \tag{2.25}$$

where y_i corresponds to the actual real-valued label and $\hat{y}_i$ is the predicted value from the model.

2. **Mean absolute error (MAE)** treats the positive and negative errors in equal measure and is given by:

$$MAE = \frac{\sum_{i=1}^{n}|y_i - \hat{y}_i|}{n} \tag{2.26}$$

3. **Root mean squared error (RMSE)** gives importance to large errors and is given by:

$$RMSE = \sqrt{\frac{\sum_{i=1}^{n}(y_i - \hat{y}_i)^2}{n}} \tag{2.27}$$

4. **Relative squared error (RSE)** is used when two errors are measured in different units:

$$RSE = \frac{\sum_{i=1}^{n}(y_i - \hat{y}_i)^2}{\sum_{i=1}^{n}(\bar{y}_i - y_i)^2} \tag{2.28}$$

5. **Coefficient of determination** (R^2) summarizes the explanatory power of the regression model and is given in terms of squared errors:

$$SSE_{residual} = \sum_{i=1}^{n}(y_i - \hat{y}_i)^2 \tag{2.29}$$

$$SSE_{total} = \sum_{i=1}^{n}(y_i - \bar{y}_i)^2 \tag{2.30}$$

$$R^2 = 1 - \frac{SSE_{residual}}{SSE_{total}} \tag{2.31}$$

2.4.4 Model Validation

Validation techniques are meant to answer the question of how to select a model(s) with the right hyperparameter values. When there are many hypotheses in the hypothesis set, then each unique hypothesis is trained on the training set D_{train} and then evaluated on the validation set D_{val}; the model(s) with the best performance metrics is then chosen. Logically, the model h^- (superscript denotes the model trained on less data) is trained on a training dataset which has fewer points M as compared to the entire set, as some data, K in all, are in the validation set. The performance on the validation set is then given as:

$$E_{Val}(h^-) = \frac{1}{K}\sum_{n=1}^{K} e(h^-(\mathbf{x}_n), y_n) \tag{2.32}$$

Using the VC bound related above, we can show that:

$$E_{out}(h^-) \leq E_{Val}(h^-) + O\left(\frac{1}{\sqrt{K}}\right) \tag{2.33}$$

This equation shows that the larger the number of validation data points K, the better the estimation of the out-of-sample error is. However, from the learning curves, we now know that the more the training data, the smaller the training error. Thus, by removing K points from the budget of training, we have theoretically increased the chances of having a larger training error.

This gives rise to an interesting learning paradox: we need a large number of validation points to have a good estimate of the out-of-sample error, while at the same time, for a model to be trained better, we need fewer data points in the validation set.

A way of addressing this paradox in practice is by training the model the model using only the training data D_{train}, using the model h^- on the validation data D_{val} to estimate the error $E_{out}(h^-)$, and then adding the data back to the training set to learn a new model h on $D_{train} + D_{val}$. This is known as the validation process and is illustrated in Fig. 2.7. Putting it all together, this allows us to obtain the following upper bound on $E_{out}(hs)$:

$$E_{out}(h) \leq E_{out}(h^-) \leq E_{Val}(h^-) + O\left(\frac{1}{\sqrt{K}}\right) \tag{2.34}$$

An important point to note is that when the validation set has been used for model performance evaluation, the estimates of out-of-sample errors derived from the validation errors are optimistic; that is, the validation set is now a biased set, because we use it indirectly to learn the hyperparameters of the models. The validation process is a simple method that can be used for model selection and is independent of the model or learning algorithm.

The only drawback with the validation process is the need to have a large number of labeled data points for creating the training set and the validation set. It is normally difficult to collect a large labeled set due to the cost or difficulty in obtaining the labels. In such cases, instead of physically separating the training set and validation set, a technique known as k-**fold cross-validation** is used. The k-fold cross-validation algorithm is shown in Fig. 2.8. First, the data is randomly divided into k sets. Then, in each of the resulting k experiments, $k - 1$ data folds are used for training and onefold is used for validation to measure the E_{Val}^k for a fold. Finally, the average of the k validation errors is used as a single estimate of the validation error E_{Val}.

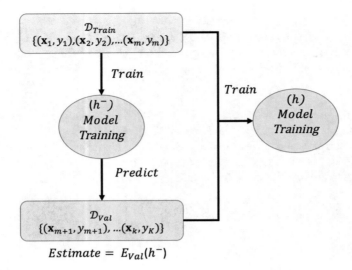

Fig. 2.7: Model training and validation process

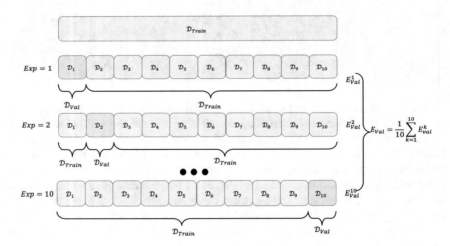

Fig. 2.8: Illustration of tenfold cross-validation

We highlight the validation process involving a hypothesis with a single parameter $\lambda = (\lambda_1, \lambda_2..\lambda_M)$ with M finite values in search of the best parameter.

Algorithm 1: FindBestParameters

Data: $D_{Train}[k]$, $D_{Val}[k]$
Result: bestParameter, lowestError
begin

    splitDataFolds: Create folds of labeled dataset D,k
    bestParameter $\leftarrow \lambda_0$
    for m $\in$ 1..M do
        $\lambda \leftarrow \lambda_m$
    for i $\in$ 1..k do
        trainModel($h(\lambda), D_{Train}[i]$) $E_{VAL}[i] \leftarrow$ testModel($h(\lambda), D_{Val}[i]$)
    $E_{VAL} \leftarrow \frac{1}{k} \sum_{k=1}^{K} E_{VAL}[k]$
    if *current value is the best seen* **then**
        lowestError $\leftarrow E_{VAL}$
        bestParameter $\leftarrow \lambda_m$

2.4.5 Model Estimation and Comparisons

Once the hypothesis or the model with the best parameters is selected and trained with the entire $D_{train} \cup D_{val}$ set, the test set D_{test} is then used to estimate the test error. Recall that the PAC equation is a function of the hypotheses set size M and dataset size N. Therefore, when only one hypothesis is considered, $M = 1$, in the presence of a sufficiently large test set, the PAC equation confirms that the test error is a good approximation of the out-of-sample error. Since the test set has not been used for either model training or model hyperparameter selection, the error remains an unbiased estimate in contrast to the training or validation error.

If a comparison needs to be made between two or more classifiers on one or more datasets, to obtain statistical estimations of differences of metrics, various statistical hypothesis testing techniques can be employed. One has to be aware of the assumptions and constraints of each technique before selecting one [JS11, Dem06, Die98]. Let us describe a few statistical hypothesis testing techniques.

Resampled t-test is a parametric test (in that it assumes a distribution) that is used to compare two classifiers on a metric such as accuracy or error via n different random resampling of training and test subsets from a single dataset. If $\hat{p}_i$ is the difference between the performance of the two classifiers and is assumed to have a Gaussian distribution, and $\bar{p}$ is the average performance difference, the t-statistic is given by:

$$t = \frac{\bar{p}\sqrt{n}}{\sqrt{\frac{\sum_{i=1}^{n}(\hat{p}_i - \bar{p})^2}{(n-1)}}} \tag{2.35}$$

McNemar's test is a popular non-parametric test used to compare two classifiers on the same test set. The test uses the counts given in Table 2.2.

The null hypothesis assumes $n_{10} \approx n_{01}$ and the statistic is given by:

Table 2.2: Basic counts of errors and correct for two classifiers on a test set

	Classifier$_2$ error	Classifier$_2$ correct
Classifier$_1$ error	n_{00}	n_{01}
Classifier$_1$ correct	n_{10}	n_{11}

$$m = \frac{(|n_{01} - n_{10}| - 1)^2}{(n_{01} + n_{10})} \qquad (2.36)$$

The *Wilcoxon signed-ranks test* is a non-parametric test that is popular when two classifiers have to be compared on multiple datasets. The test ranks the differences between metrics of the two classifiers under comparison on the i-th dataset of N datasets as $d_i = p_1^i - p_2^i$, ignoring the signs. The statistic is given by

$$T = min(R^+, R^-) \qquad (2.37)$$

where

$$R^+ = \sum_{d_i > 0} rank(d_i) + \frac{1}{2} \sum_{d_i = 0} rank(d_i)$$

and

$$R^- = \sum_{d_i < 0} rank(d_i) + \frac{1}{2} \sum_{d_i = 0} rank(d_i)$$

The *Friedman non-parametric test* with *Iman–Davenport extension* is used when there are multiple classifiers to be compared on multiple datasets, a scenario that is very common when presenting a novel classifier or aiming to select a single best classifier across many on a given dataset. If R_i is the rank of the jth classifier among K classifiers in the ith dataset among N datasets, then the statistic is given by:

$$F = \frac{(N-1)\chi_F^2}{K(K-1) - \chi_F^2} \qquad (2.38)$$

where

$$\chi_F^2 = \frac{12N}{K(K+1)} \left[\sum_{j=1}^{k} R_j^2 - \frac{K(K+1)^2}{4} \right]$$

2.4.6 Practical Tips for Machine Learning

Though not true in every scenario, there are many practical tips for machine learning practitioners. We list a few here.

- In an unbalanced dataset, when one has enough data, it is a good idea to create stratified samples of the training and test set. The general rule of thumb is to have a number of samples that is at least ten times the number of dimensions, if possible. Generally, 20% of the data is set aside for testing.
- If the labeled dataset is very sparse (that is, the number of samples is much smaller than the number of dimensions), instead of dividing the dataset into training and test, one can use the cross-validation process for both model selection and estimation. One has to be aware that the error metric will be optimistic and may not reflect the real out-of-sample error.
- To obtain a good error estimate in an unbalanced dataset, the test set should have a similar proportion of positives and negatives as the general population estimate.
- The test set needs to have similar data characteristics as the general population estimate. This includes feature statistics and distributions.
- In an unbalanced dataset, creating many samples of the same minority class with different majority class is often useful for both training and testing. The variance of the error estimate across various sets provides an important metric of the sample bias.
- Use the training set for training and use cross-validation on the training set for hyperparameter selection.
- The training sample size in an unbalanced dataset can be oversampled or undersampled. The ratio of the two classes (in binary classification) is another hyperparameter to learn.
- Always plot the learning curves on the validation set (cross-validation average with variance on folds of the training set) to evaluate the number of instances needed for a given algorithm.

2.5 Linear Algorithms

Let us now consider similar issues in the context of regression. In this chapter, we will limit our discussion to linear regression. We will give basic introduction and equations around each algorithm followed by discussion points capturing advantages and limitations.

2.5.1 Linear Regression

Linear regression is one of the simplest linear models that has been analyzed theoretically and applied practically to many domains [KK62]. The dataset assumes the labels to be numeric value or a real number; for instance, one could be interested in predicting the house price in a location where historical data of previous house sales with important features such as structure, rooms, neighborhood, and other features have been collected. Due to the popularity of linear regression, let us go over some important elements, such as concepts, facets of optimization, and others.

The hypothesis h is a linear combination of input $\mathbf{x}$ and a weight parameters $\mathbf{w}$ (that we intend to learn through training). In a d-dimensional input ($\mathbf{x} = [x_1, x_2 \ldots x_d]$), we introduce another dimension called the bias term, x_0, with value 1. Thus the input can be seen as $\mathbf{x} \in \{1\} \times \mathbb{R}^d$, and the weights to be learned are $\mathbf{w} \in \mathbb{R}^{d+1}$.

$$h(\mathbf{x}) = \sum_{i=0}^{d} w_i x_i \qquad (2.39)$$

In matrix notation, the input can be represented as a data matrix $\mathbf{X} \in \mathbb{R}^{N \times (d+1)}$, whose rows are examples from the data (e.g., $\mathbf{x}_n$), and the output is represented as a column vector $\mathbf{y} \in \mathbb{R}^N$. We will assume that the good practice of dividing the datasets into training, validation, and testing is followed, and we will represent the training error by E_{train}.

The process of learning via linear regression can be analytically represented as minimizing the squared error between the hypothesis function $h(\mathbf{x}_n)$ and the target real values y_n, as:

$$E_{train}(h(\mathbf{x}, \mathbf{w})) = \frac{1}{N} \sum_{i=0}^{d} (\mathbf{w}^\mathsf{T} \mathbf{x}_n - y_n)^2 \qquad (2.40)$$

Since the data $\mathbf{x}$ is given, we will write the equation in terms of weights $\mathbf{w}$:

$$E_{train}(\mathbf{w}) = \frac{1}{N} \|(\mathbf{X}\mathbf{w} - \mathbf{y})^2\|, \qquad (2.41)$$

where $\|(\mathbf{X}\mathbf{w} - \mathbf{y})^2\|$ is the Euclidean norm of a vector.

So, we can write

$$E_{train}(\mathbf{w}) = \frac{1}{N} (\mathbf{w}^\mathsf{T} \mathbf{X}^\mathsf{T} \mathbf{X} \mathbf{w} - 2\mathbf{w}^\mathsf{T} \mathbf{X}^\mathsf{T} \mathbf{y} + \mathbf{y}^T \mathbf{y}) \qquad (2.42)$$

We need to minimize E_{train}. This is an optimization problem, as we need to find the weights $\mathbf{w}_{opt}$ that minimize the training error. That is, we need to find:

$$\mathbf{w}_{opt} = \arg\min_{\mathbf{w} \in \mathbb{R}^{d+1}} E_{train}(\mathbf{w}) \qquad (2.43)$$

We can assume that the loss function $E_{train}(\mathbf{w})$ is differentiable. So, to obtain a solution, we take the gradient of the loss function with respect to $\mathbf{w}$, and set it to the zero vector $\mathbf{0}$:

$$\nabla E_{train}(\mathbf{w}) = \frac{2}{N} (\mathbf{X}^\mathsf{T} \mathbf{X} \mathbf{w} - \mathbf{X}^\mathsf{T} \mathbf{y}) = \mathbf{0} \qquad (2.44)$$

$$\mathbf{X}^{\mathrm{T}}\mathbf{X}\mathbf{w} = \mathbf{X}^{\mathrm{T}}\mathbf{y} \qquad (2.45)$$

We also assume that $\mathbf{X}^{\mathrm{T}}\mathbf{X}$ is invertible, and so we obtain

$$\mathbf{w}_{opt} = (\mathbf{X}^{\mathrm{T}}\mathbf{X})^{-1}\mathbf{X}^{\mathrm{T}}\mathbf{y} \qquad (2.46)$$

We can represent the *pseudo-inverse* as $\mathbf{X}^{\dagger}$, such that $\mathbf{X}^{\dagger} = (\mathbf{X}^{\mathrm{T}}\mathbf{X})^{-1}\mathbf{X}^{\mathrm{T}}$. This derivation shows that linear regression has a direct analytic formula to compute for the optimum weights, and the learning process is as simple as computing the pseudo-inverse matrix and the matrix multiplication with the label vector $\mathbf{y}$. The following algorithms implement the described optimization process.

Algorithm 2: LinearRegression train

Data: Training Dataset $(\mathbf{x_1}, y_1), (\mathbf{x_2}, y_2), ..(\mathbf{x_n}, y_n)$ such that $\mathbf{x}_i \in \mathbb{R}^d$ and $y_i \in \mathbb{R}^d$
Result: Weight vector $\mathbf{w} \in \mathbb{R}^{d+1}$
begin

> create a matrix $\mathbf{X}$ from inputs and a bias for each vector $x_0 = 1$
> create a vector $\mathbf{y}$ from labels
> compute the pseudo-inverse $\mathbf{X}^{\dagger} = (\mathbf{X}^{\mathrm{T}}\mathbf{X})^{-1}\mathbf{X}^{\mathrm{T}}$
> $\mathbf{w} = \mathbf{X}^{\dagger}\mathbf{y}$

Algorithm 3: LinearRegression predict

Data: Test Data $\mathbf{x}$ such that $\mathbf{x} \in \mathbb{R}^d$ and weight vector $\mathbf{w}$
Result: Prediction $\hat{y}$
begin

> create a vector $\mathbf{x}$ from inputs and prefix the input vector with the bias term $x_0 = 1$
> $\hat{y} = \mathbf{x}^{\mathrm{T}}\mathbf{w}$

2.5.1.1 Discussion Points

- Linear regression has an efficient training algorithm, with time complexity polynomial in the size of the training data.
- Linear regression assumes $\mathbf{X}^{\dagger}$ is invertible. Even if this is not the case, the pseudo-inverse can be employed, though doing so does not guarantee a unique optimal solution. We note that there are techniques that can compute the pseudo-inverse without inverting the matrix.
- The performance of linear regression is affected if there is correlation among the features in the training set.

2.5.2 Perceptron

Perceptrons are models based on the linear regression hypothesis composed with the sign function that provides a classification output instead of regression, as shown below, and illustrated in Fig. 2.9.

$$h(\mathbf{x}) = sign\left(\sum_{i=0}^{d} w_i x_i\right) \tag{2.47}$$

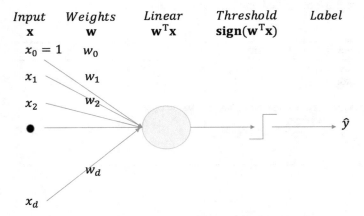

Fig. 2.9: Perceptron

The training algorithm for perceptrons in a linearly separable dataset is to initialize the weights and iterate over the training set, changing the weights only when data points are wrongly classified [Ros58]. This is an iterative process that converges only when the dataset is linearly separable. For linear but not separable datasets (having small number of labels on either side of the plane), a small modification is made to iterate only up to a maximum number of iterations and store the loss function with weights corresponding to the lowest loss function. This is known as the pocket algorithm. The perceptron training algorithm tries to find *a hyperplane* of $d - 1$ dimension in a d-dimensional dataset.

2.5.2.1 Discussion Points

- Perceptrons need not find the best hyperplane (maximum separation between points) separating the two classes and suffer from noise in the datasets.
- Outliers impact the algorithm's ability to find the best hyperplane and so addressing them is important.

Algorithm 4: Perceptron

Data: Training Dataset $(\mathbf{x_1}, y_1), (\mathbf{x_2}, y_2), ..(\mathbf{x_n}, y_n)$ such that $\mathbf{x}_i \in \mathbb{R}^d$ and $y_i \in (+1, -1)$,
         MaxIterations= T
Result: Weight vector $\mathbf{w} \in \mathbb{R}^{d+1}$
begin
    create a vector $\mathbf{x}$ from inputs and prefix the input vector with the bias term $x_0 = 1$
    create a vector $\mathbf{y}$ from labels
    initialize weight vector $\mathbf{w}_0$ to be $\mathbf{0}$
    bestWeight $\leftarrow \mathbf{w}_0$
    initialize loss($\mathbf{w}$) to be 1
    for $t \in 0..T - 1$ **do**
        for $i \in 0..N - 1$ **do**
            if $sign(\mathbf{x}_i y_i) \neq y_i$ **then**
                update the weight vector $\mathbf{w}_{(t+1)} = \mathbf{w}_{(t)} + \mathbf{x}_i y_i$
                currentLoss($\mathbf{w}$) $\leftarrow$ currentLoss($\mathbf{w}_{(t+1)}$)
                if *currentLoss*($\mathbf{w}$) $<$ *loss*($\mathbf{w}$) **then**
                    loss($\mathbf{w}$) $\leftarrow$ currentLoss($\mathbf{w}$)
                    bestWeight = $\mathbf{w}_{(t+1)}$
    return **bestWeight**

Algorithm 5: Perceptron

Data: Test Data $\mathbf{x}$ such that $\mathbf{x} \in \mathbb{R}^d$ and weight vector
Result: Prediction $\hat{y} \in (+1, -1)$
begin
    create a vector $\mathbf{x}$ from inputs and adding a bias for each vector $x_0 = 1$
    $\hat{y} = sign(\mathbf{x}^T \mathbf{w})$
    return $\hat{y}$

2.5.3 Regularization

As discussed earlier, one of the common issues in supervised machine learning is overfitting. Equation 2.15 can be seen as a penalty on the model complexity. If this penalty is taken into account in the training of the model, then learning is improved. The idea of regularization is to do just that; that is, to introduce this penalty in the training itself. Regularization can be generally considered as an application of *Occam's Razor* in that the goal is to choose a simpler hypothesis. In general, regularization is used to combat the noise inherent in the dataset.

In many weight-based machine learning algorithms, such as linear regression, perceptrons, logistic regression, and neural networks, it is common practice to put a penalty on the weights and introduce that in the loss function. The resulting *augmented* loss function that is then used for optimization is given below:

$$E_{aug}(h) = E_{train}(h) + \lambda \Omega(\mathbf{w}) \tag{2.48}$$

In the augmented loss function above, the scalar parameter λ is known as the **regularization constant**, and $\Omega(\mathbf{w})$ is known as the **regularization function**.

2.5.3.1 Ridge Regularization: L_2 Norm

One of the popular regularization functions is the L_2 norm [HK00], also known as weight decay or ridge regularization. It can be substituted in Eq. 2.49 as shown below:

$$E_{aug}(h) = E_{train}(h) + \lambda \mathbf{w}^{\mathrm{T}} \mathbf{w} \qquad (2.49)$$

Therefore, one can search for the optimal $\mathbf{w}_{opt}$, defined as:

$$\mathbf{w}_{opt} = \underset{\mathbf{w} \in \mathbb{R}^{d+1}}{\arg \min} E_{aug}(\mathbf{w}) \qquad (2.50)$$

$$\mathbf{w}_{opt} = \underset{\mathbf{w} \in \mathbb{R}^{d+1}}{\arg \min} \left(E_{train}(h) + \lambda \mathbf{w}^{\mathrm{T}} \mathbf{w} \right) \qquad (2.51)$$

The linear regression solution modified with regularization is:

$$\mathbf{w}_{opt} = (\mathbf{X}^{\mathrm{T}} \mathbf{X} + \lambda \mathbf{I})^{-1} \mathbf{X}^{\mathrm{T}} \mathbf{y} \qquad (2.52)$$

> The regularization parameter λ is normally selected using the validation technique described above for any hyperparameter and is generally a small value around 0.001. The impact of L_2 regularization is that some weights which are less relevant will have values closer to zero. In this way, L_2 regularization can be seen as conducting implicit feature selection via feature weighting. L_2 regularization is computationally efficient.

2.5.3.2 Lasso Regularization: L_1 Norm

The L_1 norm is another popular regularization used in weight-based algorithms [HTF09]:

$$\mathbf{w}_{opt} = \underset{\mathbf{w} \in \mathbb{R}^{d+1}}{\arg \min} (E_{train}(h) + \lambda |\mathbf{w}|) \qquad (2.53)$$

Due to the absolute function, the above equation does not have a closed-form solution and is generally represented as a constrained optimization problem as below:

$$\underset{\mathbf{w} \in \mathbb{R}^{d+1}}{\arg \min} (\mathbf{X}^{\mathrm{T}} \mathbf{X} \mathbf{w} - \mathbf{X}^{\mathrm{T}} \mathbf{y}) \; s.t. \; \mathbf{w} < \eta \qquad (2.54)$$

The parameter η is inversely related to the regularization parameter λ. The equation above can be shown to be a convex function, and quadratic programming is generally used to obtain the optimized weights.

As in L_2 regularization, the regularization parameter λ in L_1 regularization is selected using validation techniques. In comparison with L_2, L_1 regularization generally results in more feature weights being set to zero. Thus, L_1 regularization yields a sparse representation through implicit feature selection.

2.5.4 Logistic Regression

Logistic regression can be seen as a transformation θ on the linear combination $\mathbf{x}^T\mathbf{w}$ that allows a classifier to return a probability score [WD67]:

$$h(\mathbf{x}) = \theta(\mathbf{w}^T\mathbf{x}) \tag{2.55}$$

A *logistic function* (also known as a *sigmoid* or *softmax* function) $\theta(\mathbf{w}^T\mathbf{x})$, shown below, is generally used for the transformation:

$$h(\mathbf{x}) = \frac{\exp(\mathbf{w}^T\mathbf{x})}{1 + \exp(\mathbf{w}^T\mathbf{x})} \tag{2.56}$$

For a binary classification, where $y \in \{-1, +1\}$, the hypothesis can be seen as a likelihood of predicting $y = +1$, i.e., $P(y = +1|\mathbf{x})$. Thus, the equation can be rewritten as a log-odds ratio, and weights are learned to maximize the *conditional likelihood* given the inputs:

$$\frac{P(y = +1|\mathbf{x})}{P(y = -1|\mathbf{x})} = \mathbf{w}^T\mathbf{x} \tag{2.57}$$

The log-likelihood of the hypothesis can be written as:

$$\log h(\mathbf{x}) = \sum_{i=0}^{n} \log P(y_i|\mathbf{x}_i) \tag{2.58}$$

$$\log \mathcal{L}(h(\mathbf{x})) = \sum_{i=0}^{n} \begin{cases} \log h(\mathbf{x}_i) \ if \ y_i = +1 \\ (1 - \log h(\mathbf{x}_i)) \ if \ y_i = -1 \end{cases} \tag{2.59}$$

$$\log \mathcal{L}(h(\mathbf{x})) = \sum_{i=0}^{n} (y_i \log h(\mathbf{x}_i) + (1 - y_i)(1 - \log h(\mathbf{x}_i))) \tag{2.60}$$

In information theory, if one treats y_i and $h(\mathbf{x}_i)$ as probability distributions, the above equation is referred to as **cross-entropy** error. This cross-entropy error can be treated as our new error function E_{train}, but it cannot be solved in closed form. Instead of an analytical solution, an iterative algorithm known as gradient descent can be employed. Gradient descent is a general optimization algorithm that is used widely in machine learning, including deep learning. Let us discuss it at some length below.

2.5.4.1 Gradient Descent

Let us recall that the goal is to find weights $\mathbf{w}$ that minimize E_{train}, and that at the minimum, the gradient of E_{train} is 0. In gradient descent, the negative of the gradient is followed in an iterative process until the gradient is zero. The gradient is a vector containing partial derivatives over each of the dimension [Bry61], as shown below:

$$\mathbf{g} = \nabla E_{train}(\mathbf{w}) = \left[\frac{\partial E_{train}}{\partial w_0}, \frac{\partial E_{train}}{\partial w_1} \cdots \frac{\partial E_{train}}{\partial w_n} \right] \tag{2.61}$$

The normalized gradient $\hat{g}$ can be written as:

$$\hat{\mathbf{g}} = \frac{\nabla E_{train}(\mathbf{w})}{\|\nabla E_{train}(\mathbf{w})\|} \tag{2.62}$$

A small step size η is made in the direction of $-\hat{g}$, and the weights are updated accordingly, leading to an optimal point. Selecting a small step size is important, otherwise the algorithm oscillates and does not reach the optimum point. The algorithm can be summarized as:

Algorithm 6: Gradient descent

Data: Training Dataset $\mathcal{D}_{train} = (\mathbf{x_1}, y_1), (\mathbf{x_2}, y_2), ..(\mathbf{x_n}, y_n)$ such that $\mathbf{x}_i \in \mathbb{R}^d$ and
    $y_i \in [+1, -1]$, Loss Function $E_{train}(\mathbf{w})$, Step size η and MaxIterations T
Result: Weight vector $\mathbf{w} \in \mathbb{R}^{d+1}$
begin
    $\mathbf{w}_0 \leftarrow \text{init}(\mathbf{w})$
    for $t \in 0..T-1$ **do**
        $\mathbf{g}_t \leftarrow \nabla E_{train}(\mathbf{w}_t)$
        $\mathbf{w}_{t+1} \leftarrow \mathbf{w}_t - \eta \hat{\mathbf{g}}_t$
    return $\mathbf{w}$

The weights $\mathbf{w}$ can be initialized to the $\mathbf{0}$ vector or set to random values (each obtained from a normal distribution with 0 mean and small variance) or preset values. Another important decision in gradient descent is the termination criterion. The algorithm can be made to terminate when the number of iterations reaches a specific value or when the value of the gradient reaches a predefined threshold, close to zero.

2.5.4.2 Stochastic Gradient Descent

One of the disadvantages of gradient descent is the use of the entire training dataset when computing the gradient. This has an implication on the memory and computation speed, which increase as the number and dimension of training examples increase. Stochastic gradient descent is a version of gradient descent that, instead of utilizing the entire training dataset, picks a data point uniformly at random from the training dataset (hence the name stochastic). It has been shown that with a large number of iterations and a small step size, stochastic gradient descent generally reaches the same optimum as the batch gradient descent algorithm [BB08].

Algorithm 7: Stochastic gradient descent

Data: Training Dataset $\mathcal{D}_{train} = (\mathbf{x_1}, y_1), (\mathbf{x_2}, y_2), ..(\mathbf{x_n}, y_n)$ such that $\mathbf{x}_i \in \mathbb{R}^d$ and
  $y_i \in (+1, -1)$, Loss Function $E_{train}(\mathbf{w})$, Step size η and MaxIterations T
Result: Weight vector $\mathbf{w} \in \mathbb{R}^{d+1}$
begin
  $\mathbf{w}_0 \leftarrow \text{init}(\mathbf{w})$
  for $t \in 0..T-1$ **do**
    $d \leftarrow (\mathbf{x_i}, y_i)$
    $\mathbf{g}_t \leftarrow \nabla E_d(\mathbf{w}_t)$
    $\mathbf{w}_{t+1} \leftarrow \mathbf{w}_t - \eta \mathbf{g}_t$
  return $\mathbf{w}$

Figure 2.10 illustrates the iterative changes in the training error for (batch) gradient descent and stochastic gradient descent for a one-dimensional linear regression problem.

It can be shown that in logistic regression the gradient is:

$$\nabla E_{train}(\mathbf{w}) = -\frac{1}{N} \sum_{i=0}^{n} \frac{y_i \mathbf{x}_i}{(1 + \exp^{y_i \mathbf{w}^{\mathsf{T}} \mathbf{x}_i})} \tag{2.63}$$

The training of logistic regression using gradient descent is described by the following algorithm:

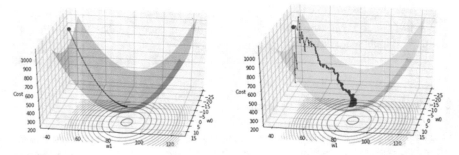

Fig. 2.10: One-dimensional regression with gradient descent and stochastic gradient descent

---

Algorithm 8: Logistic regression with gradient descent

Data: Training Dataset $(\mathbf{x_1}, y_1), (\mathbf{x_2}, y_2), ..(\mathbf{x_n}, y_n)$ such that $\mathbf{x}_i \in \mathbb{R}^d$ and $y_i \in (+1, -1)$,
MaxIterations= T and Step size η
Result: Weight vector $\mathbf{w} \in \mathbb{R}^{d+1}$
begin

create a vector $\mathbf{x}$ from inputs and adding a bias for each vector $x_0 = 1$
for $t \in 0..T-1$ **do**

$\mathbf{g}_t \leftarrow -\frac{1}{N} \sum_{i=0}^{n} \frac{y_i \mathbf{x}_i}{(1+\exp^{y_i \mathbf{w}^\mathsf{T} \mathbf{x}_i})}$

$\mathbf{w}_{t+1} \leftarrow \mathbf{w}_t - \eta \mathbf{g}_t$

return $\mathbf{w}$

2.5.5 Generative Classifiers

All algorithms we have seen so far have been **discriminative** in their approach; that is, they make no assumption about the underlying distribution of the data and focus instead on the end goal of prediction accuracy. Another popular approach in machine learning is the **generative** approach, which assumes an underlying distribution with which the data is generated and tries to find parameters of this distribution in its training.

The generative approach, though an indirect mechanism for achieving the prediction accuracy, has been very successful in real-world applications. Many machine learning algorithms, both supervised and unsupervised, naive Bayes, linear discriminant analysis, expectation maximization, and Bayes networks among others, are based on the generative approach and have a probabilistic foundation in the Bayes theorem.

Formally, given a hypothesis h and a training dataset D_{train}, the Bayes theorem helps in defining the probability of choosing the hypothesis given the data; that is, it helps define $P(h|D_{train})$ given the prior probability of the hypothesis $P(h)$, the likelihood of the data given the hypothesis $P(D_{train}|h)$, and the probability of data over all hypotheses $P(D_{train}) = \int_h P(D_{train}|h)$ as:

$$P(h|\mathcal{D}_{train}) = \frac{P(\mathcal{D}_{train}|h)P(h)}{P(\mathcal{D}_{train})} \tag{2.64}$$

If there are multiple hypotheses, the question of which one is the most probable one given the training data can be answered by the **maximum a posteriori** hypothesis as:

$$h_{MAP} = \arg\min_{h \in \mathcal{H}} P(h|\mathcal{D}_{train}) \tag{2.65}$$

$$h_{MAP} = \arg\max_{h \in \mathcal{H}} \frac{P(\mathcal{D}_{train}|h)P(h)}{P(\mathcal{D}_{train})} \tag{2.66}$$

Since $P(D_{train})$ is independent of h, we have:

$$h_{MAP} = \arg\max_{h \in \mathcal{H}} P(\mathcal{D}_{train}|h)P(h) \tag{2.67}$$

If we further assume that all the hypotheses are equally likely (i.e., $P(h_1) \approx P(h_2) \approx P(h_m)$ for m hypotheses), the equation can be reduced to:

$$h_{ML} = \arg\max_{h \in \mathcal{H}} P(\mathcal{D}_{train}|h) \tag{2.68}$$

As stated in the assumptions, if the training examples are independent and identically distributed (i.i.d), $P(\mathcal{D}_{train}|h)$ can be written in terms of the training examples as:

$$P(\mathcal{D}_{train}|h) = \prod_{i=1}^{N} P(\langle \mathbf{x}_i, y_i \rangle | h) = \prod_{i=1}^{N} P(y_i|\mathbf{x}_i; h)P(\mathbf{x}_i) \tag{2.69}$$

2.5.5.1 Naive Bayes

The hypothesis in Bayes form for a binary classification $y_i \in (0, 1)$ is:

$$h_{Bayes}(\mathbf{x}) = \arg\max_{y \in (0,1)} P(X = \mathbf{x}|Y = y)P(Y = y) \tag{2.70}$$

In *naive Bayes*, an assumption of independence between the features or attributes is made. So, for d dimensions, the equation simplifies as:

$$h_{Bayes}(\mathbf{x}) = \arg\max_{y \in (0,1)} P(Y = y) \prod_{j=1}^{d} P(X_j = x_i|Y = y) \tag{2.71}$$

As a result, training and estimating parameters of naive Bayes is just measuring two quantities, the priors for the class $P(Y = y)$ and the conditional for each feature $P(X_j = x_j|Y = y)$ given the class. It can be easily shown that the maximum likelihood estimates of these are nothing but counts in discrete datasets as shown below:

$$P(Y = y) = \frac{1}{N} \sum_{i=0}^{N} [\![y_i = y]\!] = \frac{countLabel(y)}{N} \tag{2.72}$$

$$P(X_i = x_j | Y = y) = \frac{[\![y_i = y \ and \ x_{i,j} = x]\!]}{N} \tag{2.73}$$

Prediction for new examples can be done using the estimations and Eq. 2.70.

2.5.5.2 Linear Discriminant Analysis

Linear discriminant analysis (LDA) is another generative model, where the assumption of a Gaussian distribution for $P(X|Y)$ is made along with equal priors for binary classes, i.e., $P(Y = 1) = P(Y = 0) = 1/2$. Formally, $\mu \in \mathbb{R}^d$ is the multivariate mean, and Σ is the covariance matrix. Then, we have:

$$P(X = \mathbf{x}|Y = y)P(Y = y) = \frac{1}{(2\pi)^{d/2}|\Sigma|^{1/2}} \exp\left(\frac{-1}{2}(\mathbf{x} - \mu)^{\mathrm{T}}|\Sigma|^{-1}(\mathbf{x} - \mu)\right) \tag{2.74}$$

The training of LDA, similar to naive Bayes, involves estimating the parameters, (μ and Σ here) from the training data.

2.5.6 Practical Tips for Linear Algorithms

1. It is always a good idea to scale the input real-valued features to the range $[0, 1]$ for gradient descent algorithms.
2. Binary or categorical features which are represented as one-hot vectors can be used without any transformations. In one-hot vector representation, each categorical attribute is converted into k boolean valued attributes such that only one of those k attributes has a value of one and rest zero for a given instance.
3. Grid search over a range of values spanning multiple orders of magnitude should be used to determine the learning rate and the regularization parameter.

2.6 Non-linear Algorithms

The algorithms we have seen so far, given by $sign(\mathbf{w}^{\mathrm{T}}\mathbf{x})$, are *linear* in the weights $\mathbf{w}$, as the inputs $\mathbf{x}$ are a given or constant for the training algorithm. A simple extension is to use a non-linear transform $\phi(\mathbf{x})$ applied to all the features, which transforms the points into a new space, say $\mathcal{Z}$, where a linear model given by $sign(\mathbf{w}^{\mathrm{T}}\phi(\mathbf{x}))$ can then be learned. When a prediction is needed on a new unseen data $\mathbf{x}$, first the data is transformed into the $\mathcal{Z}$ space using the transformation $\phi(\mathbf{x})$, and then linear algorithm weights are applied in the $\mathcal{Z}$ space to make the prediction.

As an example, a simple non-linear, two-dimensional training dataset can be transformed into a three-dimensional $\mathcal{Z}$ space, where the dimensions are $\mathbf{z} = \langle x_1, x_2, x_1^2 + x_2^2 \rangle$. The $\mathcal{Z}$ space is linearly separable, as shown in Fig. 2.11.

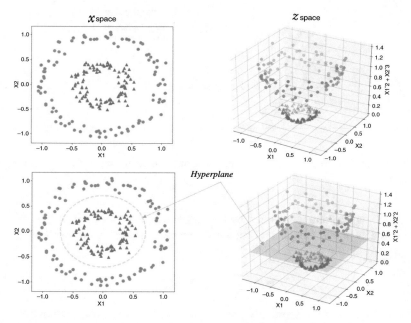

Fig. 2.11: Illustration of non-linear to linear transformation and finding a separating hyperplane in the transformed space

2.6.1 Support Vector Machines

Support vector machine (SVM) is one of the most popular non-linear machine learning algorithms that can separate both linear and non-linear data using built-in transformations known as *kernels* [Vap95]. SVMs not only separate the data but also find the hyperplane that separates the data in the most optimal way through a principle known as **maximum margin** separation, as shown in Fig. 2.12. The data points that separate the hyperplane and lie on the margin are known as *support vectors*.

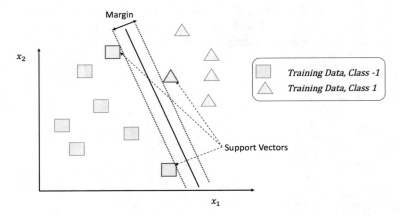

Fig. 2.12: Illustration of SVM finding the maximum margin separation between labeled data

In SVM the hyperplane is obtained by the kernel transformation $k(x, x')$, which takes any two data points x and x' and obtains the transformed real value in the inner product space:

$$y = b + \sum_i \alpha_i k(\mathbf{x}, \mathbf{x}') \tag{2.75}$$

The idea of kernels is to implicitly perform the non-linear transformation onto the $\mathcal{Z}$ space without any explicit transformation $\phi(\mathbf{x})$ through a concept known as the **kernel trick**. The **radial basis function**, also known as the **Gaussian kernel**, is one such kernel transformation:

$$k(\mathbf{x}, \mathbf{x}') = \exp^{\frac{-|\mathbf{x}-\mathbf{x}'|^2}{\sigma^2}} \tag{2.76}$$

Gaussian kernels can be shown to map the input space into an infinite-dimensional feature space. The transformation shown in Fig. 2.11 can be generalized to a **polynomial kernel** of degree σ given by:

$$k(\mathbf{x}, \mathbf{x}') = (1 + \mathbf{x}\mathbf{x}')^{\sigma} \tag{2.77}$$

2.6.2 Other Non-linear Algorithms

The *k-nearest neighbors* algorithm is another simple non-linear algorithm. It is also known as the *lazy* learner, as its core idea is to hold all the training data in memory and use the distance metric for the user-specified *k* (number of neighbors) to classify the unseen new data point. Generally, a distance metric such as **Euclidean** or **Manhattan**, generalized as the **Minkowski** distance, is used to compute the distance from the points:

$$\mathbf{dist}(\mathbf{x}, \mathbf{x}') = \Big(\sum_{d=1}^{d} |\mathbf{x} - \mathbf{x}'|^q \Big)^{\frac{1}{q}} \tag{2.78}$$

Neural networks are another extension of perceptrons that are used to create non-linear boundaries. We discuss them in detail in Chap. 4. *Decision trees* and many of its extensions, such as the *gradient boosted algorithm* and *random forest* among others, are based on the principle of finding simpler decision boundaries for the features and combining them in hierarchical trees, as shown in Fig. 2.13.

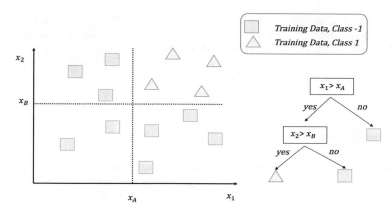

Fig. 2.13: A two-dimensional example and classifier boundaries to separate two classes using decision trees

2.7 Feature Transformation, Selection, and Dimensionality Reduction

In this section we will review some of the common techniques used for feature transformation, selection, and reduction.

2.7.1 Feature Transformation

In many algorithms, it is beneficial to have all the features in the same range, for example, in $[0, 1]$, for the algorithm to not be biased or run effectively across all features. Some commonly used transformations are as follows.

2.7.1.1 Centering or Zero Mean

Each feature can be transformed by subtracting the mean from its value, $f_{feature,i} = f_i - \bar{f}_{feature}$:

$$\bar{f}_{feature} = \frac{1}{N} \sum_{i=0}^{N} f_i \tag{2.79}$$

2.7.1.2 Unit Range

Each feature can be transformed to be in the range of $[0, 1]$. For a feature $f_{feature}$, let $f_{feature}Max$ correspond to the maximum value in the dataset, and $f_{feature}Min$ correspond to the minimum value. Then the transformation for an instance i is:

$$f_i = \frac{(f_i - f_{feature}Max)}{(f_{feature}Max - f_{feature}Min)} \tag{2.80}$$

2.7.1.3 Standardization

In this transformation, the features change to have zero mean and unit variance. The empirical variance of a feature $v_{feature}$ is calculated on a dataset by:

$$v_{feature} = \frac{1}{N} \sum_{i=0}^{N} (f_i - \bar{f}_{feature})^2 \tag{2.81}$$

The transformation for each feature is $f_i = \frac{(f_i - \bar{f}_{feature})}{\sqrt{v}}$.

2.7.1.4 Discretization

Continuous features are sometimes transformed to categorical types by defining the number of categories or the category width.

2.7.2 Feature Selection and Reduction

We have already seen that regularization with L_1 or L_2 can be considered as a feature scoring and selection mechanism and can be employed directly in an algorithm to reduce or prioritize the impact of features. There are many univariate and multivariate feature selection approaches that use information-theoretic, statistical-based, sparse learning-based, wrapping algorithms for finding features [GE03, CS14]. There are various dimensionality or feature reduction techniques to transform and reduce the feature set to smaller subset of more meaningful features. In this section, we will highlight one such statistical-based method known as *principal component analysis (PCA)*, which is also applicable to deep learning techniques.

2.7.2.1 Principal Component Analysis

PCA is a linear dimensionality reduction technique which, given an input matrix **X**, tries to find a feature matrix **W** such that the size m of the feature matrix **W** is much lower than the input dimension d ($m \ll d$) and each reduced feature in this new feature matrix captures maximum variance from the inputs [Jol86]. This can be considered as the process of finding the matrix **W** such that the weights decorrelate or minimize relationships between features (Fig. 2.14).

This can be expressed as below:

$$(\mathbf{WX})^{\mathrm{T}}(\mathbf{WX}) = (\mathbf{Z})^{\mathrm{T}}(\mathbf{Z}) = N\mathbf{I} \tag{2.82}$$

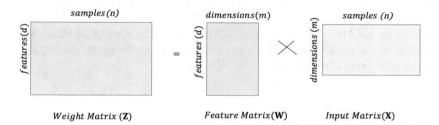

Fig. 2.14: PCA process of finding reduced dimensions from original features

$$\mathbf{WX}^{\mathrm{T}}\mathbf{XW}^{\mathrm{T}} = N\mathbf{I} \tag{2.83}$$

Solving for the diagonalization, the above equation becomes:

$$\mathbf{W}\mathrm{Cov}(\mathbf{X})\mathbf{W}^{\mathrm{T}} = \mathbf{I} \tag{2.84}$$

Covariance matrices are positive semi-definite, symmetrical, have orthogonal eigenvectors, and real-valued eigenvalues. Matrix **A** can be factorized as $\mathbf{UAU}^{\mathrm{T}} = \Lambda$, where $\mathbf{U}$ has the eigenvectors of $\mathbf{A}$ in its columns, and $\Lambda = diag(\lambda_i)$, where λ_i are the eigenvalues of $\mathbf{A}$.

Thus the solution for $\mathbf{W}\mathrm{Cov}(\mathbf{X})\mathbf{W}^T$ is a function of the eigenvectors $\mathbf{U}$ and eigenvalues Λ of the covariance matrix, i.e., $\mathrm{Cov}(\mathbf{X})$. The algorithm is shown below.

Algorithm 9: PCA

Data: Dataset $\mathbf{X} = [\mathbf{x}_1, \mathbf{x}_2 .. \mathbf{x}_N] \in \mathbb{R}^d$, Components $= m$
Result: Transformed Data $\mathbf{Y} \in \mathbb{R}^m$
begin
$\quad \mathbf{X} \leftarrow [\mathbf{x}_1 - \mu, \mathbf{x}_2 - \mu .. \mathbf{x}_N - \mu]$
$\quad \mathbf{S}_t \leftarrow \frac{1}{N} \mathbf{X}\mathbf{X}^T$
$\quad \mathbf{X}^T\mathbf{X} = \mathbf{V}\lambda\mathbf{V}^T$
$\quad \mathbf{U} \leftarrow \mathbf{X}\mathbf{V}\lambda^{-\frac{1}{2}}$
$\quad \mathbf{U}_m \leftarrow [\mathbf{u}_1, \mathbf{u}_2 .. \mathbf{u}_m]$
$\quad \mathbf{Y} \leftarrow \mathbf{U}_m^T\mathbf{X}$
$\quad$ return $\mathbf{Y}$

2.8 Sequence Data and Modeling

In many sequence data analysis problems, such as language modeling, time series analysis, and signal processing, modeling the underlying process as a **Markov** process has been very successful. Many traditional NLP tasks, such a parts-of-speech tagging, extracting information, and phrase chunking, have been modeled very successfully using **hidden Markov models (HMM)**, a special type of Markov processes. In next few sections, we will discuss some important theory, properties, and algorithms associated with Markov chains [KS+60].

2.8.1 Discrete Time Markov Chains

Markov chains are the basic building blocks for modeling many sequential processes. Consider a finite set of states modeled as $S = \{s_1, s_2, \ldots, s_n\}$, and let a variable q represent the transition at any time t as q_t. An illustration is provided in Fig. 2.15.

The Markov property states that at any time t, the probability of it being in a state s_i depends only on the previous k states rather than all the states from time 1 to $t-1$. This can be expressed as:

$$P(q_t = s_i | q_{t-1}, q_{t-2}, .. q_1) = P(q_t = s_i | q_{t-1}, q_{t-2}, .. q_{t-k}) \tag{2.85}$$

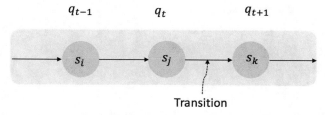

Fig. 2.15: Markov chain transitions over three time steps

The simplest Markov chain depends only on the most recent state ($k = 1$) and is represented as:

$$P(q_t = s_i | q_{t-1}, q_{t-2}, .. q_1) = P(q_t = s_i | q_{t-1}) \qquad (2.86)$$

Such Markov chain for a fixed set of states given by $S = (s_1, s_2, .. s_n)$ can be represented using an $n \times n$ transition matrix A, an $n \times n$ matrix, where each element captures the transition probability as:

$$A_{i,j} = P(q_t = s_i | q_{t-1} = s_j) \qquad (2.87)$$

and an n-dimensional vector π which contains the initial state probabilities:

$$\pi_i = P(q_1 = s_i) \ s.t. \sum_{1}^{n} \pi_i = 1 \qquad (2.88)$$

2.8.2 Discriminative Approach: Hidden Markov Models

At times, the states of the Markov chain are hidden and not observed, but they produce effects which are observable. Such Markov chains are represented using **hidden Markov models (HMM)**, as shown in Fig. 2.16, where new observed states represented by set V having fixed m elements, such as $V = (v_1, v_2, \ldots v_m)$, are added to the previous Markov chain [Rab89].

The concepts required in HMM are:

- Finite hidden states $S = (s_1, s_2, .. s_n)$ and finite observable states $V = (v_1, v_2, \ldots v_m)$.
- For a fixed state sequence transition of length T, given by $Q = q_1, q_2, \ldots qT$, the observations are given by $O = o_1, o_2, \ldots o_T$.
- The parameters of HMM are $\lambda = (\mathbf{A}, \mathbf{b}, \pi)$, where

  - The transition matrix $\mathbf{A}$ represents the transition probability from state s_i to s_j and is given by:

$$A_{i,j} = P(q_t = s_j | q_{t-1} = s_i) \qquad (2.89)$$

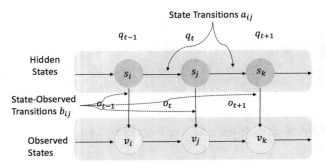

Fig. 2.16: Hidden Markov models (HMM)

- The vector **b** represents the probability of observing a state v_k given the hidden state s_i and is independent of time, given by:

$$b(k) = P(x_t = v_k | q_t = s_i) \qquad (2.90)$$

- The vector π represents the initial probability of the states and is given by:

$$\pi_i = P(q_1 = s_i) \ s.t. \sum_{1}^{n} \pi_i = 1 \qquad (2.91)$$

• The first-order HMMs have two independence assumptions:

$$P(q_t = s_i | q_{t-1}, q_{t-2}, ..q_1) = P(q_t = s_i | q_{t-1}) \qquad (2.92)$$

$$P(o_t = v_j | o_{t-1}, o_{t-2}, ..o_1 q_t, q_{t-1}, ..q_1) = P(o_t = v_j | o_t, q_t) \qquad (2.93)$$

HMMs can be used to answer various fundamental questions through different **dynamic programming**-based algorithms, of which we list a few below.

1. **Likelihood**
   Given an HMM (λ) and a sequence of observations O, what is the likelihood of the HMM generating the sequence; i.e., what is $P(O|\lambda)$? A dynamic programming-based technique known as the **forward algorithm**, which stores intermediate values of states and its probabilities to finally build up the probability of the whole sequence in an efficient manner, is generally employed.

2. **Decoding**
   Given an HMM (λ) and a sequence of observations O, what is the most likely hidden state sequence S that generated the observations? A dynamic programming-based technique known as the **Viterbi algorithm**, similar to the forward algorithm with minor changes, is used to answer this question.

3. **Learning: Supervised and Unsupervised**
   Given a sequence of observations O and the state sequences S, what are the

parameters of the HMM that could generate it? This is a supervised learning problem and can be easily obtained from the training examples by computing different probabilities using the $Count()$ function for the likelihood estimates. The individual cells $A_{i,j}$ of the transition probability matrix $\mathbf{A}$ can be estimated by counting the number of times the state s_j is followed by the state s_i as:

$$A_{i,j} = P(s_j|s_i) = \frac{Count(s_j, s_i)}{Count(s_i)} \tag{2.94}$$

The elements of the array $b(k)$ can be estimated by counting the number of times the observed state v_k happens along with the hidden state s_j and is given by:

$$b_j(k) = P(v_k|s_j) = \frac{Count(v_k, s_j)}{Count(s_j)} \tag{2.95}$$

And the initial probabilities computed as:

$$\pi_i = P(q_1 = s_i) = \frac{Count(q_1 = s_i)}{Count(q_1)} \tag{2.96}$$

If only the sequence of observations O is provided and we need to learn the model that maximizes the probability of the sequence, the unsupervised learning problem is solved using a variation of the expectation maximization (EM) algorithm known as the **Baum–Welch** algorithm.

2.8.3 Generative Approach: Conditional Random Fields

Analogous to the relation between naive Bayes and logistic regression, conditional random field (CRF) has a similar relationship to HMM in the sequence modeling world. HMMs have shortcomings in effectively modeling the dependency between the inputs or the observed states and even the overlapping relationship between them. A linear chain CRF can be considered to be an undirected graph model equivalent of a linear HMM, as shown in Fig. 2.17 [LMP01]. CRF is used mostly in supervised learning problems, though there are extensions that address unsupervised learning, whereas HMMs can be easily used for unsupervised learning.

To illustrate CRFs, let us take as input a simple sentence *Obama gave a speech at the Google campus in Mountain View*. Each input word will be assigned a tag of either Person, Organization, Location, or Other, as illustrated in Fig. 2.18. The association of these tags to words is known as the named entity recognition problem in text processing.

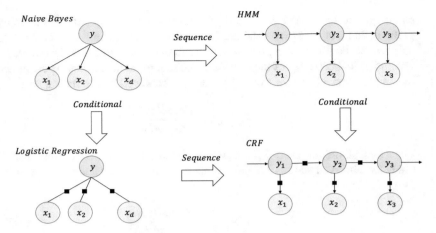

Fig. 2.17: Relationship between generative and discriminative models in non-sequence and sequence-based data

2.8.3.1 Feature Functions

Feature functions are the basic units of CRF that capture the relationship between a pair of two consecutive outputs y_{i-1}, y_i with the entire input sequence $x_1, x_2, \ldots x_n$ as a real-valued output given by $f_j(y_{i-1}, y_i, x_{1:n}, i)$. A simple binary feature function can be written for our example as:

$$f_j(y_{i-1}, y_i, \mathbf{x}, i) = \begin{cases} 1 \; if \; y_i = \; Location \;, y_{i-1} = \; Location \; and \; x_i = \; View \\ 0 \; otherwise \end{cases}$$

(2.97)

2.8.3.2 CRF Distribution

The entire labeled sequence of length n can be modeled as log-linear in terms of the feature functions $f_j(y_{i-1}, y_i, x_{1:n}, i)$ and their weights λ_j, similar to logistic regression, as given by:

$$P(\mathbf{y}|\mathbf{x}, \lambda) = \frac{1}{Z(\mathbf{x}, \lambda)} \exp\left(\sum_{i=0}^{n} \sum_{j} f_j(y_{i-1}, y_i, \mathbf{x}, i) \right)$$

(2.98)

where $Z(\mathbf{x}, \lambda)$ is known as the normalization constant or the partition function and is given by:

$$Z(\mathbf{x}, \lambda) = \sum_{y \in Y} \exp\left(\sum_{i=0}^{n} \sum_{j} f_j(y_{i-1}, y_i, \mathbf{x}, i) \right)$$

(2.99)

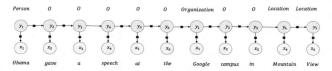

Fig. 2.18: An information extraction example given with words as the inputs and the named entity tags as the outputs of the CRF

2.8.3.3 CRF Training

Similar to logistic regression, maximum likelihood (negative log-likelihood) can be used for learning the parameters (λ) for the CRF. Considering m training sequences $\mathcal{D} = (\mathbf{x^1}, y^1), (\mathbf{x^2}, y^2) \ldots (\mathbf{x^m}, y^m)$, the total log-likelihood loss can be written as:

$$\mathcal{L}(\lambda, \mathcal{D}) = -\log \left(\prod_{k=1} mP(\mathbf{y}|\mathbf{x}, \lambda) \right) \tag{2.100}$$

$$\mathcal{L}(\lambda, \mathcal{D}) = -\sum_{k=1}^{m} \log \left(\frac{1}{Z(\mathbf{x}, \lambda)} \exp \left(\sum_{i=0}^{n} \sum_{j} f_j(y_{i-1}, y_i, \mathbf{x}, i) \right) \right) \tag{2.101}$$

The optimal parameter λ_{opt} can be estimated using the equation below, where C acts as a prior or regularization constant.

$$\lambda_{opt} = \arg\min_{\lambda} \mathcal{L}(\lambda, \mathcal{D}) + C\frac{1}{2}|\lambda|^2 \tag{2.102}$$

The above equation is convex, and solving for the optimum will guarantee obtaining the global optimum. If we rewrite the feature functions for simplicity as below and differentiate the above equation w.r.t λ_j:

$$F_j(\mathbf{y}, \mathbf{x}) = \sum_{j} f_j(y_{i-1}, y_i, \mathbf{x}, i) \tag{2.103}$$

$$\frac{\partial \mathcal{L}(\lambda, \mathcal{D})}{\partial \lambda_j} = \underbrace{\frac{-1}{m} \sum_{k=1}^{m} F_j(\mathbf{y}^k, \mathbf{x}^k)}_{observed\ mean\ feature\ value} + \underbrace{\sum_{k=1}^{m} E_{P(\mathbf{y}|\mathbf{x}^k, \lambda)}[F_j(\mathbf{y}^k, \mathbf{x}^k)]}_{expected\ feature\ value\ given\ the\ Model} \tag{2.104}$$

It can be seen that this equation is not in a closed form, hence impossible to solve analytically. Various iterative algorithms such as L-BFGS or even gradient descent (as discussed above) are generally employed to obtain the solution.

2.9 Case Study

We now take the reader through a real-world application of the concepts introduced in the chapter through a case study. The case study also equips the reader with necessary practical hands-on tools, libraries, methods, code, and analysis that will be useful for standard machine learning, as well as deep learning.

We use the Higgs Boson challenge which was hosted by Kaggle. The challenge data is now available on ATLAS Higgs Challenge 2014. The case study is to classify the events into *signals* and *background* (any other event other than the signal). This is a binary classification problem. Most Kaggle challenges or hackathons provide training data with labels. The models that get submitted are then evaluated on the blind test data on a well-known metrics. Instead of the entire dataset, we have used a sample dataset which has training data size of 10,000 and a separate testing data of size 5000 with labels on which models will be evaluated. We will also assume the best model is selected based on the classification accuracy achieved on the test data, with metrics of **accuracy**, as the data is well balanced between the two classes.

The goal of the case study is to use various techniques and methods illustrated in the chapter and compare the performance on the unseen test set. Various Python libraries such as Numpy, Scipy, Pandas, and scikit-learn, which are used extensively in machine learning, are introduced in the case study.

2.9.1 Software Tools and Libraries

First, we need to describe the main open source tools and libraries we will use for our case study.

- **Pandas** (https://pandas.pydata.org/) is a popular open source implementation for data structures and data analysis. We will use it for data exploration and some basic processing.
- **scikit-learn** (http://scikit-learn.org/) is a popular open source for various machine learning algorithms and evaluations. We will use it only for sampling and creating datasets, machine learning implementations of linear and non-linear algorithms in our case study.
- **Matplotlib** (https://matplotlib.org/) is a popular open source for visualization. We will use it to visualize performance.

2.9.2 Exploratory Data Analysis (EDA)

We use basic EDA to understand the characteristics of the data through univariate statistics, correlation analysis, and visualization.

One of the most important principles we have highlighted in the beginning of this book is to avoid the *data snooping*, i.e., letting the test set labels influence the model or process decisions. Performing distribution analysis, statistical analysis on features, and confirming that the training and test datasets look similar in the splits are all considered to be valid exploratory analysis steps. An example of exploratory data analysis follows.

1. Exploring the number of training and testing data in terms of features and number of classes per set.
2. Exploring the data types for each feature to determine whether it is categorical, continuous, ordinal, etc. and transforming them if needed based on the domain.
3. Finding if the features have missing or unknown values and transforming them as needed.
4. Understanding the distribution of each feature using scatter plots, histogram plots, box plots, etc., to see basic statistics of range, variance, and distribution for the features. This is illustrated in Fig. 2.19.
5. Understanding similarity and differences between each of these statistics and plots for the training and testing data features.
6. Calculating pairwise correlation between the features and correlation between features and the labels on training set. Plotting these and visualizing them (as in Fig. 2.19) gives a great aid to subject matter experts and data scientists.

2.9.3 Model Training and Hyperparameter Search

In this section we will go over some standard machine learning techniques performed on the data for learning effective models.

2.9.3.1 Feature Transformation and Reduction Impact

Understanding the impact of feature transformation and selection is one of the preliminaries for training a machine learning model. As discussed earlier in the chapter there are various dimensionality reduction techniques such as PCA, SVD, and others, various selection techniques, such as mutual information, chi-square, and others, and each of them have parameters that need to be tuned. These will impact the model and training algorithms, as well. The feature selection and dimensionality reduction techniques should be treated as hyperparameters that the model selection process will optimize. For the sake of brevity, in this section we will only analyze two different feature selection algorithms. We will show two different feature selection and analyze them in this section.

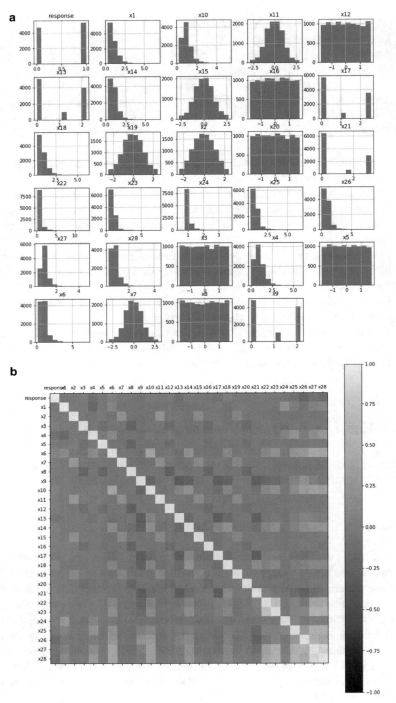

Fig. 2.19: Exploratory data analysis plots. (**a**) Histogram plot for each feature and label on training data. (**b**) Pearson correlation across features and labels on training data

We first perform PCA with two components and plot these components with labels to see if the newly reduced training dataset with two components shows improved separation. We then increase the dimensions and plot the cumulative explained variances by adding the variance captured by each transformed feature. Figure 2.20 shows the two plots and reveals that the PCA transformation and reduction may not be useful for this dataset; the transformed features need as many dimensions as the original features to capture the variances.

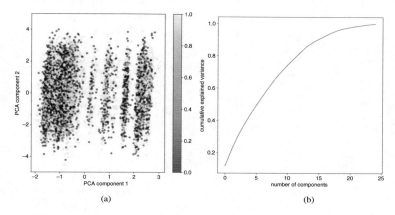

(a)   (b)

Fig. 2.20: PCA analysis. (**a**) PCA of two transformed components as scatter plot on training data. (**b**) Cumulative explained variance for first 25 dimensions

We also perform chi-square analysis on the data by first scaling the features to the range of $[0, 1]$, as chi-square needs all features to be in positive range. The plots of scores and feature names are shown in Fig. 2.21. Only 16 features have scores above a threshold 0.1, and that may be a good subset to choose if reduction is pursued.

Chi-Square Feature Selection Results

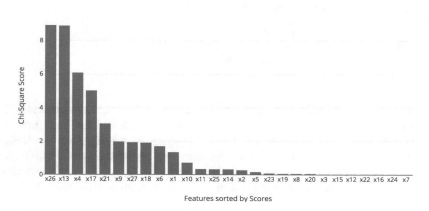

Fig. 2.21: Chi-square scores on the features plotted in the descending order

2.9.3.2 Hyperparameter Search and Validation

We choose accuracy as the metric for performing search of hyperparameters for the algorithms and for comparing algorithms, because that is the metric that the test data prediction will be evaluated on. We will use cross-validation as our validation technique in the hyperparameter search. We will use five linear and non-linear algorithms for training: (a) perceptron, (b) logistic regression, (c) linear discriminant analysis, (d) naive Bayes, and (e) support vector machines (with RBF kernel).

The code below highlights the hyperparameter search for SVM.

```
from sklearn.svm import SVC
import numpy
# gamma parameter in SVM
gammas = numpy.array([1, 0.1, 0.01, 0.001])
# C parameter for SVM
c_values = numpy.array([100, 1, 0.1, 0.01])
# grid search for gamma and C
svm_param_grid = {'gamma': gammas, 'C': c_values}
# svm with rbf kernel
svm = SVC(kernel='rbf')
scoring = 'accuracy'
# grid search
grid = GridSearchCV(estimator=svm, param_grid=svm_param_grid,
    scoring=scoring)
```

The hyperparameters found and the validation results are given in Table 2.3. It is interesting to observe that the simplest linear perceptron has the least score and as the complexity of the model is increased to completely non-linear RBF kernel SVM, the performance improves.

Table 2.3: Hyperparameters and validation scores for the classifiers

Classifier	Parameter and values	Tenfold cross-validation AUC
Perceptron	$\alpha = 0.001$, *maxIter* $= 100$	0.54
Logistic regression	penalty=L_1, C=0.1, *maxIter* $= 100$	0.61
LDA	tolerance=0.001	0.60
Naive Bayes		0.60
SVM (RBF)	$\gamma = 0.01, C = 100$	0.63

We next see if there is an impact of the feature selection techniques on a classifier by performing grid search for all the parameters of feature selection/reduction and classification. We use PCA and kbest with chi-square as the two feature selection techniques, and logistic regression as the classifier. By plotting the classification accuracy on various combinations, as shown in Fig. 2.22, we see that there is no impact of the feature selection and reduction on the validation performance.

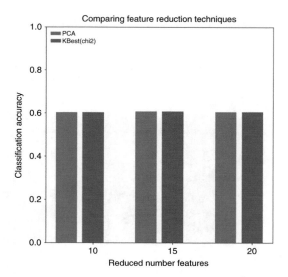

Fig. 2.22: Validation accuracy of logistic regression with different feature selection techniques

2.9.3.3 Learning Curves

To see the impact of training examples and the variance across different validation runs, learning curves are plotted. They provide useful comparison of the training and validation metrics as a function of the training set size. We plot learning curves for the tuned logistic regression and SVM, as they were high-scoring algorithms (as shown in Fig. 2.23a and b). It can be observed that the SVM validation score increases monotonically with the size of the training set, demonstrating that more examples do improve the performance. It can also be observed that SVM has low variance across runs compared to logistic regression, which indicates robustness of the SVM classifier.

2.9.4 Final Training and Testing Models

Finally, we train the best models (with the best parameters) on the entire training data and run them on the test data for estimating the out-of-sample error (Table 2.4).

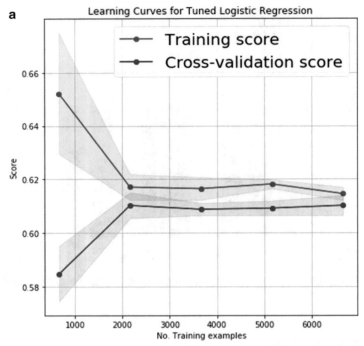

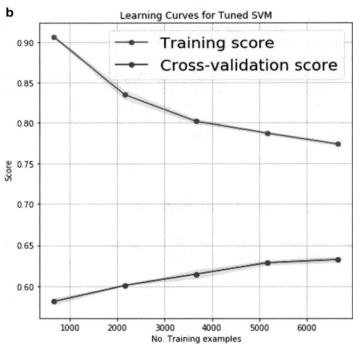

Fig. 2.23: (**a**) Learning curves for tuned logistic regression. (**b**) Learning curves for tuned SVM

Table 2.4: Hyperparameter and validation scores for classifiers

Classifier	Accuracy	Precision	Recall	F1-score
Perceptron	0.55	0.55	0.56	0.56
Logistic regression	0.61	0.61	0.62	0.61
LDA	0.61	0.61	0.61	0.61
Naive Bayes	0.60	0.61	0.60	0.60
SVM (RBF)	0.64	0.64	0.65	0.65

2.9.5 Exercises for Readers and Practitioners

Some other interesting problems readers and practitioners can attempt on their own include:

1. What is the impact of other feature transformations, such as normalization?
2. What is the impact of other univariate feature selection methods, such as mutual information (selecting high-gain features)?
3. What is the impact of multivariate feature selection, such as correlation feature selection (CFS) or minimum redundancy maximum relevance (mRmR) that consider groups of features as opposed to individual features?
4. What is the impact of wrapper-based feature selection methods like recursive feature elimination (RFE)?
5. What is the impact of other non-linear learning methods, such as decision tree, gradient boosting, and random forest?
6. What is the impact of meta-learning techniques, such as cost-based learning, ensemble learning, and others?

References

[AMMIL12]   Yaser S. Abu-Mostafa, Malik Magdon-Ismail, and Hsuan-Tien Lin. *Learning From Data*. AMLBook, 2012. ISBN: 1600490069, 9781600490064.

[BB08]   Léon Bottou and Olivier Bousquet. "The Tradeoffs of Large Scale Learning". In: *Advances in Neural Information Processing Systems*. Vol. 20. NIPS Foundation (http://books.nips.cc), 2008, pp. 161–168.

[Bry61]   A. E. Bryson. "A gradient method for optimizing multi-stage allocation processes". In: *Proc. Harvard Univ. Symposium on digital computers and their applications*. 1961.

[CS14]   Girish Chandrashekar and Ferat Sahin. "A Survey on Feature Selection Methods". In: *Comput. Electr. Eng.* 40.1 (Jan. 2014), pp. 16–28.

[Dem06]   Janez Demšar. "Statistical Comparisons of Classifiers over Multiple Data Sets". In: *J. Mach. Learn. Res.* 7 (Dec. 2006), pp. 1–30.

[Die98]   Thomas G. Dietterich. "Approximate Statistical Tests for Comparing Supervised Classification Learning Algorithms". In: *Neural Comput.* 10.7 (Oct. 1998), pp. 1895–1923.

[GE03]   Isabelle Guyon and André Elisseeff. "An Introduction to Variable and Feature Selection". In: *J. Mach. Learn. Res.* 3 (Mar. 2003), pp. 1157–1182.

[HTF09]   Trevor Hastie, Robert Tibshirani, and Jerome Friedman. *The elements of statistical learning*. Springer Series in Statistics, 2009. Chap. 15.

[HK00]   Arthur E. Hoerl and Robert W. Kennard. "Ridge Regression: Biased Estimation for Nonorthogonal Problems". In: *Technometrics* 42.1 (Feb. 2000), pp. 80–86.

[JS11]   Nathalie Japkowicz and Mohak Shah. *Evaluating Learning Algorithms: A Classification Perspective*. New York, NY, USA: Cambridge University Press, 2011.

[Jol86]   I. T. Jolliffe. *Principal Component Analysis*. Springer-Verlag, 1986.

[KS+60]   John G Kemeny, James Laurie Snell, et al. *Finite Markov chains*. Vol. 356. van Nostrand Princeton, NJ, 1960.

[KK62]   J. F. Kenney and E. S. Keeping. *Mathematics of Statistics*. Princeton, 1962, pp. 252–285.

[LMP01]   John D. Lafferty, Andrew McCallum, and Fernando C. N. Pereira. "Conditional Random Fields: Probabilistic Models for Segmenting and Labeling Sequence Data". In: *Proceedings of the Eighteenth International Conference on Machine Learning*. ICML '01. Morgan Kaufmann Publishers Inc., 2001, pp. 282–289.

[Rab89]   Lawrence R Rabiner. "A tutorial on hidden Markov models and selected applications in speech recognition". In: *Proceedings of the IEEE* 77.2 (1989), pp. 257–286.

[Ros58]   Frank Rosenblatt. "The perceptron: a probabilistic model for information storage and organization in the brain." In: *Psychological review* 65.6 (1958), p. 386.

[Vap95]   V. Vapnik. *The Nature of Statistical Learning Theory*. Springer, New York, 1995.

[WD67]   Strother H. Walker and David B. Duncan. "Estimation of the probability of an event as a function of several independent variables". In: *Biometrika* 54 (1967), pp. 167–179.

Chapter 3
Text and Speech Basics

3.1 Introduction

This chapter introduces the major topics in text and speech analytics and machine learning approaches. Neural network approaches are deferred to later chapters.

We start with an overview of **natural language** and **computational linguistics**. Representations of text that will form the basis of advanced analysis are introduced, and the core components in computational linguistics are discussed. Readers are guided through the broad range of applications that leverage these concepts. We investigate the topics of text classification and text clustering, and move onto applications in machine translation, question answering, and automated summarization. In the latter part of this chapter, acoustic models and audio representations are introduced, including MFCCs and spectrograms.

3.1.1 Computational Linguistics

Computational linguistics focuses on applying quantitative and statistical methods to understand how humans model language, as well as computational approaches to answer linguistic questions. Its beginning in the 1950s coincided with the advent of computers. **Natural language processing** (NLP) is the application of computational methods to model and extract information from human language. While the difference between the two concepts relates to underlying motivation, they are often used interchangeably.

Computational linguistics can refer to written or spoken natural language. A written language is the symbol representation of a spoken or gestural language. There are plenty of spoken natural languages without a writing system, whereas there are no written natural languages that have developed on their own without a spoken aspect. Natural language processing of written language is often called **text analytics** and of spoken language is called **speech analytics**.

© Springer Nature Switzerland AG 2019
U. Kamath et al., *Deep Learning for NLP and Speech Recognition*,
https://doi.org/10.1007/978-3-030-14596-5_3

Computational linguistics was considered in the past as a field within computer science. This has evolved considerably as computational linguistics became an interdisciplinary field of theoretical and applied science joining linguistics, psychology, neuroscience, philosophy, computer science, mathematics, and others. With the rise of social media, conversational agents, and personal assistants, computational linguistics is increasingly relevant in creating practical solutions to modeling and understanding human language.

3.1.2 Natural Language

A natural language is one that has evolved naturally through daily use by humans over time, without formal construction. They encompass a broad set that includes spoken and signed languages. By some estimates, there are about 7000 human languages currently in existence, with the top ten representing 46% of the world's population [And12] (Fig. 3.1).

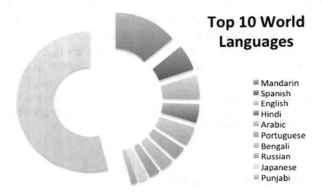

Fig. 3.1: Top 10 world languages

Natural language is inherently ambiguous, especially in its written form. To understand why this is so, consider that the English language has about 170,000 words in its vocabulary but only about 10,000 are commonly used day-to-day [And12]. Human communications have evolved to be highly efficient, allowing for reuse of shorter words whose meanings are resolved through context. This lessens the computational burden and frees up parts of the human brain for other important tasks. At the same time, this ambiguity makes it inherently hard for computers to process and understand natural language. This difficulty extends to aspects of language such as sarcasm, irony, metaphors, and humor. In any language, ambiguities exist in word sense, grammatical structure, and sentence structure. We will discuss below methods that deal with each of these ambiguities.

3.1.3 Model of Language

When we analyze a natural language, we often group language characteristics into a set of categories. For text analysis, these categories are morphology, lexical, syntax, semantics, discourse, and pragmatics. **Morphology** refers to the shape and internal structure of a word. **Lexical** refers to the segmentation of text into meaningful units like words. **Syntax** refers to the rules and principles applied to words, phrases, and sentences. **Semantics** refers to the context that provides meaning within a sentence. It is semantics that provides the efficiency of human language. **Discourse** refers to conversations and the relationships that exist among sentences. **Pragmatics** refers to external characteristics such as the intent of the speaker to convey context. For speech analysis, we typically group language characteristics into the categories of acoustics, phonetics, phonemics, and prosodics. **Acoustics** refers to the methods we use to represent sounds. **Phonetics** refers to how sounds are mapped to phonemes that serve as base units of speech. **Phonemics**, also known as phonology, refers to how phonemes are used in a language. **Prosodics** refers to non-language characteristics that accompany speech such as tone, stress, intonation, and pitch. In the following sections, we will discuss each in greater detail from a computational linguistics perspective. As we will see in the subsequent chapters, these linguistic characteristics can be used to provide a rich set of representations that are useful for machine learning algorithms (Table 3.1).

Table 3.1: Language analysis categories

Morphology	Shape and structure of words
Lexical	Segmenting text into words
Syntax	Rules for words in a sentence
Semantics	Meaning of words in a sentence
Discourse	Meaning among sentences
Pragmatics	Meaning through speaker intent
Acoustics	Representations of sound
Phonetics	Mapping sound to speech
Phonemics	Mapping speech to language
Prosodics	Stress, pitch, tone, rhythm

Because of the dependencies inherent in these categories, we often model a natural language as a hierarchical collection of linguistic characteristics as in Fig. 3.2.

This is often called a synchronic model of language—that is, a model that is based on a snapshot in time of a language [Rac14]. Some linguists have argued that such a synchronic model does not fit modern living languages that constantly evolve, preferring diachronic models that can address changes in time. The complexity of diachronic models makes them difficult to handle, however, and synchronic models as originally championed by Swiss linguist Ferdinand de Saussure at the turn of the twentieth century are widely adopted today [Sau16]. The computational and

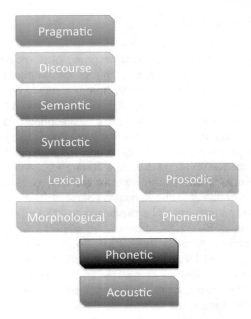

Fig. 3.2: Model of natural language

statistical methods that can be applied to each component in the synchronic model form the basis of natural language processing.

Natural language processing seeks to map language to representations that capture morphological, lexical, syntactic, semantic, or discourse characteristics that can then be processed by machine learning methods. The choice of representation can have significant impact on later tasks and will depend on the selected machine learning algorithm for analysis.

In the following sections of this chapter, we will dive into these representations to better understand their role in linguistics and purpose in natural language processing. We introduce to readers the most common text-based representations, and leave audio representations to later sections.

3.2 Morphological Analysis

All natural languages have systematic structure, even sign language. In linguistics, morphology is the study of the internal structure of words. Literally translated from its Greek roots, morphology means "the study of shape." It refers to the set of rules and conventions used to form words based on their context, such as plurality, gender, contraction, and conjugation.

Words are composed of subcomponents called morphemes, which represent the smallest unit of language that holds independent meaning. Morphemes may be components of the word that relate to its meaning, grammatical role, or derivation. Some morphemes are words by themselves, such as "run," "jump," or "hide." Other words are formed by a combination of morphemes, such as "runner," "jumps," or "unhide." Some languages, like English, have relatively simple morphologic rules for combining morphemes. Others, like Arabic, have a rich set of complex morphologic rules [HEH12].

To humans, understanding the morphological relations between the words "walk, walking, walked" is relatively simple. The plurality of possible morpheme combinations, however, makes it very difficult for computers to do so without morphological analysis. Two of the most common approaches are stemming and lemmatization, which we describe below.

3.2.1 Stemming

Often, the word ending is not as important as the root word itself. This is especially true of verbs, where the verb root may hold significantly more meaning than the verb tense. If this is the case, computational linguistics applies the process of word stemming to convert words to their root form (e.g., base morpheme in meaning). Here are some stemming examples:

$$\text{works} \rightarrow \text{work}$$
$$\text{worked} \rightarrow \text{work}$$
$$\text{workers} \rightarrow \text{work}$$

While you and I can easily recognize that each of these is related to meaning, it would be very difficult for a computer to do so without stemming. It is important to note that stemming can introduce ambiguity, as evident in the third example above where "workers" has the same stem as "works," but both words have different meanings (people versus items). On the other hand, the advantage of stemming is that it is generally robust to spelling errors, as the correct root may still be inferred correctly.

One of the most popular stemming algorithms in NLP is the **Porter stemmer**, devised by Martin Porter in 1980. This simple and efficient method uses a series of 5 steps to strip word suffixes and find word stems. Open-source implementations of the Porter stemmer are widely available.

3.2.2 Lemmatization

Lemmatization is another popular method used in computational linguistics to reduce words to base forms. It is closely related to stemming in that it is an algorithmic process that removes inflection and suffixes to convert words into their lemma (i.e., dictionary form). Some examples of lemmatization are:

$$works \rightarrow works$$
$$worked \rightarrow work$$
$$workers \rightarrow worker$$

Notice that the lemmatization results are very similar to those of stemming, except that the results are actual words. Whereas stemming is a process where meaning and context can be lost, lemmatization does a much better job as evident in the third example above. Since lemmatization requires a dictionary of lexicons and numerous lookups, stemming is faster and the generally more preferred method. Lemmatization is also extremely sensitive to spelling errors, and may require spell correction as a preprocessing step.

3.3 Lexical Representations

Lexical analysis is the task of segmenting text into its lexical expressions. In natural language processing, this means converting text into base word representations which can be used for further processing. In the next few subsections, we provide an overview of word-level, sentence-level, and document-level representations. As the reader will see, these representations are inherently sparse, in that few elements are non-zero. We leave dense representations and word embeddings to a later chapter.

Words are the elementary symbols of natural language. They are not the most elementary, as words can consist of one or more morphemes. In natural language processing, often the first task is to segment text into separate words. Note that we say "often" and not "always." As we will see later, sentence segmentation as a first step may provide some benefits, especially in the presence of "noisy" or ill-formed speech.

3.3.1 Tokens

The computational task of segmenting text into relevant words or units of meaning is called **tokenization**. Tokens may be words, numbers, or punctuation marks. In simplest form, tokenization can be achieved by splitting text using whitespace:

The rain in Spain falls mainly on the plain.
|The|, |rain|, |in|, |Spain|, |falls|, |mainly|, |on|, |the|, |plain|, |.|

This works in most cases, but fails in others:

Don't assume we're going to New York.
|Don|, |'t|, |assume|, |we|, |'|, |re|, |going|, |to|, |New|, |York|, |.|

Notice that "New York" is typically considered a single token, since it refers to a specific location. To compound problems, tokens can sometimes consist of multiple words (e.g., "he who cannot be named"). There are also numerous languages that do not use any whitespace, such as Chinese.

西班牙的降雨主要集中在平原。

Tokenization serves also to segment sentences by delineating the end of one sentence and beginning of another. Punctuation plays an important role in this task, but unfortunately punctuation is often ambiguous. Punctuation like apostrophes, hyphens, and periods can create problems. Consider the multiple use of the period in this sentence:

Dr. Graham poured 0.5ml into the beaker.
|Dr.|, |Graham poured 0.|, |5ml into the beaker.|

A simple punctuation-based sentence splitting algorithm would incorrectly segment this into three sentences. There are numerous methods to overcome this ambiguity, including augmenting punctuation-based with hand engineered rules, using regular expressions, machine learning classification, conditional random field, and slot-filling approaches.

3.3.2 Stop Words

Tokens do not occur uniformly in English text. Instead, they follow an exponential occurrence pattern known as Zipf's law, which states that a small subset of tokens occur very often (e.g., the, of, as) while most occur rarely. How rarely? Of the 884,000 tokens in Shakespeare's complete works, 100 tokens comprise over half of them [Gui+06].

In the written English language, common functional words like "the," "a," or "is" provide little to no context, yet are often the most frequently occurring words in text as seen in Fig. 3.3. By excluding these words in natural language processing, performance can be significantly improved. The list of these commonly excluded words is known as a stop word list.

3.3.3 N-Grams

While word level representations are sometimes useful, they do not capture the relationship with adjacent words that can help provide grammar or context. For instance, when working with individual tokens, there is no concept of word order.

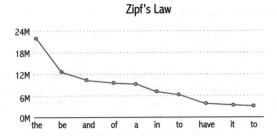

Fig. 3.3: Zipf's law as it applies to the text from the complete works of Shakespeare

One simply considers the existence and occurrence of tokens within a piece of text. This is known as a **bag-of-words** model (see Eq. (3.1)), and is based on a Markov assumption. A phrase containing L tokens would be predicted with probability:

$$P(w_1 w_2 \ldots w_L) = \prod_{i=1}^{L} P(w_i) \qquad (3.1)$$

Instead of considering individual tokens (termed unigrams), another approach would be to consider consecutive tokens. This is called a bigram approach, where a sentence with L tokens would yield $L-1$ bigrams in the form:

$$P(w_1 w_2 \ldots w_L) = \prod_{i=2}^{L} P(w_i | w_{i-1}) \qquad (3.2)$$

Notice that bigrams effectively capture the local word order of two consecutive tokens (e.g., "lion king" is not the same as "king lion"). We can extend this concept to capture lengths of n tokens, known as n-grams:

$$P(w_1 w_2 \ldots w_L) = \prod_{i=n}^{L} P(w_i | w_{i-1} w_{i-2} \ldots w_{i-n}) \qquad (3.3)$$

It is important to note that for higher values of n, n-grams become extremely infrequent. This will adversely impact count-based natural language processing methods as we see later.

3.3.4 Documents

When working with a corpus of documents, there are a number of document representation methods used in computational linguistics. Some are token based, multi-token based, or character based. Many are sparse representations, while others are dense. We discuss two of the most common ones in this section.

3.3.4.1 Document-Term Matrix

A document-term matrix is a mathematical representation of a set of documents. In the document-term matrix shown in Table 3.2, rows correspond to documents in the collection and columns correspond to unique tokens. The number of columns is equal to the unique token vocabulary across all documents. There are numerous ways to determine the value of the elements of this matrix, and we discuss two below.

3.3.4.2 Bag-of-Words

One common approach is to set each element of the document-term matrix equal to the frequency of word occurrence within each document. Imagine representing each document as a list of counts of unique words. This is known as a **bag-of-words** approach [PT13]. Obviously, there is significant information loss by simply using a document vector to represent an entire document, but this is sufficient for many computational linguistics applications. This process of converting a set of documents into a document-term matrix where each element is equal to word occurrence is commonly known as **count vectorization**.

Table 3.2: Document-term matrix

	Doc1	Doc2	Doc3	Doc4	Doc5	Doc6
Car	1	0	0	1	1	1
Bicycle	0	1	1	0	0	1
Drives	1	0	1	0	0	1
Rides	0	1	0	1	2	1
Bumpy	0	1	0	1	1	1
Smoothly	1	0	1	0	0	1
Like	0	0	1	0	2	1

You may remember that the most frequent words in the English vocabulary are generally less significant than rarer words for discerning meaning. Unfortunately, a bag-of-words model weighs words based on occurrence. In practice, a stop word filter is used to remove the most common words prior to count vectorization. Even in doing so however, we will still find that rare words occurring across a set of documents are often the most meaningful.

3.3.4.3 TFIDF

TFIDF is a method that provides a way to give rarer words greater weight by setting each document-term matrix element equal to the value w of multiplying the **term frequency** (TF) by the **inverse document frequency** (IDF) of each token [Ram99]:

$$w = tf \times idf \tag{3.4}$$

$$= (1 + \log(TF_t)) \times \log\left(\frac{N}{n_i}\right) \tag{3.5}$$

Here, we have used the definitions of term frequency as the logarithmically scaled ratio of the count of term t occurring in document d versus the total number of terms in document d, and inverse document frequency as the logarithmically scaled ratio of the total number of documents vs the count of documents with term t.

Because of the tf factor, the TFIDF value for a token increases proportionally to the number of times it appears in a document. The idf factor reduces the TFIDF value for a token based on the frequency of occurrence across all documents. Currently, TFIDF is the most popular weighting method, with over 80% of current digital libraries using it in production (Table 3.3).

Table 3.3: TFIDF matrix

	Doc1	Doc2	Doc3	Doc4	Doc5	Doc6
Car	3.50	0	0	5.91	3.21	2.82
Bicycle	0	2.79	2.51	0	0	1.73
Drives	1.21	0	0.88	0	0	0.88
Rides	0	1.26	0	3.13	2.22	0.41
Bumpy	0	1.11	0	0.45	0.61	1.23
Smoothly	0.13	0	0.12	0	0	0.92
Like	0	0	0.22	0	0.34	0.24

3.4 Syntactic Representations

Syntactic representations of natural language deal with the grammatical structure and relation of words and phrases within sentences. Grammar plays an inherent role in most languages to help provide context. Computational linguistics utilizes different approaches to extract these context clues such as part-of-speech tags, chunking, and dependency parsing. They serve well as features for downstream natural language processing tasks.

3.4.1 Part-of-Speech

A part-of-speech (POS) is a class of words with grammatical properties that play similar sentence syntax roles. It is widely accepted that there are 9 basic part-of-speech classes (you may remember them from grade school) (Table 3.4).

Table 3.4: Basic part-of-speech labels

N	Noun	Dog, cat
V	Verb	Run, hide
A	Article	The, an
ADJ	Adjective	Green, short
ADV	Adverb	Quickly, likely
P	Preposition	By, for
CON	Conjunction	And, but
PRO	Pronoun	You, me
INT	Interjection	Wow, lol

There are numerous POS subclasses in English, such as singular nouns (NN), plural nouns (NNS), proper nouns (NP), or adverbial nouns (NR). Some languages can have over 1000 parts of speech [PDM11]. Due to the ambiguity of the English language, many English words belong to more than one part-of-speech category (e.g., "bank" can be a verb, noun, or interjection), and their role depends on how they are used within a sentence. It can be difficult to identify which category the word belongs. Part-of-speech tagging is the process of predicting the part-of-speech

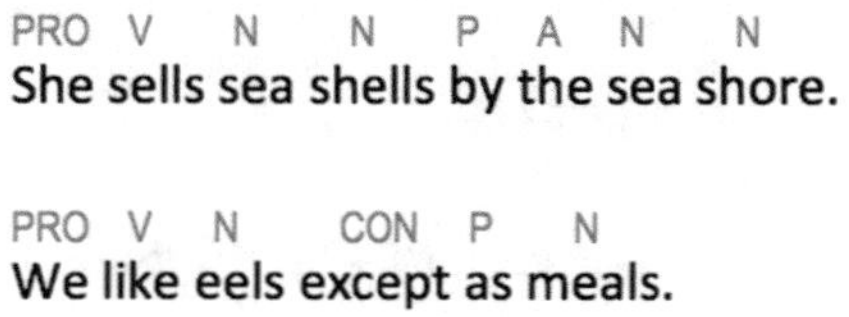

PRO V   N   N   P   A   N   N
She sells sea shells by the sea shore.

PRO V   N   CON P   N
We like eels except as meals.

Fig. 3.4: POS tagging

category for each word in the text based on its grammatical role and context within a sentence (Fig. 3.4) [DeR88]. POS tagging algorithms fall into two distinct groups: rules based and statistical methods.

The **Brown corpus** was the first major collection of English texts used in computational linguistics research. It was developed by Henry Kuč era and W. Nelson Francis at Brown University in the mid-1960s and consists of over a million words of English prose extracted from 500 randomly chosen publications of 2000 or more words. Each word in the corpus has been POS-tagged meticulously using 87 distinct POS tags. The Brown corpus is still commonly used as a gold set to measure the performance of POS-tagging algorithms.

3.4.1.1 Rules Based

The earliest part-of-speech tagging approaches were rules based and depended on dictionaries, lexicons, or regular expressions to predict possible POS labels for each word. Where ambiguities arose, ad-hoc rules were often incorporated to make POS tag decisions. This made rules-based systems brittle. For instance, a rule could declare that a word that follows an adverb and comes before a conjunction should be a noun, except it should be verb if it is not a singular common noun. The best rules-based POS tagger to date achieved only 77% accuracy on the Brown corpus [BM04].

3.4.1.2 Hidden Markov Models

Since the 1980s, hidden Markov models (HMMs) introduced in the previous chapter became popular as a better approach to POS tagging. HMMs are better able to learn and capture the sequential nature of grammar than rules-based methods. To understand this, consider the POS tagging problem, where we seek to find the most probable tag sequence $\hat{t}^n$ for a given sequence of n words w^n:

$$\hat{t}^n = \operatorname*{argmax}_{t^n} P\left(t^n | w^n\right) \tag{3.6}$$

$$\approx \operatorname*{argmax}_{t^n} \prod_{i=1}^{n} P\left(w_i | t_i\right) P\left(t_i | t_{i-1}\right) \tag{3.7}$$

The equation above represents an HMM model where the Markov states are the words w^n and the hidden states t^n are the POS tags. The transition matrices can be directly computed from text data. It turns out that assigning the most common tag to each known word can achieve fairly high accuracy. To account for more ambiguous word sequences, higher-order HMMs can also be used for larger sequences by leveraging the Viterbi algorithm. These higher-order HMMs can achieve very high accuracy, but they require significant computation load since they must explore a larger set of paths.

Beyond HMMs, machine learning methods have gained huge popularity for POS tagging tasks, including CRF, SVM, perceptrons, and maximum entropy classification approaches. Most now achieve accuracy above 97%. In the subsequent chapters, we will examine deep learning approaches that hold even greater promise to POS tag prediction.

3.4.2 Dependency Parsing

In a natural language, grammar is the set of structural rules by which words and phrases are composed. Every sentence in English follows a certain pattern. These patterns are called grammars and express a relation between a (head) word and its dependents. Most natural languages have a rich set of grammar rules, and knowledge of these rules helps us disambiguate context in a sentence. Consider the fact that without grammar, there would be practically unlimited possibilities to combine words together.

Parsing is the natural language processing task of identifying the syntactic relationship of words within a sentence, given the grammar rules of a language [Cov01]. There are two common ways to describe sentence structure in natural language. The first is to represent the sentence by its constituent phrases, recursively down to the individual word level. This is known as constituent grammar parsing, which maps a sentence to a constituent parse tree (Fig. 3.5). The other way is to link individual

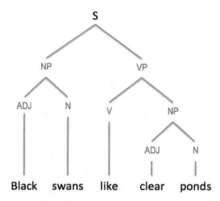

Fig. 3.5: Constituent grammar parsing

words together based on their dependency relationship. This is known as dependency grammar parsing which maps a sentence to a dependency parse tree (Fig. 3.6). Dependency is a one-to-one correspondence, which means that there is exactly

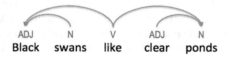

Fig. 3.6: Dependency grammar parsing

one node for every word in the sentence. Notice that the links are directional between two words in a dependency parse tree, pointing from the head word to the dependent word to convey the relationship. Constituent and dependency parse trees can be strongly equivalent. The appeal of dependency tree is that the links closely resemble semantic relationships.

Because a dependency tree contains one node per word, the parsing can be achieved with computational efficiency. Given a sentence, parsing algorithms attempt to find the most likely derivation from its grammatical rules. If the sentence exhibits structural ambiguity, more than one derivation is possible. Parsers are subdivided into two general approaches. Top-down parsers use a recursive algorithm with a back-tracking mechanism to descend from the root down to all words in the sentence. Bottom-up parsers start with the words and build up the parse tree based on a shift/reduce or other algorithm. Top-down parsers will derive trees that will always be grammatically consistent, but may not align with all words in a sentence. Bottom-up approaches will align all words, but may not be always make grammatical sense.

3.4.2.1 Context-Free Grammars

Grammar, as stated above, is the set of rules that define the syntactic structure and pattern of words in a sentence. Because these rules are generally fixed and absolute, a context-free grammar (CFG) can be used to represent the grammatical rules of a language [JM09]. Context-free grammars typically have a representation known as Backus–Naur form and are able to capture both constituency and ordering of words in a sentence.

Unfortunately, because of the inherent ambiguity of language, CFG may generate multiple possible parse derivations for a given sentence. Probabilistic context-free grammars (PCFG) deal with this issue by ranking possible parse derivations and selecting the most probable, given a set of weights learned from a distribution of text. PCFGs generally outperform CFGs, especially for languages like English.

3.4.2.2 Chunking

For some applications, a full syntactic parse with its computational expense may not be needed. Chunking, also called shallow parsing, is a natural language processing task which joins words into base syntactic units rather than generating a full parse tree. These base syntactic units are often referred to as "chunks." For example, given a sentence, we would like to identify just the base noun-phrases (i.e., phrases that serve the same grammatical function as a noun and do not contain other noun-phrases):

$[_{NP}$The winter season$]$ is depressing for $[_{NP}$many people$]$.

Chunking is often performed by a rules-based approach where regular expressions and POS-tags are used to match fixed patterns, or with machine learning algorithms such as SVM [KM01].

3.4.2.3 Treebanks

A treebank is a text corpus that has been parsed and annotated for syntactic structure. That is, each sentence in the corpus has been parsed into its dependency parse tree. Treebanks are typically generated iteratively using a parser algorithm and human review [Mar+94] [Niv+16]. Often, treebanks are built on top of a corpus that has already been annotated with part-of-speech tags. The creation of treebanks revolutionized computational linguistics, as it embodied a data-driven approach to generating grammars that could be reused broadly in multiple applications and domains. Statistical parsers trained with treebanks are able to deal much better with structural ambiguities [Bel+17].

> The **Penn Treebank** is the de facto standard treebank for parse analysis and evaluation. Initially released in 1992, it consists of a collection of articles from Dow Jones News Service written in English, of which 1 million words are POS-tagged and 1.6 million words parsed with a tagset. An improved version of the Penn Treebank was released in 1995.
>
> **Universal Dependencies** is a collection of over 100 treebanks in 60 languages, created with the goal of facilitating cross-lingual analysis [McD+13]. As its name implies, the treebanks are created with a set of universal, cross-linguistically consistent grammatical annotations. The first version was released in October of 2014.

3.5 Semantic Representations

Whereas lexical and syntactic analyses capture the form and order of language, they do not associate meaning with words or phrases. For instance, labeling "dog" as a noun gives us no clue what a "dog" is. Semantic representations give a sense of meaning to words and phrases. They attach roles to word chunks such as person, place, or amount. Semantic analysis is interested in understanding meaning primarily in terms of word and sentence relationships. There are several different kinds of semantic relations between words (see Table 3.5).

Table 3.5: Semantic relations between words

Synonymy	Words spelled differently but have the same meaning
Antonymy	Words having the opposite meanings to each other
Hyponymy	Generic term and a specific instance of it
Hypernymy	Broad category that includes other words
Meronymy	Constituent part or a member of something
Holonymy	Semantic relation between a whole and its parts
Homonymy	Words with identical forms but different meanings
Polysemy	Words with two or more distinct meanings

Homonymy and polysemy are very similar, and the key difference is that a polysemous word is one word with different meanings, while homonyms are different words that share a shape (usually both spelling and pronunciation). For example, most people would consider the noun tire (the wheels on your car) and the verb tire (what happens when you exercise) to be completely different words, even though they look and sound the same. They're homonyms. On the other hand, most people agree that there is only one word offense, but that it has various meanings which are all related: the attacking team, a criminal act, a feeling of being insulted, etc.

3.5.1 Named Entity Recognition

Named entity recognition (NER) is a task in natural language processing that seeks to identify and label words or phrases in text that refer to a person, location, organization, date, time, or quantity. It is a subtask of information extraction and is sometimes called entity extraction. Due to the reuse of words and ambiguity of natural language, entity recognition is hard. Take, for instance, the word "Washington" which may be a reference to a city, a state, or a president. It would be difficult to disambiguate this word without the context of real-world knowledge. Ambiguities can exist in two ways: different entities of the same type (George Washington and Washington Carver are both persons) or entities of different types (George Washington or Washington state) (Table 3.6).

Table 3.6: Named entities

Person	George Washington
Location	Washington State
Organization	General Motors
Date	Fourth of July
Time	Half past noon
Quantity	Four score

While regular expressions can be used to some extent for name entity recognition [HN14], the standard approach is to treat it as a sequence labeling task or HMM in similar fashion to POS-tagging or chunking [AL13]. Conditional random fields (CRFs) have shown some success in named entity recognition. However, training a CRF model typically requires a large corpus of annotated training data [TKSDM03c]. Even with a lot of data, name entity recognition is still largely unsolved.

3.5.2 Relation Extraction

Relationship extraction is the task of detecting semantic relationships of named entity mentions in text. For instance, from the following sentence,

President George Bush and his wife Laura attended the Congressional Dinner.

we can extract a set of relations between the entities: George Bush, Laura, Congressional Dinner (Table 3.7). Note that the second relation (George Bush is married

Table 3.7: Entity relations

Laura married to George Bush	Person → Person
George Bush married to Laura	Person → Person
George Bush at Congressional Dinner	Person → Location
President George Bush	Org → Person

to Laura) logically follows from the first (Laura is married to George Bush), even though it may not be explicitly stated in the text. The common approach to relation extraction is to divide it into subtasks:

1. Identify any relations between entities
2. Classify the identified relations by type
3. Derive logical/reciprocal relations.

The first subtask is typically treated as a classification problem, where a binary decision is made as to whether a relation is present between any two entities within the text. The second subtask is a multiclass prediction problem. Naive Bayes and SVM models have been successfully applied to both subtasks [BB07, Hon05]. The last subtask is a logical inference task. Relation extraction plays an important role in question answering tasks.

3.5.3 Event Extraction

Events are mentions within text that have a specific location and instance or interval in time associated with them. The task of event detection is to detect the mentions of events in text and to identify the class to which they belong. Some examples of events are: the Superbowl, The Cherry Blossom festival, and our 25th wedding anniversary celebration. Both rules-based and machine learning approaches for event detection are similar to those for relationship extraction [Rit+12, MSM11]. Such approaches have had mixed success due to the need for external context and the importance of temporal relations.

3.5.4 Semantic Role Labeling

Semantic role labeling (SRL), also known as thematic role labeling or shallow semantic parsing, is the process of assigning labels to words and phrases that indicate their semantic role in the sentence. A semantic role is an abstract linguistic construct that refers to the role that a subject or object takes on with respect to a verb. These roles include: agent, experiencer, theme, patient, instrument, recipient, source, beneficiary, manner, goal, or result.

Semantic role labeling can provide valuable context [GJ02], whereas syntactic parsing can only provide grammatical structure. The most common approach to SRL is to parse a set of target sentences to identify predicates [PWM08]. For each of these predicates, a machine learning classifier trained on a dataset such as PropBank or FrameNet is used to predict a semantic role label. These labels serve as highly useful features for further tasks such as text summarization or question answering [JN08, BFL98b].

> **PropBank** (the Proposition Bank) is a corpus of Penn Treebank sentences fully annotated with semantic roles, where each of the roles is specific to an individual verb sense. Each verb maps to a single instance in PropBank. The corpus was released in 2005.

> **FrameNet** is another corpus of sentences annotated with semantic roles. Whereas PropBank roles are specific to individual verbs, FrameNet roles are specific to semantic frames. A frame is the background or setting in which a semantic role takes place—it provides a rich set of contexts for the roles within the frame. FrameNet roles have much finer grain than those of PropBank. FrameNet contains over 1200 semantic frames, 13,000 lexical units, and 202,000 example sentences.

3.6 Discourse Representations

Discourse analysis is the study of the structure, relations, and meaning in units of text that are longer than a single sentence. More specifically, it investigates the flow of information and meaning by a collection of sentences taken as a whole. Discourse presumes a sender, receiver, and message. It encompasses characteristics such as the document/dialogue structure, topics of discussion, cohesion, and coherence of the text. Two popular tasks in discourse analysis are coreference resolution and discourse segmentation.

3.6.1 Cohesion

Cohesion is a measure of the structure and dependencies of sentences within discourse. It is defined as the presence of information elsewhere in the text that supports presuppositions within the text. That is, cohesion provides continuity in word and sentence structure. It is sometimes called "surface level" text unity, since it provides the means to link structurally unrelated phrases and sentences together [BN00]. There are six types of cohesion within text: coreference, substitution, ellipsis, conjunction, reiteration, and collocations. Of these, coreference is by far the most popular, as observed in the relation between "Jack" and "He" in the two sentences:

> _Jack_ ran up the hill.
> _He_ walked back down.

3.6.2 Coherence

Coherence refers to the existence of semantic meaning to tie phrases and sentences together within text. It can be defined as continuity in meaning and context, and usually requires inference and real-world knowledge. Coherence is often based on conceptual relationships implicitly shared by both the sender and receiver that are used to construct a mental representation of the discourse [WG05]. An example of coherence can be seen in the following example which presumes knowledge that a bucket holds water:

> _Jack carried the bucket._
> _He spilled the water._

3.6.3 Anaphora/Cataphora

Anaphora refers to the relation between two words or phrases where the interpretation of one, called an anaphor, is determined by the interpretation of a word that came before, called an antecedent. Cataphora is where the interpretation of a word is determined by another word that came after in the text. Both are important characteristics of cohesion in discourse.

> *Anaphora: <u>The court</u> cleared <u>its</u> docket before adjoining.*
> *Cataphora: Despite <u>his</u> carefulness, <u>Jack</u> spilled the water.*

3.6.4 Local and Global Coreference

The linguistics process by which anaphors are linked with their antecedents is known as coreference resolution. It is a well-studied problem in discourse. When this occurs within a document, it is commonly termed local coreference. If this occurs across documents, it is termed global coreference. Essential when disambiguating pronouns and connecting them with the right individual mentions within text, coreference also plays an important role in entity resolution [Lee+13, Sin+13].

Coreference resolution can be considered a classification task, and algorithms for resolving coreference range in accuracy from 70% for named entities to 90% for pronouns [PP09].

3.7 Language Models

A statistical language model is a probability distribution over sequences of words. Given such a sequence, say of length L, it assigns a probability to the whole sequence. In other words, it tries to assign a probability to each possible sequence of words or tokens. Given a set of L tokens $w_1, w_2, \ldots, w_L$, a language model will predict the probability $P(W)$:

$$P(W) = P(w_1 w_2 \ldots w_L) \tag{3.8}$$

How is this useful? A language model is one that tries to predict how frequent a phrase occurs within the natural use of a language. Having a way to estimate the relative likelihood of different phrases is useful in many natural language processing applications, especially ones that generate text as an output. For instance, language models can be used for spell correction by predicting a word w_L given all of the previous words before it:

$$P\left(w_L \mid w_{L-1} w_{L-2} \ldots w_1\right) \tag{3.9}$$

Language modeling is used in speech recognition, machine translation, part-of-speech tagging, parsing, handwriting recognition, information retrieval, and other applications.

3.7.1 N-Gram Model

We can extend this to the general case of n-grams. We assume that the probability of observing the ith word w_i in the context history of the preceding words can be approximated by the probability of observing it in the shortened context history of the preceding words (nth order Markov property).

A unigram model used in information retrieval can be treated as the combination of several one state finite automata. It splits the probabilities of different terms in a context, e.g., from:

$$P(w_1 w_2 \ldots w_L) = \prod_{i=1}^{L} P(w_i) \tag{3.10}$$

The words bigrams and trigrams denote n-gram model language models with $n = 2$ and $n = 3$, respectively. The conditional probability can be calculated from n-gram model frequency counts:

$$P(w_1 w_2 \ldots w_L) = \prod_{i=n}^{L} P(w_i | w_{i-1} w_{i-2} \ldots w_{i-n}) \tag{3.11}$$

3.7.2 Laplace Smoothing

The sparsity of n-grams can become a problem, especially if the set of documents used to create the n-grams language model is small. In those cases, it is not uncommon for certain n-grams to have zero counts in the data. The language model would assign zero probability to these n-grams. This creates a problem when these n-grams occur in test data. Because of the Markov assumption, the probability of a sequence is equal to the product of the individual probabilities of the n-grams. A single zero probability n-gram would set the probability of the sequence to be zero.

To overcome this problem, it is common to use a technique called smoothing. The simplest smoothing algorithm initializes the count of every possible n-gram at 1. This is known as Laplace or add-one smoothing, and guarantees that there will always be a small probability that any n-gram occurs. Unfortunately, as n-gram sparsity grows, this approach becomes less useful as it dramatically shifts occurrence probabilities.

If a word was never seen in the training data, then the probability of that sentence is zero. Clearly this is undesirable, so we apply Laplacian smoothing to help deal with that. We add 1 to every count so it's never zero. To balance this, we add the number of possible words to the divisor, so the division will never be greater than 1.

Laplace smoothing is a simple, inelegant approach that provides modest improvements to results for like text classification. In general, we can use a pseudocount parameter $\alpha > 0$:

$$\vartheta_i = \frac{x_i + \alpha}{N + \alpha d} \tag{3.12}$$

A more effective and wisely used method is Kneser–Ney smoothing, due to its use of absolute discounting by subtracting a fixed value from the probability's lower order terms to omit n-grams with lower frequencies:

$$P_{abs}(w_i|w_{i-1}) = \frac{\max(c(w_{i-1}w_i) - \delta, 0)}{\sum_w c(w_{i-1}w)} + \alpha p_{abs}(w_i) \tag{3.13}$$

3.7.3 Out-of-Vocabulary

Another serious problem for language models arise when the word is not in the vocabulary of the model itself. Out-of-vocabulary (OOV) words create serious problems for language models. In such a scenario, the n-grams that contain an out-of-vocabulary word are ignored. The n-gram probabilities are smoothed over all the words in the vocabulary even if they were not observed.

To explicitly model the probability of out-of-vocabulary words, we can introduce a special token (e.g., <unk>) into the vocabulary. Out-of-vocabulary words in the corpus are effectively replaced with this special <unk> token before n-grams counts are accumulated. With this option, it is possible to estimate the transition probabilities of n-grams involving out-of-vocabulary words. By doing so, however, we treat all OOV words as a single entity, ignoring the linguistic information.

Another approach is to use approximate n-gram matching. OOV n-grams are mapped to the closest n-gram that exists in the vocabulary, where proximity is based on some semantic measure of closeness (we will describe word embeddings in more detail in a later chapter).

A simpler way to deal with OOV n-grams is the practice of backoff, based on the concept of counting smaller n-grams with OOV terms. If no trigram is found, we instead count bigrams. If no bigram found, use unigrams.

3.7.4 Perplexity

In information theory, perplexity measures how well a probability distribution predicts a sample. Perplexity is a commonly used measure to evaluate the performance of a language model. It measures the intrinsic quality of an n-gram model as a function of the probability $P(W)$ that the model predicts a test sequence $W = w_1 w_2 \ldots w_N$ can occur, given by:

$$P(W) = P(w_1 w_2 \ldots w_N)^{-\frac{1}{N}} \tag{3.14}$$

$$= \sqrt[N]{\frac{1}{P(w_1 w_2 \ldots w_N)}} \tag{3.15}$$

If the model is based on bigrams, perplexity reduces to the expression:

$$P(W) = \sqrt[N]{\prod_{N}^{i=1} \frac{1}{P(w_i|w_{i-1})}} \tag{3.16}$$

Lower measures of perplexity imply that the model predicts the test data better, while higher perplexity values imply lower prediction quality. Note that it is important for the test sequence to be comprised of the same n-grams as was used to train the language model, or else the perplexity will be very high.

3.8 Text Classification

Text classification is a core task in many applications such as information retrieval, spam detection, or sentiment analysis. The goal of text classification is to assign documents to one or more categories. The most common approach to building classifiers is through supervised machine learning whereby classification rules are learned from examples [SM99, CT94, Seb02]. We provide a brief overview of the process by which these classifiers are created. Readers can refer to the previous chapter for the details of the machine learning algorithms.

3.8.1 Machine Learning Approach

Most problems in computational linguistics end up as text classification problems that can be addressed with a supervised machine learning approach. Text classification consists of document representation, feature selection, application of machine learning classifier, and finally the evaluation of classifier performance. Feature selection can leverage any of the morphological, lexical, syntactic, semantic, or discourse representations introduced in the previous sections.

Given a set $\mathcal{D}_{labeled}$ of n documents, the first step is to construct representations of these documents in a feature space. The common method is to use a bag-of-words approach with n-gram frequency or TFIDF to create document vectors $\mathbf{x_i}$ and their labeled categories y_i:

$$\mathcal{D}_{labeled} = (\mathbf{x_1}, y_1), (\mathbf{x_2}, y_2), \ldots, (\mathbf{x_n}, y_n) \tag{3.17}$$

With this data, we can train a classification model to predict labels of un-annotated text samples. Popular machine learning algorithms for text classification include K-nearest neighbor, decision trees, naive Bayes, support vector machines, and logistic regression. The general text classification pipeline can be summarized as:

Algorithm 1: Text classification pipeline

Data: A set of documents $\mathcal{D}_{labeled}$
Result: A trained model $h(x)$
begin
    preprocess documents (e.g., tokenize)
    create document representations $\mathbf{x_i}$
    split into train, validation, test sets
    for $\mathbf{x_i} \in \mathbf{X}$ **do**
        train machine learning classifier model on train set
        tune model on dev set;
    evaluate tuned model on test set

3.8.2 Sentiment Analysis

Sentiment analysis is a task that evaluates written or spoken language to determine if linguistic expressions are favorable, unfavorable, or neutral, and to what degree. It has widespread uses in business that include discerning customer feedback, gauging overall mood and opinion, or tracking human behavior. Sentiment encompasses both the affective aspects of text—how one's emotions affect our communication—and subjective aspects of text—the expression of our emotions, opinions, and beliefs. Textual sentiment analysis is the task of detecting type and strength of one's attitudes in sentences, phrases, or documents.

3.8.2.1 Emotional State Model

Models of emotion have been researched for several decades. An emotional state model is one that captures the human states of emotion. The Mehrabian and Russell model, for instance, decomposes human emotional states into three dimensions (Table 3.8). There are other emotional state models used in sentiment analysis, in-

Table 3.8: Mehrabian and Russell model of emotion

Valence	Measures the pleasurableness of an emotion
	also known as polarity
	Ambivalence is the conflict between positive and negative valence
Arousal	Measures the intensity of emotion
Dominance	Measures the dominion of an emotion over others

cluding Plutchik's wheel of emotions and Russell's two-dimensional emotion circumplex model (Fig. 3.7).

The simplest computational approach to sentiment analysis is to take the set of words that describe emotional states and vectorize them with the dimensional values of the emotional state model [Tab+11]. The occurrence of these words is computed

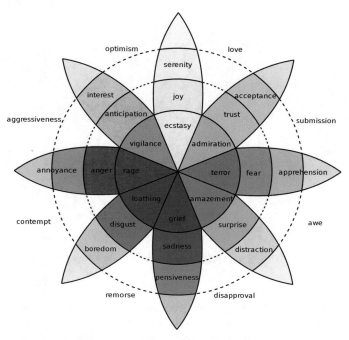

Fig. 3.7: Plutchik's wheel of emotions

within a document, and the sentiment of the document is equal to the aggregated scores of the words. This lexical approach is very fast, but suffers from the inability to effectively model subtlety, sarcasm, or metaphor [RR15]. Negation (e.g., "not nice" vs. "nice") is also problematic with pure lexical approaches.

The affective norms for English words (ANEW) dataset is a lexicon created by Bradley and Lang containing 1000 words scored for emotional ratings of valence, dominance, and arousal (Fig. 3.8). ANEW is very useful for longer texts and newswire documents. Another model is the SentiStrength model for short informal text developed by Thelwall et al., which has been applied successfully to analyze text and Twitter messages.

3.8.2.2 Subjectivity and Objectivity Detection

A closely related task in sentiment analysis is to grade the subjectivity or objectivity of a particular piece of text. The ability to separate subjective and objective parts, followed by sentiment analysis on each part, can be very useful for analysis.

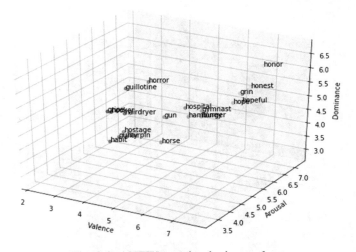

Fig. 3.8: ANEW emotion lexicon subset

Objectivity detection could help identify personal bias, track hidden viewpoints, and alleviate the "fake news" problem existing today [WR05].

One approach to objectivity detection is to use *n*-grams or shallow parsing and pattern matching with a set of learned lexical–syntactic patterns. Another is to use lexical–syntactic features in combination with conversational discourse features to train a classifier for subjectivity.

3.8.3 Entailment

Textual entailment is the logical concept that truth in one text fragment leads to truth in another text fragment. It is a directional relation, analogous to the "if-then" clause. Mathematically, given text fragments X and Y, entailment is given by:

$$P(Y|X) > P(X) \tag{3.18}$$

where $P(Y|X)$ is considered the entailment confidence. Note that the relation X entails Y does not give any certainty that Y entails X (logical fallacy).

Entailment is considered a text classification problem. It has widespread use in many NLP applications (e.g., question answering). Initial approaches toward entailment were logical-form based methods that required a many axioms, inference rules, and a large knowledge base. These theorem-proving methods performed poorly in comparison to other statistical NLP approaches [HMM16].

Currently, the most popular entailment approaches are syntax based [AM10]. Parse trees are used to generate and compare similarity scores that are combined with an SVM or LR classifier to detect entailment. Such approaches are quite capable of capturing shallow entailment but do poorly on more complex text (such as text that switches between active and passive voice).

Recent semantic approaches have shown better ability to generalize by incorporating semantic role labeling in addition to lexical and syntactic features [Bur+07]. Even so, the gap between human level entailment and computational approaches is still significant. Entailment remains an open research topic.

3.9 Text Clustering

While text classification is the usual go-to approach for text analytics, we are often presented with a large corpus of unlabeled data in which we seek to find texts that share common language and/or meaning. This is the task of text clustering [Ber03].

The most common approach to text clustering is via the k-means algorithm [AZ12]. Text documents are tokenized, sometimes stemmed or lemmatized, stop words are removed, and text is vectorized using bag-of-words or TFIDF. K-means is applied to the resulting document-term matrix for different k values.

Algorithm 2: Text clustering pipeline

Data: A set of documents $\mathbb{D}_{unlabeled}$
Result: k text clusters
begin
    preprocess documents (e.g., tokenize)
    create document representations $\mathbf{x_i}$
    for *values of k* **do**
        apply k-means algorithm
    choose best k value

There are two main considerations when using k-means. The first is the notion of distance between two text fragments. For k-means, this is the Euclidean distance, but other measures like cosine distance could theoretically be used. The second is determining the value of k—how many different clusters of text exist within a corpus. As in standard k-means, the elbow method is most widely used for determining the value of k.

3.9.1 Lexical Chains

Traditional approaches relying on bag-of-words ignore semantic relationships between words in a document and do not capture meaning. A method that can incorporate this semantic information is lexical chains. These chains originate from the linguistic concept of textual cohesion, where a sequence of related words are known to contain a semantic relation. For instance, the following words form a lexical chain:

$$\text{car} \rightarrow \text{automobile} \rightarrow \text{sedan} \rightarrow \text{roadster}$$

Usually, a lexical database like WordNet is utilized to both predict lexical chains and to cluster the resulting concepts. Lexical chains are useful for higher-order tasks such as text summarization and discourse segmentation [MN02, Wei+15].

3.9.2 Topic Modeling

Often, we have a collection of documents and want to broadly know what is discussed within the collection. Topic modeling provides us the ability to organize, understand, and summarize large collections of text. A topic model is a statistical model used to discover abstract "topics" within in a collection of documents. It is a form of text mining, seeking to identify recurring patterns of words in discourse.

3.9.2.1 LSA

Latent semantic analysis (LSA) is a technique that seeks to identify relationships between a set of documents and words based on the implicit belief that words close in meaning will occur in similar pieces of text. It is one of the oldest methods for topic modeling [Bir+08]. It uses a mathematical technique named singular value decomposition (SVD) to convert the document-term matrix of a text corpus into two lower-rank matrices: a document-topic matrix that maps topics to documents, and a topic-word matrix that maps words to topics. In doing so, LSA acts to reduce the dimensionality of the corpus vector space while identifying higher-order patterns within the corpus. To measure relatedness, LSA utilizes the cosine distance measure between two term vectors.

LSA is very easy to train and tune, and the two matrices derived from LSA can be reused for other tasks as they contain semantic information. Unfortunately for large collections of documents, LSA can be quite slow.

3.9.2.2 LDA

Latent Dirichlet allocation (LDA) is a model that also acts to decompose a document-term matrix into a lower-order document-topic matrix and topic-word matrix. It differs from LSA in that it takes a stochastic, generative model approach and assumes topics to have a sparse Dirichlet prior. This is equivalent to the belief that only a small set of topics belong to any particular document and that topics mostly contain small sets of frequent words. As compared to LSA, LDA does better at disambiguation of words and identifies topics with finer details.

3.10 Machine Translation

Machine translation (abbreviated MT) refers to the process of translating text from a source language to a different target language. Language translation is hard even for humans to be able to fully capture meaning, tone, and style. Languages can have significantly different morphology, syntax, or semantic structure. For instance, it will be rare to find English words with more than 4 morphemes, but it is quite common in Turkish or Arabic. German sentences commonly follow the subject-verb-object syntactic structure, while Japanese mostly follows a subject-object-verb order, and Arabic prefers a verb-subject-object order. With machine translation, we typically focus on two measures:

- Faithfulness = preserving the meaning of text in translation
- Fluency = natural sounding text or speech to a native speaker.

3.10.1 Dictionary Based

In simplest form, machine translation can be achieved by a direct translation of each word using a bilingual dictionary. A slight improvement may be to directly translate word phrases instead of individual words [KOM03]. Because of the lack of syntactic or semantic context, direct translation tends to do poorly in all but the simplest machine translation tasks [Dod02].

Another classical method for machine translation is based on learning lexical and syntactic transfer rules from the source to the target language. These rules provide a means to map the parse trees between languages, potentially altering the structure in the transformation. Due to the need for parsing, transfer methods are generally complex and difficult to manage, especially for large vocabularies. For this reason, classic machine translation systems usually take a combined approach, using direct translation for simple text structure and lexical/syntactic transfer for more messy cases.

3.10.2 Statistical Translation

Statistical machine translation adopts a probabilistic approach to map from one language to another. Specifically, it builds two types of models by treating the problem as one similar to a Bayesian noisy channel problem in communications:

- Language model (fluency) = $P(X)$
- Translation model (faithfulness) = $P(Y|X)$.

The language model measures the probability that any sequence of words X is an actual sentence—that is, there is consistency within a language. The translation model measures the conditional probability that a sequence of words Y in the target language is a true translation of a sequence of words X in the source language. A statistical machine translation model will find the best translation to the target language Y by optimizing for:

$$\hat{Y} = \underset{Y}{\operatorname{argmax}} P(X|Y)P(Y) \tag{3.19}$$

Statistical models are based on the notion of **word alignment**, which is a mapping of a sequence of words from the source language to those of a target language. Because of differences between languages, this mapping will almost never be one-to-one. Furthermore, the order of words may be quite different.

> BLEU (bilingual evaluation understudy) is a common method to measure the quality of machine translation [Pap+02]. It measures the similarity between phrase-based model translations and human-created translations averaged over an entire corpus. Similar to precision, it is normally expressed as a value between 0 and 1 but sometimes scaled by a factor of 10.

3.11 Question Answering

Question answering (QA) is the NLP task of answering questions in natural language. It can leverage expert system, knowledge representation, and information retrieval methods. Traditionally, question answering is a multi-step process where relevant documents are retrieved, useful information is extracted from these documents, possible answers are proposed and scored against evidence, and a short text answer in natural language is generated as a response.

Question: Who won the Tournament of Champions on Jeopardy in 2011?

Answer: IBM Watson debuted a system named DeepQA in 2011 that went on to win first place against the legendary champions on Jeopardy.

Early question answering systems focused only on answering a predefined set of topics within a particular domain [KM11]. These were known as closed-domain QA systems, as opposed to open-domain QA systems that attempt to answer queries in any topic. Closed-domain systems often avoided the complexity of dialog processing and produced structured, pattern-based answers derived directly from expert systems. Modern open-domain QA systems provide much richer capability and in theory can leverage an unlimited set of knowledge sources (e.g., the internet) to answer questions through statistical processing.

Question decomposition is the first step in any QA system, where a question is processed to form a query. In simple versions, questions would be parsed to find keywords which served as queries to an expert system to produce answers. This is known as query formation, where keywords are extracted from the question to formulate a relevant query. Sometimes, query expansion is used to identify additional query terms similar to those within a question [CR12]. In more advanced versions, syntactic processing (e.g., noun-phrases) and semantic processing (e.g., extracting entities) can be used to enrich extraction. Another method is query reformation, where the entities in the question are extracted along with its semantic relation. For instance, the following sentence and semantic relation:

Who invented the telegraph? → Invented (Person, telegraph)

An answer module can pattern-match this relation against semantic databases and knowledge bases to retrieve a set of candidate answers. The candidates are scored against evidence and the one with the highest confidence is returned as a natural language response. Some questions are easier to answer. For instance, it is much easier to determine the date or year of an event (e.g., when was Superbowl XX) than it is to relate entities in particular contexts (e.g., which city is most like Toronto). The former would require only a small, targeted search while the latter search space is much larger.

3.11.1 Information Retrieval Based

Web-based question answering systems like Google Search are based on information retrieval (IR) methods that leverage the web. These text-based systems seek to answer questions by finding short texts from the internet or some other large collection of documents. Typically, they map queries into a bag-of-words and use methods like LSA to retrieve a set of relevant documents and extract passages within them. Depending on the question type, answer strings can be generated with a pattern-extraction approach or *n*-gram tiling methods. IR-based QA systems are entirely statistical in nature and are unable to truly capture meaning beyond distributional similarity.

3.11.2 Knowledge-Based QA

Knowledge-based question answering systems, on the other hand, take a semantic approach. They apply semantic parsing to map questions into relational queries over a comprehensive database. This database can be a relational database or knowledge base of relational triples (e.g., subject-predicate-object) capturing real-world relationships such as DBpedia or Freebase. Because of their ability to capture meaning, knowledge-based methods are more applicable for advanced, open-domain question-answering applications as they can bring in external information in the form of knowledge bases [Fu+12]. At the same time, they are constrained by the set relations of those knowledge bases (Fig. 3.9).

DBpedia is a free semantic relation database with 4.6 million entities extracted from Wikipedia pages in multiple languages. It contains over 3 billion relational triples expressed in the resource description framework (RDF) format. DBpedia is often considered the foundation for the semantic web, also known as the linked open data cloud. First released in 2007, DBpedia continues to evolve through crowdsourced updates in similar fashion to Wikipedia.

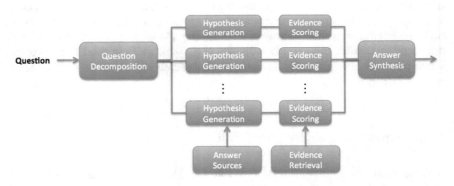

Fig. 3.9: Open-domain QA system

3.11.3 Automated Reasoning

Recent QA systems have begun to incorporate automated reasoning (AR) to extend beyond the semantic relations of knowledge-based systems. Automated reasoning is a field in artificial intelligence that explores methods in abductive, probabilistic, spatial and temporal reasoning by computer systems. By creating a set of first-order logic clauses, QA systems can enhance a set of semantic relations and evidence retrieved in support of answer hypotheses [FGP10]. Prolog is a common declarative language approach used to maintain this set of clauses.

IBM Watson's DeepQA is an example of a question answering system that incorporates a variety of IR-based, knowledge-based, and automated reasoning methods. By leveraging reportedly 100 different approaches and knowledge base sources to generate candidate answers that are evidence-scored and merged [Wan+12], DeepQA was able to exceed human level performance in the game of Jeopardy in 2011. IBM has since deployed DeepQA into a variety of other domains with varying success.

A common metric used to measure question answering system performance is **mean reciprocal rank** (MRR). It is based on using a gold set of questions that have been manually labeled by humans with correct answers. To evaluate a QA system, the set of ranked answers of the system would be compared with the gold set labels of a corpus of N questions, and the MRR is given by:

$$MRR = \frac{1}{N} \sum_{i=1}^{N} \frac{1}{rank_i} \qquad (3.20)$$

Current state-of-the-art QA systems exceed $MRR = 0.83$ on the commonly used TREC-QA benchmark.

3.12 Automatic Summarization

Automatic summarization is a useful NLP task that identifies the most relevant information in a document or group of documents and creates a summary of the content. It can be an extraction task that takes the most relevant phrases or sentences in original form and uses them to generate the summary, or an abstraction task that generates natural language summaries from the semantic content [AHG99, BN00]. Both approaches mirror how humans tend to summarize text, though the former extracts text while the latter paraphrases text.

3.12.1 Extraction Based

Extraction-based summarization is a content selection approach to distilling documents. In most implementations, it simply extracts a subset of sentences deemed most important. One method to measure importance is to count informative words based on lexical measures (e.g., TFIDF). Another is to use discourse measures (e.g., coherence) to identify key sentences. Centroid-based methods evaluate word probability relative to the background corpus to determine importance. A creative approach called TextRank takes a graph-based approach to assign sentence scores based on lexical similarity of words. As long as plagiarism is not a concern, extraction-based summarization is the more popular approach.

3.12.2 Abstraction Based

Unlike extraction-based copying, abstraction-based approaches take a semantic approach. One method is to use entity recognition and semantic role labeling to identify relations. These can be fed into standard templates (e.g., mad-lib approach) or a natural language generation engine to create synopses. The use of lexical chains can aid in the identification of central themes, where the strongest chain is indicative of the main topic [SM00].

Automatic summarization remains a difficult task. State-of-the-art methods are around the 35% precision level, with performance differing greatly by underlying document type [GG17]. Deep learning methods hold significant promise, as we will see in a later chapter.

3.13 Automated Speech Recognition

Automatic speech recognition (ASR) is the NLP task of real-time computational transcription of spoken language. ASR has been at the forefront in the study of human–computer interfaces since the 1950s. With the advent of personal AI assistants like Siri, Alexa, or Cortana, the importance of ASR has skyrocketed in recent years. The ultimate goal of ASR is human-level (near 100%) speech transcription. Current ASR in perfect conditions can only approach 95% [Bak+09]. Evolution has given us the ability to recognize speech in a variety of conditions (e.g., noise, accents, diction, and tone) that computers cannot yet deal with, and much room for improvement in ASR remains. In the next sections, some background on the computational representation of speech and the classical approaches to ASR are provided.

3.13.1 Acoustic Model

An acoustic model is a representation of the sounds in an audio signal used in automatic speech recognition. Its main purpose is to map acoustic waves to the statistical properties of phonemes, which are elementary linguistic units of sound that distinguish one word from another in a language. Consider an audio signal as a sequence of short, consecutive time frames $S = s_1, s_2, \ldots s_T$. Let a sequence of M phonemes be represented by $F = f_1, f_2, \ldots f_M$ and a sequence of N words be represented by $W = w_1, w_2, \ldots w_N$. In speech recognition, the goal is to predict the set of words W from the audio input S:

$$\hat{W} = \underset{W}{\operatorname{argmax}} P(W|S) \tag{3.21}$$

$$\hat{W} \approx \underset{W}{\operatorname{argmax}} P(S|F) P(F|W) P(W) \tag{3.22}$$

Here, $P(W)$ represents the probability that a string of words is an English sentence—that is, $P(W)$ is the language model. The quantity $P(S|F)$ is known as the pronunciation model and the quantity $P(F|W)$ is the acoustic model.

3.13.1.1 Spectrograms

A spectrogram is a visual representation of the frequencies of an acoustic signal over a period of time, where the horizontal axis is time, the vertical axis is frequency, and the intensity of the audio signal is represented by the color at each point. A spectrogram is generated using a sliding time window in which a short-time Fourier transform is performed. As a time-frequency visualization of a speech signal, spectrograms are useful for both speech representations and for evaluation of text to speech systems (Fig. 3.10).

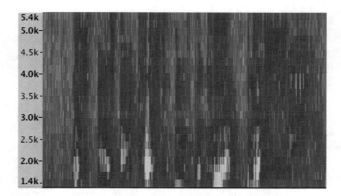

Fig. 3.10: Spectrogram

3.13.1.2 MFCC

Mel-frequency cepstral coefficients (MFCCs) are another useful representation of speech signals. MFCCs transform continuous audio signals into feature vectors, each representing a small window in time. Consider the cepstrum, which is the inverse fast-Fourier transform of the log of the fast-Fourier transform of an audio signal (Fig. 3.11):

$$C = \left| F^{-1} \left(\log F \left(f(t) \right) \right) \right|^2 \tag{3.23}$$

MFCCs is similar to the cepstrum and is given by taking the discrete cosine transform of the log of the fast-Fourier transform of an audio signal where a triangular filter of Mel frequency banks has been applied. MEL filters are placed linearly for frequencies less than 1000Hz and on a log scale for frequencies above 1000Hz, closely corresponding to the response of the human ear:

$$C = DCT \left(\log \left(MEL \left(F \left(f(t) \right) \right) \right) \right) \tag{3.24}$$

MFCCs contain both time and frequency information about the audio signal. They are particularly useful for ASR because cepstral features are effectively orthogonal to each other and robust to noise.

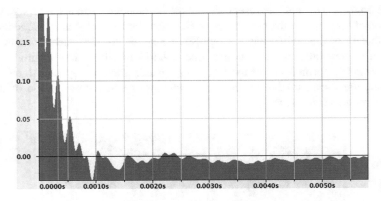

Fig. 3.11: Cepstrum

3.14 Case Study

To provide further insight on applications of natural language processing, we present the following case study to guide readers through an application of text clustering, topic modeling, and text classification principles. The case study is based on the Reuters-21578 dataset, a collection of 21578 newswire stories from 1987. We begin by cleaning the dataset and transforming it into a format that permits easier analysis. Through exploratory data analysis, we will examine corpus structure and identify if text clusters exist and to what degree. We will model topics within the corpus, and compare our findings with the annotations provided in the dataset. Finally, we will explore various methods to classify the documents by topic. Hopefully, this case study will reinforce the fundamental principles of text analytics as well as identify key gaps in classical NLP.

3.14.1 Software Tools and Libraries

For this case study, we will use Python and the following libraries:

- **Pandas** (https://pandas.pydata.org/) is a popular open source implementation for data structures and data analysis. We will use it for data exploration and some basic processing.
- **scikit-learn** (http://scikit-learn.org/) is a popular open source for various machine learning algorithms and evaluations. We will use it only for sampling, creating datasets, and machine learning implementations of linear and non-linear algorithms in our case study.
- **NLTK** (https://www.nltk.org/) is a suite of text and natural language processing tools. We will use it to convert text into vectors for processing.
- **Matplotlib** (https://matplotlib.org/) is a popular open source library for visualization. We will use it to visualize performance.

3.14.2 EDA

Our first task is to take a close look at the dataset by loading and performing exploratory data analysis. To do so, we must extracting metadata and the text body from each document in the corpus. If we take a close look at the corpus, we find (Figs. 3.12, 3.13 and 3.14):

1. There are 11,367 documents that have one or more topic annotations.
2. The greatest number of topics in a single document is 16.
3. There are a total of 120 distinct topic labels in the corpus.
4. There are 147 distinct place and 32 organization labels.

So far so good. But before we perform any NLP analysis, we will want to perform some cursory text normalization:

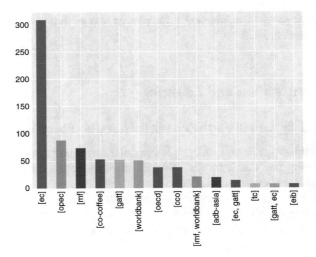

Fig. 3.12: Document count by organization

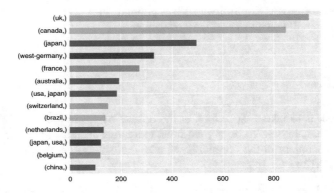

Fig. 3.13: Document count by non-US location

1. Transform to lower case
2. Remove punctuation and numbers
3. Stem verbs
4. Remove stopwords.

To do so, we define a SimpleTokenizer method that will be useful when creating document representations.

```
import re
import nltk
from nltk import word_tokenize
from nltk.corpus import stopwords
from nltk.stem.porter import PorterStemmer
from sklearn.preprocessing.label import MultiLabelBinarizer
```

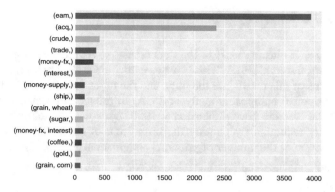

Fig. 3.14: Document count by topic

```
7  from sklearn.feature_extraction.text import TfidfVectorizer
8
9  nltk.download("punkt")
10 nltk.download("stopwords","data")
11 nltk.data.path.append('data')
12
13 labelBinarizer = MultiLabelBinarizer()
14 data_target = labelBinarizer.fit_transform(data_set[u'topics'
      ])
15
16 stopWords = stopwords.words('english')
17 charfilter = re.compile('[a-zA-Z]+');
18
19 def SimpleTokenizer(text):
20   words = map(lambda word: word.lower(), word_tokenize(text))
21   words = [word for word in words if word not in stopWords]
22   tokens = (list(map(lambda token: PorterStemmer().stem(token),
        words)))
23   ntokens = list(filter(lambda token:charfilter.match(token),
      tokens))
24   return ntokens
25
26 vec = TfidfVectorizer(tokenizer=SimpleTokenizer,
27                       max_features=1000,
28                       norm='12')
29
30
31 mytopics = [u'cocoa',u'trade',u'money-supply',u'coffee',u'gold
      ']
32 data_set = data_set[data_set[u'topics'].map(set(mytopics).
      intersection)
33   .apply( lambda x: len(x)>0 )]
34 docs = list(data_set[u'body'].values)
35
36 dtm = vec.fit_transform(docs)
```

3.14.3 Text Clustering

We want to see if clusters exist in the documents, so let's create some document
representations through TFIDF. This gives us a document-term matrix, but typically
the dimensions of this matrix are too large and the representations are sparse. Let's
first apply principal component analysis (PCA) to reduce the dimensionality. The
original TFIDF vectors have dimension = 1000. Let's take a look at the effect of
dimensionality reduction by plotting the proportion of explained variance of the
data as a function of the number of principal components (Fig. 3.15):

```
from sklearn.decomposition import PCA

explained_var = []
for components in range(1,100,5):
  pca = PCA(n_components=components)
  pca.fit(dtm.toarray())
  explained_var.append(pca.explained_variance_ratio_.sum())

plt.plot(range(1,100,5),explained_var,"ro")
plt.xlabel("Number of Components")
plt.ylabel("Proportion of Explained Variance")
```

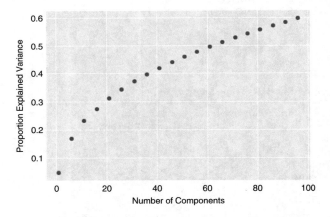

Fig. 3.15: Explained variance by number of PCA components

The graph above shows that half of the variance can be explained by 60 components.
Let's apply this to the dataset, and visualize the results by plotting the first two PCA
components of each document (Fig. 3.16).

```
from sklearn.decomposition import PCA
import seaborn as sns

components = 60

```

```
 6  palette = np.array(sns.color_palette("hls", 120))
 7
 8  pca = PCA(n_components=components)
 9  pca.fit(dtm.toarray())
10  pca_dtm = pca.transform(dtm.toarray())
11
12  plt.scatter(pca_dtm[:,0],pca_dtm[:,1],
13            c=palette[data_target.argmax(axis=1).astype(int)])
14
15  explained_variance = pca.explained_variance_ratio_.sum()
16  print("Explained variance of the PCA step: {}%".format(
17   int(explained_variance * 100)))
```

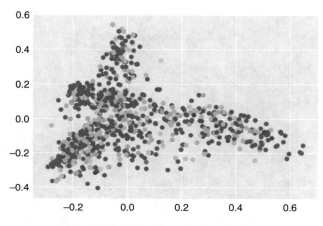

Fig. 3.16: PCA document projection

We know that there are 5 distinct topics (though some documents might have overlap), so let's run the k-means algorithm with $k = 5$ to examine document grouping (Fig. 3.17).

```
 1  from sklearn.cluster import KMeans
 2  palette = np.array(sns.color_palette("hls", 5))
 3
 4  model = KMeans(n_clusters=5,max_iter=100)
 5  clustered = model.fit(pca_dtm)
 6  centroids = model.cluster_centers_
 7  y = model.predict(pca_dtm)
 8
 9  ax = plt.subplot()
10  sc = ax.scatter(pca_dtm[:,0],pca_dtm[:,1],
11            c=palette[y.astype(np.int)])
```

How does this compare with the manually annotated labels? (Fig. 3.18)

```
 1  palette = np.array(sns.color_palette("hls", 5))
 2
```

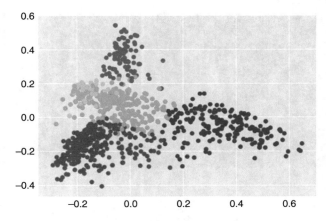

Fig. 3.17: K-means clusters

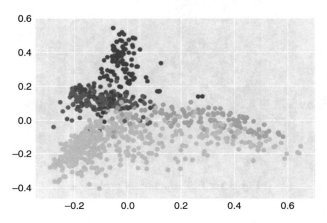

Fig. 3.18: Manually labeled clusters

```
gold_labels = data_set['topics'].map(set(mytopics).
    intersection)
  .(lambda x: x.pop()).apply(lambda x: mytopics.index(x))

ax = plt.subplot()
sc = ax.scatter(pca_dtm[:,0],pca_dtm[:,1],c=palette[
    gold_labels])
```

3.14.4 Topic Modeling

In addition to the lexical clustering of documents, let's see if we can discern any natural topic structure within the corpus. We apply the LSA and LDA algorithms, which will associate words to a set of topics, and topics to our set of documents.

3.14.4.1 LSA

We start with the LSA algorithm and set the number of dimensions to 60 (Fig. 3.19):

```
from sklearn.decomposition import TruncatedSVD
import seaborn as sns

components = 60

palette = np.array(sns.color_palette("hls", 120))

lsa = TruncatedSVD(n_components=components)
lsa.fit(dtm)
lsa_dtm = lsa.transform(dtm)

plt.scatter(lsa_dtm[:,0],lsa_dtm[:,1],
            c=palette[data_target.argmax(axis=1).astype(int)
    ])

explained_variance = lsa.explained_variance_ratio_.sum()
print("Explained variance of the SVD step: {}%".format(
    int(explained_variance * 100)))
```

As with PCA, let's apply k-means with $k = 5$ clusters (Fig. 3.20).

```
from sklearn.cluster import KMeans
palette = np.array(sns.color_palette("hls", 8))

model = KMeans(n_clusters=5,max_iter=100)
clustered = model.fit(lsa_dtm)
centroids = model.cluster_centers_
y = model.predict(lsa_dtm)

ax = plt.subplot()
sc = ax.scatter(lsa_dtm[:,0],lsa_dtm[:,1],c=palette[y.astype(
    np.int)])
```

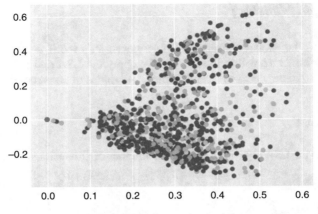

Fig. 3.19: LSA topic model

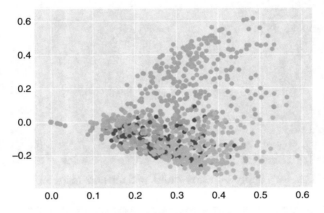

Fig. 3.20: *k*-means on LSA

Let's examine the documents of one of these clusters:

232	Talks on the possibility of reintroducing...
235	Indonesia's agriculture sector will grow...
249	The International Coffee Organization...
290	Talks on coffee export quotas at the...
402	Coffee quota talks at the International...
42	International Coffee Organization, ICO,...
562	Talks at the extended special meeting of...
75	International Coffee Organization (ICO)...
754	Efforts to break an impasse between...
842	A special meeting of the International Coffee...

3.14.4.2 LDA

Let's see if the LDA algorithm can do better as a Bayesian approach to document clustering and topic modeling. We set the number of topics to the known number of topics = 5.

```python
import numpy as np
import seaborn as sns
from sklearn.cluster import KMeans
from sklearn.decomposition import LatentDirichletAllocation

components = 5
n_top_words = 10

palette = np.array(sns.color_palette("hls", 120))

def print_top_words(model, feature_names, n_top_words):
    for topic_idx, topic in enumerate(model.components_):
        message = "Topic #%d: " % topic_idx
        message += " ".join([feature_names[i]
            for i in topic.argsort()[:-n_top_words - 1:-1]])
        print(message)
    print()

lda = LatentDirichletAllocation(n_components=components,
    max_iter=5, learning_method='online')
lda.fit(dtm)
lda_dtm = lda.transform(dtm)

vec_feature_names = vec.get_feature_names()
print_top_words(lda, vec_feature_names, n_top_words)
```

Topic 0	said trade u.s. deleg quota brazil export year coffe market
Topic 1	gold mine ounc ton said ltd compani ore feet miner
Topic 2	fed volcker reserv treasuri bank borrow pct rate growth dlr
Topic 3	said trade u.s. export japan coffe would ec market offici
Topic 4	billion dlr mln pct januari februari rose bank fell year

The LDA results are encouraging, and we can easily discern 4 of the 5 original topics from the list of words associated with each topic.

3.14.5 Text Classification

Now let's see if we can build classifiers to possibly identify the topics above. We first randomize and split our dataset into train and test sets.

```python
from sklearn.model_selection import train_test_split
```

```
3  data_set['label'] = gold_labels
4
5  X_train, X_test, y_train, y_test = train_test_split(data_set,
       gold_labels, test_size=0.2, random_state=10)
6  print("Train Set = ",len(X_train))
7  print("Test Set = ",len(X_test))
8
9  X_train = X_train[u'body']
10 X_test = X_test[u'body']
```

We then create a pipeline that builds classifiers based on 5 models: naive Bayes, logistic regression, SVM, K-nearest neighbor, and random forest.

```
1  from sklearn.naive_bayes import MultinomialNB
2  from sklearn.linear_model import LogisticRegression
3  from sklearn.svm import LinearSVC
4  from sklearn.neighbors import KNeighborsClassifier
5  from sklearn.ensemble import RandomForestClassifier
6
7  models = [('multinomial_nb', MultinomialNB()),
8            ('log_reg', LogisticRegression()),
9            ('linear_svc', LinearSVC()),
10           ('knn', KNeighborsClassifier(n_neighbors=6)),
11           ('rf', RandomForestClassifier(n_estimators=6))]
```

We then train each model on the training set and evaluate on the test set. For each model, we want to see the precision, recall, F1 score, and support (number of samples) for each topic class.

```
1  from sklearn.pipeline import Pipeline
2  from sklearn.metrics import classification_report
3
4  for m_name, model in models:
5      pipeline = Pipeline([('vec', TfidfVectorizer(tokenizer=
       SimpleTokenizer)),(m_name,model)])
6      pipeline.fit(X_train,y_train)
7      test_y = pipeline.predict(X_test)
8      print(classification_report(y_test,test_y,digits=6))
```

The results seem to indicate that a linear SVM model seems to perform the best, with random forest a close second. This is a bit misleading, since we didn't tune any of these models to obtain our results. Hyperparameter tuning can significantly affect how well a classifier performs. Let's try tuning the LinearSVC model. We want to tune parameters by using grid search with cross-validation. Note that cross-validation is important as we do not want to tune with our test set, which we will use only at the end to assess performance. Note also that this can take a while!

```
1  from sklearn.model_selection import GridSearchCV
2
3  pipeline = Pipeline([('vec', vectorizer),
4   ('model', model)])
```

```
5
6  parameters = {'vec__ngram_range': ((1, 1), (1, 2)),
7               'vec__max_features': (500, 1000),
8               'model__loss': ('hinge', 'squared_hinge'),
9               'model__C': (1, 0.9)}
10
11 grid_search = GridSearchCV(pipeline, parameters, verbose=1)
12 grid_search.fit(X_train, y_train)
13
14 test_y = grid_search.best_estimator_.predict(X_test)
15 print(classification_report(y_test, test_y, digits=6))
```

As you see, the SVM model typically outperforms other machine learning algorithms, and often provides state-of-the-art quality (Fig. 3.21). Unfortunately, SVM suffers from several major drawbacks, including the inability to scale to large datasets. As we will learn in later chapters, neural networks can bypass the limitations of SVMs.

	Test Set Precision	Test Set Recall	Test Set F1
Naïve Bayes	0.8262	0.7361	0.7048
Logistic Regression	0.8929	0.8704	0.8606
Linear SVM	**0.9567**	**0.9537**	**0.9541**
K Nearest Neighbors	0.5802	0.3981	0.3959
Random Forest	0.8854	0.8843	0.8803

Fig. 3.21: Classification results

3.14.6 Exercises for Readers and Practitioners

Here are further exercises for the reader to consider:

1. Instead of TFIDF, what other document representations can we try?
2. How can we incorporate syntactic information to enhance the text clustering task?
3. What semantic representations could be useful for text classification?
4. What are some other ways to cluster documents?
5. Can we combine classification models to improve prediction accuracy?

References

[AZ12]   Charu C. Aggarwal and ChengXiang Zhai. "A Survey of Text Cluster-
              ing Algorithms." In: *Mining Text Data*. Springer, 2012, pp. 77–128.
[And12]   S.R. Anderson. *Languages: A Very Short Introduction*. OUP Oxford,
              2012.
[AM10]   Ion Androutsopoulos and Prodromos Malakasiotis. "A Survey of
              Paraphrasing and Textual Entailment Methods". In: *J. Artif. Int. Res.*
              38.1 (May 2010), pp. 135–187.
[AL13]   Samet Atdag and Vincent Labatut. "A Comparison of Named En-
              tity Recognition Tools Applied to Biographical Texts". In: *CoRR*
              abs/1308.0661 (2013).
[AHG99]   Saliha Azzam, Kevin Humphreys, and Robert Gaizauskas. "Using
              Coreference Chains for Text Summarization". In: *in ACL Workshop
              on Coreference and its Applications*. 1999.
[BB07]   Nguyen Bach and Sameer Badaskar. "A Review of Relation Extrac-
              tion". 2007.
[BFL98b]   Collin F. Baker, Charles J. Fillmore, and John B. Lowe. "The Berke-
              ley FrameNet Project". In: *Proceedings of the 36th Annual Meeting of
              the Association for Computational Linguistics and 17th International
              Conference on Computational Linguistics - Volume 1*. ACL '98. As-
              sociation for Computational Linguistics, 1998, pp. 86–90.
[Bak+09]   Janet Baker et al. "Research Developments and Directions in Speech
              Recognition and Understanding, Part 1". In: *IEEE Signal Processing
              Magazine* 26 (2009), pp. 75–80.
[BM04]   Michele Banko and Bob Moore. "Part of Speech Tagging in Context".
              In: International Conference on Computational Linguistics, 2004.
[Bel+17]   Anya Belz et al. "Shared Task Proposal: Multilingual Surface Re-
              alization Using Universal Dependency Trees". In: *Proceedings of
              the 10th International Conference on Natural Language Generation*.
              2017, pp. 120–123.
[Ber03]   Michael Berry. *Survey of Text Mining : Clustering Classification, and
              Retrieval*. Springer, 2003.
[Bir+08]   Istvan Biro et al. "A Comparative Analysis of Latent Variable Mod-
              els for Web Page Classification". In: *Proceedings of the 2008 Latin
              American Web Conference*. LA-WEB '08. IEEE Computer Society,
              2008, pp. 23–28.
[BN00]   Branimir K. Boguraev and Mary S. Neff. "Lexical Cohesion, Dis-
              course Segmentation and Document Summarization". In: *Content-
              Based Multimedia Information Access - Volume 2*. RIAO '00. 2000,
              pp. 962–979.
[Bur+07]   Aljoscha Burchardt et al. "A Semantic Approach to Textual Entail-
              ment: System Evaluation and Task Analysis". In: *Proceedings of the
              ACL-PASCAL Workshop on Textual Entailment and Paraphrasing*.
              Association for Computational Linguistics, 2007, pp. 10–15.

[CR12]   Claudio Carpineto and Giovanni Romano. "A Survey of Automatic Query Expansion in Information Retrieval". In: *ACM Comput. Surv.* 44.1 (Jan. 2012), 1:1–1:50.

[CT94]   William B. Cavnar and John M. Trenkle. "N-Gram-Based Text Categorization". In: *Proceedings of SDAIR-94, 3rd Annual Symposium on Document Analysis and Information Retrieval.* 1994, pp. 161–175.

[Cov01]   Michael A. Covington. "A fundamental algorithm for dependency parsing". In: *In Proceedings of the 39th Annual ACM Southeast Conference.* 2001, pp. 95–102.

[DeR88]   Steven J. DeRose. "Grammatical Category Disambiguation by Statistical Optimization". In: *Comput. Linguist.* 14.1 (Jan. 1988), pp. 31–39.

[Dod02]   George Doddington. "Automatic Evaluation of Machine Translation Quality Using N-gram Co-occurrence Statistics". In: *Proceedings of the Second International Conference on Human Language Technology Research.* HLT '02. Morgan Kaufmann Publishers Inc., 2002, pp. 138–145.

[Fu+12]   Linyun Fu et al. "Towards Better Understanding and Utilizing Relations in DBpedia". In: *Web Intelli. and Agent Sys.* 10.3 (July 2012), pp. 291–303.

[FGP10]   Ulrich Furbach, Ingo Glöckner, and Björn Pelzer. "An Application of Automated Reasoning in Natural Language Question Answering". In: *AI Commun.* 23.2–3 (Apr. 2010), pp. 241–265.

[GG17]   Mahak Gambhir and Vishal Gupta. "Recent Automatic Text Summarization Techniques: A Survey". In: *Artif. Intell. Rev.* 47.1 (Jan. 2017), pp. 1–66.

[GJ02]   Daniel Gildea and Daniel Jurafsky. "Automatic Labeling of Semantic Roles". In: *Comput. Linguist.* 28.3 (Sept. 2002), pp. 245–288.

[Gui+06]   Yves Guiard et al. "Shakespeare's Complete Works As a Benchmark for Evaluating Multiscale Document Navigation Techniques". In: *Proceedings of the 2006 AVI Workshop on BEyond Time and Errors: Novel Evaluation Methods for Information Visualization.* ACM, 2006, pp. 1–6.

[HMM16]   Mohamed H, Marwa M.A., and Ahmed Mohammed. "Different Models and Approaches of Textual Entailment Recognition". In: 142 (May 2016), pp. 32–39.

[HEH12]   Nizar Habash, Ramy Eskander and Abdelati Hawwari. "A Morphological Analyzer for Egyptian Arabic". In: *Proceedings of the Twelfth Meeting of the Special Interest Group on Computational Morphology and Phonology.* Association for Computational Linguistics, 2012, pp. 1–9.

[HN14]   Kazi Saidul Hasan and Vincent Ng. "Automatic keyphrase extraction: A survey of the state of the art". In: *In Proc. of the 52nd Annual Meeting of the Association for Computational Linguistics (ACL).* 2014.

[Hon05]   Gumwon Hong. "Relation Extraction Using Support Vector Machine". In: *Proceedings of the Second International Joint Conference on Natural Language Processing*. Springer-Verlag, 2005, pp. 366–377.

[JN08]   Richard Johansson and Pierre Nugues. "Dependency-based Semantic Role Labeling of PropBank". In: *Proceedings of the Conference on Empirical Methods in Natural Language Processing*. EMNLP '08. Association for Computational Linguistics, 2008, pp. 69–78.

[JM09]   Daniel Jurafsky and James H. Martin. *Speech and Language Processing (2Nd Edition)*. Prentice-Hall, Inc., 2009.

[KOM03]   Philipp Koehn, Franz Josef Och, and Daniel Marcu. "Statistical Phrase-based Translation". In: *Proceedings of the 2003 Conference of the North American Chapter of the Association for Computational Linguistics on Human Language Technology - Volume 1*. Association for Computational Linguistics, 2003, pp. 48–54.

[KM11]   Oleksandr Kolomiyets and Marie-Francine Moens. "A Survey on Question Answering Technology from an Information Retrieval Perspective". In: *Inf. Sci.* 181.24 (Dec. 2011), pp. 5412–5434.

[KM01]   Taku Kudo and Yuji Matsumoto. "Chunking with Support Vector Machines". In: *Proceedings of the Second Meeting of the North American Chapter of the Association for Computational Linguistics on Language Technologies*. Association for Computational Linguistics, 2001, pp. 1–8.

[Lee+13]   Heeyoung Lee et al. "Deterministic Coreference Resolution Based on Entity-centric, Precision-ranked Rules". In: *Comput. Linguist.* 39.4 (Dec. 2013), pp. 885–916.

[Mar+94]   Mitchell Marcus et al. "The Penn Treebank: Annotating Predicate Argument Structure". In: *Proceedings of the Workshop on Human Language Technology*. Association for Computational Linguistics, 1994, pp. 114–119.

[MSM11]   David McClosky, Mihai Surdeanu, and Christopher D. Manning. "Event Extraction As Dependency Parsing". In: *Proceedings of the 49th Annual Meeting of the Association for Computational Linguistics: Human Language Technologies - Volume 1*. Association for Computational Linguistics, 2011, pp. 1626–1635.

[McD+13]   Ryan T. McDonald et al. "Universal Dependency Annotation for Multilingual Parsing." In: The Association for Computer Linguistics, 2013, pp. 92–97.

[MN02]   Dan Moldovan and Adrian Novischi. "Lexical Chains for Question Answering". In: *Proceedings of the 19th International Conference on Computational Linguistics - Volume 1*. Association for Computational Linguistics, 2002, pp. 1–7.

[Niv+16]   Joakim Nivre et al. "Universal Dependencies v1: A Multilingual Treebank Collection". In: *LREC*. 2016.

[PT13]   Georgios Paltoglou and Mike Thelwall. "More than Bag-of-Words: Sentence-based Document Representation for Sentiment Analysis." In: *RANLP*. RANLP 2013 Organising Committee / ACL, 2013, pp. 546–552.

[Pap+02]   Kishore Papineni et al. "BLEU: A Method for Automatic Evaluation of Machine Translation". In: *Proceedings of the 40th Annual Meeting on Association for Computational Linguistics*. ACL '02. Association for Computational Linguistics, 2002, pp. 311–318.

[PDM11]   Slav Petrov, Dipanjan Das, and Ryan McDonald. "A universal part-of-speech tagset". In: *IN ARXIV:1104.2086*. 2011.

[PP09]   Simone Paolo Ponzetto and Massimo Poesio. "State-of-the-art NLP Approaches to Coreference Resolution: Theory and Practical Recipes". In: *Tutorial Abstracts of ACL-IJCNLP 2009*. Association for Computational Linguistics, 2009, pp. 6–6.

[PWM08]   Sameer Pradhan, Wayne Ward, and James H. Martin. "Towards robust semantic role labeling". In: *Computational Linguistics* (2008).

[Rac14]   Jiří Raclavský "A Model of Language in a Synchronic and Diachronic Sense". In: *Lodź Studies in English and General Linguistic 2: Issues in Philosophy of Language and Linguistic*. Łodź University Press, 2014, pp. 109–123.

[Ram99]   Juan Ramos. *Using TF-IDF to Determine Word Relevance in Document Queries*. 1999.

[RR15]   Kumar Ravi and Vadlamani Ravi. "A Survey on Opinion Mining and Sentiment Analysis". In: *Know.-Based Syst.* 89.C (Nov. 2015), pp. 14–46.

[Rit+12]   Alan Ritter et al. "Open Domain Event Extraction from Twitter". In: *Proceedings of the 18th ACM SIGKDD International Conference on Knowledge Discovery and Data Mining*. ACM, 2012, pp. 1104–1112.

[Sau16]   Ferdinand de Saussure. *Cours de Linguistique Générale* Payot, 1916.

[SM99]   Sam Scott and Stan Matwin. "Feature engineering for text classification". In: *Proceedings of ICML-99, 16th International Conference on Machine Learning*. Morgan Kaufmann Publishers, San Francisco, US, 1999, pp. 379–388.

[Seb02]   Fabrizio Sebastiani. "Machine Learning in Automated Text Categorization". In: *ACM Comput. Surv.* 34.1 (Mar. 2002), pp. 1–47.

[SM00]   H. Gregory Silber and Kathleen F. McCoy. "Efficient Text Summarization Using Lexical Chains". In: *Proceedings of the 5th International Conference on Intelligent User Interfaces*. IUI '00. ACM, 2000, pp. 252–255.

[Sin+13]   Sameer Singh et al. "Joint Inference of Entities, Relations, and Coreference". In: *Proceedings of the 2013 Workshop on Automated Knowledge Base Construction*. ACM, 2013, pp. 1–6.

[Tab+11]   Maite Taboada et al. "Lexicon-based Methods for Sentiment Analysis". In: *Comput. Linguist.* 37.2 (June 2011), pp. 267–307.

[TKSDM03c] Erik F. Tjong Kim Sang and Fien De Meulder. "Introduction to the CoNLL-2003 Shared Task: Language-independent Named Entity Recognition". In: *Proceedings of the Seventh Conference on Natural Language Learning at HLT-NAACL 2003 - Volume 4*. Association for Computational Linguistics, 2003, pp. 142–147.

[Wan+12]   Chang Wang et al. "Relation Extraction and Scoring in DeepQA". In: *IBM Journal of Research and Development* 56.3/4 (2012), 9:1–9:12.

[Wei+15]   Tingting Wei et al. "A semantic approach for text clustering using WordNet and lexical chains". In: *Expert Systems with Applications* 42.4 (2015), pp. 2264–2275.

[WR05]   Janyce Wiebe and Ellen Riloff. "Creating Subjective and Objective Sentence Classifiers from Unannotated Texts". In: *Proceedings of the 6th International Conference on Computational Linguistics and Intelligent Text Processing*. Springer-Verlag, 2005, pp. 486–497.

[WG05]   Florian Wolf and Edward Gibson. "Representing Discourse Coherence: A Corpus-Based Study". In: *Comput. Linguist.* 31.2 (June 2005), pp. 249–288.

Part II
Deep Learning Basics

Chapter 4
Basics of Deep Learning

4.1 Introduction

One of the most talked-about concepts in machine learning both in the academic community and in the media is the evolving field of deep learning. The idea of neural networks, and subsequently deep learning, gathers its inspiration from the biological representation of the human brain (or any brained creature for that matter).

The perceptron is loosely inspired by biological neurons (Fig. 4.1), connecting multiple inputs (signals to dendrites), combining and accumulating these inputs (as would take place in the cell body proper), and producing an output signal that resembles an axon.

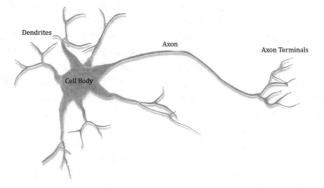

Fig. 4.1: Diagram of a biological neuron

Neural networks extend this analogy, combining a network of artificial neurons to create a neural network where information is passed between neurons (synapses), as

© Springer Nature Switzerland AG 2019   141
U. Kamath et al., *Deep Learning for NLP and Speech Recognition*,
https://doi.org/10.1007/978-3-030-14596-5_4

illustrated in Fig. 4.2. Each of these neurons learns a different function of its input, giving the network of neurons an extremely diverse representational power.

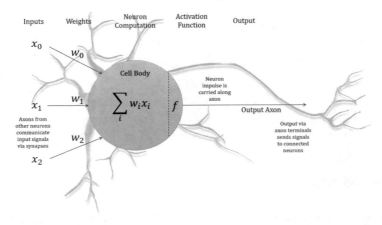

Fig. 4.2: Diagram of an artificial neuron (perceptron)

The last 6–7 years have seen exponential growth in the popularity and application of deep learning. Although the foundations of neural networks can be traced back to the late 1960s [Iva68], the AlexNet architecture [KSH12c] ushered in an explosion of interest in the deep learning when it handily won the 2012 Imagenet image classification competition [Den+09b] with a 5-layer convolutional neural network. Since then deep learning has been applied to a multitude of domains and has achieved state-of-the-art performance in most of these areas.

The purpose of this chapter is to introduce the reader to deep learning. By the end of this chapter, the reader should be able to understand the basics of neural networks and how to train them. We begin this chapter with a review of the perceptron algorithm that was introduced in Chap. 2, where neural networks found their origin. We then introduce the multilayer perceptron (MLP) classifier, the most simplistic form of feed-forward neural networks. Following this is a discussion of the essential components of training an MLP. This section contains an introduction to both forward and back propagation and explains the overall training process for neural networks. We then move toward an exploration of the essential architectural components: activation functions, error metrics, and optimization methods. After this section, we broaden the MLP concept to the deep learning domain, where we introduce additional considerations when training deep neural networks, such as computation time and regularization. Finally, we conclude with a practical discussion of common deep learning framework approaches.

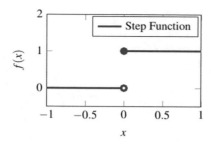

Fig. 4.3: The step function performs perfectly adequately for the perceptron; however, the lack of a non-zero gradient makes it useless for neural networks

4.2 Perceptron Algorithm Explained

Deep learning in its simplest form is an evolution of the perceptron algorithm, trained with a gradient-based optimizer. Chapter 2 introduced the perceptron algorithm. This section propounds the importance of the perceptron algorithm as one of the building blocks of deep learning.

The perceptron algorithm is one of the earliest supervised learning algorithms, dating back to the 1950s. Much like a biological neuron, the perceptron algorithm acts as an artificial neuron, having multiple inputs, and weights associated with each input, each of which then yields an output. This is illustrated in Fig. 4.6b.

The basic form of the perceptron algorithm for binary classification is:

$$y(x_1,\ldots,x_n) = f(w_1x_1 + \ldots + w_nx_n). \tag{4.1}$$

We individually weigh each x_i by a learned weight w_i to map the input $\mathbf{x} \in \mathbb{R}^n$ to an output value y, where $f(x)$ is defined as the step function shown below and in Fig. 4.3.

$$f(v) = \begin{cases} 0 & \text{if } v < 0.5 \\ 1 & \text{if } v \geq 0.5 \end{cases} \tag{4.2}$$

The step function takes a real number input and yields a binary value of 0 or 1, indicating a positive or negative classification if it exceeds the threshold of 0.5.

4.2.1 Bias

The perceptron algorithm learns a hyperplane that separates two classes. However, at this point, the separating hyperplane cannot shift away from the origin, as shown in Fig. 4.4a. Restricting the hyperplane in this fashion causes issues, as we can see in Fig. 4.4b.

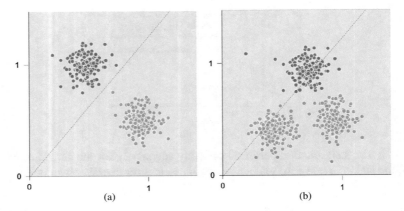

Fig. 4.4: (**a**) The perceptron algorithm is able to separate the two classes with the line passing through the origin. (**b**) Although the data is linearly separable, the perceptron algorithm is not able to separate the data. This is due to the restriction of the separating plane needing to pass through the origin

One solution is to ensure that our data is learnable if we normalize the method to center around the origin as a potential solution to alleviate this issue or add a bias term b to Eq. 4.1, allowing the classification hyperplane to move away from the origin, as shown in Fig. 4.5.

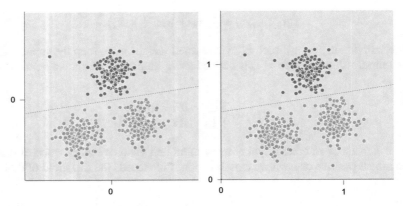

Fig. 4.5: (**a**) The perceptron algorithm is able to separate the two classes after centering the data at the origin. Note the location of the origin in the figure. (**b**) The bias allows the perceptron algorithm to relocate the separating plane, allowing it to correctly classify the data points

We can write the perceptron with a *bias* term as:

$$y(x_1,\ldots,x_n) = f(w_1 x_1 + \ldots + w_n x_n + b) \tag{4.3}$$

Alternatively, we can treat b as an additional weight w_0 tied to a constant input of 1 as shown in Fig. 4.6b and write it as:

$$y(x_1,\ldots,x_n) = f(w_1 x_1 + \ldots + w_n x_n + w_0) \tag{4.4}$$

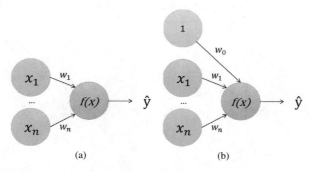

(a) (b)

Fig. 4.6: Perceptron classifier diagram. (**a**) Perceptron classifier diagram drawn without the bias. (**b**) Perceptron diagram including the bias

Some authors describe this as adding an input constant $x_0 = 1$, allowing the learned value for $b = w_0$ to move the decision boundary away from the origin. We will continue to write the bias term for now as a reminder of its importance; however, the bias term is implicit even when not written, which is commonly the case in academic literature. Switching to vector notation, we can rewrite Eq. 4.3 as:

$$y(\mathbf{x}) = f(\mathbf{w}\mathbf{x} + b). \tag{4.5}$$

> The bias term is a learned weight that removes the restriction that the separating hyperplane must pass through the origin.

The learning process for the perceptron algorithm is to modify the weights $\mathbf{w}$ to achieve 0 error on the training set. For example, suppose we need to separate sets of points A and B. Starting with random weights $\mathbf{w}$, we incrementally improve the boundary through each iteration with the aim of achieving $E(\mathbf{w},b) = 0$. Thus, we would minimize the error of the following function over the entire training set.

$$E(\mathbf{w}) = \sum_{\mathbf{x} \in A} (1 - f(\mathbf{w}\mathbf{x} + b)) + \sum_{\mathbf{x} \in B} f(\mathbf{w}\mathbf{x} + b) \tag{4.6}$$

4.2.2 Linear and Non-linear Separability

Two sets of data are linearly separable if a single decision boundary can separate them. For example, two sets, A and B, are linearly separable if, for some decision threshold t, every $x_i \in A$ satisfies the inequality $\sum_i w_i x_i \geq t$ and every $y_j \in B$ satisfies $\sum_i w_i y_i < t$. Conversely, two sets are not linearly separable if separation requires a non-linear decision boundary.

If we apply the perceptron to a non-linearly separable dataset, like the dataset shown in Fig. 4.7a, then we are unable to separate the data as shown in 4.7b since we are only able to learn three parameters, w_1, w_2, and b.

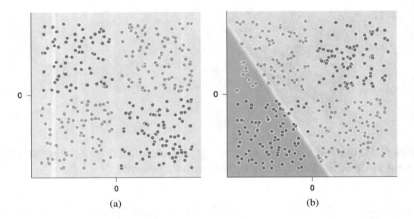

      (a)   (b)

Fig. 4.7: (**a**) Non-linearly separable dataset (generalization of the XOR function). (**b**) Result of training the perceptron algorithm on the non-linearly separable dataset in (**a**). The linear boundary is incapable of classifying the data correctly

Unfortunately, most data that we tend to encounter in NLP and speech is highly non-linear. One option (as we saw in Chap. 2) is to create non-linear combinations of the input data and use them as features in the model. Another option is to learn non-linear functions of the raw data, which is the principal aim of neural networks.

4.3 Multilayer Perceptron (Neural Networks)

The multilayer perceptron (MLP) links multiple perceptrons (commonly referred to as neurons) together into a network. Neurons that take the same input are grouped into a layer of perceptrons. Instead of using the step function, as seen previously, we substitute a differentiable, non-linear function. Applying this non-linear function, commonly referred to as an **activation function** or non-linearity, allows the output value to be a non-linear, weighted combination of its inputs, thereby creating non-linear features used by the next layer. In contrast, using a linear function as the activation function restricts the network to only being able to learn linear transforms of the input data. Furthermore, it is shown that any number of layers with a linear activation function can be reduced to a 2-layer MLP [HSW89].

The MLP is composed of interconnected neurons and is, therefore, a neural network. Specifically, it is a feed-forward neural network, since there is one direction to the flow of data through the network (no cycles—recurrent connections). Figure 4.8 shows the simplest multilayer perceptron. An MLP must contain an input and output layer and at least one hidden layer. Furthermore, the layers are also "fully connected," meaning that the output of each layer is connected to each neuron of the next layer. In other words, a weight parameter is learned for each combination of input neuron and output neuron between the layers.

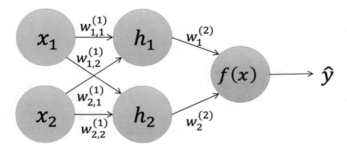

Fig. 4.8: Illustration of the multilayer perceptron network with an input layer, one hidden layer containing two neurons, and an output layer. The hidden layer, h, is the result of $\mathbf{h} = g(\mathbf{W}^{(1)}\mathbf{x})$, where $g(x)$ is the activation function. The output of the network $\hat{y} = f(\mathbf{W}^{(2)}\mathbf{h})$, where $f(x)$ is the output function, such as the step or sigmoid function

The hidden layer provides two outputs, h_1 and h_2, which may be non-linear combinations of their input values x_1 and x_2. The output layer weighs its inputs from the hidden layer, now a potential non-linear mapping, and makes its prediction.

4.3.1 Training an MLP

Training the weights of the MLP (and by extension, a neural network) relies on four main components.

Steps to train a neural network:

1. **Forward propagation**: Compute the network output for an input example.
2. **Error computation**: Compute the prediction error between the network prediction and the target.
3. **Backpropagation**: Compute the gradients in reverse order with respect to the input and the weights.
4. **Parameter update**: Use stochastic gradient descent to update the weights of the network to reduce the error for that example.

We will walk through each of these components with the network shown in Fig. 4.8.

4.3.2 Forward Propagation

The first step in training an MLP is to compute the output of the network for an example from the dataset. We use the sigmoid function, represented by $\sigma(x)$, as the activation function for the MLP. It can be thought of as a smooth step function and is illustrated in Fig. 4.14. Additionally, it is continuously differentiable, which is a desirable property for backpropagation, as is shown momentarily. The definition of the sigmoid function is:

$$\sigma(x) = \frac{1}{1 + e^{-x}}. \tag{4.7}$$

The forward propagation step is very similar to steps 3 and 4 of the perceptron algorithm. The goal of this process is to compute the current network output for a particular example x, with each output connected as the input to the next layer's neuron(s).

For notational and computational convenience, the layer's weights are combined into a single weight matrix, $\mathbf{W}_l$, representing the collection of weights in that layer, where l is the layer number. The linear transform performed by the layer computation for each weight is an inner product computation between $\mathbf{x}$ and $\mathbf{W}_l$. This type is regularly referred to as a "fully connected," "inner product," or "linear" layer because a weight connects each input to each output. Computing the prediction $\hat{y}$ for an example x where h_1 and h_2 represent the respective layer outputs becomes:

$$f(v) = \sigma(v)$$

$$\begin{aligned}
\mathbf{h_1} &= f(\mathbf{W}_1 \mathbf{x} + \mathbf{b}_1) \\
h_2 &= f(\mathbf{W}_2 \mathbf{h}_1 + b_2) \\
\hat{y} &= h_2.
\end{aligned} \tag{4.8}$$

Note the bias $\mathbf{b}_1$ is a vector because there is a bias value associated with each neuron in the layer. There is only one neuron in the output layer, so the bias b_2 is a scalar.

By the end of the forward propagation step, we have an output prediction for our network. Once the network is trained, a new example is evaluated through forward propagation.

4.3.3 Error Computation

The error computation step verifies how well our network performed on the example given. We use mean squared error (MSE) as the loss function used in this example (treating the training as a regression problem). MSE is defined as:

$$E(\hat{\mathbf{y}}, \mathbf{y}) = \frac{1}{2n} \sum_{i=1}^{n} (\hat{y}_i - y_i)^2. \tag{4.9}$$

The $\frac{1}{2}$ simplifies backpropagation. With a single output this quantity is reduced to:

$$E(\hat{y}, y) = \frac{1}{2}(\hat{y} - y)^2. \tag{4.10}$$

Error functions will be explored more in Sect. 4.4.2.

This error function is commonly used for regression problems, measuring the average of the square errors for the target. The squaring function forces the error to be non-negative and functions as a quadratic loss with the values closer to zero, yielding a polynomially smaller error than values further from zero.

The error computation step produces a scalar error value for the training example. We will talk more about error functions in Sect. 4.4.2.

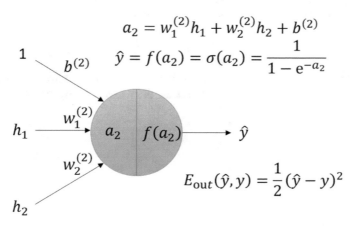

Fig. 4.9: Output neuron of Fig. 4.12 showing the full computation of the pre-activation and post-activation output

Figure 4.9 shows the forward propagation step and error propagation for the output neuron of Fig. 4.8.

4.3.4 Backpropagation

During forward propagation, an output prediction $\hat{y}$ is computed for the input x and the network parameters θ. To improve our prediction, we can use SGD to decrease the error of the whole network. Determining the error for each of the parameters can be done via the chain rule of calculus. We can use the chain rule of calculus to compute the derivatives of each layer (and operation) in the reverse order of forward propagation as seen in Fig. 4.10.

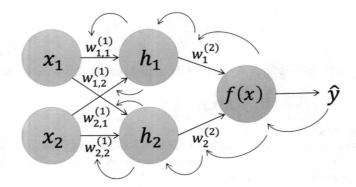

Fig. 4.10: Visualization of backward propagation

In our previous example, the prediction $\hat{y}$ was dependent on $\mathbf{W}_2$. We can compute the prediction error with respect to $\mathbf{W}_2$, by using the chain rule:

$$\frac{\partial E}{\partial \mathbf{W}_2} = \frac{\partial E}{\partial \hat{y}} \cdot \frac{\partial \hat{y}}{\partial \mathbf{W}_2} \tag{4.11}$$

The chain rule allows us to compute the gradient of the error for each of the learnable parameters θ, allowing us to update the network using stochastic gradient descent.

We begin by computing the gradient on the output layer with respect to the prediction.

$$\nabla_{\hat{y}} E(\hat{y}, y) = \frac{\partial E}{\partial \hat{y}} = (\hat{y} - y) \tag{4.12}$$

We can then compute error with respect to the layer 2 parameters.

We currently have the "post-activation" gradient, so we need to compute the pre-activation gradient:

$$
\begin{aligned}
\nabla_{\mathbf{a_2}} E = \frac{\partial E}{\partial \mathbf{a_2}} &= \frac{\partial E}{\partial \hat{y}} \cdot \frac{\partial \hat{y}}{\partial \mathbf{a_2}} \\
&= \frac{\partial E}{\partial \hat{y}} \odot f'(\mathbf{W}_2 \mathbf{h}_1 + \mathbf{b}_2)
\end{aligned}
\tag{4.13}
$$

Now we can compute the error with respect to $\mathbf{W}_2$ and $\mathbf{b}_2$.

$$
\begin{aligned}
\nabla_{\mathbf{W_2}} E = \frac{\partial E}{\partial \mathbf{W_2}} &= \frac{\partial E}{\partial \hat{y}} \cdot \frac{\partial \hat{y}}{\partial \mathbf{a_2}} \cdot \frac{\partial \mathbf{a_2}}{\partial \mathbf{W_2}} \\
&= \frac{\partial E}{\partial \mathbf{a_2}} \mathbf{h}_1^{\mathsf{T}}
\end{aligned}
\tag{4.14}
$$

$$
\begin{aligned}
\nabla_{\mathbf{b_2}} E = \frac{\partial E}{\partial \mathbf{b_2}} &= \frac{\partial E}{\partial \hat{y}} \cdot \frac{\partial \hat{y}}{\partial \mathbf{a_2}} \cdot \frac{\partial \mathbf{a_2}}{\partial \mathbf{b_2}} \\
&= \frac{\partial E}{\partial \mathbf{a_2}}
\end{aligned}
\tag{4.15}
$$

We can also compute the error for the input to layer 2 (the post-activation output of layer 1).

$$
\begin{aligned}
\nabla_{\mathbf{h_1}} E = \frac{\partial E}{\partial \mathbf{h_1}} &= \frac{\partial E}{\partial \hat{y}} \cdot \frac{\partial \hat{y}}{\partial \mathbf{a_2}} \cdot \frac{\partial \mathbf{a_2}}{\partial \mathbf{h_1}} \\
&= \mathbf{W}_2^{\mathsf{T}} \frac{\partial E}{\partial \mathbf{a_2}}
\end{aligned}
\tag{4.16}
$$

We then repeat this process to calculate the error for layer 1's parameters $\mathbf{W}_1$ and $\mathbf{b}_1$, thus propagating the error backward throughout the network.

Figure 4.11 shows the backward propagation step for the output neuron of the network shown in Fig. 4.8. We leave numerical exploration and experimentation for our notebook exercises.

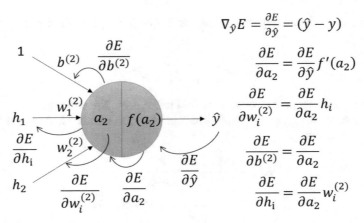

$$\nabla_{\hat{y}} E = \frac{\partial E}{\partial \hat{y}} = (\hat{y} - y)$$

$$\frac{\partial E}{\partial a_2} = \frac{\partial E}{\partial \hat{y}} f'(a_2)$$

$$\frac{\partial E}{\partial w_i^{(2)}} = \frac{\partial E}{\partial a_2} h_i$$

$$\frac{\partial E}{\partial b^{(2)}} = \frac{\partial E}{\partial a_2}$$

$$\frac{\partial E}{\partial h_i} = \frac{\partial E}{\partial a_2} w_i^{(2)}$$

Fig. 4.11: Backpropagation through the output neuron

4.3.5 Parameter Update

The last step in the training process is the parameter update. After obtaining the gradients with respect to all learnable parameters in the network, we can complete a single SGD step, updating the parameters for each layer according to the learning rate α.

$$\theta = \theta - \alpha \nabla_{\theta} E \tag{4.17}$$

The simplicity of the SGD update rule presented here does come at a cost. The value of α is particularly vital in SGD and affects the speed of convergence, the quality of convergence, and even the ability for the network to converge at all. Too small of a learning rate and the network converges very slowly and can potentially get stuck in local minima near the random weight initialization. If the learning rate is too large, the weights may grow too quickly, becoming unstable and failing to converge at all. Furthermore, the selection of the learning rate depends on a combination of factors such as network depth and normalization method. The simplicity of the network presented here alleviates the tedious nature of selecting a learning rate, but for deeper networks, this process can be much more difficult. The importance of choosing a good learning rate has led to an entire area of research around gradient descent optimization algorithms. We discuss some of these techniques more in Sect. 4.4.3.

The overall process is described in Algorithm 1.

Algorithm 1: Neural network training

Data: Training Dataset $\mathcal{D} = \{(\mathbf{x}_1, \mathbf{y}_1), (\mathbf{x}_1, \mathbf{y}_2), \ldots, (\mathbf{x}_n, \mathbf{y_n})\}$
Neural network with l layers with learnable parameters
$\theta = (\{\mathbf{W}_1, \ldots \mathbf{W}_l\}, \{\mathbf{b}_1, \ldots \mathbf{b}_l\})$
Activation function $f(\mathbf{v})$
Learning rate α
Error function $E(\hat{\mathbf{v}}, \mathbf{v})$
Initialize neural network parameters $\theta = (\{\mathbf{W}_1, \ldots \mathbf{W}_l\}, \{\mathbf{b}_1, \ldots \mathbf{b_l}\})$
for $e \leftarrow 1$ *to e epochs* **do**
   for $(\mathbf{x}, \mathbf{y})$ *in* $\mathcal{D}$ **do**
      for $i \leftarrow 1$ *to l* **do**
         if *i=1* **then**
            $\mathbf{h}_{i-1} = \mathbf{x}$
         $\mathbf{a}_i = \mathbf{W}_i \mathbf{h}_{i-1} + \mathbf{b}_i$
         $\mathbf{h}_i = f(\mathbf{a}_i)$

      $\hat{\mathbf{y}} = \mathbf{h}_l$
      error $= E(\hat{\mathbf{y}}, \mathbf{y})$
      $\mathbf{g}_{\mathbf{h}_{i+1}} = \nabla_{\hat{\mathbf{y}}} E(\hat{\mathbf{y}}, \mathbf{y})$

      for $i \leftarrow l$ *to 1* **do**
         $\mathbf{g}_{\mathbf{a}_i} = \nabla_{\mathbf{a}_i} E = \mathbf{g}_{\mathbf{h}_{i+1}} \circ f'(\mathbf{a}_i)$
         $\nabla_{\mathbf{W}_i} E = \mathbf{g}_{\mathbf{a}_i} \mathbf{h}_{i-1}^\mathsf{T}$
         $\nabla_{\mathbf{b}_i} E = \mathbf{g}_{\mathbf{a}_i}$
         $\mathbf{g}_{\mathbf{h}_i} = \nabla_{\mathbf{h}_{i-1}} E = \mathbf{W}_i^\mathsf{T} \mathbf{g}_{\mathbf{a}_i}$

      $\theta = \theta - \alpha \nabla_\theta E$

4.3.6 Universal Approximation Theorem

Neural network architectures are applied to a variety of problems because of their representational power. The universal approximation theorem [HSW89] has shown that a feed-forward neural network with a single layer can approximate any continuous function with only limited restrictions on the number of neurons in the layer.[1] This theorem often gets summarized as "neural networks are universal approximators." Although this is technically true, the theorem does not provide any guarantees on the likelihood of learning a particular function.

---

[1] The universal approximation theorem was initially proved for neural network architectures using the sigmoid activation function, but was subsequently shown to apply to all fully connected networks [Cyb89b, HSW89].

The topography of the parameter space becomes more varied as machine learning problems become more complex. It is typically non-convex with many local minima. A simple gradient descent approach may struggle to learn the specific function. Instead, multiple layers of neurons are stacked consecutively and trained jointly with backpropagation. The network of layers then learns multiple non-linear functions to fit the training dataset. Deep learning refers to many neural network layers connected in sequence.

4.4 Deep Learning

The term "deep learning" is somewhat ambiguous. In many circles deep learning is a re-branding term for neural networks or is used to refer to neural networks with many consecutive (deep) layers. However, the number of layers to distinguish a deep network from a shallow network is relative. For example, would the neural network shown in Fig. 4.12 be considered deep or shallow?

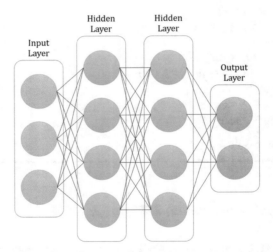

Fig. 4.12: Feed-forward neural network with two hidden layers

In general, deep networks are still neural networks (trained with backpropagation, learning hierarchical abstractions of the input, optimized using gradient-based learning), but typically with more layers. The distinguishing characteristic of deep learning is its application to problems previously infeasible to traditional methods and smaller neural networks, such as the MLP shown in Fig. 4.8. Deeper networks allow for more layers of hierarchical abstractions to be learned for the input data, thus becoming capable of learning higher-order functions in more complex domains. For this book however we utilize the term deep learning as described above—a neural network with more than one hidden layer.

The flexibility of neural networks is what makes them so compelling. Neural networks are applied to many types of problems given the simplicity and effectiveness of backpropagation and gradient-based optimization methods. In this section, we introduce additional methods and considerations that impact the architecture design and model training for deep neural networks (DNN). In particular, we focus on activation functions, error functions, optimization methods, and regularization approaches.

4.4.1 Activation Functions

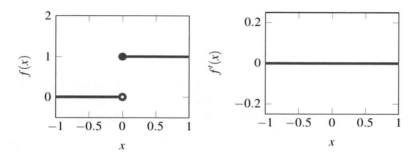

Fig. 4.13: The step function performed perfectly adequate for the perceptron, however its derivative makes it bad for gradient descent methods

When computing the gradient of the output layer, it becomes apparent that the step function is not exactly helpful when trying to compute a gradient. As shown in Fig. 4.13, the derivative is 0 everywhere which means any gradient descent is useless. Therefore we wish to use a non-linear activation function that provides a meaningful derivative in the backpropagation process.

4.4.1.1 Sigmoid

A better function to use as an activation function is the logistic sigmoid:

$$\sigma(x) = \frac{1}{1 + e^{-x}} \tag{4.18}$$

The sigmoid function is a useful activation for a variety of reasons. As we can see from the graph in Fig. 4.14, this function acts as a continuous squashing function that bounds its output in the range $(0, 1)$. It is similar to the step function but has a

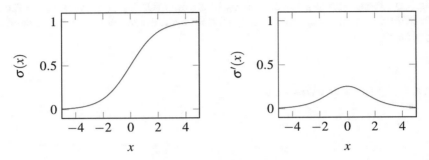

Fig. 4.14: Sigmoid activation function and its derivative

smooth, continuous derivative ideal for gradient descent methods. It is also zero-centered, creating a simple decision boundary for binary classification tasks, and the derivative of the sigmoid function is mathematically convenient:

$$\sigma'(x) = \sigma(x)(1 - \sigma(x)). \tag{4.19}$$

There are, however, some undesirable properties of the sigmoid function.

- Saturation of the sigmoid gradients at the ends of the curve (very close to $\sigma(x) \leftarrow 0$ or $\sigma(x) \leftarrow 1$) will cause the gradients to be very close to 0. As backpropagation continues for subsequent layers, the small gradient is multiplied by the post-activation output of the previous layer, forcing it smaller still. Preventing this can require careful initialization of the network weights or other regularization strategies.
- The outputs of the sigmoid are not centered around 0, but instead around 0.5. This introduces a discrepancy between the layers because the outputs are not in a consistent range. This is often referred to as "internal covariate shift" which we will talk more about later.

4.4.1.2 Tanh

The *tanh* function is another common activation function. It also acts as a squashing function, bounding its output in the range $(-1, 1)$ as shown in Fig. 4.15.

$$f(x) = tanh(x) \tag{4.20}$$

It can also be viewed as a scaled and shifted sigmoid.

$$tanh(x) = 2 * \sigma(2x) - 1 \tag{4.21}$$

The tanh function solves one of the issues with the sigmoid non-linearity because it is zero-centered. However, we still have the same issue with the gradient saturation at the extremes of the function, shown in Fig. 4.16.

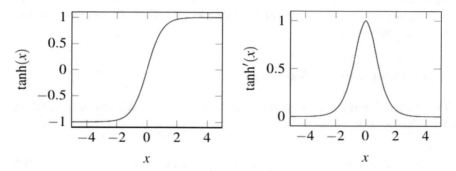

Fig. 4.15: Tanh activation function and its derivative

4.4.1.3 ReLU

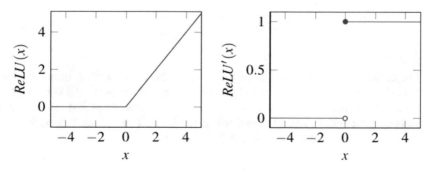

Fig. 4.16: ReLU activation function and its derivative

The rectified linear unit (ReLU) is a simple, fast activation function typically found in computer vision. The function is a linear threshold, defined as:

$$f(x) = \max(0, x). \tag{4.22}$$

This simple function has become popular because it has shown faster convergence compared to sigmoid and tanh, possibly due to its non-saturating gradient in the positive direction.

In addition to faster convergence, the ReLU function is much faster computationally. The sigmoid and tanh functions require exponentials which take much longer than a simple max operation.

One drawback from the simplicity of the gradient updates being 0 or 1 is that it can lead to neurons "dying" during training. If a large gradient is backpropagated through a neuron, the neuron's output can become so affected that the update prevents the neuron from ever updating again. Some have shown that as many as 40% of the neurons in a network can "die" with the ReLU activation function if the learning rate is set too high.

4.4.1.4 Other Activation Functions

Other activation functions have been incorporated to limit the effects of those previously described, displayed in Fig. 4.17.

- Hard tanh
  The hard tanh function is computationally cheaper than the tanh. It does, however, re-introduce the disadvantage of gradient saturation at the extremes.

$$f(x) = \max(-1, \min(1, x)) \tag{4.23}$$

- Leaky ReLU
  The Leaky ReLU introduces an α parameter that allows small gradients to be backpropagated when the activation is not active, thus eliminating the "death" of neurons during training.

$$f(x) = \begin{cases} x & \text{if } x \geq 0 \\ \alpha x & \text{if } x < 0 \end{cases}. \tag{4.24}$$

- PRELU
  The parametric rectified linear unit, similar to the Leaky ReLU, uses an α parameter to scale the slope of the negative portion of the input; however, an alpha parameter is learned for each neuron (doubling the number of learned weights). Note that when the value of $\alpha = 0$ this is the ReLU function and when the α is fixed, it is equivalent to the Leaky ReLU.

$$f(x) = \begin{cases} x & \text{if } x \geq 0 \\ \alpha x & \text{if } x < 0 \end{cases}. \tag{4.25}$$

- ELU

  The ELU is a modification of the ReLU that allows the mean of activations to push closer to 0, which therefore potentially speeds up convergence.

  $$f(x) = \begin{cases} x & \text{if } x > 0 \\ \alpha(e^x - 1) & \text{if } x \leq 0 \end{cases}. \tag{4.26}$$

- Maxout

  The maxout function takes a different approach to activation functions. It differs from the element-wise application of a function to each neuron output. Instead, it learns two weight matrices and takes the highest output for each element.

  $$f(x) = \max(w_1 x + b_1, w_2 x + b_2) \tag{4.27}$$

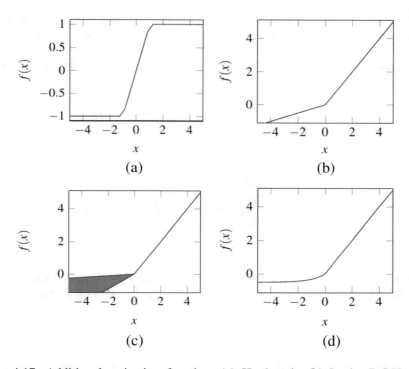

Fig. 4.17: Additional activation function. (**a**) Hard tanh. (**b**) Leaky ReLU. (**c**) PReLU. (**d**) ELU

4.4.1.5 Softmax

The squashing concept of the sigmoid function is extended to multiple classes by way of the softmax function. The softmax function allows us to output a categorical probability distribution over K classes.

$$f(x_i) = \frac{e^{x_i}}{\sum_j e^{x_j}} \tag{4.28}$$

We can use the softmax to produce a vector of probabilities according to the output of that neuron. In the case of a classification problem that has $K = 3$ classes, the final layer of our network will be a fully connected layer with an output of three neurons. If we apply the softmax function to the output of the last layer, we get a probability for each class by assigning a class to each neuron. The softmax computation is shown in Fig. 4.18.

The softmax probabilities can become very small, especially when there are many classes and the predictions become more confident. Most of the time a log-based softmax function is used to avoid underflow errors. The softmax function is a particular case for activation functions, in that it is rarely seen as an activation that

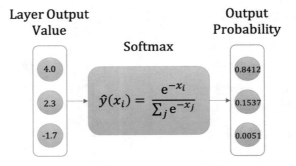

Fig. 4.18: The output of a neural network can be mapped to a multi-class classification task (three classes shown here). The softmax function maps the real-value network output to a probability distribution over the number of classes, where the number of classes equals the number of neurons in the final layer

occurs between layers. Therefore, the softmax is often treated as the last layer of a network for multiclass classification rather than an activation function.

4.4.1.6 Hierarchical Softmax

As the number of classes begins to grow, as is often the case in language tasks, the computation of the softmax function can become expensive to compute. For example, in a language modeling task, our output layer may be trying to predict which word will be next in the sequence. Therefore, the output of the network would be a probability distribution over the number of terms in our vocabulary, which could be thousands or hundreds of thousands. The Hierarchical Softmax [MB05] approximates the softmax function by representing the function as a binary tree with the depth yielding less probable class activations. The tree must be balanced as the network is training, but it will have a depth of $\log_2(K)$ where K is the number of classes, which means only $\log_2(K)$ states need to be evaluated to compute the output probability of a class.

4.4.2 Loss Functions

Another important aspect of training neural networks is the choice of error functions often referred to as the criteria/criterion. The selection of the error function depends on the type of problem being addressed. For a classification problem, we want to predict a probability distribution over a set of classes. In regression problems, however, we want to predict a specific value rather than a distribution. We present the basic, most commonly used loss functions here.

4.4.2.1 Mean Squared (L_2) Error

Mean squared error(MSE) computes the squared error between the classification prediction and the target. Training with it minimizes the difference in magnitude. One drawback to MSE is that it is susceptible to outliers since the difference is squared.

$$E(\hat{\mathbf{y}}, \mathbf{y}) = \frac{1}{n} \sum_{i=1}^{n} (y_i - \hat{y}_i)^2 \tag{4.29}$$

So far, we have been using the MSE or L_2 for its simplicity as the loss for a binary classification problems, classifying it as a 0 if $\hat{y} \geq 0.5$ or 1 if $\hat{y} < 0.5$; however, it is typically used for regression problems and could be easily extended for the simple problems that we have been working with.

4.4.2.2 Mean Absolute (L_1) Error

Mean absolute error gives a measure of the absolute difference between the target value and prediction. Using it minimizes the magnitude of the error without considering direction, making it less sensitive to outliers.

$$E(\hat{\mathbf{y}}, \mathbf{y}) = \frac{1}{n} \sum_{i=1}^{n} |y_i - \hat{y}_i| \tag{4.30}$$

4.4.2.3 Negative Log Likelihood

Negative log likelihood (NLL), is the most common loss function used for multi-class classification problems. It is also known as the multiclass cross-entropy loss. The softmax provides a probability distribution over the output classes. The entropy computation is a weighted-average log probability over the possible events or classifications in a multiclass classification problem. This causes the loss to increase as the probability distribution of the prediction diverges from the target label.

$$E(\hat{\mathbf{y}}, \mathbf{y}) = -\frac{1}{n} \sum_{i=1}^{n} (y_i \log(\hat{y}_i) - (1 - y_i) \log(1 - \hat{y}_i)) \tag{4.31}$$

4.4.2.4 Hinge Loss

The hinge loss is a max-margin loss classification taken from the SVM loss. It attempts to separate data points between classes by maximizing the margin between them. Although it is not differentiable, it is convex, which makes it useful to work with as a loss function.

$$E(\hat{\mathbf{y}}, \mathbf{y}) = \sum_{i=1}^{n} \max(0, 1 - y_i \hat{y}_i) \tag{4.32}$$

4.4.2.5 Kullback–Leibler (KL) Loss

Additionally, we can optimize on functions, such as the KL-divergence, which measures a distance metric in a continuous space. This is useful for problems like generative networks with continuous output distributions. The KL-divergence error can be described by:

$$\begin{aligned}
E(\hat{\mathbf{y}}, \mathbf{y}) &= \frac{1}{n} \sum_{i=1}^{n} \mathcal{D}_{KL}(y_i || \hat{y}_i) \\
&= \frac{1}{n} \sum_{i=1}^{n} \left(y_i \cdot \log(y_i) \right) - \frac{1}{n} \sum_{i=1}^{n} \left(y_i \cdot \log(\hat{y}_i) \right)
\end{aligned} \tag{4.33}$$

4.4.3 Optimization Methods

The training process of neural networks is based on gradient descent methods, specifically SGD. However, as we have seen in the previous section, SGD can cause many undesirable difficulties during the training process. We will explore additional optimization methods in addition to SGD and the benefits associated with them. We consider all learnable parameters including weights and biases as θ.

4.4.3.1 Stochastic Gradient Descent

As presented in Chap. 2, stochastic gradient descent is the process of making updates to a set of weights in the direction of the gradient to reduce the error. In Algorithm 7, SGD's update rule was the simple form:

$$\theta_{t+1} = \theta_t - \alpha \nabla_\theta E. \tag{4.34}$$

where θ represents the learnable parameters, α is the learning rate, and $\nabla_\theta E$ is the gradient of the error with respect to the parameters.

4.4.3.2 Momentum

One issue that commonly arises with SGD is that there are areas of feature space that have long shallow ravines, leading up to the minima. SGD will oscillate back and forth across the ravine because the gradient will point down the steepest gradient on one of the sides rather than in the direction of the minima. Thus, SGD can yield slow convergence.

Momentum is one modification of SGD to move the objective more quickly to the minima. The parameter update equation for momentum is

$$v_t = \gamma v_{t-1} + \eta \nabla_\theta E$$
$$\theta_{t+1} = \theta_t - v_t \tag{4.35}$$

where θ_t represents a parameter at iteration t.

Momentum, taking its inspiration from physics computes a velocity vector capturing the cumulative direction that previous gradients have yielded. This velocity vector is scaled by an additional hyper-parameter η, which suggests how heavily the cumulative velocity can contribute to the update.

4.4.3.3 Adagrad

Adagrad [DHS11] is an adaptive gradient-based optimization method. It adapts the learning rate to each of the parameters in the network, making more substantial updates to infrequent parameters, and smaller updates to frequent ones. This makes it particularly useful for learning problems with sparse data [PSM14]. Perhaps the most significant benefit of adagrad is that it removes the need to tune the learning rate manually. This does, however, come at the cost of having an additional parameter for every parameter in the network.

The adagrad equation is given by:

$$g_{t,i} = \nabla_\theta E(\theta_{t,i})$$
$$\theta_{t+1,i} = \theta_{t,i} - \frac{\eta}{\sqrt{G_{t,ii} + \varepsilon}} \circ g_{t,i} \tag{4.36}$$

where g_t is the gradient at time t along each component of θ, G_t is the diagonal matrix of the sum of up to t time steps of past gradients w.r.t. to all parameters θ on the diagonal, η is the general learning rate, and ε is a smoothing term (usually $1e-8$) that keeps the equation from dividing by zero.

The main drawback to adagrad is that the accumulation of the squared gradients is positive, causing the sum to grow, shrinking the learning rate, and stopping the model from further learning. Additional variants, such as Adadelta [Zei12], have been introduced to alleviate this problem.

4.4.3.4 RMS-Prop

RMS-prop [TH12] developed by Hinton was also introduced to solve the inadequacies of adagrad. It also divides the learning rate by an average of squared gradients, but it also decays this quantity exponentially.

$$\mathbb{E}[g^2]_t = \rho \mathbb{E}[g^2]_{t-1} + (1-\rho)g_t^2$$
$$\theta_{t+1} = \theta_t - \frac{\eta}{\sqrt{\mathbb{E}[g^2]_t + \varepsilon}} g_t \tag{4.37}$$

where $\rho = 0.9$ and the learning rate $\eta = 0.001$ is suggested in the presented lecture.

4.4.3.5 ADAM

Adaptive moment estimation, referred to as Adam [KB14] is another adaptive optimization method. It too computes learning rates for each parameter, but in addition to keeping an exponentially decaying average of the previous squared gradients, similar to momentum, it also incorporates an average of past gradients m_t.

$$m_t = \beta_1 m_{t-1} + (1-\beta_1)g_t$$
$$v_t = \beta_2 v_{t-1} + (1-\beta_2)g_t^2$$
$$\hat{m}_t = \frac{m_t}{1-\beta_1^t}$$
$$\hat{v}_t = \frac{v_t}{1-\beta_2^t} \tag{4.38}$$
$$\theta_{t+1} = \theta_t - \frac{\eta}{\sqrt{\hat{v}_t} + \varepsilon} \hat{m}_t$$

Empirical results show that Adam works well in practice in comparison with other gradient-based optimization techniques.

While Adam has been a popular technique, some criticisms of the original proof have surfaced showing convergence to sub-optimal minima in some situations [BGW18, RKK18]. Each work proposes a solution to the issue, however the subsequent methods remain less popular than the original Adam technique.

4.5 Model Training

Achieving the best generalization error (best performance on the test set) is the main goal for machine learning, which requires finding the best position on the spectrum between overfitting and underfitting. Deep learning is more prone to overfitting. With many free parameters, it can be relatively easy to find a path to achieve $E = 0$. It has been shown that many standard deep learning architectures can be trained on random labeling of the training data and achieve $E = 0$ [Zha+16].

In contrast to overfitting, for many complex functions there are diverse local minima that may not be the optimal solution, and it is common to settle in a local minima. Deep learning relies on finding a solution to a non-convex optimization problem which is NP-complete for a general non-convex function [MK87]. In practice, we see that computing the global minimum for a well-regularized deep network is mostly irrelevant because local minima are usually roughly similar and get closer to the global minimum as the complexity of the model increases [Cho+15a]. In a poorly regularized network, however, the local minima may yield a high loss, which is undesirable.

The best model is one that achieves the smallest gap between its training loss and validation loss; however, selecting the correct architecture configuration and training technique can be taxing. Here we discuss typical training and regularization techniques to improve model generalization.

4.5.1 Early Stopping

One of the more practical ways that we can prevent a model from overfitting is "early stopping." Early stopping hinges on the assumption: "As validation error decreases, test error should also decrease." When training we compute the validation error at distinct points (usually at the end of each epoch) and keep the model with the lowest validation error, as shown in Fig. 4.19.

The learning curve shows that the training error will continue to decrease towards 0. However, the model begins to perform worse on the validation set as it overfits to the training data. Therefore, to maintain the generalization of the model on the test set, the model (learned parameters of model) that performed best on our validation set would be selected. It is also important to point out here that this requires a dataset that is split into training, validation, and testing sets with no overlap. The test set should be kept separate from the training and validation, as, otherwise, this compromises the integrity of the model.

The simplicity of early stopping makes it the most commonly used form of regularization in deep learning.

4.5.2 Vanishing/Exploding Gradients

When training neural networks with many layers with backpropagation, the issue of vanishing/exploding gradients arises. During backpropagation, we are multiplying the gradient by the output of each successive layer. This means that the gradient can

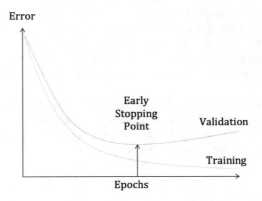

Fig. 4.19: Early stopping point is shown when validation error begins to diverge from training error

get larger and larger if $\nabla E > 1$ or $\nabla E < 1$ and smaller if the gradient is $1 < \nabla E < 0$ as it is multiplied by each successive layer. Practically, this means, in the case of vanishing gradients, very little of the error is propagated back to the earlier layers of the network causing learning to be very slow or nonexistent. For exploding gradients, this causes the weights to eventually overflow which prevents learning. The deeper a neural network becomes, the greater a problem this becomes.

In the case of exploding gradients a simple, practical solution is to "clip" the gradients setting a maximum for the gradient values at each backpropagation step to control the growth of the weights. We revisit this topic when addressing recurrent neural networks.

4.5.3 Full-Batch and Mini-Batch Gradient Decent

Batch gradient decent is a variant of gradient descent that evaluates the error on the whole dataset before updating the model by accumulating the error after each example. This alleviates some of the problems of SGD, such as the noise introduced from each example, but the frequency of the updates can cause a higher variance between training epochs, which can create significant differences in the models. This approach is rarely used in practice with deep learning.

A suitable compromise between these two strategies is mini-batch gradient descent. Mini-batch gradient descent splits the dataset into batches, and the model accumulates the error over a mini-batch before making an update. This approach provides a variety of advantages, including:

- Reduced noise in each model update due to accumulating the gradients from multiple training examples
- Greater efficiency than SGD
- Faster training by taking advantages of matrix operations to reduce IO time

One downside of mini-batch gradient descent is the addition of the mini-batch size as a hyperparameter. The mini-batch size, often just called "batch" size for convenience, is usually set based on the model's hardware limitations to not exceed the memory of either the CPU or GPU. Additionally, batch sizes are typically powers of 2 (8, 16, 32, etc.) due to common hardware implementations. In general, it is desirable to strike a balance with a small batch size yielding a quicker convergence and a larger batch size which converges more slowly but with more accurate estimates. It is recommended to review the learning curves of a few different batch sizes to decide on the best size.

4.5.4 Regularization

Practically, controlling the generalization error is achieved by creating a large model that is appropriately regularized [GBC16a, Bis95]. Regularization can take many

forms. Some methods focus on reducing the capacity of the models by penalizing
the abnormal parameters in the objective function by adding a regularization term

$$E(\mathbf{W}; \hat{\mathbf{y}}, \mathbf{y}) = E(\hat{\mathbf{y}}, \mathbf{y}) + \Omega(\mathbf{W}) \tag{4.39}$$

where $\mathbf{W}$ is the weights of the network. Some approaches focus on limiting the information provided to the network (e.g., dropout) or normalizing the output of layers
(batch normalization), while others may make changes to the data directly. Here we
will explore a variety of regularization methods, and it is typically suggested to
incorporate multiple into every problem.

4.5.4.1 L_2 Regularization: Weight Decay

One of the most common regularization methods is the L_2 regularization method,
commonly referred to as weight decay. Weight decay adds a regularization term
to the error function that pushes the weights towards the origin, penalizing high
weight variations. Weight decay introduces a scalar α that penalizes weights moving
away from the origin. This functions as a zero-mean Gaussian prior on the training
objective, limiting the freedom of the network to learn large weights that might be
associated with overfitting. The setting of this parameter becomes quite important
because if the model is too constrained, it may be unable to learn.

L_2 regularization is defined as:

$$\Omega(\mathbf{w}) = \frac{\alpha}{2} \mathbf{W}^\mathsf{T} \mathbf{W} \tag{4.40}$$

The loss function can then be described as:

$$E(\mathbf{W}; \hat{\mathbf{y}}, \mathbf{y}) = \frac{\alpha}{2} \mathbf{W}^\mathsf{T} \mathbf{W} + E(\hat{\mathbf{y}}, \mathbf{y}). \tag{4.41}$$

With the gradient being:

$$\nabla_\mathbf{W} E(\mathbf{W}; \hat{\mathbf{y}}, \mathbf{y}) = \alpha \mathbf{W} + \nabla_\mathbf{W} E(\hat{\mathbf{y}}, \mathbf{y}) \tag{4.42}$$

And the parameter update becomes:

$$\mathbf{W} = \mathbf{W} - \varepsilon(\alpha \mathbf{W} + \nabla_\mathbf{W} E(\hat{\mathbf{y}}, \mathbf{y})), \tag{4.43}$$

where ε is the learning rate.

4.5.4.2 L_1 Regularization

A less common regularization method is L_1 regularization. This technique also functions as a weight penalization. The regularizer is a sum of the absolute values of the
weights:

$$\Omega(\mathbf{w}) = \alpha \sum |w_i| \tag{4.44}$$

As training progresses many of the weights will become zero, introducing sparsity into the model weights. This is often used in feature selection but is not always a desirable quality with neural networks.

4.5.4.3 Dropout

Perhaps the second-most common regularization method in deep learning is Dropout [Sri+14]. Dropout has been a simple and highly effective method to reduce overfitting of neural networks. It stems from the idea that neural networks can have very fragile connections from the input to the output. These learned connections may work for the training data but do not generalize to the test data. Dropout aims to correct this tendency by randomly "dropping out" connections in the neural network training process so that a prediction cannot depend on any single neuron during training, as illustrated in Fig. 4.20.

Applying dropout to a network involves applying a random mask sampled from a Bernoulli distribution with a probability of p. This mask matrix is applied element-wise (multiplication by 0) during the feed-forward operation. During the backpropagation step, the gradients for each parameter and the parameters that were masked the gradient are set to 0 and other gradients are scaled up by $\frac{1}{1-p}$.

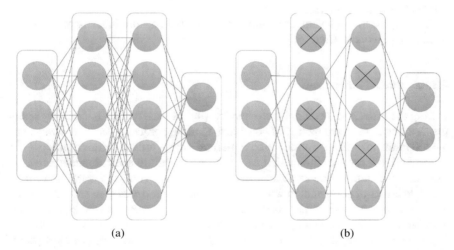

(a)   (b)

Fig. 4.20: Dropout when applied to a fully connected neural network. (**a**) Standard 2-layer (hidden) neural network. (**b**) Standard 2-layer (hidden) neural network with dropout

4.5.4.4 Multitask Learning

In all machine learning tasks, we are optimizing for a specific error metric or function. Therefore, to perform well on various tasks simultaneously, we usually train a model for each metric and then ensemble, linearly combine, or connect them in some other meaningful way to perform well on our overall collection of tasks. Because deep learning is achieved via gradient-based computation and descent, we can simultaneously optimize for a variety of optimization functions. This allows our underlying representation to learn a general representation that can accomplish multiple tasks. Multitask learning has become a widely used approach recently. The addition of auxiliary tasks can help improve the gradient signal to the learned parameters leading to better quality on the overall task [Rud17a].

4.5.4.5 Parameter Sharing

Another form of regularization is parameter sharing. So far we have only considered fully connected neural networks, which learn an individual weight for every input. In some tasks the inputs are similar enough that it is undesirable to learn a different set of parameters for each task, but rather share the learnings in multiple places. This can be accomplished by sharing a set of weights across different inputs. Parameter sharing is not only useful as a regularizer, but also provides multiple training benefits such as reduced memory (one copy of a set of weights) and a reduced number of unique model parameters.

One approach that leverages parameter sharing is a convolutional neural network, which we explore in Chap. 6.

4.5.4.6 Batch Normalization

During the process of training, there may be a lot of variation in the training examples leading to the introduction of noise in the training process. One of the ways that we recommended in the introduction is normalizing our data before training. Normalization reduces the amount that weights need to shift to accommodate a specific example, maintaining the same distribution properties. With deep learning we have multiple layers of computation with hidden values that are passed to subsequent layers. The output of each of these layers is likely to be a non-normalized input, and the distribution is likely to change frequently during the training process. This process is commonly referred to as "internal covariate shift." Batch normalization [IS15] aims to reduce internal covariate shift in a network by normalizing the outputs of intermediate layers during training. This speeds the training process and allows for higher learning rates without risking divergence.

Batch normalization achieves this by normalizing the output of the previous hidden layer by the batch's (mini-batch's) mean and variance. This normalization, however, would affect the inference phase, and, thus, batch normalization captures a moving average of the mean and variance and fixes them at inference time.

For an input mini-batch $\beta = \{x_{1:m}\}$, we learn parameters γ and β via:

$$
\begin{aligned}
\mu_\beta &= \frac{1}{m} \sum_{i=1}^{m} x_i \\
\sigma_\beta^2 &= \frac{1}{m} \sum_{i=1}^{m} (x_i - \mu_\beta)^2 \\
\hat{x}_i &= \frac{x_i - \mu_\beta}{\sqrt{\sigma_\beta^2 + \varepsilon}} \\
y_i &= \gamma \hat{x}_i + \beta.
\end{aligned}
\tag{4.45}
$$

4.5.5 Hyperparameter Selection

Most learning techniques and regularization methods have some form of training configuration parameters associated with them. Learning rate, momentum, dropout probability, and weight decay, for example, all need to be selected for each model. Selecting the best combination of these **hyperparameters** can be a challenging task.

4.5.5.1 Manual Tuning

Manual hyperparameter tuning is recommended when applying an existing model to a new dataset to an existing model or new model to an existing dataset. Manual selection helps provide intuition about the network. This can be useful to understand if a particular set of parameters will cause the network to overfit or underfit. It is advised to monitor the norm of the gradients, and how quickly a model's loss converges or diverges. In general, the learning rate is the most important hyperparameter, having the most impact on the effective capacity of the network [GBC16b]. Selecting the right learning rate for a model will allow good convergence, and early stopping will prevent the model from overfitting to the training set. If the learning rate is too high, large gradients can cause the network to diverge preventing future learning in some cases (even when the learning rate becomes lower). If the learning rate is too low, small updates will slow the learning process and can also cause the model to settle into a local minimum with a high training and generalization error.

4.5.5.2 Automated Tuning

Automatic hyperparameter selection is a much faster and robust method for optimizing a training configuration. Grid search, introduced in Chap. 2, is the most common and straightforward technique. In a grid search, uniform or logarithmic samples are provided for each parameter to be optimized, and a model is trained for every combination of parameters. This approach is effective, however it does require a significant amount of computation time to train the set of models. Typically, this cost can be reduced by investigating large ranges first and then narrowing the set of parameters or ranges, performing another grid search with the new ranges.

Random hyperparameter search is sometimes more robust to the nuances of training, as some combinations of hyperparameters can have cumulative effect. Similar to a grid search, random search randomly samples values in the range of the grid search rather than evenly spaced samples. This has shown to consistently outperform grid search as there are spaces of the hyperparameter grid that are unexplored (given the same number of parameter combinations).

Typically, the majority of the models explored with grid search and random search are subject to poor combinations. This can be alleviated to some degree by setting appropriate bounds for the search gleaned from manual exploration, however ideally the performance of the model can be used to determine the next set of parameters. Various conditioned and Bayesian hyperparameter selection procedures have been introduced to accomplish this [SLA12].

4.5.6 Data Availability and Quality

Regularization is the most common technique to prevent overfitting, but it can also be accomplished with increasing the amount of data. Data is the most important component of any machine learning model. Although it may seem obvious, this is often one of the most overlooked components in real-world scenarios. Abstractly, neural networks are learning from the experiences they encounter. In binary classification, for example, the positive example-label pairs are encouraged, while negative pairs are discouraged. Tuning the neural network's hyperparameters is typically the best appropriate step to improve generalization error. If a performance gap still exists between the training and generalization error, it may be necessary to increase the amount of data (or quality in some cases).

Neural networks can be robust to some amount of noise in a dataset, and during the training process, the effects of outliers are typically lessened. However, erroneous data can cause many issues. Poor model performance in real-world applications can be caused by consistently incorrect labels or insufficient data.

> In real-world applications, if there seems to be odd behavior throughout the training process, it may be a sign of data inconsistencies.

This typically manifests itself in one of two ways: overfitting or poor convergence. In the case of overfitting, the model may learn an anomaly of the data (such as the presence of a user name in many negative sentiment reviews).

Deep learning in particular benefits more from larger datasets than other machine learning algorithms. Much of the quality improvements achieved by deep learning are directly attributable to the increase in the size of datasets used. Large datasets can act as a regularization technique to prevent a model from overfitting to specific examples.

4.5.6.1 Data Augmentation

One of the easiest ways to improve model performance is to introduce more training data. Practically, this can be expensive, but if the data can be augmented in a meaningful way, this method can be quite useful. This technique can be particularly beneficial to reduce over-fitting to specific anomalies in a dataset.

In the case of images, we can imagine rotating and horizontal flipping as creating a different $(\mathbf{X}, y)$ pair, without having to re-label any data. This, however, would not be the case for handwritten numbers, where a horizontal flip might corrupt the interpretation of the label (think of the 5 and the 2). When incorporating data augmentation, make sure to keep the constraints of the example and target relationship in mind.

4.5.6.2 Bagging

Bagging is another technique commonly used in machine learning. This technique is based on the idea that we can reduce the ability for models to overfit by training multiple models on different portions of the training set. The bagging technique samples from the original dataset (with replacement), creating subtraining sets on which models are trained. The models should learn different features since they are learning different portions of the data, leading to a lower generalization error after combining the results from each model. This strategy tends to be used less often in practice due to the computation time of deep learning models, the large data requirements of deep models, and the introduction of other regularization methods (like Dropout).

4.5.6.3 Adversarial Training

Adversarial examples are examples designed to cause a classifier to misclassify the example. The free parameter space of neural networks means that we can find specific input examples that can take advantage of the specific set of trained parameters within a model [GSS14].

Because of the properties of adversarial examples, we can use the techniques used to create adversarial examples to produce training data for the network to reduce the likelihood of success of a particular attack, as well as improve the robustness of the network by providing training examples that focus on the areas of uncertainty in the parameter space.

4.5.7 Discussion

Broadly speaking, there are typically four pillars in tension when configuring and training neural networks:

- Data availability (and quality)
- Computation speed
- Memory requirements
- Quality

In practice, it is generally a good idea to establish the end goal and work backwards to figure out the boundaries for each of the constraints.

Generally speaking, the initial stage of model selection ensures the model has the capacity to learn reliably. This inevitably leads to overfitting on the training dataset, at which point regularization is introduced to decrease the gap between the training loss and validation loss. In practice, it is usually not necessary (nor feasible) to start from scratch for each new model type or task. However, we believe introducing complexity gradually is best with highly dynamic systems. It is common to start with empirically verified architecture sizes and apply regularization directly from the beginning, however it is best to remove complexity when unexpected situations arise.

4.5.7.1 Computation and Memory Constraints

While numerous advancements made deep learning possible, one of the most significant contributors to the recent growth in adoption is undoubtedly hardware improvements, particularly specialized computer architectures (GPUs). The processing speeds accomplished with GPUs have been among one of the most significant contributing factors to the popularity and practicality of deep learning. Speed advantages through matrix optimizations and the ability to batch compute make the problems of deep learning ideal for GPU architectures. This development made it possible to move beyond shallow architectures to the deep, complex architectures that we see today.

Large datasets and deep learning architectures have led to significant quality improvements; however, the computational cost of deep learning models is typically higher other machine learning methods, which needs to be considered in limited resource environments (such as mobile devices). The model requirements also impact the amount of hyperparameter optimization that can be done. It is unlikely that a full grid search can be performed for models that take days or weeks to train.

The same reasoning applies to memory concerns, with larger models requiring more space. Although, many quantization techniques are being introduced to shrink model sizes, such as quantizing parameters or using hashing parameter values [Jou+16b].

4.6 Unsupervised Deep Learning

So far, we have examined examples of feed-forward neural networks for supervised learning. We will now look at some other architectures that extend neural networks and deep learning to unsupervised tasks by looking at three common unsupervised architectures: Restricted Boltzmann machines (RBM), deep belief networks, and autoencoders. We will build on our current knowledge by analyzing some simple architectures that accomplish tasks other than classification.

As discussed in Chap. 2, unsupervised models learn representations, and these features form data without labels. This is usually a very desirable property because unlabeled data is readily available at large volumes.

4.6.1 Energy-Based Models

Energy-based models (EBMs) gain their inspiration from physics. The free energy in a system can be correlated with the probability of an observation. High energies are associated with a low probability observation and low energies are associated with a high probability observation. Thus, in EBMs, the aim is to learn an energy function that results in low energies for observed examples from the dataset and higher energies for unobserved examples [LeC+06].

For an energy-based model, a probability distribution is defined through the energy function, similar to:

$$p(x) = \frac{e^{-E(x)}}{Z} \tag{4.46}$$

where Z is the normalization constant, commonly referred to as the partition function.

$$Z = \sum_x e^{-E(x)} \tag{4.47}$$

The partition function is intractable for many algorithms, as it requires an exponential sum over all the possible combinations of the input x as defined by the distribution P. However, it can be approximated as we will see in the case of RBMs.

Learning useful features requires learning the weight of our input x and also a hidden portion h. Thus, the probability of an observation x can be written as:

$$P(x) = \sum_h P(x,h) = \sum_h \frac{e^{-E(x)}}{Z} \tag{4.48}$$

Free energy is defined as:

$$F(x) = -\log \sum_h e^{-E(x,h)}. \tag{4.49}$$

and the negative log-likelihood gradient:

$$-\frac{\log p(x)}{\partial \theta} = \frac{\partial F(x)}{\partial \theta} = -\sum_{\hat{x}} p(\hat{x}) \frac{\partial F(x)}{\partial \theta}. \tag{4.50}$$

This function yields a negative log-likelihood gradient with two parts, commonly referred to as the positive phase and negative phase. The positive phase increases the probability of training data.

4.6.2 Restricted Boltzmann Machines

The restricted Boltzmann machine [HS06] is a technique for using log-linear Markov random field (MRF) to model the energy function for unsupervised learning. The RBM is, as the name suggests, a restricted form of the Boltzmann machine [HS83], which provides some useful constraints on the architecture to improve the tractability and convergence of the algorithm. The RBM limits the connectivity of the network, as shown in Fig. 4.21, allowing only visible-hidden connections. This modification allows for more efficient training algorithms, such as gradient-based Contrastive Divergence.

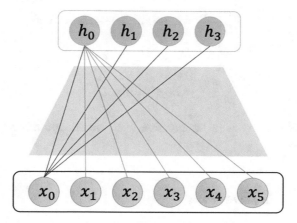

Fig. 4.21: Illustration of an RBM. Note that this can be seen as a fully connected layer as shown previously with only visible-hidden connections in the network. Connections are only shown for one visible neuron and one hidden neuron for the sake of clarity

The energy function of the RBM is defined as:

$$E(\mathbf{x}, \mathbf{h}) = -\mathbf{h}^\mathsf{T}\mathbf{W}\mathbf{x} - \mathbf{c}^\mathsf{T}\mathbf{x} - \mathbf{b}^\mathsf{T}\mathbf{h} \tag{4.51}$$

where W represents the weight matrix connecting the visible units and the hidden units, b is the bias of the hidden unit, and c is the bias of the probability for each x_i.

We then get the probability from the energy function:

$$p(\mathbf{x}, \mathbf{h}) = \frac{e^{-E(\mathbf{x}, \mathbf{h})}}{Z} \tag{4.52}$$

Furthermore, if $x, h \in \{0, 1\}$, we can further reduce the equation to:

$$\begin{aligned} p(h_i = 1|\mathbf{x}) &= \sigma(b_i + \mathbf{W}_i\mathbf{x}) \\ p(x_j = 1|\mathbf{h}) &= \sigma(c_j + \mathbf{W}_j^\mathsf{T}\mathbf{h}) \end{aligned} \tag{4.53}$$

where σ is the sigmoid function.

The free energy formula therefore, becomes:

$$F(\mathbf{x}) = -\mathbf{c}^\mathsf{T}\mathbf{x} - \sum_i \log(1 + e^{b_i + \mathbf{W}_i\mathbf{x}}) \tag{4.54}$$

We can then compute the gradients for the RBM as:

$$-\frac{\partial \log p(\mathbf{x})}{\partial W_{ij}} = E_x[p(h_i|\mathbf{x})x_j] - \sigma(c_i + \mathbf{W}_i\mathbf{x})$$

$$-\frac{\partial \log p(\mathbf{x})}{\partial b_i} = E_x[p(h_i|\mathbf{x})] - \sigma(\mathbf{W}_i\mathbf{x}) \qquad (4.55)$$

$$-\frac{\partial \log p(\mathbf{x})}{\partial c_j} = E_x[p(x_i|\mathbf{h})] - x_j$$

Once we have samples of the function $p(x)$, we can run a Markov chain with Gibbs sampling.

4.6.3 Deep Belief Networks

The effectiveness of RBMs showed that these architectures can be stacked and trained together to create a deep belief network (DBN) [HOT06b]. Each sub-network is trained in isolation, with a hidden layer serving as the visible layer to the next network. The concept of this layer-by-layer training led to one of the first effective approaches to deep learning. A deep belief network is shown in Fig. 4.22.

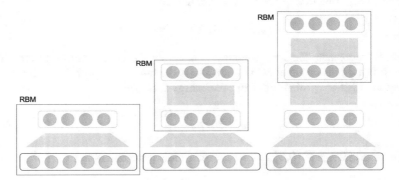

Fig. 4.22: Illustration of a three layer, deep belief network. Each rbm layer is trained individually, starting with the lowest layer

4.6.4 Autoencoders

The autoencoder is an unsupervised deep learning approach to perform dimensionality reduction on each set of data. The aim is to learn a lower dimensional representation of the input data by training one encoder to reduce the dimensionality of the data and another decoder to reproduce the input. The autoencoder is a neural network that is trained to reproduce its input rather than predict a class. The learned representation contains the same information as the input in a smaller, compressed vector, learning what is most important for the reconstruction to minimize a reconstruction error.

The autoencoder is split into two components, the encoder and the decoder. The encoder converts the input, $\mathbf{x}$ into an embedding,[2] $\mathbf{z}$. The decoder maps the encoding, $\mathbf{z}$, back to the original input $\mathbf{x}$. Thus, for a neural network encoder, $\mathrm{Enc}(x)$, and decoder, $\mathrm{Dec}(z)$, the loss $\mathcal{L}$ (mean squared error) is minimized by:

$$\mathbf{z} = \mathrm{Enc}(\mathbf{x})$$
$$\hat{\mathbf{x}} = \mathrm{Dec}(\mathbf{z}) \tag{4.56}$$
$$\mathcal{L}(\mathbf{x}, \hat{\mathbf{x}}) = \|\mathbf{x} - \hat{\mathbf{x}}\|^2.$$

An illustration of the autoencoder architecture is shown in Fig. 4.23.

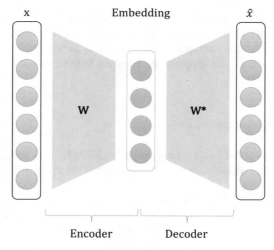

Fig. 4.23: Architecture diagram of an autoencoder with six input values and an embedding of size 4

Training an autoencoder is very similar to other neural network architectures for classification, except for the loss function. Whereas the softmax was previously used to predict a distribution over a set of classes, we now want to produce a real-valued output that can be compared with the input. This is exactly what we accomplished with the MSE objective function that we used previously and is primarily used for autoencoders.[3]

The training of this network is the same as defined in Algorithm 1 with an occasional difference. Many times in autoencoders, it is beneficial to tie the weights

---

[2] This output of the encoder is sometimes referred to as the code, encoding or embedding.

[3] If the task has real-valued inputs between 0 and 1, then Bernoulli cross-entropy is a better choice for the objective function.

together between the encoder and decoder, with the decoder weights $\mathbf{W}^* = \mathbf{W}^T$. In this scenario, the gradients for the weights W will be the sum of two gradients, one from the encoder and one from the decoder.

In general, there are four types of autoencoders:

- Undercomplete autoencoders (standard)
- Sparse
- Denoising autoencoder
- Variational autoencoders (VAE)

with variants of each depending on the application.

4.6.4.1 Undercomplete Autoencoders

An undercomplete autoencoder is the most common type. As shown in Fig. 4.23, the encoder narrows the network to produce an encoding that is smaller than the input. This operates as a learned dimensionality reduction technique. Ideally, the encoder learns to compress the most essential information into the encoding, so the decoder can reconstruct the input.

4.6.4.2 Denoising Autoencoders

A denoising autoencoder takes a noisy input and attempts to decode to a noiseless output. The learned representation will be less sensitive to noise perturbations in the input.

For a noise function[4] $N(x)$, the autoencoder can be described as:

$$
\begin{aligned}
\mathbf{x}' &= N(\mathbf{x}) \\
\mathbf{z} &= \mathrm{Enc}(\mathbf{x}') \\
\hat{\mathbf{x}}' &= \mathrm{Dec}(\mathbf{z}) \\
\mathcal{L}(\mathbf{x}, \hat{\mathbf{x}}') &= \|\mathbf{x} - \hat{\mathbf{x}}'\|^2 .
\end{aligned}
\tag{4.57}
$$

4.6.4.3 Sparse Autoencoders

Sparse autoencoders rely on a minimum threshold of the activations to enforce sparsity in the encoding, rather than relying on a bottleneck of the encoder. In this scenario, the encoder can have larger hidden layers than the input, and sparsity can be achieved by setting a minimum threshold for a neuron, zeroing the outputs for neurons below the threshold.

One way to train a sparse autoencoder is to add a term to the loss such as L_1 to penalize the output activations in the encoder. For a single layer encoder, the loss

---

[4] Note: there are no learned parameters in the noise function presented here.

function can be described as $\mathcal{L}(\mathbf{x}, \hat{\mathbf{x}}) = \|\mathbf{x} - \hat{\mathbf{x}}\|^2 + \lambda \sum_i |z_i|$, where λ sets the weight of the sparsity.

4.6.4.4 Variational Autoencoders

Variational autoencoders describe the latent space in terms of probability distributions. The encoding that has been learned by autoencoders so far describes a sample drawn from some latent space, determined by the encoder. Instead of each value of the encoding being represented by a single value as the other autoencoders have done so far, the variational autoencoder learns to represent the encoding as latent distributions. The parameters are typically learned with respect to the Gaussian distribution in that two parameters must be learned: the mean, μ, and the standard deviation, σ. The decoder is trained on samples, referred to as "sampled latent vectors," drawn from a random distribution parameterized by the learned μ and σ values. A diagram of a VAE is shown in Fig. 4.24.

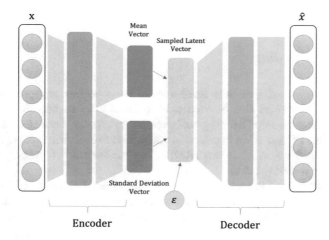

Fig. 4.24: The variational autoencoder learns a vector of means, μ, and a vector of standard deviations, σ. The sampled latent vector, $\mathbf{z}$, is computed by $\mathbf{z} = \mu + \sigma \circ \varepsilon$ where ε is sampled from a normal distribution, $\mathcal{N}(0, 1)$

A problem arises when trying to backpropagate through a stochastic operation of sampling from the Gaussian distribution. The computation is in the path of the forward propagation, and the gradient for the sampling must be computed to obtain gradients for the encoder; however, the stochastic operation does not have a well-defined gradient. The reparameterization trick [JGP16] offers a way to rewrite the sampling procedure to make the stochastic element independent of the learned μ and σ parameters. The sampling of latent variable z is changed from:

$$z = \mathcal{N}(\mu, \sigma^2) \tag{4.58}$$

to the reparameterized:

$$z = \mu + \sigma\varepsilon, \tag{4.59}$$

where ε is sampled from a normal distribution, $\mathcal{N}(0,1)$. Now, although ε is still stochastic, μ and σ do not depend on it for backpropagation.

Training the VAE requires optimizing two components in the loss function. The first component is the reconstruction error that we have optimized for normal autoencoders and the second part is the KL-divergence. The KL-divergence loss ensures the learned mean and variance parameters stay close to $\mathcal{N}(0,1)$.

The overall loss is defined as:

$$\mathcal{L}(x,\hat{x}) + \sum_j D_{KL}(q_j(z|x) \| p(z)), \tag{4.60}$$

where D_{KL} is the KL-divergence, $p(z)$ is prior distribution, and $q_j(z|x)$ is the learned distribution.

4.6.5 Sparse Coding

Sparse coding [Mai+10] aims to learn a set of basis vectors to represent the data. These basis vectors can then be used to form linear combinations to represent the input x. The technique of learning basis vectors to represent our data is similar to techniques like PCA that we explored in Chap. 2. However, with sparse coding, we instead learn an over-complete set that will allow the learning of a variety of patterns and structures within the data.

Sparse coding itself is not a neural network algorithm, but we can add a penalty to our network to enforce sparsity of an autoencoder that creates a sparse autoencoder. This is merely the addition of an L_1 penalty to the loss function that forces most of weights to be 0.

4.6.6 Generative Adversarial Networks

Generative adversarial networks (GAN) [Goo+14a] are an unsupervised technique that structures the learning procedure like a zero-sum game. The technique uses two neural networks referred to as the generator and the discriminator. The generator provides a generated example to the discriminator network, often drawn from a latent space or distribution. The discriminator must discern whether the provided example is a generated (fake) example or an actual example from dataset/ distribution. An illustration of a GAN is shown in Fig. 4.25.

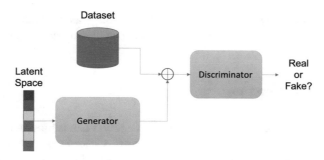

Fig. 4.25: Illustration of a generative adversarial network

At training time both true and generated examples are provided to the discriminator. The discriminator and generator are trained jointly, with the generator's objective to increase the error of the discriminator, and the discriminator's objective to decrease its error. This is related to the minimax decision rule used in statistics and decision theory in zero-sum games. This technique has been used as both a regularization technique and a way to generated synthetic data.

For a generator G and a discriminator D the objective function is given by:

$$\min_{G} \max_{D} \mathbb{E}_{\mathbf{x} \sim \mathbb{P}_r}[\log(D(\mathbf{x}))] + \mathbb{E}_{\widetilde{\mathbf{x}} \sim \mathbb{P}_g}[\log(1 - D(\widetilde{\mathbf{x}}))], \quad (4.61)$$

where $\mathbb{P}_r$ and $\mathbb{P}_g$ represent the real data distribution and generated data distribution, respectively, and $\widetilde{\mathbf{x}} = G(\mathbf{z})$, where $\mathbf{z}$ is drawn from a noise distribution such as the Gaussian distribution.

GANs tend to be used more commonly in computer vision rather than NLP. For example, some amount of Gaussian noise can be added to an image, while still maintaining the overall structure and meaning of the image's content. Sentences are typically mapped to a discrete space instead of a continuous space, as a word is discrete (present or not), where noise cannot be readily applied without changing the meaning. However, a form of character-level language modeling was accomplished in [Gul+17] by using a latent vector to generate 32 one-hot character vectors through a convolutional neural network.

4.7 Framework Considerations

The majority of the architectural and algorithmic considerations that have been discussed are already implemented in deep learning frameworks, with CPU and GPU support. Many of the differences center on the implementation language, target users, and abstractions. The most common implementation language is C++ with a Python interface. The target users can vary broadly and with that variation, the decisions on abstractions. A key abstraction is how deep networks are composed.

Early abstractions focused on layers as blocks of computation that could be linked together, while more recent frameworks rely on a computational graph approach.

4.7.1 Layer Abstraction

Earlier, we briefly introduced the concept of the layer abstraction, referring to the linear transformation operation as a "linear layer." Conceptually, we can continue the layer abstraction to include all portions of the neural network, representing the MLP in Fig. 4.8 as three layers with one hidden layer as shown in Fig. 4.26.

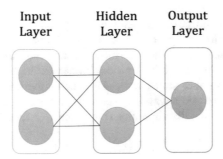

Fig. 4.26: Layer representation of an MLP

Note that although we have represented the inputs, non-linearities, and output as layers this is still a single hidden layer network.

This makes it easier to split a neural network into logical blocks that can be composed together. Early deep learning frameworks took this approach for composing neural networks. One could create any layer by implementing a minimal set of functions, namely the forward propagation step and backward propagation step. The layers are connected to form a neural network.

This abstraction is useful when constructing standard neural networks, with defined behavior and has been a common approach for frameworks. It is reasonably straightforward to reason about the interaction of the layers and make guarantees about the computational requirements. A downside to this approach, as we will see, the layer abstraction becomes difficult when dealing with complex network structures. For example, if we wanted recursive connections in a network, we usually have to implement all recurrent computation in a single layer block (we will explore this more in the Chap. 7).

4.7.2 Computational Graphs

Many frameworks have since moved beyond the layer abstraction to computational graphs. The computational graph approach is similar in concept to abstract symbol trees (AST) in compilers. A dependency graph of inputs and outputs can be represented with symbols in a tree. This allows a compiler to generate assembly instructions linking libraries and functions for an executable model. Data flows through the AST based on the dependencies present in the graph.

In deep learning, a computational graph is a directed graph which defines the order of computations. The nodes of the graph correspond to operations or variables. The inputs of a specific node into the graph are the dependencies present in the computational graph. Subsequently, the backpropagation process can readily be determined by following the operations in the reverse order from which they were computed in the forward propagation step. An example of a neural network computational graph is shown in Fig. 4.27.

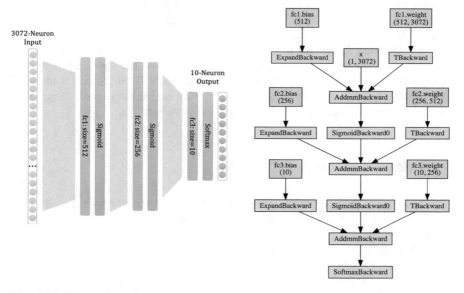

Fig. 4.27: Illustration of a 3-layer neural network with the corresponding backward computation graph. Notice how certain operations can still be combined programmatically for optimization (e.g., Addmm combines the addition and the multiplication into a single operation.) (**a**) Illustration of a 3-layer neural network with sigmoid activation functions and a softmax output for 10 classes. (**b**) Computational graph built from the network shown in (**a**)

4.7.3 Reverse-Mode Automatic Differentiation

Not only is the computational graph approach convenient for complex functions, it can be extended to allow for simpler gradient approximations in a complex neural network. Gradient computation is central in neural networks. One of the most difficult parts of programming deep neural networks is the gradient computation for a specific layer or operation. However, the graph-based approach to deep learning allows for the efficient and automatic computation of gradients in the reverse mode over the computational graph.

Computational graphs make it much easier to leverage reverse-mode automatic differentiation methods. Automatic differentiation (AD) [GW08] is a method used to compute the derivative of a function numerically. AD leverages the concept that, in computers, all mathematical computation is executed as a sequence of basic mathematical operations (addition, subtraction, multiplication, exp, log, sin, cos, etc.). The AD approach leverages the chain rule of differentiation to decompose a function into the differentials for each basic operation in the function. This allows derivatives to be applied automatically and accurately (within a small precision of the theoretical derivative). This approach is typically straightforward to implement achieving much simpler implementations for complex architectures.

The algorithm of reverse mode AD [Spe80] is the select approach to AD for deep learning, because it differentiates a single scalar loss. The forward propagation operation can be seen through the computational graph. This graph can be finely decomposed to the primitive operations, and, during the backward pass, the gradient for the output can be computed with respect to the scalar error.

4.7.4 Static Computational Graphs

Static computational graphs are graphs that have been created with a static view of memory. The static structure allows for the optimization of the graph before it is computed, allowing parallel computation and optimal sequencing of operations. For example, fusing certain operations may reduce the time needed for memory IO or efficient optimization of the computation across a collection of GPUs that may improve the overall performance. This upfront optimization cost is beneficial when there are resource constraints such as in embedded applications, or when the network architecture is relatively rigid, as it repeatedly executes the same graph with little variability in the input.

One of the disadvantages of static computational graphs is that once they are created, they cannot be modified. Any modifications would eliminate potential advantages in the applied optimization strategy.

4.7.5 Dynamic Computational Graphs

Dynamic computational graphs take a different approach where the operations are computed dynamically at run-time. This is useful in situations where you do not know what the computation will be beforehand or where we would like to execute different computations on given data points. A clear example of this is recursive computation in recurrent neural networks that are based on time sequence inputs of often variable-length. Dynamic computation is often desirable in NLP applications where sentence lengths differ and similarly in ASR with the variable lengths of audio files.

Each of these approaches has trade-offs, much like comparing dynamic typed programming languages with statically typed languages. Two current examples of each of these approaches are TensorFlow [Aba+15] and PyTorch [Pas+17]. Tensor-Flow relies on static computational graphs while PyTorch utilizes dynamic computational graphs.

4.8 Case Study

In this section, we will apply the concepts of this chapter to the common Free Spoken Digit Dataset[5] (FSDD). FSDD is a collection of 1500 recordings of spoken digits, 0–9, from 3 speakers. We increase the number of files by performing data augmentation. We discuss this in the next section.

The spoken words are relatively short (most less than 1.5 s). In its raw form, audio is a single series of samples in the time domain, however it is typically more useful to convert it to the frequency domain using an FFT. We convert each audio file to a logMel Spectrogram.

A spectrogram shows the features in a two-dimensional representation with the intensity of a frequency at a point in time. These representations will be discussed more in Chap. 8. A set of logMel spectrogram samples from the FSDD dataset are shown in Fig. 4.28.

4.8.1 Software Tools and Libraries

In these sections, we will use PyTorch for our example code. We find that the code used for PyTorch mixes effortlessly with Python, making it easier to focus on the deep learning concepts rather than the syntax associated with other frameworks. In addition to PyTorch, we also use librosa to perform the audio manipulation and augmentation.

---

[5] https://github.com/Jakobovski/free-spoken-digit-dataset.

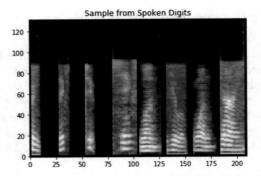

Fig. 4.28: FSDD sample, showing logMel spectrograms for spoken digits

4.8.2 Exploratory Data Analysis (EDA)

The original FSDD dataset contains 1500 examples, with no dedicated validation or testing set. This is a relatively small number of examples, when considering deep learning, so we scale the dataset by using data augmentation. We focus on two types of augmentation time stretching and pitch shifting. Time stretching either increases or decreases the length of the file, while pitch shifting moves the frequencies higher or lower. For time stretching we move the file 25% faster or slower, and with pitch shifting we shift up or down one half-step. Every combination of these is applied to each file, yielding $13,500$ examples, a $9\times$ increase in the amount of data.

```
samples, sample_rate = librosa.load(file_path)
for ts in [0.75,1,1.25]:
    for ps in [-1,0,+1]:
        samples_new = librosa.effects.time_stretch(samples,
        rate=ts)
        y_new = librosa.effects.pitch_shift(samples_new,
        sample_rate, n_steps=ps)
```

The neural networks described so far are only able to take fixed length inputs. The temporal nature of speech makes that difficult, as some of the files are longer than others. In order to alleviate this constraint, we choose to trip all files to a maximum duration of 1.5 s. This allows us to work with a fixed representation for all files. This also helps when batching, as all files in a batch should typically be the same length for computational efficiency.

After increasing the total amount of data and limiting the length, we randomly split into training, validation, and testing sets. 80% of the data is used for training, 10% for validation, and 10% for testing.

We use librosa to obtain the logMel Spectrogram, with 128 mel filters applied (typically < 40 is still fine).

```
1  max_length = 1.5   # Max length in seconds
2  samples, sample_rate = librosa.load(file_path)
3  short_samples = librosa.util.fix_length(samples, sample_rate*
       max_length)
4  melSpectrum = librosa.feature.melspectrogram(short_samples.
       astype(np.float16), sr=sample_rate, n_mels=128)
5  logMelSpectrogram = librosa.power_to_db(melSpectrum, ref=np.
       max)
```

In addition to saving the audio files in a raw, wav format, we also save them as numpy arrays. Loading numpy arrays is much faster during training, especially if we are applying any augmentation. The input data will be the scaled pixel input from the spectrograms. The dimensionality of the input will be $d \times t$, where d is the number of mel features extracted and t, the number of time steps. At loading time, we normalize the logMel spectrogram to be between 0 and 1. Converting the data range from the power decibel range $[-80, 0]$ to be continuous in the range $[0, 1]$ alleviates the need for the network to learn higher weights in the early stages of training. This typically makes training more stable, as there is less internal covariate shift.

In theory, scaling and normalization is not necessarily required in neural networks. Any normalization can be converted by changing the weights and bias associated with the input to achieve the same outcome. However, some gradient descent methods are very sensitive to scaling, standardizing the input data reduces the need for the network to learn extreme values for outliers. This typically improves training times, because it reduces the dependency on the scale of the initial weights.

The next thing that we would like to look for in our data is if there is a class or dataset imbalance. If there is a substantial class imbalance, then we would want to ensure that we have a representative sample across our datasets. Figure 4.29 shows a histogram for our dataset splits. From the histograms, we can see that each class is well represented in each of our sets, and that all classes are relatively balanced in the number of examples per class. This usually is true for academic datasets, but is infrequently the case in practice.

Now that we have a good representation of our data, we will show an example of a supervised classification problem with a neural network as well as an unsupervised learning method using an autoencoder.

4.8.3 Supervised Learning

A supervised classifier first requires us to define an error function that we optimize. We use the cross-entropy loss for our model with a softmax output. In practice, the log of the softmax is used to prevent underflow if the probabilities of one class become very low.

The second step is to define our network architecture. The architecture is often obtained experimentally, considering computational resources, and representational power. In our example, we initially choose a small, 2-hidden layer network with 128 neurons in each layer with a ReLU activation function after each hidden layer. This network is shown in Fig. 4.30.

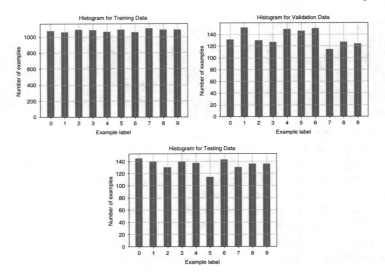

Fig. 4.29: Histograms for the FSDD training, validation, and testing sets. Each example has a spoken label of 0–9. The distribution between the classes is roughly consistent across the datasets

The PyTorch network definition is shown below:

```
import torch.nn as nn

# PyTorch Network Definition
class Model(nn.Module):
    def __init__(self):
        super(Model, self).__init__()
        self.fc1 = nn.Linear(3072, 128)
        self.fc2 = nn.Linear(128, 128)
        self.fc3 = nn.Linear(128, 10)

    def forward(self, x):
        x = x.view((-1, 3072))   # Converts 2D data to 1D
        h = self.fc1(x)
        h = torch.relu(h)

        h = self.fc2(h)
        h = torch.relu(h)

        h = self.fc3(h)
        out = torch.log_softmax(h, dim=1)
        return out
```

In the network definition, we only need to instantiate the learned layers, and the forward function then defines the order of computation that will be executed.

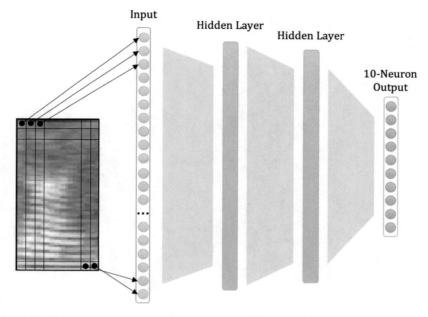

Fig. 4.30: 3-layer neural network for FSDD classification. A ReLU layer is used as the activation function after the first two hidden layers and a log-softmax transformation after the output layer

Linear layers expect the input to be represented in a 1-dimensional form. Thus we include a call to the "view" function, which converts the input from the 2-dimensional input into 1-dimension.[6]

The gradients are paired with each learnable parameter, thus for each step in the forward pass, memory is reserved for the gradient at that step. After passing data through our network we will have an output tensor of size $[n, 1, 1, 10]$. We can then compute our loss by using our error metric, cross entropy. This function takes in two tensors of the same size and computes the scalar loss. The backward function on the loss then computes the gradient of all parameters that contributed to the loss in reverse order using backpropagation. Once the backward pass has been performed we call a single step for our optimizer which takes one step in the direction of the gradient (with respect to our learning rate and other hyperparameters). We repeat this process for the entire dataset for e epochs. The Python code for the training function is shown below.

```
import torch.optim as optim
use_cuda = torch.cuda.is_available()  # Run on GPU if
    available

```

---

[6] Note, PyTorch can still train in mini-batch mode. The view function converts the input tensor into the dimensions $[n, 1, 1, 3072]$, where n is the mini-batch size.

```
 4  # Neural Network Training in PyTorch
 5  model = Model()
 6  model.train()
 7  if use_cuda:
 8      model.cuda()
 9  optimizer = optim.Adam(model.parameters(), lr=0.01)
10  n_epoch = 40
11  for epoch in range(n_epoch):
12      for data, target in train_loader:
13          # Get Samples
14          if use_cuda:
15              data, target = data.cuda(), target.cuda()
16
17          # Clear gradients
18          optimizer.zero_grad()
19
20          # Forward Propagation
21          y_pred = model(data)
22
23          # Error Computation
24          loss = torch.cross_entropy(y_pred, target)
25
26          # Backpropagation
27          loss.backward()
28
29          # Parameter Update
30          optimizer.step()
```

This code snippet is not complete because it does not incorporate validation evaluation during the training process. A more robust example is given in the accompanying notebook. It is left to the reader to experiment with different hyper-parameter
configurations in the exercises. During the training process, we save a copy of the
model with the best validation loss. This model is used to compute the error on the
test set. The training curves and test set results are shown in Fig. 4.31.

We can additionally modify our network to include some of the regularization
techniques and activation functions that we discussed previously, such as batch normalization, dropout, and ReLUs. Incorporating these features is a simple modification of the model architecture described previously. The training graph for this
model is also given in Fig. 4.31.

```
 1  # PyTorch Network Definition
 2  class Model(nn.Module):
 3      def __init__(self):
 4          super(Model, self).__init__()
 5          self.fc1 = nn.Linear(3072, 128)
 6          self.bc1 = nn.BatchNorm1d(128)
 7
 8          self.fc2 = nn.Linear(128, 128)
 9          self.bc2 = nn.BatchNorm1d(128)
10
11          self.fc3 = nn.Linear(128, 10)
```

```
12
13  def forward(self, x):
14      x = x.view((-1, 3072))
15      h = self.fc1(x)
16      h = self.bc1(h)
17      h = torch.relu(h)
18      h = F.dropout(h, p=0.5, training=self.training) #
    Disabled during evaluation
19
20      h = self.fc2(h)
21      h = self.bc2(h)
22      h = torch.relu(h)
23      h = F.dropout(h, p=0.2, training=self.training) #
    Disabled during evaluation
24
25      h = self.fc3(h)
26      out = torch.log_softmax(h, dim=1)
27      return out
```

4.8.4 Unsupervised Learning

For the unsupervised example, we will train a simple autoencoder on the FSDD dataset. This autoencoder learns a low-dimensional encoding of the input data that the decoder is able to produce examples, and the architecture that we will use in this example is shown in Fig. 4.32.

Because this is an unsupervised task, we will use the MSE error function comparing our input with the output of our decoder. The output of our network must be the same size as our input, $d = 3072$, thus the final layer of our network must ensure that the dimensionality matches the input.

The network architecture is a very simple definition with four linear layers learned for each of the encoder and the decoder. The PyTorch autoencoder definition is show below.

```
1  import torch.nn as nn
2  import torch.nn.functional as F  # In place operations for non
       -linearities
3
4  # PyTorch Network Definition
5  class autoencoder(nn.Module):
6      def __init__(self):
7          super(autoencoder, self).__init__()
8
9          self.e_fc1 = nn.Linear(3072, 512)
10         self.e_fc2 = nn.Linear(512, 128)
11         self.e_fc3 = nn.Linear(128, 64)
12         self.e_fc4 = nn.Linear(64,64)
13
14         self.d_fc1 = nn.Linear(64, 64)
15         self.d_fc2 = nn.Linear(64, 128)
16         self.d_fc3 = nn.Linear(128, 512)
17         self.d_fc4 = nn.Linear(512, 3072)
```

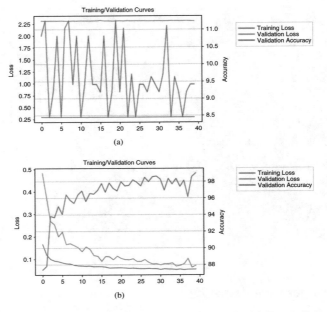

Fig. 4.31: Learning curve for a 40 epoch run with two different architecture definitions. Notice the stability of the regularized architecture in (**b**) compared to (**a**). (**a**) Learning curve for a 40 epoch run of the 2-hidden layer network shown in Fig. 4.30. On the test set, the best performing validation model achieves a loss of 2.3050 an accuracy of 10%, statistically the same as random guessing. (**b**) Learning curve for a 40 epoch run of the 2-hidden layer network shown in Fig. 4.30 with the incorporation of batch normalization, and dropout. On the test set, the best performing validation model achieves a loss of 0.0825 an accuracy of 98%

```
18
19     def forward(self, x):
20         # Encoder
21         h = F.relu(self.e_fc1(x))
22         h = F.relu(self.e_fc2(h))
23         h = F.relu(self.e_fc3(h))
24         h = self.e_fc4(h)
25
26         # Decoder
27         h = F.relu(self.d_fc1(h))
28         h = F.relu(self.d_fc2(h))
29         h = F.relu(self.d_fc3(h))
30         h = self.d_fc4(h)
31         out = F.tanh(h)
32
33         return out
```

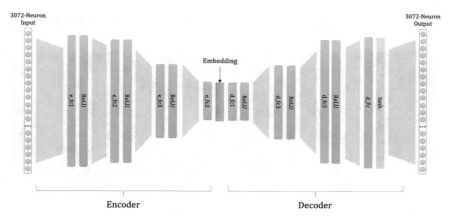

Fig. 4.32: Autoencoder for the FSDD dataset. Note: layer sizes define the output size of that layer

The training algorithm is very similar to the one that was introduced for the classification example. We use the Adam optimizer and add a weight decay term for regularization. Additionally, as we will be using the same size input as output, we will move the 2D to 1D transformation outside of the model. The rest of the algorithm is the same as previously shown. The training algorithm is shown below.

```python
import torch.optim as optim
import torch.nn.functional as F

# Neural Network Training in PyTorch
model = autoencoder()
optimizer = optim.Adam(
    model.parameters(), lr=learning_rate, weight_decay=1e-5)

for epoch in range(n_epoch):
    for data, _ in train_loader:
        # Get samples
        input = data.view(-1,3072)  # We will reuse the
        formatted input as our target

        # Forward Propagation
        output = model(input)

        # Error Computation
        loss = F.mse_loss(output, input)

        # Clear gradients
        optimizer.zero_grad()

        # Backpropagation
        loss.backward()

        # Parameter Update
        optimizer.step()
```

A sample of the decoded output of an input example is shown in Fig. 4.33.

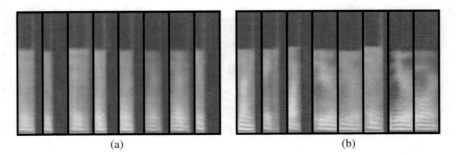

(a)   (b)

Fig. 4.33: Autoencoder output after n epoch(s) on the training data. Notice how the horizontal lines in the spectrogram are starting to form differently for separate inputs. (**a**) Autoencoder reconstruction of its input after 1 epoch. (**b**) Autoencoder reconstruction of its input after 100 epochs

When examining the reconstructed inputs, we notice that they appear to be less sharp than the examples shown in Fig. 4.28. This is mainly due to the MSE loss function. Because it is computing the squared error, it tends to pull all values toward the mean prioritizing the average over specific areas of the input.

4.8.5 Classifying with Unsupervised Features

The RBM learns unsupervised features during the training process. Once these un-supervised features are learned, we can create a low-dimensional, labeled dataset using these features to be used in a supervised classifier. In our example, we train a RBM and then use the learned features as input to a logistic regression classifier.

We can define an RBM with the following code:

```
class RBM(nn.Module):
    def __init__(self, n_vis=3072, n_hin=128, k=5):
        super(RBM, self).__init__()
        self.W = nn.Parameter(torch.randn(n_hin, n_vis)*1e-2)
        self.v_bias = nn.Parameter(torch.zeros(n_vis))
        self.h_bias = nn.Parameter(torch.zeros(n_hin))
        self.k = k

    def sample_from_p(self, p):
        return F.relu(torch.sign(p - Variable(torch.rand(p.size()))))

    def v_to_h(self, v):
        p_h = F.sigmoid(F.linear(v, self.W, self.h_bias))
        sample_h = self.sample_from_p(p_h)
        return p_h, sample_h
```

```
17     def h_to_v(self,h):
18         p_v = F.sigmoid(F.linear(h,self.W.t(),self.v_bias))
19         sample_v = self.sample_from_p(p_v)
20         return p_v,sample_v
21
22     def forward(self,v):
23         pre_h1,h1 = self.v_to_h(v)
24
25         h_ = h1
26         for _ in range(self.k):
27             pre_v_,v_ = self.h_to_v(h_)
28             pre_h_,h_ = self.v_to_h(v_)
29
30         return v,v_
31
32     def free_energy(self,v):
33         vbias_term = v.mv(self.v_bias)
34         wx_b = F.linear(v,self.W,self.h_bias)
35         hidden_term = wx_b.exp().add(1).log().sum(1)
36         return (-hidden_term - vbias_term).mean()
```

We train the model with Adam. The sample code to do this is as follows:

```
1  rbm = RBM(n_vis=3072, n_hin=128, k=1)
2
3  train_op = optim.Adam(rbm.parameters(), 0.01)
4  for epoch in range(epochs):
5      loss_ = []
6      for _, (data,target) in enumerate(train_loader):
7          data = Variable(data.view(-1, 3072))
8          sample_data = data.bernoulli()
9
10         v,v1 = rbm(sample_data)
11         loss = rbm.free_energy(v) - rbm.free_energy(v1)
12         loss_.append(loss.data[0])
13         train_op.zero_grad()
14         loss.backward()
15         train_op.step()
```

After training our RBM features, we can create a logistic regression classifier to classify our examples based on the unsupervised features we have learned.

```
1  from sklearn.linear_model import LogisticRegression
2
3  clf = LogisticRegression()
4  clf.fit(train_features, train_labels)
5  predictions = clf.predict(test_features)
```

The classifier achieves an accuracy of 71.04% on the dataset, 128-dimensional features from the RBM. A confusion matrix for the classifier is shown in Fig. 4.34.

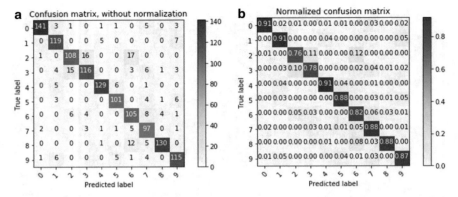

Fig. 4.34: Confusion matrices for a logistic regression classifier with RBM features on the FSDD dataset. (**a**) Confusion matrix for FSDD. (**b**) Normalized confusion matrix for FSDD

4.8.6 Results

Combining the conclusions from previous sections, we compare the methods of classification in Table 4.1.

4.8.7 Exercises for Readers and Practitioners

Some other interesting problems readers and practitioners can try on their own include:

Table 4.1: End-to-end speech recognition performance on FSDD test set. Highlighted result indicates best performance

Approach	Accuracy
2-layer MLP	10.38
2-layer MLP (with regularization)	98.44
RBM + Logistic Regression	71.04

1. What is the effect of training the FSDD classifier with each of the learning rates $[0.001, 0.1, 1.0, 10]$? What is the effect when switching the optimization method?
2. What is the result of a learning rate of 0.1 for the FSDD autoencoder?
3. How would the architecture change if we want to learn a set of sparse features instead of a low-dimensional encoding of the handwritten digits?
4. What effect does batch size have on the learning process? Does it effect the learning rate?

5. What additional data augmentations could be applied to the audio for the system to be more robust?
6. Train a classifier with the trained autoencoder's encoding as the features. How does the accuracy compare to the supervised model?
7. Change the autoencoder to a variational autoencoder. Does it improve the visible quality of the generated output? Vary the inputs to the decoder to understand the features that have been learned.
8. Extend the RBM to create a deep belief network for classifying the FSDD dataset.

References

[Aba+15]   Martín Abadi et al. *TensorFlow: Large-Scale Machine Learning on Heterogeneous Systems*. 2015.

[Bis95]   Christopher M Bishop. "Regularization and complexity control in feed-forward networks". In: (1995).

[BGW18]   Sebastian Bock, Josef Goppold, and Martin Weiß. "An improvement of the convergence proof of the ADAM-Optimizer". In: *arXiv preprint arXiv:1804.10587* (2018).

[Cho+15a]   Anna Choromanska et al. "The loss surfaces of multilayer networks". In: *Artificial Intelligence and Statistics*. 2015, pp. 192–204.

[Cyb89b]   George Cybenko. "Approximation by superpositions of a sigmoidal function". In: *Mathematics of control, signals and systems* 2.4 (1989), pp. 303–314.

[Den+09b]   Jia Deng et al. "Imagenet: A large-scale hierarchical image database". In: *Computer Vision and Pattern Recognition, 2009. CVPR 2009. IEEE Conference on*. IEEE. 2009, pp. 248–255.

[DHS11]   John Duchi, Elad Hazan, and Yoram Singer. "Adaptive subgradient methods for online learning and stochastic optimization". In: *Journal of Machine Learning Research* 12.Jul (2011), pp. 2121–2159.

[GBC16a]   Ian Goodfellow, Yoshua Bengio, and Aaron Courville. *Deep Learning*. MIT Press, 2016.

[GBC16b]   Ian Goodfellow, Yoshua Bengio, and Aaron Courville. "Deep learning (adaptive computation and machine learning series)". In: *Adaptive Computation and Machine Learning series* (2016), p. 800.

[Goo+14a]   Ian Goodfellow et al. "Generative adversarial nets". In: *Advances in neural information processing systems*. 2014, pp. 2672–2680.

[GSS14]   Ian J Goodfellow, Jonathon Shlens, and Christian Szegedy. "Explaining and harnessing adversarial examples". In: *arXiv preprint arXiv:1412.6572* (2014).

[GW08]   Andreas Griewank and Andrea Walther. *Evaluating derivatives: principles and techniques of algorithmic differentiation*. SIAM, 2008.

[Gul+17]   Ishaan Gulrajani et al. "Improved training of Wasserstein GANs". In: *Advances in Neural Information Processing Systems*. 2017, pp. 5767–5777.

[HOT06b]   Geoffrey E Hinton, Simon Osindero, and Yee-Whye Teh. "A fast learning algorithm for deep belief nets". In: *Neural computation* 18.7 (2006), pp. 1527–1554.

[HS06]   Geoffrey E Hinton and Ruslan R Salakhutdinov. "Reducing the dimensionality of data with neural networks". In: *science* 313.5786 (2006), pp. 504–507.

[HS83]   Geoffrey E Hinton and Terrence J Sejnowski. "Optimal perceptual inference". In: *Proceedings of the IEEE conference on Computer Vision and Pattern Recognition*. Citeseer. 1983, pp. 448–453.

[HSW89]   Kurt Hornik, Maxwell Stinchcombe, and Halbert White. "Multilayer feedforward networks are universal approximators". In: *Neural networks* 2.5 (1989), pp. 359–366.

[IS15]   Sergey Ioffe and Christian Szegedy. "Batch Normalization: Accelerating Deep Network Training by Reducing Internal Covariate Shift". In: *CoRR* abs/1502.03167 (2015).

[Iva68]   Aleksey Grigorievtch Ivakhnenko. "The group method of data handling - a rival of the method of stochastic approximation". In: *Soviet Automatic Control* 13.3 (1968), pp. 43–55.

[JGP16]   Eric Jang, Shixiang Gu, and Ben Poole. "Categorical reparameterization with gumbel-softmax". In: *arXiv preprint arXiv:1611.01144* (2016).

[Jou+16b]   Armand Joulin et al. "Fasttext. zip: Compressing text classification models". In: *arXiv preprint arXiv:1612.03651* (2016).

[KB14]   Diederik Kingma and Jimmy Ba. "Adam: A method for stochastic optimization". In: *arXiv preprint arXiv:1412.6980* (2014).

[KSH12c]   Alex Krizhevsky, Ilya Sutskever, and Geoffrey E Hinton. "Imagenet classification with deep convolutional neural networks". In: *Advances in neural information processing systems*. 2012, pp. 1097–1105.

[LeC+06]   Yann LeCun et al. "A tutorial on energy-based learning". In: *Predicting structured data* 1.0 (2006).

[Mai+10]   Julien Mairal et al. "Online learning for matrix factorization and sparse coding". In: *Journal of Machine Learning Research* 11.Jan (2010), pp. 19–60.

[MB05]   Frederic Morin and Yoshua Bengio. "Hierarchical Probabilistic Neural Network Language Model." In: *Aistats*. Vol. 5. Citeseer. 2005, pp. 246–252.

[MK87]   Katta G Murty and Santosh N Kabadi. "Some NP-complete problems in quadratic and nonlinear programming". In: *Mathematical programming* 39.2 (1987), pp. 117–129.

[Pas+17]   Adam Paszke et al. "Automatic differentiation in PyTorch". In: (2017).

[PSM14]   Jeffrey Pennington, Richard Socher, and Christopher Manning. "Glove: Global vectors for word representation". In: *Proceedings of the 2014 conference on empirical methods in natural language processing (EMNLP)*. 2014, pp. 1532–1543.

[RKK18]   Sashank J Reddi, Satyen Kale, and Sanjiv Kumar. "On the convergence of Adam and beyond". In: (2018).

[Rud17a]   Sebastian Ruder. "An Overview of Multi-Task Learning in Deep Neural Networks". In: *CoRR* abs/1706.05098 (2017).

[SLA12]   Jasper Snoek, Hugo Larochelle, and Ryan P Adams. "Practical Bayesian optimization of machine learning algorithms". In: *Advances in neural information processing systems*. 2012, pp. 2951–2959.

[Spe80]   Bert Speelpenning. *Compiling fast partial derivatives of functions given by algorithms*. Tech. rep. Illinois Univ., Urbana (USA). Dept. of Computer Science, 1980.

[Sri+14]   Nitish Srivastava et al. "Dropout: a simple way to prevent neural networks from overfitting." In: *Journal of machine learning research* 15.1 (2014), pp. 1929–1958.

[TH12]   Tijmen Tieleman and Geoffrey Hinton. "Lecture 6.5-rmsprop: Divide the gradient by a running average of its recent magnitude". In: *COURSERA: Neural networks for machine learning* 4.2 (2012), pp. 26–31.

[Zei12]   Matthew D. Zeiler. "ADADELTA: An Adaptive Learning Rate Method". In: *CoRR* abs/1212.5701 (2012).

[Zha+16]   Chiyuan Zhang et al. "Understanding deep learning requires rethinking generalization". In: *CoRR* abs/1611.03530 (2016).

Chapter 5
Distributed Representations

5.1 Introduction

In this chapter, we introduce the notion of **word embeddings** that serve as core representations of text in deep learning approaches. We start with the **distributional hypothesis** and explain how it can be leveraged to form semantic representations of words. We discuss the common distributional semantic models including **word2vec** and **GloVe** and their variants. We address the shortcomings of embedding models and their extension to document and concept representation. Finally, we discuss several applications to natural language processing tasks and present a case study focused on language modeling.

5.2 Distributional Semantics

Distributional semantics is a subfield of natural language processing predicated on the idea that word meaning is derived from its usage. The *distributional hypothesis* states that words used in similar contexts have similar meanings. That is, if two words often occur with the same set of words, then they are semantically similar in meaning. A broader notion is the *statistical semantic hypothesis*, which states that meaning can be derived from statistical patterns of word usage. Distributional semantics serve as the fundamental basis for many recent computational linguistic advances.

5.2.1 Vector Space Model

Vector space models (VSMs) represent a collection of documents as points in a hyperspace, or equivalently, as vectors in a vector space (Fig. 5.1). They are based

© Springer Nature Switzerland AG 2019
U. Kamath et al., *Deep Learning for NLP and Speech Recognition*,
https://doi.org/10.1007/978-3-030-14596-5_5

on the key property that the proximity of points in the hyperspace is a measure of the semantic similarlity of the documents. In other words, documents with similar vector representations imply that they are semantically similar. VSMs have found widespread adoption in information retrieval applications, where a search query is achieved by returning a set of nearby documents sorted by distance. We have already seem VSMs in the form of the bag-of-words term-frequency or TFIDF example back in Chap. 3.

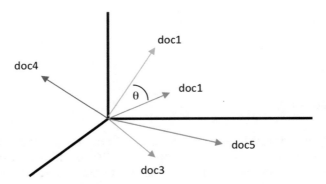

Fig. 5.1: Vector space model representation for documents

5.2.1.1 Curse of Dimensionality

VSMs can suffer from a major drawback if they are based on high-dimensional sparse representations. Here, sparse means that a vector has many dimensions with zero values. This is termed the **curse of dimensionality**. As such, these VSMs require large memory resources and are computationally expensive to implement and use. For instance, a term-frequency based VSM would theoretically require as many dimensions as the number of words in the dictionary of the entire corpus of documents. In practice, it is common to set an upper bound on the number of words and hence, dimensionality of the VSM. Words that are not within the VSM are termed **out-of-vocabulary** (OOV). This is a meaningful gap with most VSMs in that they are unable to attribute semantic meaning to new words that they haven't seen before and are OOV.

The distributional hypothesis says that the meaning of a word is derived from the context in which it is used, and words with similar meaning are used in similar contexts.

5.2.2 Word Representations

One of the earliest use of word representations dates back to 1986. Word vectors explicitly encode linguistic regularities and patterns. Distributional semantic models can be divided into two classes, co-occurrence based and predictive models. Co-occurrence based models must be trained over the entire corpus and capture global dependencies and context, while predictive models capture local dependencies within a (small) context window. The most well-known of these models, word2vec and GloVe, are known as word models since they model word dependencies across a corpus. Both learn high-quality, dense word representations from large amounts of unstructured text data. These word vectors are able to encode linguistic regularities and semantic patterns, which lead to some interesting algebraic properties.

5.2.2.1 Co-occurrence

The distributional hypothesis tells us that co-occurrence of words can reveal much about their semantic proximity and meaning. Computational linguistics leverages this fact and uses the frequency of two words occurring alongside each other within a corpus to identify word relationships. **Pointwise Mutual Information** (PMI) is a commonly used information-theoretic measure of co-occurrence between two words w_1 and w_2:

$$PMI(w_1, w_2) = \log \frac{p(w_1, w_2)}{p(w_1)p(w_2)} \tag{5.1}$$

where $p(w)$ is the probability of the word occurring, and $p(w_1, w_2)$ is joint probability of the two words co-occurring. High values of PMI indicate collocation and co-incidence (and therefore strong association) between the words. It is common to estimate the single and joint probabilities based on word frequency and co-occurrence within the corpus. PMI is a useful measure for word clustering and many other tasks.

5.2.2.2 LSA

Latent semantic analysis (LSA) is a technique that effectively leverages word co-occurrence to identify topics within a set of documents. Specifically, LSA analyzes word associations within a set of documents by forming a document-term matrix (see Fig. 5.2), where each cell can be the frequency of occurrence or TFIDF of a term within a document. As this matrix can be very large (with as many rows as words in the vocabulary of the corpus), a dimensionality reduction technique such as *singular-value decomposition* is applied to find a low-rank approximation. This low-rank space can be used to identify key terms and cluster documents or for information retrieval (as discussed in Chap. 3).

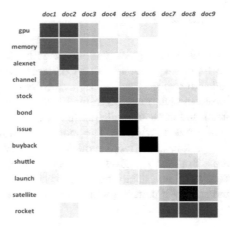

Fig. 5.2: LSA document-term matrix

5.2.3 Neural Language Models

Recall that language models seek to learn the joint probability function of sequences of words. As stated above, this is difficult due to the curse of dimensionality—the sheer size of the vocabulary used in the English language implies that there could be an impossibly huge number of sequences over which we seek to learn. A language model estimates the conditional probability of the next word w_T given all previous words w_t:

$$p(w_T) = \prod_{t=1}^{T} p(w_t|w_1,\ldots,w_{t-1}) \qquad (5.2)$$

Many methods exist for estimating continuous representations of words, including latent semantic analysis (LSA) and latent Dirichlet allocation (LDA). The former fails to preserve linear linguistic regularities while the latter requires huge computational expense for anything beyond small datasets. In recent years, different neural network approaches have been proposed to overcome these issues (Fig. 5.3), which we introduce below. The representations learned by these neural network models are termed **neural embeddings** or simply **embeddings** and will be referenced as such in the rest of this book.

5.2.3.1 Bengio

In 2003, Bengio et al. [Ben+03] presented a neural probabilistic model for learning a distributed representation of words. Instead of sparse, high-dimensional representations, the Bengio model proposed representing words and documents in lower-dimensional continuous vector spaces by using a multilayer neural network

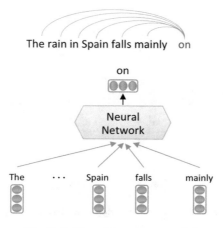

Fig. 5.3: Neural language model

to predict the next word given the previous ones. This network is iteratively trained to maximize the conditional log-likelihood J over the training corpus using back-propagation:

$$J = \frac{1}{T} \sum_{t=1}^{T} \log f(\mathbf{v}(w_t), \mathbf{v}(w_{t-1}), \dots, \mathbf{v}(w_{t-n+1}); \theta) + R(\theta) \tag{5.3}$$

where $\mathbf{v}(w_t)$ is the feature vector for word w_t, f is the mapping function representing the neural network, and $R(\theta)$ is the regularization penalty applied to weights θ of the network. In doing so, the model concurrently associates each word with a distributed word feature vector as well as learning the joint probability function of word sequences in terms of the feature vectors of the words in the sequence. For instance, with a corpus of vocabulary size of 100,000, a one-hot encoded 100,000-dimensional vector representation, the Bengio model can learn a much smaller 300-dimensional continuous vector space representation (Fig. 5.4).

5.2.3.2 Collobert and Weston

In 2008, Collobert and Weston [CW08] applied word vectors to several NLP tasks and showed that word vectors could be trained in an unsupervised manner on a corpus and used to significantly enhance NLP tasks. They used a multilayer neural network trained in an end-to-end fashion. In the process, the first layer in the network learned distributed word representations that are shared across tasks. The output of this word representation layer was passed to downstream architectures that were able to output part-of-speech tags, chunks, named entities, semantic roles, and sentence likelihood. The Collobert and Weston's model is an example of multitask learning enabled through the adoption of dense layer representations.

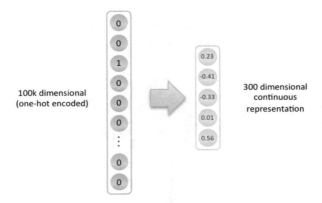

Fig. 5.4: Sparse vs. dense representations

> Neural language models can be trained by stochastic gradient descent and thereby avoid the heavy computational and memory burden of storing co-occurrence matrices in memory.

5.2.4 word2vec

In 2013, Mikolov et al. [Mik+13b] proposed a set of neural architectures could compute continuous representations of words over large datasets. Unlike other neural network architectures for learning word vectors, these architectures were highly computationally efficient, able to handle even billion-word vocabularies, since they do not involve dense matrix multiplications. Furthermore, the high-quality representations learned by these models possessed useful translational properties that provided semantic and syntactic meaning. The proposed architectures consisted of the **continuous bag-of-words** (CBOW) model and the **skip-gram model**. They termed the group of models **word2vec**. They also proposed two methods to train the models based on a hierarchical softmax approach or a negative-sampling approach.

The translational properties of the vectors learned through word2vec models can provide highly useful linguistic and relational similarities. In particular, Mikolov et al. revealed that vector arithmetic can yield high-quality word similarities and analogies. They showed that the vector representation of the word *queen* can be recovered from representations of *king*, *man*, and *woman* by searching for the nearest vector based on cosine distance to the vector sum:

$$\mathbf{v}(queen) \approx \mathbf{v}(king) - \mathbf{v}(man) + \mathbf{v}(woman)$$

Vector operations could reveal both semantic relationships such as:

$$\mathbf{v}(\text{Rome}) \approx \mathbf{v}(\text{Paris}) - \mathbf{v}(\text{France}) + \mathbf{v}(\text{Italy})$$
$$\mathbf{v}(\text{niece}) \approx \mathbf{v}(\text{nephew}) - \mathbf{v}(\text{brother}) + \mathbf{v}(\text{sister})$$
$$\mathbf{v}(\text{Cu}) \approx \mathbf{v}(\text{Zn}) - \mathbf{v}(\text{zinc}) + \mathbf{v}(\text{copper})$$

as well as syntactic relationships such as:

$$\mathbf{v}(\text{biggest}) \approx \mathbf{v}(\text{smallest}) - \mathbf{v}(\text{small}) + \mathbf{v}(\text{big})$$
$$\mathbf{v}(\text{thinking}) \approx \mathbf{v}(\text{read}) - \mathbf{v}(\text{reading}) + \mathbf{v}(\text{think})$$
$$\mathbf{v}(\text{mice}) \approx \mathbf{v}(\text{dollars}) - \mathbf{v}(\text{dollar}) + \mathbf{v}(\text{mouse})$$

In the next sections, we present the intuition behind the CBOW and skip-gram models and their training methodologies. Notably, people have found that CBOW models are better able to capture syntactic relationships, whereas skip-gram models excel at encoding semantic relationships between words.

> Note that word2vec models are fast—they can quickly learn vector representations of much larger corpora than previous methods.

5.2.4.1 CBOW

The **CBOW** architecture is based on a projection layer that is trained to predict a target word given a context window of c words to the left and right side of the target word (Fig. 5.5). The input layer maps each context word through an embedding matrix $\mathbf{W}$ to a dense vector representation of dimension k, and the resulting vectors of the context words are averaged across each dimension to yield a single vector of k dimension. The embedding matrix $\mathbf{W}$ is shared for all context words. Because word order of the context words is irrelevant in the summation, this model is analogous to a bag-of-words model, except that a continuous representation is used. The CBOW model objective seeks to maximize the average log probability:

$$\frac{1}{T} \sum_{t=1}^{T} \sum_{-c < j < c, j \neq 0} \log \left(p \left(w_t | w_{t+j} \right) \right) \tag{5.4}$$

where c is the number of context words to each side of the target word (Fig. 5.6). For the simple CBOW model, the average vector representation from the output of the projection layer is fed into a softmax that predicts over the entire vocabulary of the corpus, using backpropagation to maximize the log probability objective:

$$p(w_t | w_{t+j}) = \frac{\exp \left(\mathbf{v}'_{w_t}{}^{\mathsf{T}} \mathbf{v}_{w_{t+j}} \right)}{\sum_{w=1}^{V} \exp \left(\mathbf{v}'_w{}^{\mathsf{T}} \mathbf{v}_{w_{t+j}} \right)} \tag{5.5}$$

where V is the number of words in the vocabulary. Note that after training, the matrix $\mathbf{W}$ are the learned word embeddings of the model.

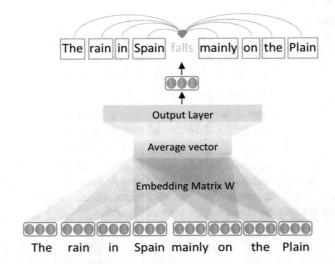

Fig. 5.5: Continuous bag-of-words model (context window = 4)

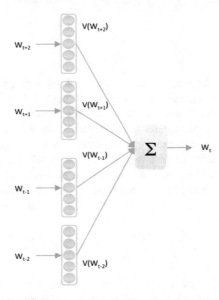

Fig. 5.6: CBOW vector construction (context window = 2)

5.2.4.2 Skip-Gram

Whereas the CBOW model is trained to predict a target word based on the nearby context words, the **skip-gram** model is trained to predict the nearby context words based on the target word (Fig. 5.7). Once again, word order is not considered. For a context size c, the skip-gram model is trained to predict the c words around the target word. The objective of the skip-gram model is to maximize the average log

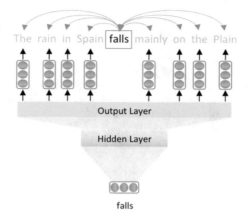

Fig. 5.7: Skip-gram model (context window = 4)

probability:

$$\frac{1}{T} \sum_{t=1}^{T} \sum_{-c < j < c, j \neq 0} \log\left(p\left(w_{t+j} | w_t\right)\right) \tag{5.6}$$

where c is the size of the training context (Fig. 5.8). Higher values of c result in more training examples and thus can lead to a higher accuracy, at the expense of the training time. The most simple skip-gram formulation utilizes the softmax function:

$$p(w_{t+j} | w_t) = \frac{\exp\left(\mathbf{v}'_{w_{t+j}}{}^{\mathsf{T}} \mathbf{v}_{w_t}\right)}{\sum_{w=1}^{V} \exp\left(\mathbf{v}'_{w}{}^{\mathsf{T}} \mathbf{v}_{w_t}\right)} \tag{5.7}$$

where V is the number of words in the vocabulary.

It is interesting to note that shorter training contexts result in vectors that capture syntactic relationships well, while larger context windows better capture semantic relationships. The intuition behind this is that syntactic information is typically dependent on the immediate context and word order, whereas semantic information can be non-local and require larger window sizes.

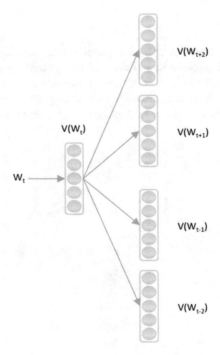

Fig. 5.8: Skipgram vector construction (context window $= 2$)

5.2.4.3 Hierarchical Softmax

The simple versions of CBOW and skip-gram use a full softmax output layer, which can be computationally expensive when the vocabulary is large. A more computationally efficient approximation to the full softmax is the hierarchical softmax, which uses a binary tree representation of the output layer. Each word w can be reached by an appropriate path from the root of the tree:

$$p(w|w_t) = \prod_{j=1}^{L(w)-1} \sigma\left(b\left(n\left(w,j+1\right) = ch\left(n\left(w,j\right)\right)\right)\right) {v'_{n(w,j)}}^{\mathsf{T}} v_{w_t} \qquad (5.8)$$

where

$$\sigma(x) = \frac{1}{1+e^{-x}} \qquad (5.9)$$

$$\sum_{w=1}^{V} p(w|w_t) = 1 \qquad (5.10)$$

In practice, it is common to use a binary Huffman tree, which assigns short codes to the frequent words and results in fast training as it requires only calculating over $\log_2(V)$ words instead of V words for the softmax.

5.2.4.4 Negative Sampling

Mikolov et al. [Mik+13b] proposed an even better alternative to the hierarchical softmax based on **noise contrastive estimation** (NCE). NCE is premised on the notion that a good model should be able to differentiate data from noise via logistic regression. Negative sampling is a simplification of NCE that seeks separate true context words from randomly selected words by maximizing the modified log probability:

$$\log\left(\sigma\left(\mathbf{v}'_{W_O}{}^\mathsf{T}\mathbf{v}_{W_I}\right)\right) + \sum_{i=1}^{k}\mathbb{E}_{w_i\sim p_n(w)}\left[\log\left(\sigma\left(-\mathbf{v}'_{w_i}{}^\mathsf{T}\mathbf{v}_{W_I}\right)\right)\right] \tag{5.11}$$

> When choosing the number of negative samples k, note that word2vec's performance will decrease as this number increases in most cases. In practice, k in the range of 5–20 can be used.

The main difference between negative sampling and NCE is that NCE needs both samples and the numerical probabilities of the noise distribution, while negative sampling uses only samples.

$$\sum_{t=1}^{\mathsf{T}}\sum_{c\in c_t}\ell(s(w_t,w_c)) + \sum_{n\in\mathcal{N}_{t,c}}\ell(-s(w_t,n)) \tag{5.12}$$

$$s(w,c) = \sum_{g\in G_w}\mathbf{z}_g^\mathsf{T}\mathbf{v}_c \tag{5.13}$$

$$s(w_t,w_c) = \mathbf{u}_{w_t}^\mathsf{T}\mathbf{v}_{w_c} \tag{5.14}$$

We have previously noted the need to remove stop words when using count-based methods, as these frequent words can occur at very high rates but convey little semantic information. When training word vectors, they can have a similar disproportionate effect. A common way to deal with this effect is to subsample frequent words. During training, each word w_i is potentially discarded with probability:

$$p(w_i) = 1 - \sqrt{\frac{t}{f(w_i)}} \tag{5.15}$$

Subsampling can considerably speed up training times as well as increase the accuracy of the learned vectors of rare words.

5.2.4.5 Phrase Representations

Previous word representations are limited by their inability to satisfy **composition-ality**—that is, they cannot infer the meaning of phrases from the individual words. Many phrases have a meaning that is not a simple composition of the meanings of its individual words. For example, the meaning of *New England Patriots* is not the sum of the meanings of each word.

To deal with this, one approach is to represent phrases by replacing the words with a single token (e.g., *New_England_Patriots*). This process can be automated using a scoring mechanism:

$$score(w_i, w_j) = \log \frac{count(w_i, w_j)}{count(w_i)count(w_j)} \tag{5.16}$$

such that words are combined and replaced by a single token whenever the score rises above a threshold value. This equation is an approximation to the pointwise-mutual information.

Interestingly, word2vec models and in particular the skip-gram model have shown the vector compositionality—the ability to use simple vector additions can often produce meaningful phrases. Adding the vectors for *Philadelphia* and *Eagles* can yield a vector that is in closest proximity to other sports teams.

5.2.4.6 word2vec CBOW: Forward and Backward Propagation

We will derive equations for forward and backward propagation for CBOW to give the readers insight into the training mechanisms and how the weights are updated. Let a single input word be represented as a one-hot vector given by $\mathbf{x} \in \mathbb{R}^V$ where V is the vocabulary and many such word vectors given by $\{\mathbf{x}_1, \mathbf{x}_2, \cdots, \mathbf{x}_C\}$ of size C form the context. Let the vectors flow into a single hidden layer $\mathbf{h} \in \mathbb{R}^D$, where D is the dimension of the embeddings to be learned through training, with the identify activation function and the input values are averaged across the context words. Let $\mathbf{W} \in \mathbb{R}^{V \times D}$ be the weight matrix that captures weights between input and the hidden layer. Figure 5.9 shows the different layers and the connections as described above.

The hidden layer can be given as:

$$\mathbf{h} = \mathbf{W}^\mathsf{T} \left(\frac{1}{C} \sum_{c=1}^{C} \mathbf{x}_c \right) \tag{5.17}$$

We will represent the $\frac{1}{C} \sum_{c=1}^{C} \mathbf{x}_c$ as the average input vector given by $\bar{\mathbf{x}}$. Thus:

$$\mathbf{h} = \mathbf{W}^\mathsf{T} \bar{\mathbf{x}} \tag{5.18}$$

The hidden layer $\mathbf{h} \in \mathbb{R}^D$ is mapped to a single output layer $\mathbf{u} \in \mathbb{R}^V$ with weights $\mathbf{W}' \in \mathbb{R}^{D \times V}$. This is given by:

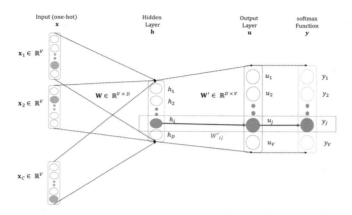

Fig. 5.9: word2vec CBOW with one-hot encoded inputs, single hidden layer, output layer and a softmax layer

$$\mathbf{u} = \mathbf{W}'^{\mathsf{T}} \mathbf{h} \tag{5.19}$$

$$\mathbf{u} = \mathbf{W}'^{\mathsf{T}} \mathbf{W}^{\mathsf{T}} \bar{\mathbf{x}} \tag{5.20}$$

The output layer is then mapped to a softmax output layer $\mathbf{y} \in \mathbb{R}^V$ given by:

$$\mathbf{y} = \text{softmax}(\mathbf{u}) = \text{softmax}(\mathbf{W}'^{\mathsf{T}} \mathbf{W}^{\mathsf{T}} \bar{\mathbf{x}}) \tag{5.21}$$

When we are training the model with target (w_t)-context words$(w_{c,1}, w_{c,2}, \cdots, w_{c,C})$, the output value should match the target in the one-hot encoded representation, i.e. at position $j*$ the output has value 1 and 0 elsewhere. The loss in terms of conditional probability of target word given context words is given by:

$$\mathcal{L} = -\log P(w_t | w_{c,1}, w_{c,2}, \cdots, w_{c,C}) = -\log(y_j*) = -\log(\text{softmax}(u_j*)) \tag{5.22}$$

$$\mathcal{L} = -\log(y_j*) = -\log\left(\frac{\exp(u_j*)}{\sum_i \exp(u_i)}\right) \tag{5.23}$$

$$\mathcal{L} = -u_j* + \log \sum_i \exp(u_i) \tag{5.24}$$

The idea of training through gradient descent as discussed in Chap. 4 is to find values of $\mathbf{W}$ and $\mathbf{W}'$ that minimize the loss function given by Eq. (5.24). The loss function depends on $\mathbf{W}$ and $\mathbf{W}'$ through the output variable $\mathbf{u}$. So to find the values we differentiate the loss function $\mathcal{L}$ with respect to both $\mathbf{W}$ and $\mathbf{W}'$. Since $\mathcal{L} = \mathcal{L}(\mathbf{u}(\mathbf{W}, \mathbf{W}'))$ the two derivatives can be written as:

$$\frac{\partial \mathcal{L}}{\partial W'_{ij}} = \sum_{k=1}^{V} \frac{\partial \mathcal{L}}{\partial u_k} \frac{\partial u_k}{\partial W'_{ij}} \tag{5.25}$$

$$\frac{\partial \mathcal{L}}{\partial W_{ij}} = \sum_{k=1}^{V} \frac{\partial \mathcal{L}}{\partial u_k} \frac{\partial u_k}{\partial W_{ij}} \tag{5.26}$$

Let us consider Eq. (5.25) where W'_{ij} is the connection between hidden layer i and output layer j and since the output is one-hot encoded affects only at value $k = j$ and will be 0 in all other places. Thus the equation reduces to:

$$\frac{\partial \mathcal{L}}{\partial W'_{ij}} = \frac{\partial \mathcal{L}}{\partial u_j} \frac{\partial u_j}{\partial W'_{ij}} \tag{5.27}$$

Now $\frac{\partial \mathcal{L}}{\partial u_j}$ can be written as:

$$\frac{\partial \mathcal{L}}{\partial u_j} = -\delta_{jj*} + y_j = e_j \tag{5.28}$$

where $-\delta_{jj*}$ is the Kronecker delta where the value is 1 if $j = j*$ and 0 elsewhere. This can be represented in the vector form as $\mathbf{e} \in \mathbb{R}^V$.

The other term $\frac{\partial u_j}{\partial W'_{ij}}$ can be written in terms of W_{ij} and average input vector $\bar{x}_k$ as:

$$\frac{\partial u_j}{\partial W'_{ij}} = \sum_{k=1}^{V} W_{ki} \bar{x}_k \tag{5.29}$$

Thus combining:

$$\frac{\partial \mathcal{L}}{\partial W'_{ij}} = (-\delta_{jj*} + y_j) \left(\sum_{k=1}^{V} W_{ki} \bar{x}_k \right) \tag{5.30}$$

This can be written as:

$$\frac{\partial \mathcal{L}}{\partial \mathbf{W'}} = (\mathbf{W}^\mathsf{T} \bar{\mathbf{x}}) \otimes \mathbf{e} \tag{5.31}$$

Next, $\mathbf{u}$ written as the expanded form becomes:

$$u_k = \sum_{d=1}^{D} \sum_{l=1}^{V} W'_{mk} \left(\frac{1}{C} \sum_{c=1}^{C} W_{lm} x_l^c \right) \tag{5.32}$$

For Eq. (5.26) after we fix the input, the output y_j at node j depends on all the connections from that input and thus

$$\frac{\partial \mathcal{L}}{\partial W_{ij}} = \sum_{k=1}^{V} \frac{\partial \mathcal{L}}{\partial u_k} \frac{\partial}{\partial W_{ij}} \left(\frac{1}{C} \sum_{d=1}^{D} \sum_{l=1}^{V} W'_{mk} \sum_{c=1}^{C} W_{lm} x_l^c \right) \tag{5.33}$$

$$\frac{\partial \mathcal{L}}{\partial W_{ij}} = \frac{1}{C} \sum_{k=1}^{V} \sum_{c=1}^{C} (-\delta_{kk*} + y_k) W'_{jk} x_i^c \qquad (5.34)$$

This can be written as:

$$\frac{\partial \mathcal{L}}{\partial \mathbf{W}} = \bar{\mathbf{x}} \otimes (\mathbf{W'e}) \qquad (5.35)$$

Thus the new values $\mathbf{W}_{new}$ and $\mathbf{W'}_{new}$ using a learning rate η is given by:

$$\mathbf{W}_{new} = \mathbf{W}_{old} - \eta \frac{\partial \mathcal{L}}{\partial \mathbf{W}} \qquad (5.36)$$

and

$$\mathbf{W'}_{new} = \mathbf{W'}_{old} - \eta \frac{\partial \mathcal{L}}{\partial \mathbf{W'}} \qquad (5.37)$$

5.2.4.7 word2vec Skip-gram: Forward and Backward Propagation

As we have defined that the skip-gram model is the inverse of the CBOW, i.e., the center word is given in the input and the context words are predicted at the output as shown in Fig. 5.10. We will derive the skip-gram equations on the same lines using a simple network similar to CBOW above.

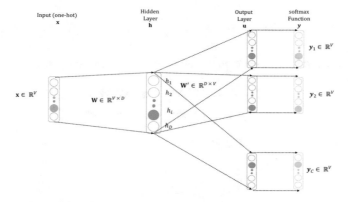

Fig. 5.10: word2vec skip-gram with input word, single hidden layer, generating C context words as output that maps to the softmax function generating the one-hot representation for each

The input $\mathbf{x} \in \mathbb{R}^V$ goes into the hidden layer $\mathbf{h} \in \mathbb{R}^D$ through the weights $\mathbf{W} \in \mathbb{R}^{V \times D}$ and unit activation function. The hidden layer then generates C context word vectors $\mathbf{u}_c \in \mathbb{R}^{D \times V}$ as the output and that can be mapped to a softmax function to generate one-hot representation $\mathbf{y} \in \mathbb{R}^V$ that maps a word in the vocabulary for each embedding output.

$$\mathbf{h} = \mathbf{W}^\mathsf{T} \mathbf{x} \tag{5.38}$$

$$\mathbf{u}_c = \mathbf{W'}^\mathsf{T} \mathbf{h} \tag{5.39}$$

$$\mathbf{u}_c = \mathbf{W'}^\mathsf{T} \mathbf{W}^\mathsf{T} \mathbf{x} \quad c = 1, \cdots, C \tag{5.40}$$

$$\mathbf{y}_c = \mathrm{softmax}(\mathbf{u}_c) = \mathrm{softmax}(\mathbf{W'}^\mathsf{T} \mathbf{W}^\mathsf{T} \mathbf{x}) \quad c = 1, \cdots, C \tag{5.41}$$

The loss function for skip-gram can be written as:

$$\mathcal{L} = -\log P(w_{c,1}, w_{c,2}, \cdots, w_{c,C} | w_i) \tag{5.42}$$

where the word w_i is the input word and $w_{c,1}, w_{c,2}, \cdots, w_{c,C}$ are the output context words.

$$\mathcal{L} = -\log \prod_{c=1}^{C} P(w_{c,i} | w_i) \tag{5.43}$$

Similar to CBOW, this can be further written as:

$$\mathcal{L} = -\log \prod_{c=1}^{C} \left(\frac{\exp(u_{c,j*})}{\sum\limits_{j=1}^{V} \exp(u_{c,j})} \right) \tag{5.44}$$

$$\mathcal{L} = \sum_{c=1}^{C} \exp(u_{c,j*}) + \sum_{c=1}^{C} \log \sum_{j=1}^{V} \exp(u_{c,j}) \tag{5.45}$$

The loss function is dependent on $\mathbf{u}_c$ and each $\mathbf{u}$ dependent on $(\mathbf{W}, \mathbf{W'})$. This can be expressed as:

$$\mathcal{L} = \mathcal{L}(\mathbf{u}_1(\mathbf{W}, \mathbf{W'}) \cdots \mathbf{u}_C(\mathbf{W}, \mathbf{W'})) \tag{5.46}$$

$$\mathcal{L} = \mathcal{L}(u_{1,1}(\mathbf{W}, \mathbf{W'}) \cdots u_{C,V}(\mathbf{W}, \mathbf{W'})) \tag{5.47}$$

The chain-rule applied to skip-gram gives:

$$\frac{\partial \mathcal{L}}{\partial W'_{ij}} = \sum_{k=1}^{V} \sum_{c=1}^{C} \frac{\partial \mathcal{L}}{\partial u_{c,k}} \frac{\partial u_{c,k}}{\partial W'_{ij}} \tag{5.48}$$

$$\frac{\partial \mathcal{L}}{\partial W_{ij}} = \sum_{k=1}^{V} \sum_{c=1}^{C} \frac{\partial \mathcal{L}}{\partial u_{c,k}} \frac{\partial u_{c,k}}{\partial W_{ij}} \tag{5.49}$$

Similar to CBOW we can write:

$$\frac{\partial \mathcal{L}}{\partial W'_{ij}} = \sum_{k=1}^{V} \sum_{c=1}^{C} \frac{\partial \mathcal{L}}{\partial u_{c,k}} \frac{\partial u_{c,k}}{\partial W'_{ij}} = \sum_{c=1}^{C} \frac{\partial \mathcal{L}}{\partial u_{c,j}} \frac{\partial u_{c,j}}{\partial W'_{ij}} \tag{5.50}$$

$$\frac{\partial \mathcal{L}}{\partial W'_{ij}} = \sum_{c=1}^{C} (-\delta_{jj_{c*}} + y_{c,j}) \left(\sum_{k=1}^{V} W_{k,i} x_k \right) \tag{5.51}$$

where $\frac{\partial \mathcal{L}}{\partial u_{c,j}} = -\delta_{jj_{c*}} + y_{c,j} = e_{c,j}$

This can be simplified as:

$$\frac{\partial \mathcal{L}}{\partial \mathbf{W'}} = (\mathbf{W^T x}) \otimes \sum_{c=1}^{C} \mathbf{e}_c \tag{5.52}$$

Similarly we can write Eq. (5.49) as:

$$\frac{\partial \mathcal{L}}{\partial W_{ij}} = \sum_{k=1}^{V} \sum_{c=1}^{C} \frac{\partial \mathcal{L}}{\partial u_{c,k}} \frac{\partial}{\partial W_{ij}} \left(\sum_{m=1}^{D} \sum_{l=1}^{V} W'_{mk} W_{ld} x_l \right) \tag{5.53}$$

$$\frac{\partial \mathcal{L}}{\partial W_{ij}} = \sum_{k=1}^{V} \sum_{c=1}^{C} (-\delta_{kk*_c} + y_{c,k}) W'_{jk} x_i \tag{5.54}$$

Now, representing $-\delta_{kk*_c} + y_{c,k} = e_{ck}$ we can simplify it as

$$\frac{\partial \mathcal{L}}{\partial \mathbf{W}} = \mathbf{x} \otimes \left(\mathbf{W'} \sum_{c=1}^{C} \mathbf{e}_c \right) \tag{5.55}$$

5.2.5 GloVe

The global co-occurrence based models can be the alternative to predictive, local-context window methods like word2vec. Co-occurrence methods are usually very high dimensional and require much storage. When dimensionality reduction methods are used like in LSA, the resulting representations typically perform poorly in capturing semantic word regularities. Furthermore, frequent co-occurrence terms tend to dominate. Predictive methods like word2vec are local-context based and generally perform poorly in capturing the statistics of the corpus. In 2014, Pennington et al. [PSM14] proposed a log-bilinear model that combines both global co-occurrence and shallow window methods. They termed this the **GloVe** model, which is play on the words **Glo**bal and **Ve**ctor. The GloVe model is trained via least squares using the cost function:

$$J = \sum_{i=1,j=1}^{V} f(X_{ij}) \left(\mathbf{u}_i^\mathsf{T} \mathbf{v}_j - \log\left(X_{ij}\right) \right)^2 \tag{5.56}$$

where V is the size of the vocabulary, X_{ij} is the count of times that words i and j co-occur in the corpus (Fig. 5.11), f is a weighting function that acts to reducethe im-

	where	in	the	sacred	river	ran	man	to	sunlit	sea
where	0	2	1	0	0	0	1	2	0	0
in	2	0	0	1	0	0	1	0	1	0
the	1	0	0	4	3	1	5	0	2	1
sacred	0	1	4	0	2	0	1	0	0	1
river	0	0	3	2	0	3	0	1	0	0
ran	0	0	1	0	3	0	3	3	0	0
man	1	1	5	1	0	3	0	1	0	2
to	2	0	0	0	1	3	1	0	1	0
sunlit	0	1	2	0	0	0	0	1	0	2
sea	0	0	1	1	0	0	2	0	2	0

Fig. 5.11: GloVe co-occurrence matrix (context window = 3)

pact of frequent counts, and $\mathbf{u}_i$ and $\mathbf{v}_j$ are word vectors. Typically, a clipped power-law form is assumed for weighting function f:

$$f(X_{ij}) = \begin{cases} \left(\dfrac{X_{ij}}{X_{\max}} \right)^a & \text{if } X_{ij} < X_{\max} \\ 1 & \text{otherwise} \end{cases} \tag{5.57}$$

with $X_{\max}$ is set at training time based on the corpus. Note that the model trains context vectors U and word vectors V separately, and GloVe embeddings are the given by the sum of these two vector representations $\mathbf{U} + \mathbf{V}$. Similar to word2vec, GloVe embeddings can express semantic and syntactic relationships through vector addition and subtraction [SL14]. Furthermore, word embeddings generated by GloVe are superior to word2vec in performance over many NLP tasks, especially in situations where global context is important such as named entity recognition.

GloVe outperforms word2vec when the corpus is small or where insufficient data may be available to capture local context dependencies.

5.2.6 Spectral Word Embeddings

Spectral approaches based on eigen-decomposition are another family of methods to generate dense word embeddings. One of these, canonical correlation analysis, has shown significant potential. This method overcomes many shortcomings of previous methods including scale invariance and providing for better sample complexity of rare words.

Canonical correlation analysis (CCA) is analogous to **principal component analysis** (PCA) for pairs of matrices. Whereas PCA calculates the directions of maximum covariance within a single matrix, CCA calculates the direction of maximum correlation between two matrices. CCA exhibits desirable properties for use in learning word embeddings in that it is scale invariant to linear transformations and provides better sample complexity.

The CCA model learns embeddings by first computing the dominant canonical correlations between target words and a context of c words nearby [DFU11]. The goal is to find vectors ϕ_w and ϕ_c so that linear combinations are maximally correlated:

$$\max_{\phi_w, \phi_c} \frac{\phi_w^{\mathsf{T}} \mathbf{C}_{wc} \phi_c}{\sqrt{\phi_w^{\mathsf{T}} \mathbf{C}_{ww} \phi_w} \sqrt{\phi_c^{\mathsf{T}} \mathbf{C}_{cc} \phi_c}} \tag{5.58}$$

Similar to LSA, this is accomplished by applying SVD to a scaled co-occurrence matrix of counts of words with their context. Thus, the optimization objective can be cast as:

$$\max_{\mathbf{g}_{\phi_w}, \mathbf{g}_{\phi_c}} \mathbf{g}_{\phi_w}^{\mathsf{T}} \mathbf{D}_{wc} \mathbf{g}_{\phi_c} \tag{5.59}$$

where

$$\mathbf{g}_{\phi_w}^{\mathsf{T}} \mathbf{g}_{\phi_w} = \mathbf{I} \tag{5.60}$$

$$\mathbf{g}_{\phi_c}^{\mathsf{T}} \mathbf{g}_{\phi_c} = \mathbf{I} \tag{5.61}$$

$$\mathbf{D}_{wc} = \Lambda_w^{-1/2} \mathbf{V}_w^{\mathsf{T}} \mathbf{C}_{wc} \mathbf{V}_c \Lambda_c^{-1/2} \tag{5.62}$$

An **eigenword** dictionary is created from which word embeddings are extracted. By using explicit left and right contexts, CCA possesses a "multi-view" capability that can allow it to implicitly account for word order in contrast to word2vec or GloVe. This "multi-view" capability can be leveraged to induce context-specific embeddings that can significantly improve certain NLP tasks. This is especially true if a mixture of short and long contexts are applied which can capture both short- and long-range dependencies as necessary in NLP tasks such as word sense disambiguation or entailment.

5.2.7 Multilingual Word Embeddings

It is well known that the distributional hypothesis holds for most human languages. This implies that we can train word embedding models in many languages
[Cou+16, RVS17], and companies such as Facebook and Google have released
pre-trained word2vec and GloVe vectors for up to 157 languages [Gra+18]. These
embedding models are monolingual—they are learned on a single language. Several languages exist with multiple written forms. For instance, Japanese possesses
three distinct writing systems (Hiragana, Katakana, Kanji). Mono-lingual embedding models cannot associate the meaning of a word across different written forms.
The term *word alignment* is used to describe the NLP process by which words are
related together across two written forms across languages (translational relationships) (Fig. 5.12) [Amm+16]. Embedding models have provided a path for deep
learning to make major breakthroughs in word alignment tasks, as we will learn in
Chap. 6 and beyond.

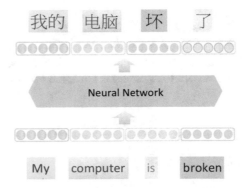

Fig. 5.12: Word alignment

5.3 Limitations of Word Embeddings

Embedding models suffer from a number of well-known limitations. These include
out-of-vocabulary words, antonymy, polysemy, and bias. We explore these in detail
in the next sections.

5.3.1 Out of Vocabulary

The Zipfian distributional nature of the English language is such that there exists
a huge number of infrequent words. Learning representations for these rare words
would require huge amounts of (possibly unavailable) data, as well as potentially

excessive training time or memory resources. Due to practical considerations, a word embedding model will contain only a limited set of the words in the English language. Even a large vocabulary will still have many **out-of-vocabulary** (OOV) words. Unfortunately, many important domain-specific terms tend to occur infrequently and can contribute to the number of OOV words. This is especially true with domain-shifts. As a result, OOV words can have crucial role in the performance NLP tasks.

With models such as word2vec, the common approach is to use a "UNK" representation for words deemed too infrequent to include in the vocabulary. This maps many rare words to an identical vector (zero or random vectors) in the belief that their rarity implies they do not contribute significantly to semantic meaning. Thus, OOV words all provide an identical context during training. Similarly, OOV words at test time are mapped to this representation. This assumption can break down for many reasons, and a number of methods have been proposed to address this shortfall.

Ideally, we would like to be able to somehow predict a vector representation that is semantically similar to either words that are outside our training corpus or that occurred too infrequently in our corpus. Character-based or subword (char-n-gram) embedding models are compositional approaches that attempt to derive a meeting from parts of a word (e.g., roots, suffixes) [Lin+15, LM16, Kim+16]. Subword approaches are especially useful for foreign languages that are rich in morphology such as Arabic or Icelandic [CJF16]. Byte-pair encoding is a character-based, bottom-up method that iteratively groups frequent character pairs and subsequently learning embeddings on the final groups [KB16]. Other methods that leverage external knowledgebases (e.g., WordNet) have also been explored, including the copy mechanism that take into account word position and alignment, but tend to be less resilient to shifts in domain [Gu+16, BCB14].

5.3.2 Antonymy

Another significant limitation is an offshoot of the fundamental principle of distributional similarity from which word models are derived—that words used in similar contexts are similar in meaning. Unfortunately, two words that are **antonyms** of each other often co-occur with the same sets of word contexts:

> I really hate spaghetti on Wednesdays.
> I really love spaghetti on Wednesdays.

While word embedding models can capture synonyms and semantic relationships, they fail notably to distinguish antonyms and overall polarity of words. In other words, without intervention, word embedding models cannot differentiate between synonyms and antonyms and it is common to find antonyms closely co-located within a vector-space model.

An adaptation to word2vec can be made to learn word embeddings that disambiguate polarity by incorporating thesauri information [OMS15]. Consider the skip-gram model that optimizes for an objective function:

$$J(\theta) = \sum_{w \in V} \sum_{c \in V} \{ \#(w,c) \log \sigma \left(sim(w,c) \right)$$
$$+ k\#(w)P_o(c) \log \sigma \left(-sim(w,c) \right) \} \tag{5.63}$$

where the first term are the co-occurrence pairs within a context window and the second term represents negative sampling. Given a set of synonyms $\mathbb{S}_w$ and antonyms $\mathbb{A}_w$ of a word w, we can modify the skip-gram model objective function to the form:

$$J(\theta) = \sum_{w \in V} \sum_{s \in \mathbb{S}_w} \log \sigma(sim(w,s)) + \alpha \sum_{w \in V} \sum_{a \in \mathbb{A}_w} \log \sigma(-sim(w,s))$$
$$+ \sum_{w \in V} \sum_{c \in V} \{ \#(w,c) \log \sigma(sim(w,c)) k \log \sigma \left(-sim(w,c) \right) \} \tag{5.64}$$

This objective can be optimized to learn embeddings that can distinguish synonyms from antonyms. Studies have shown that embeddings learned in this manner to incorporate both distributional and thesauri information perform significantly better in tasks such as question-answering.

5.3.3 Polysemy

In the English language, words can sometimes have several meanings. This is known as **polysemy**. Sometimes these meanings can be very different or complete opposites of each other. Look up the meaning of the word *bad* and you might find up to 46 distinct meanings. As models such as word2vec or GloVe associate each word with a single vector representation, they are unable to deal with homonyms and polysemy. Word sense disambiguation is possible but requires more complex models.

In linguistics, word sense relates to the notion that, in the English language and many other languages, words can take on more than one meaning. **Polysemy** is the concept that a word can have multiple meanings. **Homonymy** is a related concept where two words are spelled the same but have different meanings. For instance, compare the usage of the word *play* in the sentences below:

> *She enjoyed the play very much.*
> *She likes to play cards.*
> *She made a play for the promotion.*

For NLP applications to differentiate between the meanings of a polysemous word, it would require separate representations to be learned for the same word, each associated with a particular meaning [Nee+14]. This is not possible with word2vec or GloVe embedding models since they learn a single embedding for a word. Embedding models must be extended in order to properly handle word sense.

Humans do remarkably well in distinguishing the meaning of a word based on context. In the sentences above, it is relatively easy for us to distinguish the different meanings of the word *play* based on the part-of-speech or surrounding word context. This gives rise to multi-representation embedding models that can leverage surrounding context (cluster-weighted context embeddings) or part-of-speech (sense2vec). We briefly discuss each in the following sections, including other model variants.

5.3.3.1 Clustering-Weighted Context Embeddings

One approach to deal with word sense disambiguation is to start by building an inventory of senses for words within a corpus. Each instance of a word w_i is associated with a representation based on context words surrounding it. These representations, termed *context embeddings*, are then clustered together. The centroid of each cluster is the representation S_{w_i} for the different senses of the word:

$$sense(w_i) = \underset{j:s_j \in S_{w_i}}{\arg\min} d(\mathbf{c}_i, \mathbf{s}_j) \tag{5.65}$$

where d is a distance metric (usually cosine distance). This can be implemented as the multi-sense skip-gram model (Fig. 5.13) where each word is associated with a vector $\mathbf{v}$ with context vectors $\mathbf{c}$ and each sense of the word is associated with a representation μ. Given a target word, a word sense is predicted based on $\mathbf{v}_{context}$:

$$s_t = \underset{k=1,2,...K}{\arg\max} sim(\mu(w_t, k), \mathbf{v}_{context}(\mathbf{c}_t)) \tag{5.66}$$

where $sim(\mathbf{a}, \mathbf{b})$ is a similarity function. The multi-sense word embeddings are learned from a training set by maximizing the objective function:

$$J(\theta) = \sum_{(w_t, c_t) \in D^+} \sum_{c \in c_t} \log P(D = 1 | \mathbf{v}_s(w_t, s_t), \mathbf{v}_g(c))$$
$$+ \sum_{(w_t, c_t') \in D^-} \sum_{c' \in c_t'} \log P(D = 0 | \mathbf{v}_s(w_t, s_t), \mathbf{v}_g(c')) \tag{5.67}$$

5.3.3.2 Sense2vec

Multi-sense word embedding models are more computationally expensive to train and apply in relation to single-sense models [CP18]. **Sense2vec** is a simpler method to achieve world-sense disambiguation that leverages supervised labeling such as part-of-speech [TML15]. It is an efficient method that eliminates the need for clustering during training as seen in context embeddings. For instance, the meanings of the word *plant* are distinct based on its use as a verb or noun:

verb: He planted the tree.
noun: He watered the plant.

The sense2vec model can learn different word senses of this word by combining a single-sense embedding model with POS labels (Fig. 5.14). Given a corpus, sense2vec will create a new corpus for each word for each sense by concatenating a word with its POS label. The new corpus is then trained using word2vec's CBOW or skip-gram to create word embeddings that incorporate word sense (as it relates to their POS usage). Sense2vec has been shown to be effective for many NLP tasks beyond word-sense disambiguation (Fig. 5.15).

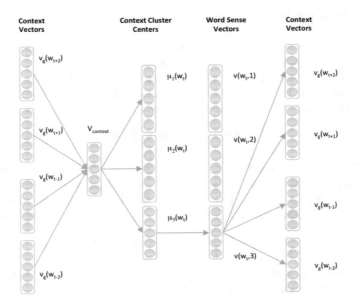

Fig. 5.13: Cluster-weighted context embeddings

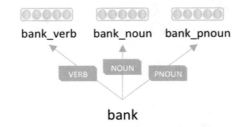

Fig. 5.14: Sense2vec with POS supervised labeling

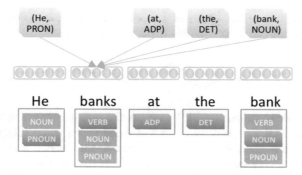

Fig. 5.15: Sense2vec

5.3.4 Biased Embeddings

Recently, we have become aware of the potential **biases** that may implicitly exist within embedding models. Learned word representations are only as good as the data that they were trained on—that is, they will capture the semantic and syntactic context inherent in the training data. For instance, recent studies have revealed racial and gender biases within popular word embedding models such as GloVe and word2vec trained on a broad news corpus:

$$\mathbf{v}(\text{nurse}) \approx \mathbf{v}(\text{doctor}) - \mathbf{v}(\text{father}) + \mathbf{v}(\text{mother})$$
$$\mathbf{v}(\text{Leroy}) \approx \mathbf{v}(\text{Brad}) - \mathbf{v}(\text{happy}) + \mathbf{v}(\text{angry})$$

5.3.5 Other Limitations

A further limitation of word embedding models relates to the batch nature of training and the practicality of **augmenting an existing model** with new data or expanded vocabulary. Doing so requires us to retrain an embedding model with both the original data and new data—the entire data needs to be available and embeddings recomputed. An online learning approach to word embeddings would allow them to be more practical.

5.4 Beyond Word Embeddings

Recent interest in word embedding models has led to practical adaptations that can leverage word compositionality (**subword embeddings**) and address memory constraints (**word2bits**). Others have extended word2vec to learn distributed representations of sentences, documents (**DM** and **DBOW**), and concepts (**RDF2Vec**). Interest has also given rise to Bayesian approaches that map words to latent probability densities (**Gaussian embeddings**) as well as hyperbolic space (**Poincaré embeddings**). We examine these innovations in the next sections.

5.4.1 Subword Embeddings

Methods such as word2vec or GloVe ignore the internal structure of words and associate each word (or word sense) to a separate vector representation. For morphologically rich languages, there may be a significant number of rare word forms such that either a very large vocabulary must be maintained or a significant number of words are treated as out-of-vocabulary (OOV). As previously stated, out-of-vocabulary words can significantly impact performance due to the loss of context from rare words [Bak18]. An approach that can help deal with this limitation is the use of subword embeddings [Boj+16], where vector representations z_g are associated with character n-grams g and words w_i are represented by the sum of the n-gram vectors (Fig. 5.16).

$$\mathbf{w}_i = \sum_{g \in \mathbb{G}_w} z_g \tag{5.68}$$

For instance, the vector for the word *indict* consists of the sum of the vectors for the n-grams $\{ind, ndi, dic, ict, indi, ndic, dict, indic, ndict, indict\}$ when $n \in (3,6)$. Thus, the set of n-grams is a superset of the vocabulary of the corpus (Fig. 5.17). As n-grams are shared across words, this allows for representation of even unseen words since an OOV word will still consist of n-grams that will have representations. Subword embeddings can significantly boost NLP tasks such as language modeling and text classification.

5.4.2 Word Vector Quantization

Even for a small vocabulary, word models can require a significant amount of memory and storage. Consider a 150,000-word vocabulary. A 300-dimensional continuous 64-bit representation of these words can easily occupy over 360 megabytes. It is possible to learn a compact representation by applying quantization to word vectors. In some cases, compression ratios of 8x-16x are possible relative to full-precision word vectors while maintaining comparable performance [Lam18]. Furthermore, the quantization function can act as a regularizer that can improve generalization [Lam18].

Word2Bits is an approach that adapts the word2vec CBOW method by introducing a quantization element to its loss function:

$$J_{quantized}\left(\mathbf{u}_o^{(q)}, \hat{\mathbf{v}}_c^{(q)}\right) = -\log\left(\sigma\left((\mathbf{u}_o^{(q)})^\intercal \hat{\mathbf{v}}_c^{(q)}\right)\right)$$
$$- \sum_{i=1}^{k} \log\left(\sigma\left((-\mathbf{u}_i^{(q)})^\intercal \hat{\mathbf{v}}_c^{(q)}\right)\right) \tag{5.69}$$

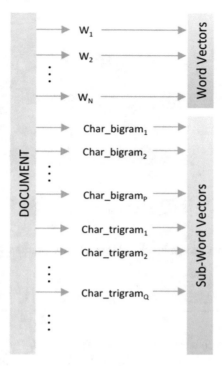

Fig. 5.16: Word and subword vectors

where

$$\mathbf{u}_o^{(q)} = Q_{bitlevel}(\mathbf{u}_o) \tag{5.70}$$

$$\hat{\mathbf{v}}_c^{(q)} = \sum_{-w+i \leq i \leq w+o, i \neq o} Q_{bitlevel}(\mathbf{v}_i) \tag{5.71}$$

Here, w is the context window width, $Q_{bitlevel}$ is the quantization function, $\mathbf{u}_o$ and $\hat{\mathbf{v}}_c$ are the target and context word vectors, respectively, and $\mathbf{u}_o^{(q)}$ and $\hat{\mathbf{v}}_c^{(q)}$ are their quantized equivalents. The Heaviside step function is commonly chosen as the quantization function $Q_{bitlevel}$. Similar to the standard CBOW algorithm, the loss function is optimized over the target $\mathbf{u}_i$ and context $\mathbf{v}_j$ over the corpus. The gradient updates for the target word $\mathbf{u}_o$, negative sampling word $\mathbf{u}_i$, and context word $\mathbf{v}_i$ are given by:

$$\mathbf{u}_o : \frac{\partial J_{quantized}\left(\mathbf{u}_o^{(q)}, \hat{\mathbf{v}}_c^{(q)}\right)}{\partial \mathbf{u}_o} = \frac{\partial J_{quantized}\left(\mathbf{u}_o^{(q)}, \hat{\mathbf{v}}_c^{(q)}\right)}{\partial \mathbf{u}_o^{(q)}} \tag{5.72}$$

$$\mathbf{u}_i : \frac{\partial J_{quantized}\left(\mathbf{u}_o^{(q)}, \hat{\mathbf{v}}_c^{(q)}\right)}{\partial \mathbf{u}_i} = \frac{\partial J_{quantized}\left(\mathbf{u}_o^{(q)}, \hat{\mathbf{v}}_c^{(q)}\right)}{\partial \mathbf{u}_i^{(q)}} \tag{5.73}$$

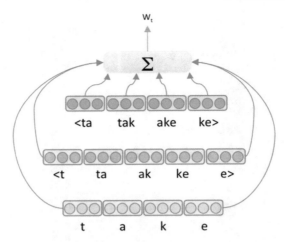

Fig. 5.17: Sub-word embeddings (character n-grams with $n = 1, 2, 3$)

$$\mathbf{v}_i : \frac{\partial J_{quantized}\left(\mathbf{u}_o^{(q)}, \hat{\mathbf{v}}_c^{(q)}\right)}{\partial \mathbf{v}_i} = \frac{\partial J_{quantized}\left(\mathbf{u}_o^{(q)}, \hat{\mathbf{v}}_c^{(q)}\right)}{\partial \mathbf{v}_i^{(q)}} \tag{5.74}$$

The final vector for each word is expressed as $Q_{bitlevel}(\mathbf{u}_i + \mathbf{v}_j)$ whose elements can take on one of $2^{bitlevel}$ values and requires only *bitlevel* bits to represent in comparison with full-precision 32/64 bits. Studies have shown that quantized vectors can perform comparably on word similarity tasks and question answering tasks even with 16x compression.

5.4.3 Sentence Embeddings

While word embedding models capture semantic relationships between words, they lose this ability at the sentence level. Sentence representations are usually expressed the sum of the word vectors of the sentence. This bag-of-words approach has a major flaw in that different sentences can have identical representations as long as the same words are used. To incorporate word order information, people have attempted to use bag-of-n-grams approaches that can capture short order contexts. However, at the sentence level, they are limited by data sparsity and suffer from poor generalization due to high dimensionality.

Le and Mikolov in 2014 [LM14] proposed an unsupervised algorithm to learn useful representations of sentences that capture word order information. Their approach was inspired by Word2Vec for learning word vectors and is commonly known as *doc2vec*. It generates fixed-length feature representations from variable-length pieces of text, making it useful for application to sentences, paragraphs, sections, or entire documents. The key to the approach is to associate every paragraph with a unique *paragraph vector* $\mathbf{u}^i$, which is averaged with the word vectors $\mathbf{w}^i_j$ of the J words in the paragraph to yield a representation of the paragraph $\mathbf{p}^i$:

$$\mathbf{p}^i = \mathbf{u}^i + \sum_{j=1,J} \mathbf{w}^i_j \qquad (5.75)$$

Note that the term *paragraph* can refer to a sentence or document as well. This approach is termed a **distributed memory** model (DM) (Fig. 5.18). The paragraph vector $\mathbf{u}^i$ can be thought of acting as a memory that remembers word order context.

During training, a sliding window of context words $\mathbb{C}$ and the paragraph vector $\mathbf{p}^i$ are used to predict the next word in the paragraph context. Both paragraph vectors and word vectors are trained via backpropagation. While the paragraph vector is unique to each paragraph and shared across all contexts generated from the same paragraph, the word vectors are shared across the entire corpus. It is notable that the DM architecture resembles the CBOW architecture of word2vec, except with the added paragraph context vector. Le and Mikolov also presented an architecture they

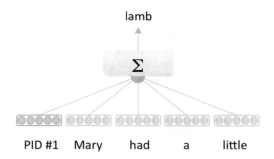

Fig. 5.18: Distributed memory architecture for paragraph vectors

called **distributed bag-of-words** (DBOW) which used only the paragraph context vector to predict the words in the paragraph (Fig. 5.19). This simple model is analogous to the skip-gram version of word2vec, except the paragraph vector is used to predict all the words paragraph instead of using the target word to predict the context words. As in the skip-gram model, DBOW is very computationally and memory efficient. Empirical results have shown that both DM and DBOW outperform bag-of-words and bag-of-n-gram models for text representations. Furthermore, averaging the DM and DBOW vector representations often yields the best performance overall.

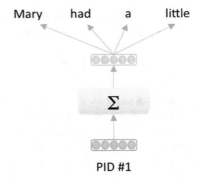

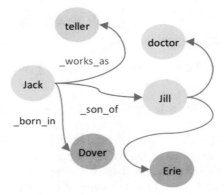

Fig. 5.19: Distributed bag-of-words architecture for paragraph vectors

5.4.4 Concept Embeddings

A key characteristic of embedding models is their ability to capture semantic relationships using simple vector arithmetic. Leveraging this idea, embedding models have recently been developed to map ontological concepts into a vector space [Als+18]. These embeddings can reflect the entity types, semantics, and relationships of a knowledge graph. **RDF2Vec** is an approach for learning embeddings of

Fig. 5.20: Knowledge graph

entities in knowledge graphs (Fig. 5.20). An RDF is a statement with three constituent parts: a subject, predicate, and object. A collection of these can be used to build a knowledge graph. RDF2Vec converts RDF graphs into a set of sequences using graph walks/subtree graph kernels and then applies the word2vec algorithm to map entities to latent representations. In the resulting embedding space, entities that share a background concept are clustered close to each other, such that entities such as "New York" are close to entities such as "city."

TransE was proposed as a general method that aims to specifically represent relationships between entities as translations in an embedding space. The key notion is that, given a set of relationships in the from (head, label, tail), the vector of the tail entity should be close to the vector of the head entity plus the vector of the label:

$$\mathbf{v}_{tail} \approx v\mathbf{v}_{head} + \mathbf{v}_{label} \tag{5.76}$$

TransE is trained in similar manner to negative sampling by minimizing the loss function over a set of triplets S using stochastic gradient descent:

$$J(\theta) = \sum_{(h,l,t)\in S} \sum_{(h,'l,t')\in S'} \max\left(d(h+l,t) - d(h'+l,t'),0\right) + R(\theta) \tag{5.77}$$

where d is a dissimilarity measure and R is a regularizer (typically L_2 norm). Figure 5.21 illustrates vector translations as relationships from the knowledge graph

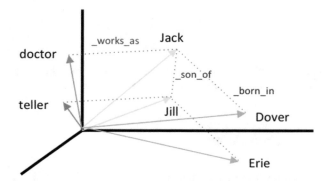

Fig. 5.21: Relationships mapped to vector translations by TransE method

in Fig. 5.18 embedded by the TransE method. For instance, the following translations hold:

$$\mathbf{v}(teller) \approx \mathbf{v}(doctor) — \mathbf{v}(Jill) + \mathbf{v}(Jack)$$
$$\mathbf{v}(Jill) \approx \mathbf{v}(Jack) — textbfv(Dover) + \mathbf{v}(Erie)$$

With the ability to scale to large datasets, TransE and related methods [Bor+13] are useful for both relation extraction and linked prediction as well as NLP tasks.

5.4.5 Retrofitting with Semantic Lexicons

To take advantage of relational information contained in lexical databases such as WordNet or FrameNet, Faruqui et. al. [Far+14] proposed a method to refine word embeddings such that lexically linked words have similar vector representations. This refinement method, commonly called *retrofitting* with semantic lexicons, makes no assumptions on how these vector representations are learned and is applicable across spectral and neural approaches. Given a vocabulary of words $(w_1, w_2, \ldots, w_n)$, a set of semantic relations expressed as an undirected graph with edges (w_i, w_j), and a set of learned word vectors $\hat{\mathbf{q}}_i$ for each w_i, the goal is to infer a new set of word vectors $\mathbf{q}_i$ such that they are close in distance to their counterparts in $\hat{\mathbf{q}}_i$ and to adjacent vertices w_j. With a Euclidean distance measure, this is equivalent to minimizing the objective function:

$$J = \sum_{i=1}^{n} \left[\alpha_i ||\mathbf{q}_i - \hat{\mathbf{q}}_i||^2 + \sum_{(i,j) \in E} \beta_{ij} ||\mathbf{q}_i - \hat{\mathbf{q}}_j||^2 \right] \tag{5.78}$$

where α_i and β_{ij} reflect the relative strength of associations. Retrofitting can be accomplished iteratively with the following update:

$$\mathbf{q}_i = \frac{\sum_{j:(i,j) \in E} \beta_{ij} \mathbf{q}_i + \alpha_i \hat{\mathbf{q}}_i}{\sum_{j:(i,j) \in E} \beta_{ij} + \alpha_i} \tag{5.79}$$

Retrofitting has led to substantial improvements in many lexical semantic evaluation tasks and is useful where external knowledge can be leveraged.

5.4.6 Gaussian Embeddings

Rather than assuming that embedding models map words to point vectors in latent representation space, words can be mapped to continuous probability densities. This gives rise to several interesting advantages, as they can inherently capture uncertainty and asymmetry of semantic relationships between words.

5.4.6.1 Word2Gauss

One such approach is **Word2Gauss**, which maps each word w in the vocabulary D and context word c in the dictionary C to a Gaussian distribution over a latent embedding space. The vectors of this space are termed *word types* and the words observed are word instances. Word2Gauss presents two methods for generating embeddings. The first way is to replace the notion of cosine distance for point vectors in latent density space by inner product E between two Gaussian densities:

$$E(P_i, P_j) = \int_{x \in \mathbb{R}^n} \mathcal{N}(x; \mu_i, \Sigma_i) \mathcal{N}(x; \mu_j, \Sigma_j) dx = \mathcal{N}(0; \mu_i - \mu_j, \Sigma_i + \Sigma_j) \tag{5.80}$$

where $\mathcal{N}(x; \mu_i, \Sigma_i)$ and $\mathcal{N}(x; \mu_j, \Sigma_j)$ are the densities of a target and context word. This is a symmetric measure that is computationally efficient, but it cannot model asymmetry relationships between words. The second, more expressive method, is to model similarity through the notion of KL divergence and train to optimize on a loss function:

$$D_{KL}(\mathcal{N}_j||\mathcal{N}_i) = \int_{x \in \mathbb{R}^n} \mathcal{N}(x; \mu_i, \Sigma_i) \log \frac{\mathcal{N}(x; \mu_j, \Sigma_j)}{\mathcal{N}(x; \mu_i, \Sigma_i)} dx \qquad (5.81)$$

This KL-divergence method enables Gaussian embeddings to incorporate the notion of entailment, as low KL-divergence from w to c implies that c entails w. Furthermore, as KL-divergence is asymmetric, these embeddings can encode asymmetric similarity in the word types.

Word2Gauss has been shown to perform significantly better at asymmetric tasks such as entailment [VM14]. Still, unimodal Gaussian densities do not adequately deal with polysemous words, and computational complexity during training is an important consideration.

5.4.6.2 Bayesian Skip-Gram

A recent approach that builds upon the notion of words embeddings as probability densities takes a generative Bayesian approach. The **Bayesian skip-gram** model (BSG) models each word representation in the form of a Bayesian model generated from prior densities associated with each occurrence of a given word in the corpus. By incorporating context, the BSG model can overcome the polysemy limitation of Word2Gauss. In fact, it can potentially model an infinite set of continuous word senses.

For a target word w and a set of context words $\mathbf{c}$, the BSG model assumes a prior distribution $p_\theta(\mathbf{z}|w)$ and posterior distribution $q_\theta(\mathbf{z}|\mathbf{c}, w)$ in the form of a Gaussian distribution:

$$p_\theta(\mathbf{z}|w) = \mathcal{N}(\mathbf{z}|\mu_w, \Sigma_w) \qquad (5.82)$$

$$q_\theta(\mathbf{z}|\mathbf{c}, w) = \mathcal{N}(\mathbf{z}|\mu_q, \Sigma_q) \qquad (5.83)$$

with diagonal covariance matrices Σ_w and Σ_q and z are latent vectors drawn from the prior (Fig. 5.22). The larger the covariance matrix Σ_q values, the more uncertain the meaning of target word w in context $\mathbf{c}$.

The BSG model aims to maximize the probability $p_\theta(\mathbf{c}|w)$ of words in a context window given a target word w, which is analogous to the skip-gram model. It is trained by taking a target word w, sampling its latent meaning z from the prior, and drawing context words $\mathbf{c}$ from $p_\theta(\mathbf{c}|z)$. The goal is to maximize the log-likelihood function:

$$\log p_\theta(\mathbf{c}|w) = \log \int \prod_{j=1}^{C} p_\theta(\mathbf{c}_j|z) p_\theta(z|w) dz \qquad (5.84)$$

where C is the context window size and c_j are the context words for target word w. For computational simplicity, the BSG model is trained by optimizing on the lower bound of the log-likelihood, given by:

$$J(\theta) = \sum_{(j,k)} \left(D_{KL}\left[q_\phi || \mathcal{N}(\mathbf{z}; \mu_{\tilde{c}_k}, \Sigma_{\tilde{c}_k})\right] - D_{KL}\left[q_\phi || \mathcal{N}(\mathbf{z}; \mu_{c_j}, \Sigma_{c_j})\right] \right)$$
$$- D_{KL}\left[q_\phi || p_\theta(\mathbf{z}|w)\right] \tag{5.85}$$

where the sum is over pairs of positive c_j and negative $\tilde{c}_k$ context words. The re-

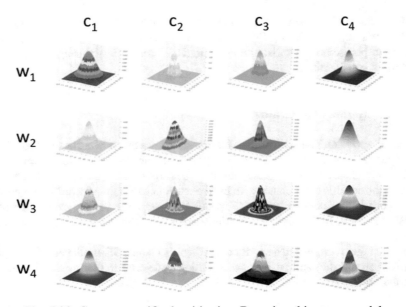

Fig. 5.22: Context specific densities in a Bayesian skip-gram model

sult of training are embeddings associated with the prior p_θ that represent a word type and associated with the posterior q_θ that encode dynamic context. In comparison with Word2Gauss, BSG can provide better context-sensitivity, as in the case of polysemy.

5.4.7 Hyperbolic Embeddings

The ability of embedding models to model complex patterns is constrained by the dimensionality of the embedding space. Furthermore, it has been shown that embedding models do poorly in capturing latent hierarchical relationships. In an effort to overcome these limitations, Nickel and Keila [NK17] proposed **Poincaré**

embeddings as a method to effectively increase the representation capacity while learning latent hierarchy. Their approach is based on learning representations in hyperbolic space instead of Euclidean space. Hyperbolic geometry can effectively and efficiently model hierarchical structures such as trees (Fig. 5.23). In fact, trees can be thought of as instances of discrete hyperbolic spaces. It is notable that the dimensionality needed to represent trees grows linearly in hyperbolic space while quadratically in Euclidean space.

The Poincaré embedding model learns hierarchical representations by mapping words to an n-dimensional unit ball $\mathbb{B}^d = \{\mathbf{x} \in \mathbb{R}^d \mid ||\mathbf{x}|| < 1\}$, where $||x||$ is the Euclidean norm. In hyperbolic space, the distance between two points $\mathbf{u}, \mathbf{v} \in \mathbb{R}^d$ is:

$$d(\mathbf{u}, \mathbf{v}) = \cosh^{-1}\left(1 + 2\frac{||\mathbf{u} - \mathbf{v}||^2}{(1 - ||\mathbf{u}||^2)(1 - ||\mathbf{v}||^2)}\right) \tag{5.86}$$

Note that as $||\mathbf{x}||$ approaches 1, the distance to other points grows exponentially. The

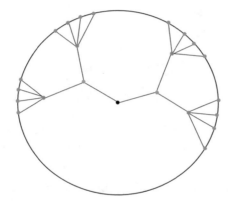

Fig. 5.23: Embedding a tree within a hyperbolic space

notion of straight lines in Euclidean space map to geodesics in $\mathbb{B}^d$ (Fig. 5.24). Such a formulation allows the modeling of trees by placing the root node at/near the origin and leaf nodes at the boundary. During training, the model learn representations Θ:

$$\Theta = \{\theta_i\}_{i=1}^n \quad \text{where } \theta_i \in \mathbb{B}^d, \tag{5.87}$$

by minimizing the loss function $L(\theta)$ using a negative sampling approach on a set of data $\mathbb{D} = \{(\mathbf{u}, \mathbf{v})\}$:

$$L(\Theta) = \sum_{(\mathbf{u}, \mathbf{v}) \in \mathbb{D}} \log \frac{e^{-d(\mathbf{u}, \mathbf{v})}}{\sum_{\mathbf{v}' \in N(\mathbf{u})} e^{-d(\mathbf{u}, \mathbf{v}')}} \tag{5.88}$$

where $N(\mathbf{u}) = \{\mathbf{v}'|(\mathbf{u}, \mathbf{v}') \notin \mathbb{D}\}$ is the set of negative samples. Nickel and Keila formulation required the use of stochastic Riemannian optimization methods to induce embeddings. These methods like Riemannian stochastic gradient descent suffer from several limitations and require an extra projection step to bring the embeddings back into the unit hyperball. Furthermore, they are computationally expensive to train, which make them less feasible for a large text corpus. Recently, Dhingra

Fig. 5.24: Whereas shortest paths are straight lines in Euclidean space, they are curved lines within a hyperbolic space and are called geodesics

et al. generalized **hyperbolic embeddings** by incorporating a parametric approach based on learning encoder functions f_θ that map word sequences to embeddings on the Poincaré ball $\mathbb{B}^d$ [Dhi+18]. The method is predicated on the notion that semantically general concepts occur in a wider range of contexts, while semantically specific concepts occur in a narrower range. By using a simple parameterization of the direction and norm of the hyperbolic embeddings and applying a sigmoid function to the norm, this method allows embeddings to be induced using popular optimization methods with only a modified distance metric and loss function.

5.5 Applications

Word embedding models have led to the improvement of state-of-the-art scores in a wide range of NLP tasks. In many applications, traditional methods have been almost completely replaced by word embedding approaches. Their ability to map variable-length sequences to fix-length representations has opened the door for the application of deep learning to natural language processing. In the next sections, we provide simple examples of how word embeddings can be applied to NLP, while leaving deep learning approaches to later chapters.

5.5.1 Classification

Text classification forms the basis of many important tasks in NLP. Traditional linear-classifier bag-of-word approaches such as Naive Bayes or logistic regression can perform well for text classification. However, they suffer from the inability to generalize to words and phrases unseen in the training data. Embedding models provide the ability to overcome this shortcoming. By leveraging *pretrained embeddings*—learning word representations on a separate large corpus—we can build classifiers that generalize across text.

The **FastText** model proposed by Joulin et al. [Jou+16a] is an example of a text-classification model that leverages word embeddings (Fig. 5.25). The first phase of FastText learns word representations on a large corpus, in effect capturing the semantic relationships on a wide range of vocabulary. These embeddings are then used during the classifier training phase where words of a document are mapped to vectors using these embeddings and these vectors are subsequently averaged together to form a latent representation of the document. These latent representations and their labels form the training set for a softmax or hierarchical softmax classifier. FastText, as its name implies, is computationally efficient, and reportedly able to train on a billion words in 10 min while achieving near state-of-the-art performance [Jou+16a].

5.5.2 Document Clustering

Traditional document clustering based on bag-of-words leads often to excessively high dimensionality and data sparsity. Topic modeling methods such as latent semantic analysis (LSA) and latent Dirichlet allocation (LDA) can be applied to document clustering, but either ignore word co-occurrence or suffer from computational scalability.

We have seen how word embeddings can be used to create latent representations of documents. These representations capture the semantic information within the documents, and it is fairly easy to perform k-means or another conventional clustering method to identify document clusters. Empirical evidence has shown the superiority of using embeddings to perform document clustering [LM14] over bag-of-words or topic model approaches. Use of pre-trained embeddings can enhance the semantic information available to cluster documents.

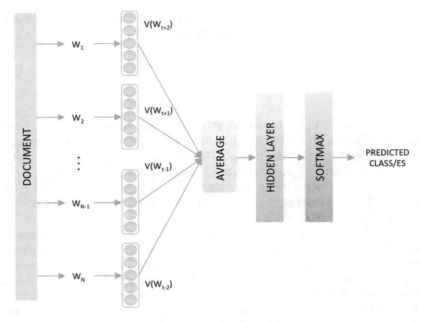

Fig. 5.25: FastText model

5.5.3 Language Modeling

As noted previously, language models are strongly related to the training of embedding models, given that both predict a target word given a set of context words. An n-gram language model predicts a word w_t given the previous words $w_{t-1}, w_{t-2}, \ldots, w_{t-n}$. Training of an n-gram language model is equivalent to maximizing the negative log-likelihood:

$$J(\theta) = \sum_{t=1}^{T} \log p(w_t | w_{t-1}, w_{t-2}, \ldots, w_{t-n+1}) \qquad (5.89)$$

In comparison, training of the CBOW word2vec model is equivalent to maximizing the objective function:

$$J(\theta) = \frac{1}{T} \sum_{t=1}^{T} \log p(w_t | w_{t-n}, \ldots, w_{t-1}, w_{t+1}, \ldots, w_{t+n}) \qquad (5.90)$$

So the language model predicts a target word based on the previous n context words, while CBOW predicts a target word based on the n context words on each side.

Embedding methods excel and language modeling tasks and have led to deep neural network approaches leading the state-of-the-art in performance [MH09].

5.5.4 Text Anomaly Detection

Anomaly detection plays an important part in many applications. Unfortunately, anomaly detection of text is generally difficult to model due to data sparsity and the extremely high dimensionality nature of text. Existing methods on structured data fall into distance-based methods, density based methods, and subspace methods [Kan+17]. However, these methods do not generalize easily to unstructured text data. While matrix factorization and topic modeling approaches can bridge this gap, they can still suffer from high dimensionality and noise as many words tend to be topically irrelevant to the context of a document. Embedding models can map text

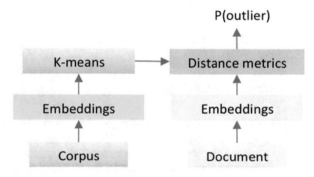

Fig. 5.26: Embedding-based outlier detection

sequences into dense representations that permit the application of distance-based and density-based methods [Che+16]. An example of an embedding approach is illustrated in Fig. 5.26. At training time, the model learns text representations and clusters entities via k-means, such that clusters and dense regions within the latent entity space are identified. At prediction time, a document can be mapped to its latent representation $\mathbf{v}_d$. A distance-based approach could calculate the distance of this representation $\mathbf{v}_d$ to the cluster centroids $\mathbf{c}_j$ identified at training time, and flag the document if the distance exceeds a threshold T (Fig. 5.27):

$$\min_{\mathbf{c}_j} ||\mathbf{v}_d - \mathbf{c}_j|| > T \rightarrow anomaly \tag{5.91}$$

A density-based approach could count the number of entities within a small neighborhood of $\mathbf{v}_d$ and flag the document if the count fell below a threshold T.

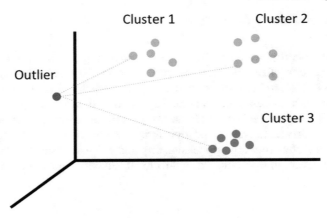

Fig. 5.27: Outlier detection model

5.5.5 *Contextualized Embeddings*

In the past year, a number of new methods leveraging **contextualized embeddings** have been proposed. These are based on the notion that embeddings for words should be based on contexts in which they are used. This context can be the position and presence of surrounding words in the sentence, paragraph, or document. By generatively pre-training contextualized embeddings and language models on massive amounts of data, it became possible to discriminatively fine-tune models on a variety of tasks and achieve state-of-the-art results. This has been commonly referred to as "NLP's ImageNet moment" [HR18].

One of the notable methods is the **Transformer** model, an attention-based stacked encoder–decoder architecture (see Chap. 7) that is pre-trained at scale. Vaswani et al. [Vas+17a] applied this model to the task of machine translation and broke performance records.

Another important method is **ELMo**, short for Embeddings from Language Models, which generates a set of contextualized word representations that effectively capture syntax and semantics as well as polysemy. These representations are actually the internal states of a bidirectional, character-based LSTM language model that is pre-trained on a large external corpus (see Chap. 10).

Building on the power of Transformers, a method has recently been proposed called **BERT**, short for Bidirectional Encoder Representations from Transformers. BERT is a transformer-based, masked language model that is bidirectionally trained to generate deep contextualized word embeddings that capture left-to-right and right-to-left contexts. These embeddings require very little fine-tuning to excel at downstream complex tasks such as entailment or question-answering [Dev+18]. BERT has broken multiple performance records and represents one of the bright breakthroughs in language representations today.

5.6 Case Study

We start off by a detailed look into the word2vec algorithm and examine a python implementation of the skip-gram model with negative sampling. Once the concepts underpinning word2vec are examined, we will use the Gensim package to speed up training time and investigate the translational properties of word embeddings. We will examine GloVe embeddings as an alternative to word2vec. Both methods, however, are unable to handle antonymy, polysemy, and word-sense disambiguation. We consider document clustering by using an embeddings approach. Lastly, we study how an embedding method like sense2vec can better handle word sense disambiguation.

5.6.1 Software Tools and Libraries

In this case study, we will be examining the inner operations of word2vec's skip-gram and negative sampling approach as well as GloVe embeddings with python. We will also leverage the popular nltk, gensim, glove, and spaCy libraries for our analysis. NLTK is a popular open-source toolkit for natural language processing and text analytics. The gensim library is an open-source toolkit for vector space modeling and topic modeling implemented in Python with Cython performance acceleration. The glove library is an efficient open-source implementation of GloVe in python. SpaCy is a fast open-source NLP library written in Python and Cython for part-of-speech tagging and named entity recognition.

For our analysis, we will leverage the Open American National Corpus, which consists of roughly 15 million spoken and written words from a variety of sources. Specifically, we will be using the subcorpus which consists of 4531 Slate magazine articles from 1996 to 2000 (approximately 4.2 million words).

5.6.2 Exploratory Data Analysis

Let's take a look at some basic statistics on this dataset, such as document length and sentence length (Figs. 5.28 and 5.29). By examining word-frequency by looking at the top 1000 terms in this corpus (Fig. 5.30), we see that the top 100 terms are what we typically consider stop-words (Table 5.1). They are common across most sentences and do not capture much, if any, semantic meaning. As we move further down the list, we start to see words that play a more important role in conveying the meaning within a sentence or document.

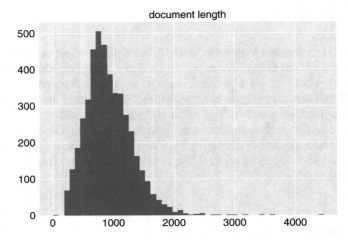

Fig. 5.28: Document length

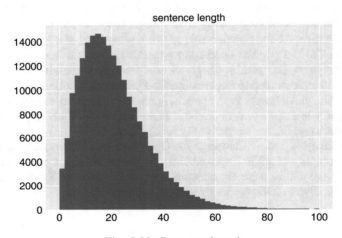

Fig. 5.29: Sentence length

5.6.3 Learning Word Embeddings

Our goal is to train a set of word embeddings for the corpus above. Let's build a skip-gram model with negative sampling, followed by a GloVe model. Before we train either model, we see that there are 77,440 unique words in the preprocessed 4.86 million word corpus.

Table 5.1: Word frequency

	Word	Frequency
0	the	266,007
1	of	115,973
2	–	114,156
3	to	107,951
4	a	100,993
5	and	96,375
6	in	74,561
7	that	64,448
8	is	51,590
9	it	38,175
⋮	⋮	⋮
990	Eyes	500
991	Troops	499
992	Raise	499
993	Pundits	499
994	Calling	498
995	de	498
996	Sports	498
997	Strategy	497
998	Numbers	496
999	Argues	496

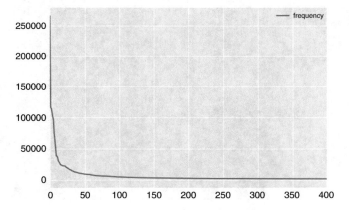

Fig. 5.30: Word frequency histogram

5.6.3.1 Word2Vec

We are now ready to train the neural network of the word2vec model. Let's define our model parameters:

- dim = dimension of the word vectors
- win = context window size (number of tokens)
- start_alpha = starting learning rate
- neg = number of samples for negative sampling
- min_count = minimum mentions to be included in vocabulary

We can reduce the size of this vocabulary by filtering out rare words. If we apply a minimum count threshold of 5 mentions in the corpus, we find that our vocabulary size drops down to 31,599, such that 45,842 words will be considered OOV. We will be mapping all of these words to a special out-of-vocabulary token.

```
truncated = []
truncated.append(VocabWord('<unk>'))
unk_hash = 0

count_unk = 0
for token in vocab_items:
    if token.count < min_count:
        count_unk += 1
        truncated[unk_hash].count += token.count
    else:
        truncated.append(token)

truncated.sort(key=lambda token : token.count, reverse=True)

vocab_hash = {}
for i, token in enumerate(truncated):
    vocab_hash[token.word] = i

vocab_items = truncated
vocab_hash = vocab_hash
vocab_size = len(vocab_items)
print('Unknown vocab size:', count_unk)
print('Truncated vocab size: %d' % vocab_size)
```

5.6.3.2 Negative Sampling

To speed up training, let's create a negative sampling lookup table that we will use during training.

```
power = 0.75
norm = sum([math.pow(t.count, power) for t in vocab_items])

table_size = int(1e8)
```

```
5  table = np.zeros(table_size , dtype=np.int)
6
7  p = 0
8  i = 0
9  for j, unigram in enumerate(vocab_items):
10     p += float(math.pow(unigram.count, power))/norm
11     while i < table_size and float(i) / table_size < p:
12         table[i] = j
13         i += 1
14
15 def sample(table ,count):
16     indices = np.random.randint(low=0, high=len(table), size=
       count)
17     return [table[i] for i in indices]
```

5.6.3.3 Training the Model

We are now ready to train the word2vec model. The approach is to train a two-layer (syn0, syn1) neural network by iterating over the sentences in the corpus and adjusting lawyer weights to maximize the probabilities of context words given a target word (skip-gram) with negative sampling. After completion, the weights of the hidden layer syn0 are the word embeddings that we seek.

```
1  tmp = np.random.uniform(low=-0.5/dim, high=0.5/dim, size=(
       vocab_size , dim))
2  syn0 = np.ctypeslib.as_ctypes(tmp)
3  syn0 = np.array(syn0)
4
5  tmp = np.zeros(shape=(vocab_size , dim))
6  syn1 = np.ctypeslib.as_ctypes(tmp)
7  syn1 = np.array(syn1)
8
9  current_sent = 0
10 truncated_vocabulary = [x.word for x in vocab_items]
11 corpus = df['text'].tolist()
12
13 while current_sent < df.count()[0]:
14     line = corpus[current_sent]
15     sent = [vocab_hash[token] if token in truncated_vocabulary
           else vocab_hash['<unk>']
16             for token in [['<bol>'] + line.split() + ['<eol>']]]
17     for sent_pos , token in enumerate(sent):
18
19         current_win = np.random.randint(low=1, high=win+1)
20         context_start = max(sent_pos --current_win , 0)
21         context_end = min(sent_pos + current_win+1, len(sent))
22         context = sent[context_start:sent_pos] + sent[sent_pos
           +1:context_end]
23
```

```
24      for context_word in context:
25          embed = np.zeros(DIM)
26          classifiers = [(token, 1)] + [(target, 0) for
       target in table.sample(neg)]
27          for target, label in classifiers:
28              z = np.dot(syn0[context_word], syn1[target])
29              p = sigmoid(z)
30              g = alpha * (label──p)
31              embed += g * syn1[target]
32              syn1[target] += g * syn0[context_word]
33          syn0[context_word] += embed
34
35      word_count += 1
36  current_sent += 1
37  if current_sent % 2000 == 0:
38      print("\rReading sentence %d" % current_sent)
39
40 embedding = dict(zip(truncated_vocabulary,syn0))
```

The semantic translation properties of these embeddings are noteworthy. Let's examine the cosine similarity between two similar words (man, woman) and two dissimilar words (candy, social). We would expect the similar words to exhibit higher similarity.

- dist(man, woman) = 0.01258108
- dist(candy, social) = 0.05319491

5.6.3.4 Visualize Embeddings

We can visualize the word embeddings using the T-SNE algorithm to map the embeddings to 2D space. Note that T-SNE is a dimensionality reduction technique that preserves notions of proximity within a vector space (points close together in 2D are close in proximity in higher dimensions). The figure below shows the relationships of a 300-word sample from the vocabulary (Fig. 5.31).

5.6.3.5 Using the Gensim package

The python code above is useful for understanding principles, but is not the fastest to run. The original word2vec package was written in C++ to facilitate rapid training speed over multiple cores. The gensim package provides an API to the word2vec library, as well as several useful methods to examine vectors neighborhoods. Let's see how we can use gensim to train on the sample data corpus. Gensim expects us to provide a set of documents as a list of list of tokens. We will call the simple_preprocess() method of gensim to remove punctuation, special and uppercase characters. With the wrapper API provided by the gensim package, training word2vec is as simple as defining a model and passing the set of training documents.

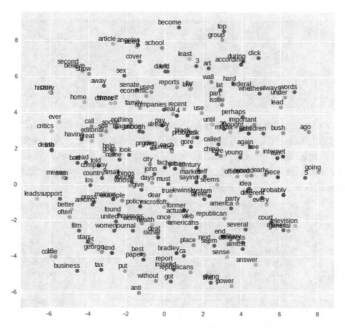

Fig. 5.31: word2vec embeddings visualized using T-SNE

```
documents = [gensim.utils.simple_preprocess(df['text'].iloc[i
    ]) for i in range(len(df))]
model = gensim.models.Word2Vec(documents,
                               size=100,
                               window=10,
                               min_count=2,
                               workers=10)
model.train(documents, total_examples=len(documents), epochs
    =10)
```

5.6.3.6 Similarity

Let's assess the quality of the learned word embeddings by examining word neighborhoods. If we look at the most similar words to "man" or "book'," we find highly similar words in their neighborhoods. So far so good.

```
model.wv.most_similar("man",topn=5)
[('guy', 0.6880463361740112),
 ('woman', 0.6301935315132141),
 ('person', 0.6296881437301636),
 ('soldier', 0.5808842182159424),
 ('someone', 0.5552011728286743)]
```

```
model.wv.most_similar("book",topn=5)
[('books', 0.7232613563537598),
 ('novel', 0.6448987126350403),
 ('biography', 0.6039375066757202),
 ('memoir', 0.6010321378707886),
 ('chapter', 0.5646576881408691)]
```

Let's look at some polysemous words. The similar words to the word "bass" reflect
the music definition of bass. That is, they only capture a single word sense (there
are no words related to the aquatic definition of bass). Similarly, words similar to
"bank" all reflect its financial word sense, but no seashores or riverbeds. This is one
of the major shortcomings of word2vec.

```
model.wv.most_similar("bass",topn=5)
[('guitar', 0.6996911764144897),
 ('solo', 0.6786242723464966),
 ('blazer', 0.6665750741958618),
 ('roars', 0.6658747792243958),
 ('corduroy', 0.6525936126708984)]
```

```
model.wv.most_similar("bank",topn=5)
[('banks', 0.6580432653427124),
 ('bankers', 0.5862468481063843),
 ('imf', 0.5782995223999023),
 ('reserves', 0.5546875),
 ('loans', 0.5457302331924438)]
```

We can examine the semantic translation properties in more detail with some vector algebra. If we start with the word "son" and subtract "man" and add "woman,"
we indeed find that "daughter" is the closest word to the resulting sum. Similarly,
if we invert the operation and start with the word "daughter" and subtract "woman"
and add "man," we find that "son" is closest to the sum. Note that reciprocity is not
guaranteed with word2vec.

```
model.wv.similar_by_vector(model.wv['son']-model.wv['man']
                                              +model.wv['woman'],
   topn=5)
[('daughter', 0.7489624619483948),
 ('sister', 0.7321654558181763),
 ('mother', 0.7243343591690063),
 ('boyfriend', 0.7229076623916626),
 ('lover', 0.7120637893676758)]
```

```
model.wv.similar_by_vector(model.wv['daughter']
                             -model.wv['woman']
                             +model.wv['man'],topn=5)
[('son', 0.7144862413406372),
 ('daughter', 0.6668421030044556),
 ('man', 0.6652499437332153),
 ('grandfather', 0.5896619558334351),
```

```
8  ('father',  0.585667073726654)]
```

We can also see that word2vec captures geographic similarities by taking the word "paris,"subtracting "france" and adding "russia." There resulting sum is close to what we expect—"moscow."

```
1  model.wv.similar_by_vector(model.wv['paris']
2                              −model.wv['france']
3                              +model.wv['russia'],topn=5)
4  [('russia',  0.7788714170455933),
5   ('moscow',  0.6269053220748901),
6   ('brazil',  0.6154285669326782),
7   ('japan',  0.592476487159729),
8   ('gazeta',  0.5799405574798584)]
```

We have previously discussed that word embeddings generated by word2vec are unable to distinguish antonyms, as these words often share the same context words in normal usage and consequentially have learned embeddings close to each other. For instance, the most similar word to "large" is "small," or the most similar word to "hard" is "easy." Antonymy is hard!

```
1  model.wv.most_similar("large",topn=5)
2  [('small',  0.726446270942688),
3   ('enormous',  0.5439934134483337),
4   ('huge',  0.5070887207984924),
5   ('vast',  0.5017688870429993),
6   ('size',  0.48968151211738586)]
```

```
1  model.wv.most_similar("hard",topn=5)
2  [('easy',  0.6564798355102539),
3   ('difficult',  0.6085934638977051),
4   ('tempting',  0.5201482772827148),
5   ('impossible',  0.5099537372589111),
6   ('easier',  0.4868208169937134)]
```

5.6.3.7 GloVe Embeddings

Whereas word2vec captures the local context of words within sentences, GloVe embeddings can additionally account for global context across the corpus. Let's take a deeper dive on how to calculate GloVe embeddings. We begin by building a vocabulary dictionary from the corpus.

```
1  from collections import Counter
2
3  vocab_count = Counter()
4  for line in corpus:
5      tokens = line.strip().split()
```

```
 6    vocab_count.update(tokens)
 7  vocab = {word: (i, freq) for i, (word, freq) in enumerate(
        vocab_count.items())}
```

5.6.3.8 Co-occurrence Matrix

Let's build the word co-occurrence matrix from the corpus. Note that word occurrences go both ways, from the main word to context, and vice versa. For smaller values of the context window, this matrix is expected to be sparse.

```
 1  # Build co-occurrence matrix
 2  from scipy import sparse
 3
 4  min_count = 10
 5  window_size = 5
 6
 7  vocab_size = len(vocab)
 8  id2word = dict((i, word) for word, (i, _) in vocab.items())
 9  occurrence = sparse.lil_matrix((vocab_size, vocab_size),dtype=
        np.float64)
10
11  for i, line in enumerate(corpus):
12      tokens = line.split()
13      token_ids = [vocab[word][0] for word in tokens]
14
15      for center_i, center_id in enumerate(token_ids):
16          context_ids=token_ids[max(0, center_i——window_size) :
        center_i]
17          contexts_len = len(context_ids)
18
19          for left_i, left_id in enumerate(context_ids):
20              distance = contexts_len——left_i
21              increment = 1.0 / float(distance)
22              occurrence[center_id, left_id] += increment
23              occurrence[left_id, center_id] += increment
24      if i % 10000 == 0:
25          print("Processing sentence %d" % i)
26
27  def occur_matrix(vocab, coccurrence, min_count):
28    for i, (row, data) in enumerate(zip(coccurrence.rows,
        coccurrence.data)):
29      if min_count is not None and vocab[id2word[i]][1] <
        min_count:
30              continue
31      for data_idx, j in enumerate(row):
32          if min_count is not None and vocab[id2word[j]][1] <
        min_count
33              :continue
34          yield i, j, data[data_idx]
```

5.6.3.9 GloVe Training

We can now train the embeddings by iterating over the documents (sentences) in the corpus.

```python
from random import shuffle
from math import log
import pickle

iterations = 30
dim = 100
learning_rate = 0.05
x_max = 100
alpha = 0.75

vocab_size = len(vocab)
W = (np.random.rand(vocab_size * 2, dim)---0.5)/float(dim + 1)
biases = (np.random.rand(vocab_size * 2)---0.5)/float(dim + 1)

gradient_squared = np.ones((vocab_size * 2, dim), dtype=np.float64)
gradient_squared_biases = np.ones(vocab_size * 2, dtype=np.float64)

data = [(W[i_main], W[i_context + vocab_size],
        biases[i_main : i_main + 1],
        biases[i_context + vocab_size : i_context + vocab_size + 1],
        gradient_squared[i_main], gradient_squared[i_context + vocab_size],
        gradient_squared_biases[i_main : i_main + 1],
        gradient_squared_biases[i_context + vocab_size
                                : i_context + vocab_size + 1],
        cooccurrence)
        for i_main, i_context, cooccurrence in comatrix]

for i in range(iterations):
    global_cost = 0
    shuffle(data)
    for (v_main, v_context, b_main, b_context, gradsq_W_main, gradsq_W_context,
        gradsq_b_main, gradsq_b_context, cooccurrence) in data:

        weight = (cooccurrence / x_max) ** alpha if cooccurrence < x_max else 1

        cost_inner = (v_main.dot(v_context)
                    + b_main[0] + b_context[0]
                    ---log(cooccurrence))
        cost = weight * (cost_inner ** 2)
        global_cost += 0.5 * cost
```

```
42    grad_main = weight * cost_inner * v_context
43    grad_context = weight * cost_inner * v_main
44    grad_bias_main = weight * cost_inner
45    grad_bias_context = weight * cost_inner
46
47    v_main -= (learning_rate * grad_main / np.sqrt(
      gradsq_W_main))
48    v_context -= (learning_rate * grad_context / np.sqrt(
      gradsq_W_context))
49
50    b_main -= (learning_rate * grad_bias_main / np.sqrt(
      gradsq_b_main))
51    b_context -= (learning_rate * grad_bias_context / np.
      sqrt(
52          gradsq_b_context))
53
54    gradsq_W_main += np.square(grad_main)
55    gradsq_W_context += np.square(grad_context)
56    gradsq_b_main += grad_bias_main ** 2
57    gradsq_b_context += grad_bias_context ** 2
```

The learned weight matrix consists of two sets of vectors, one if the word is in the
main word position and one for the context word position. We will average them to
generate the final GloVe embeddings for each word.

```
1  def merge_vectors(W, merge_fun=lambda m, c: np.mean([m, c],
     axis=0)):
2
3    vocab_size = int(len(W) / 2)
4    for i, row in enumerate(W[:vocab_size]):
5        merged = merge_fun(row, W[i + vocab_size])
6        merged /= np.linalg.norm(merged)
7        W[i, :] = merged
8
9    return W[:vocab_size]
10
11 embedding = merge_vectors(W)
```

5.6.3.10 GloVe Vector Similarity

Let's examine the translational properties of these vectors. We define a simple function that returns the 5 most similar words to the word "man."

```
1  most_similar(embedding, vocab, id2word, 'man,' 5)
2  ('woman', 0.9718018808969603)
3  ('girl', 0.9262655177669397)
4  ('single', 0.9222400016708986)
5  ('dead', 0.9187203648559261)
6  ('young', 0.9081009733127359)
```

Interestingly, the similarity results fall into two categories. Whereas "woman" and "girl" have similar semantic meaning to "man," the words "dead" and "young" do not. But these words do co-occur often, with phrases such as "young man" or "dead man." GloVe embeddings can capture both contexts. We can see this when we visualize the embeddings using T-SNE (Fig. 5.32).

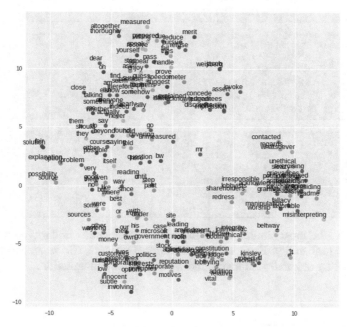

Fig. 5.32: GloVe embeddings visualized using T-SNE

5.6.3.11 Using the Glove Package

While useful, our python implementation is too slow to run with a large corpus. The glove library is a python package that implements the GloVe algorithm efficiently. Let's retrain our embeddings using the glove package.

```
from glove import Corpus, Glove

corpus = Corpus()
corpus.fit(documents, window=5)

glove = Glove(no_components=100, learning_rate=0.05)
glove.fit(corpus.matrix, epochs=30, no_threads=4, verbose=True)
glove.add_dictionary(corpus.dictionary)
```

Let's assess the quality of these embeddings by examining a few words.

```
glove.most_similar('man', number=6)
[('woman', 0.9417155142176431),
 ('young', 0.8541752252243202),
 ('guy', 0.8138920634188781),
 ('person', 0.8044470112897205),
 ('girl', 0.793038798219135)]
```

```
glove.most_similar('nice', number=6)
[('guy', 0.7583150809899194),
 ('very', 0.7071106359169386),
 ('seems', 0.7048211092737807),
 ('terrible', 0.697033427158236),
 ('fun', 0.6898111303194308)]
```

```
glove.most_similar('apple', number=6)
[('industry', 0.6965166116455955),
 ('employee', 0.6724064797672178),
 ('fbi', 0.6280345651329606),
 ('gambling', 0.6276268857034702),
 ('indian', 0.6266591982382662)]
```

Once again, the most similar words exhibit both semantic similarity and high co-occurrence probability. Even with the additional context, GloVe embeddings still lack the ability to handle antonyms and word sense disambiguation.

5.6.4 Document Clustering

The use of word embeddings provides a useful and efficient means for document clustering in comparison with traditional approaches such as LSA or LDA. The simplest approach is a bag-of-words method where a document vector is created by averaging the vectors of each of the words in the document. Let's take our Slate corpus and see what we can find with this approach.

5.6.4.1 Document Vectors

We create a set of document vectors by adding the vectors of each word in the document and dividing by the total number of words.

```
documents=[gensim.utils.simple_preprocess(ndf['text'].iloc[i])
    for i in range(len(ndf))]
corpus = Corpus()
corpus.fit(documents, window=5)
glove = Glove(no_components=100, learning_rate=0.05)
glove.fit(corpus.matrix, epochs=10, no_threads=4, verbose=True)
```

```
 6  glove . add_dictionary ( corpus . dictionary )
 7  print ("Glove embeddings trained .")
 8
 9  doc_vectors = []
10  for doc in documents :
11      vec = np . zeros (( dim ,) )
12      for token in doc :
13          vec += glove . word_vectors [ glove . dictionary [ token ]]
14      if len ( doc ) > 0:
15          vec = vec / len ( doc )
16      doc_vectors . append ( vec )
17
18  print ("Processed documents = ", len ( doc_vectors ) )
```

If we visualize these embeddings using T-SNE, we can see there are several pro-nounced clusters (Fig. 5.33).

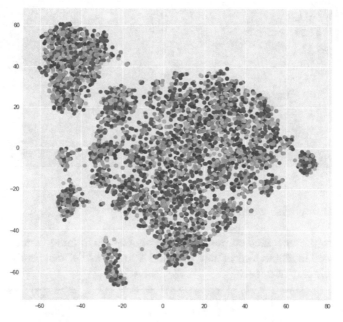

Fig. 5.33: Document vectors visualized using T-SNE

5.6.5 Word Sense Disambiguation

Word sense disambiguation is an important task in computational linguistics. How-ever, word2vec or GloVe embeddings map words to a single embedding vector, and therefore lack the ability to disambiguate between multiple senses of words.

The sense2vec algorithm is an improved approach that can deal with polysemy or antonymy through supervised disambiguation. Moreover, sense2vec is computationally inexpensive and can be implemented as a preprocessing task prior to training a word2vec or GloVe model. To see this, let's apply the sense2vec algorithm to our corpus by leveraging the spaCy library to generate part-of-speech labels that will serve as our supervised disambiguation labels.

5.6.5.1 Supervised Disambiguation Annotations

Let's process the sentences in our corpus using the spaCy NLP annotations. We create a separate corpus where each word is augmented by its part-of-speech label. For instance, the word *he* is mapped to *he_PRON*.

```
import spacy
nlp = spacy.load('en',disable=['parser', 'ner'])
corpus = df['text'].tolist()
print("Number of docs = ",len(corpus))

docs = []
count = 0
for item in corpus:
  docs.append(nlp(item))
  count += 1
  if count % 10000 == 0:
    print("Processed document #",count)

sense_corpus = [[x.text+"_"+x.pos_ for x in y] for y in docs]
```

5.6.5.2 Training with word2vec

With the new preprocessed corpus, we can proceed with training word2vec. We can use this trained model to look at how words like "run" or "lie" can be disambiguated based on their part-of-speech.

```
model.wv.most_similar("run_NOUN",topn=5)
[('runs_NOUN', 0.5418172478675842),
 ('term_NOUN', 0.5085563063621521),
 ('ropy_VERB', 0.5027114152908325),
 ('distance_NOUN', 0.49787676334381104),
 ('sosa_NOUN', 0.4942496120929718)]
```

```
model.wv.most_similar("run_VERB",topn=5)
[('put_VERB', 0.6089274883270264),
 ('work_VERB', 0.599068284034729),
 ('hold_VERB', 0.5984195470809937),
 ('break_VERB', 0.5887631177902222),
 ('get_VERB', 0.5873323082923889)]
```

```
model.wv.most_similar("lie_NOUN",topn=5)
[('truth_NOUN',  0.6057517528533936),
 ('guilt_NOUN',  0.5678446888923645),
 ('sin_NOUN',  0.565475344657898),
 ('perjury_NOUN',  0.5402902364730835),
 ('madness_NOUN',  0.5183135867118835)]
```

```
model.wv.most_similar("lie_VERB",topn=5)
[('talk_VERB',  0.662897527217865),
 ('expose_VERB',  0.64887535572052),
 ('testify_VERB',  0.6263021230697632),
 ('commit_VERB',  0.6155776381492615),
 ('leave_VERB',  0.5946056842803955)]
```

5.6.6 Exercises for Readers and Practitioners

Word embedding algorithms can be extended in a number of interesting ways, and the reader is encouraged to investigate:

1. Training embeddings based on character n-grams, byte-pairs, or other subword approaches.
2. Applying an embeddings approach to cluster named entities.
3. Using embeddings as input features for a classifier.

In subsequent chapters, the reader will realize that embeddings are fundamental to the application of neural networks to text and speech. Furthermore, embeddings enable transfer learning and are an important consideration in any deep learning algorithm.

References

[Als+18]   Faisal Alshargi et al. "Concept2vec: Metrics for Evaluating Quality of Embeddings for Ontological Concepts." In: *CoRR* abs/1803.04488 (2018).

[Amm+16]   Waleed Ammar et al. "Massively Multilingual Word Embeddings." In: *CoRR* abs/1602.01925 (2016).

[BCB14]   Dzmitry Bahdanau, Kyunghyun Cho, and Yoshua Bengio. "Neural machine translation by jointly learning to align and translate". In: *CoRR* abs/1409.0473 (2014).

[Bak18]   Amir Bakarov. "A Survey of Word Embeddings Evaluation Methods". In: *CoRR* abs/1801.09536 (2018).

[Ben+03]   Yoshua Bengio et al. "A neural probabilistic language model". In: *JMLR* (2003), pp. 1137–1155.

[Boj+16]   Piotr Bojanowski et al. "Enriching Word Vectors with Subword Infor-
              mation". In: *CoRR* abs/1607.04606 (2016).

[Bor+13]   Antoine Bordes et al. "Translating Embeddings for Modeling Mul-
              tirelational Data." In: *NIPS*. 2013, pp. 2787–2795.

[CP18]   José Camacho-Collados and Mohammad Taher Pilehvar. "From Word
              to Sense Embeddings: A Survey on Vector Representations of Mean-
              ing". In: *CoRR* abs/1805.04032 (2018).

[Che+16]   Ting Chen et al. "Entity Embedding-Based Anomaly Detection for
              Heterogeneous Categorical Events." In: *IJCAI*. IJCAI/AAAI Press,
              2016, pp. 1396–1403.

[CW08]   Ronan Collobert and Jason Weston. "A Unified Architecture for Nat-
              ural Language Processing: Deep Neural Networks with Multitask
              Learning". In: *Proceedings of the 25th International Conference on
              Machine Learning*. ACM, 2008, pp. 160–167.

[CJF16]   Marta R. Costa-Jussà and José A. R. Fonollosa. "Character-based
              Neural Machine Translation." In: *CoRR* abs/1603.00810 (2016).

[Cou+16]   Jocelyn Coulmance et al. "Trans-gram, Fast Cross-lingual Word em-
              beddings". In: *CoRR* abs/1601.02502 (2016).

[Dev+18]   Jacob Devlin et al. "BERT: Pre-training of Deep Bidirectional Trans-
              formers for Language Understanding." In: *CoRR* abs/1810.04805
              (2018).

[DFU11]   Paramveer S. Dhillon, Dean Foster, and Lyle Ungar. "Multiview
              learning of word embeddings via cca". In: *In Proc. of NIPS*. 2011.

[Dhi+18]   Bhuwan Dhingra et al. "Embedding Text in Hyperbolic Spaces".
              In: *Proceedings of the Twelfth Workshop on Graph-Based Methods
              for Natural Language Processing (TextGraphs-12)*. Association for
              Computational Linguistics, 2018, pp. 59–69.

[Far+14]   Manaal Faruqui et al. *Retrofitting Word Vectors to Semantic Lexicons*.
              2014.

[Gra+18]   Edouard Grave et al. "Learning Word Vectors for 157 Languages". In:
              CoRR abs/1802.06893 (2018).

[Gu+16]   Jiatao Gu et al. *Incorporating Copying Mechanism in Sequence-to-
              Sequence Learning*. 2016.

[HR18]   Jeremy Howard and Sebastian Ruder. "Universal Language Model
              Fine-tuning for Text Classification". In: Association for Computa-
              tional Linguistics, 2018.

[Jou+16a]   Armand Joulin et al. "Bag of Tricks for Efficient Text Classification".
              In: *CoRR* abs/1607.01759 (2016).

[Kan+17]   Ramakrishnan Kannan et al. "Outlier Detection for Text Data: An
              Extended Version." In: *CoRR* abs/1701.01325 (2017).

[Kim+16]   Yoon Kim et al. "Character-Aware Neural Language Models". In:
              AAAI. 2016.

[KB16]   Anoop Kunchukuttan and Pushpak Bhattacharyya. "Learning variable
              length units for SMT between related languages via Byte Pair Encod-
              ing." In: *CoRR* abs/1610.06510 (2016).

[Lam18]   Maximilian Lam. "Word2Bits - Quantized Word Vectors". In: *CoRR* abs/1803.05651 (2018).

[LM14]   Quoc V. Le and Tomas Mikolov. "Distributed Representations of Sentences and Documents". In: *CoRR* abs/1405.4053 (2014).

[Lin+15]   Wang Ling et al. "Finding Function in Form: Compositional Character Models for Open Vocabulary Word Representation." In: *CoRR* abs/1508.02096 (2015).

[LM16]   Minh-Thang Luong and Christopher D. Manning. "Achieving Open Vocabulary Neural Machine Translation with Hybrid Word-Character Models." In: *CoRR* abs/1604.00788 (2016).

[Mik+13b]   Tomas Mikolov et al. "Distributed Representations of Words and Phrases and their Compositionality". In: *Advances in Neural Information Processing Systems 26*. 2013, pp. 3111–3119.

[MH09]   Andriy Mnih and Geoffrey E Hinton. "A scalable hierarchical distributed language model". In: *Advances in neural information processing systems*. 2009, pp. 1081–1088.

[Nee+14]   Arvind Neelakantan et al. "Efficient Non-parametric Estimation of Multiple Embeddings per Word in Vector Space." In: *EMNLP*. ACL, 2014, pp. 1059–1069.

[NK17]   Maximillian Nickel and Douwe Kiela. "Poincaré Embeddings for Learning Hierarchical Representations". In: *Advances in Neural Information Processing Systems 30*. Curran Associates, Inc., 2017, pp. 6338–6347.

[OMS15]   Masataka Ono, Makoto Miwa, and Yutaka Sasaki. "Word Embedding based Antonym Detection using Thesauri and Distributional Information." In: *HLT-NAACL*. 2015, pp. 984–989.

[PSM14]   Jeffrey Pennington, Richard Socher, and Christopher D. Manning. "GloVe: Global Vectors for Word Representation". In: *Empirical Methods in Natural Language Processing (EMNLP). 2014*, pp. 1532–1543.

[RVS17]   Sebastian Ruder, Ivan Vulic, and Anders Sogaard. *A Survey Of Crosslingual Word Embedding Models*. 2017.

[SL14]   Tianze Shi and Zhiyuan Liu. "Linking GloVe with word2vec." In: *CoRR* abs/1411.5595 (2014).

[TML15]   Andrew Trask, Phil Michalak, and John Liu. "sense2vec - A Fast and Accurate Method for Word Sense Disambiguation In Neural Word Embeddings." In: *CoRR* abs/1511.06388 (2015).

[Vas+17a]   Ashish Vaswani et al. "Attention is all you need". In: *Advances in Neural Information Processing Systems*. 2017, pp. 5998–6008.

[VM14]   Luke Vilnis and Andrew McCallum. "Word Representations via Gaussian Embedding." In: *CoRR* abs/1412.6623 (2014).

Chapter 6
Convolutional Neural Networks

6.1 Introduction

In the last few years, convolutional neural networks (CNNs), along with recurrent neural networks (RNNs), have become a basic building block in constructing complex deep learning solutions for various NLP, speech, and time series tasks. LeCun first introduced certain basic parts of the CNN frameworks as a general NN framework to solve various high-dimensional data problems in computer vision, speech, and time series [LB95]. ImageNet applied convolutions to recognize objects in images; by improving substantially on the state of the art, ImageNet revived interest in deep learning and CNNs. Collobert et al. pioneered the application of CNNs to NLP tasks, such as POS tagging, chunking, named entity resolution, and semantic role labeling [CW08b]. Many changes to CNNs, from input representation, number of layers, types of pooling, optimization techniques, and applications to various NLP tasks have been active subjects of research in the last decade.

The initial sections of this chapter describe CNNs, starting with the basic operations, and demonstrate how these networks address the reduction in parameters while creating an inductive bias towards local patterns. Later sections derive the forward and backward pass equations for the basic CNN. Applications of CNNs and their adaptations to text inputs are introduced next. Classic CNN frameworks are then presented, followed by modern frameworks, in order to provide readers with examples of the diversity of ways in which CNNs are used in different domains. Special attention is paid to popular applications of CNNs to various NLP tasks. This chapter also describes specific algorithms that allow deep CNN frameworks to run more efficiently on modern GPU-based hardware. To provide readers with a practical, hands-on experience, the chapter concludes with a detailed case study of airline tweet sentiment analysis using many of the discussed CNN frameworks using Keras and TensorFlow for implementation. In this case study, readers are provided with a detailed exploratory data analysis, preprocessing, training, validation, and evaluation, similar to what one can experience in a real-world project.

© Springer Nature Switzerland AG 2019
U. Kamath et al., *Deep Learning for NLP and Speech Recognition*,
https://doi.org/10.1007/978-3-030-14596-5_6

6.2 Basic Building Blocks of CNN

The next few sections introduce fundamental concepts and building blocks of CNNs. Note that since CNNs originated in computer vision applications, many of the terms and examples in their building blocks refer to images or two-dimensional (2d) matrices. As the chapter continues, these will be mapped to one-dimensional (1d) text input data.

6.2.1 Convolution and Correlation in Linear Time-Invariant Systems

6.2.1.1 Linear Time-Invariant Systems

In signal processing or time series analysis, a transformation or a system that is linear and time-invariant is called a *linear time-invariant system* (LTI); that is, if $y(t) = T(x(t))$, then $y(t-s) = T(x(t-s))$, where $x(t)$ and $y(t)$ are the inputs and the outputs, while $T()$ is the transformation.

A linear system possesses the following two properties:

1. Scaling: $T(ax(t)) = aT(x(t))$
2. Superposition: $T(x_1(t) + x_2(t)) = T(x_1(t)) + T(x_2(t))$

6.2.1.2 The Convolution Operator and Its Properties

Convolution is a mathematical operation performed on LTI systems in which an input function $x(t)$ is combined with a function $h(t)$ to give a new output that signifies an overlap between $x(t)$ and the reverse translated version of $h(t)$. The function $h(t)$ is generally known as a **kernel** or **filter** transformation. In the continuous domain, this can be defined as:

$$y(t) = (h \times x)(t) = \int_{-\infty}^{\infty} h(\tau)x(t-\tau)d\tau \tag{6.1}$$

In the discrete domain, in one dimension, this can be defined as:

$$y(i) = (h \times x)(i) = \sum_{n} h(n)x(i-n) \tag{6.2}$$

Similarly in two dimensions, mostly used in computer vision with still images:

$$y(i,j) = (h \times x)(i,j) = \sum_{n}\sum_{m} h(m,n)x(i-m,i-n) \tag{6.3}$$

This can be also written as cross-correlation or flipped or rotated kernel:

$$y(i,j) = (h \times x)(i,j) = \sum_{n}\sum_{m} x(i+m,i+n)h(-m,-n) \tag{6.4}$$

$$y(i,j) = (h \times x)(i,j) = x(i+m,i+n) \times \text{rotate}_{180}\{h(m,n)\} \tag{6.5}$$

Convolution exhibits the general commutative, distributive, associative, and differentiable properties.

6.2.1.3 Cross-Correlation and Its Properties

Cross-correlation is a mathematical operation very similar to convolution and is a measure of similarity or of the strength of the correlation between two signals $x(t)$ and $ht(t)$. It is given by:

$$y(t) = (h \otimes x)(t) = \int_{-\infty}^{\infty} h(\tau)x(t+\tau) \tag{6.6}$$

In the discrete domain, in one dimension, this can be defined as:

$$y(i) = (h \otimes x)(i) = \sum_n h(n)x(i+n) \tag{6.7}$$

Similarly in two dimensions:

$$y(i,j) = (h \otimes x)(i,j) = \sum_n \sum_m h(m,n)x(i+m,i+n) \tag{6.8}$$

It is important to note that cross-correlation is very similar to convolution but does not exhibit commutative and associative properties.

Many CNNs employ the cross-correlation operator, but the operation is called convolution. We will use the terms synonymously, as the main idea of both is to capture similarity in input signals. Many of the terms used in CNNs have their roots in image processing.

In a regular NN, the transformation between two subsequent layers involves multiplication by the weight matrix. By contrast, in CNNs, the transformation instead involves the convolution operation.

6.2.2 Local Connectivity or Sparse Interactions

In a basic NN, all the units of the input layer connect to all the units of the next layer. This connectivity negatively impacts both computational efficiency and the ability to capture certain local interactions. Consider an example shown in the Fig. 6.1, where the input layer has $m = 9$ dimensions, and the hidden layer has $n = 4$ dimensions; in a fully connected NN, as shown in Fig. 6.1a, there will be $m \times n = 36$ connections (thus, weights) that the NN has to learn. On the other hand, if we allow only $k = 3$ spatially proximal inputs to connect to a single unit of the hidden layer, as shown in Fig. 6.1b, the number of connections reduces to $n \times k = 12$. Another advantage of limiting the connectivity is that restricting the hidden layer connections to spatial-proximal inputs forces the feed-forward system to learn local features through backpropagation. We will refer to the matrix of dimension k as our **filter**, or **kernel**. The spatial extent of the connectivity or the size of the filter (width and height) is generally called the **receptive field**, due to its computer vision heritage. In a 3d input space, the depth of the filter is always equal to the depth of the input, but the width and height are the hyperparameters that can be obtained via search.

6.2.3 Parameter Sharing

Parameter sharing or **tied weights** is a concept in which the same parameters (weights) are reused across all the connections between two layers. Parameter sharing helps reduce the parameter space and hence, memory usage. As shown in Fig. 6.2, instead of learning $n \times k = 12$ parameters, as would happen in Fig. 6.2a, if local connections share the same set of k weights, as shown in Fig. 6.2b, there is a reduction in memory usage by a factor of n. Note that the feed-forward computation still requires $n \times k$ operations. Parameter sharing can also result in the transformation property called **equivariance**, in which the mapping preserves the structure. A function $f()$ is equivariant to a function $g()$, if $f(g(x)) = g(f(x))$ for input x.

> The combination of local connectivity and parameter sharing results in filters that capture common features acting as building blocks across all inputs. In image processing, these common features learned through filters can be basic edge detection or higher-level shapes. In NLP or text mining, these common features can be combinations of n-grams that capture associations of words or characters that are over-represented in the training corpus.

6.2.4 Spatial Arrangement

Note that the number, arrangement, and connections of the filters are hyperparameters in a model. The **depth** of the CNN is the number of layers, whereas the number of filters in a layer determines the "depth" of the subsequent layer. How the filters get moved, that is, how many inputs get skipped before the next instance of the filter, is known as the **stride** of the filter. For example, when the stride is 2, two inputs are

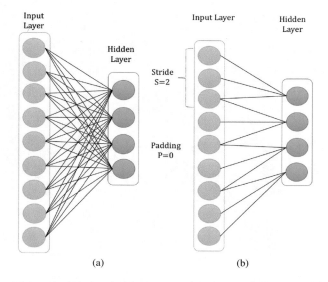

Fig. 6.1: Local connectivity and sparse interactions. (**a**) Fully connected layers. (**b**) Locally connected layers

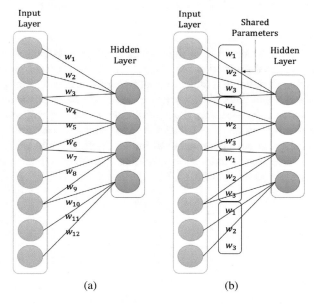

Fig. 6.2: Local connectivity and parameter sharing. (**a**) Locally connected layers. (**b**) Locally connected layers with parameter sharing

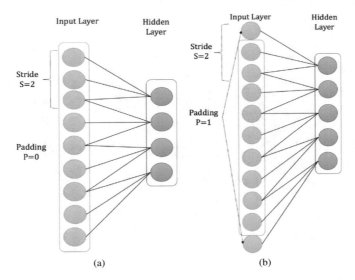

Fig. 6.3: Spatial arrangement and relationship between hyperparameters. (**a**) Spatial arrangement leading to $N = \frac{9-3+0}{2} + 1 = 4$ neurons in the next layer. (**b**) Spatial arrangement leading to $N = \frac{9-3+2}{2} + 1 = 5$ neurons in the next layer

skipped before the next instance of the filter is connected to the subsequent layer. The inputs can be **zero-padded** at the edges, so that filters can fit around the edge units. The number of padding units is another hyperparameter for tuning. Figure 6.3 illustrates different paddings between layers. Adding paddings to the edges with values of 0 is called **zero-padding**; the convolution performed with zero-padding is called **wide convolution**, and one without is called **narrow convolution**.

The relationship between the number of inputs (input volume) W, receptive field (filter size) F, stride size S, and the zero-padding P leads to the number of neurons in the next layer, as provided in Eq. (6.9).

$$N = \frac{W - F + 2P}{S} + 1 \qquad (6.9)$$

Figure 6.4 illustrates how a 2d filter convolves with the input generating the final output for that layer, with the convolution steps broken down. Figure 6.4 visualizes all of the above spatial arrangements and illustrates how changing the padding affects the number of neurons according to the equation above.

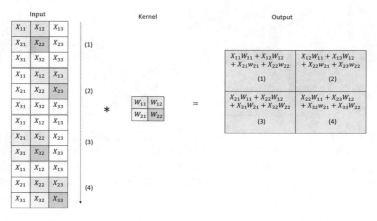

Fig. 6.4: A 2d filter of size 2×2 on a 2d input of size 3×3 gives a convolved output of 2×2 with a stride of 1 and no zero-padding

Figure 6.5 extends the convolution process to a 3d input with three channels (similar to RGB channels in an image) and two filters, showing how the linear convolution reduces the volume between layers. As shown in Fig. 6.6, these filters act as a complex feature detection mechanism. Given a large volume of training data, a CNN can learn filters, such as horizontal edge, vertical edge, outline, and more in the classification process.

The number of channels in the filters should match the number of channels in its input.

Most practical toolboxes or libraries that implement CNNs throw exceptions or safely handle the relationships when hyperparameters violate the constraints given by Eq. (6.9).

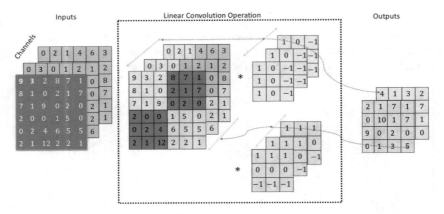

Fig. 6.5: An illustration of an image with dimensions of $6 \times 6 \times 3$ (height × width × channels) convolving with two filters each of size $3 \times 3 \times 3$, no padding and stride 1, resulting in the output of $4 \times 4 \times 2$. The two filters can be thought of as working in parallel, resulting in two outputs on the right of the diagram

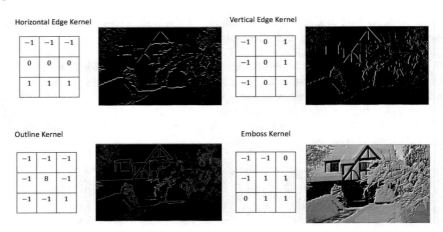

Fig. 6.6: Four filters capturing basic image features

6.2.5 Detector Using Nonlinearity

The output of a convolution is an affine transformation that feeds into a nonlinear layer or transformation known as a **detector layer/stage**. This is very similar to the activation function studied in Chap. 4, where the affine transformation of weights and inputs passes through a nonlinear transformation function. The detector layer normally uses the sigmoid $f(x) = \frac{1}{1+e^{-x}}$, hyperbolic tangent $f(x) = \tan h(x)$, or ReLU $f(x) = \max(0,x)$ as the nonlinear function. As discussed in Chap. 4, ReLU is the most popular function because of its easy computation and simple differentiation. ReLU has also been shown to lead to better generalization when used in CNNs.

6.2.6 Pooling and Subsampling

The outputs of one layer can be further downsampled to capture summary statistics of local neurons or sub-regions. This process is called **pooling**, **subsampling** or **downsampling** based on the context. Usually, the outputs from the detector stage become the inputs for the pooling layer.

Pooling has many useful effects, such as reduction of overfitting and reduction of the number of parameters. Specific types of pooling can also result in **invariance**. Invariance allows identification of a feature irrespective of its precise location and is an essential property in classification. For example, in a face detection problem, the presence of features that indicate an eye is not only essential but it also exists irrespective of its location in an image.

There are different pooling methods, each with its own benefits, and the particular choice depends on the task at hand. When the bias of the pooling method matches the assumptions made in a particular CNN application, such as in the example of face detection, one can expect significant improvement in the results. Some of the more popular pooling methods are listed below.

6.2.6.1 Max Pooling

As the name suggests, the max pooling operation chooses the maximum value of neurons from its inputs and thus contributes to the invariance property discussed above. This is illustrated in Fig. 6.7. Formally, for a 2d output from the detection stage, a max pooling layer performs the transformation:

$$h_{i,j}^l = \max_{p,q} h_{i+p,j+q}^{l-1} \tag{6.10}$$

where p and q denote the coordinates of the neuron in its local neighborhood and l represents the layer. In k-max pooling, k values are returned instead of a single value in the max pooling operation.

6.2.6.2 Average Pooling

In average or mean pooling, the local neighborhood neuron values are averaged to give the output value, as illustrated in Fig. 6.7. Formally, average pooling performs the transformation:

$$h_{i,j}^l = \frac{1}{m^2} \sum_{p,q} h_{i+p,j+q}^{l-1} \tag{6.11}$$

where $m \times m$ is the dimension of the kernel.

6.2.6.3 L2-Norm Pooling

L2-norm pooling is a generalization of the average pooling and is given by:

$$h_{i,j}^l = \sqrt{\sum_{p,q} {h_{i+p,j+q}^{l-1}}^2} \tag{6.12}$$

There are indeed many variants of pooling, such as k-max pooling, dynamic pooling, dynamic k-max pooling, and others.

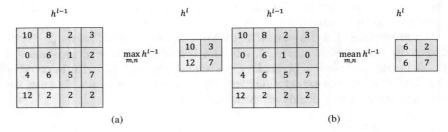

(a)   (b)

Fig. 6.7: Examples of pooling operations. (**a**) Max pooling. (**b**) Average pooling

6.2.6.4 Stochastic Pooling

In stochastic pooling, instead of picking the maximum, the picked neuron is drawn from a multinomial distribution. Stochastic pooling works similar to dropout in addressing the issue of the overfitting [ZF13b].

6.2.6.5 Spectral Pooling

In spectral pooling, the spatial input is transformed into a frequency domain through a discrete Fourier transform (DFT) to capture important signals in the lower dimension. For example, if the input is $\mathbf{x} \in \mathbb{R}^{m \times m}$ and has to be reduced to the size $h \times w$, a DFT operation is performed on the input, so that the frequency representation maintains the central $h \times w$ submatrix; then, an inverse DFT is performed to transform back into the spatial domain [RSA15]. This transformation has the effect of dimensionality reduction on the space and can be very effective in certain applications.

6.3 Forward and Backpropagation in CNN

Now that all basic components are covered, they will be connected. This section will also go through the step-by-step process to clearly understand the different operations involved in the forward and the backward pass in a CNN.

In the interest of clarity, we will consider a basic CNN block consisting of one convolutional layer, a nonlinear activation function, such as ReLU that performs a non-linear transformation, and a pooling layer. In real-world applications, multiple blocks like this are stacked together to form network layers. The output of these blocks is then **flattened out** and connected to a fully connected output layer. The **flattening out** process converts a multidimensional tensor to a mono-dimensional vector, for example a three-dimensional (W, H, N) to a vector of dimension $d = W \times H \times N$.

Let us start the derivation for the layer l, which is the output of convolution on layer $l - 1$. Layer l has height h, width w, and channels c. Let us assume that there is one channel, i.e. $c = 1$, and has iterators i, j for the input dimensions. We will consider the filter $k_1 \times k_2$ dimensions with iterators m, n for the convolution operation. The weight matrix with weights $W_{m,n}^l$ and the bias b^l transforms the previous layer $l - 1$ into the layer l via the convolution operation. The convolution layer is followed by a non-linear activation function, such as ReLU $f(\cdot)$. The output for layer l is denoted by $O_{i,j}^l$.

Thus forward propagation for layer l can be written as:

$$X_{i,j}^l = \text{rotate}_{180}\{W_{m,n}^l\} \times O_{i,j}^{l-1} + b^l \tag{6.13}$$

This can be expanded as:

$$X_{i,j}^l = \sum_m \sum_n W_{m,n}^l O_{i+m,j+n}^{l-1} + b^l \tag{6.14}$$

and

$$O_{i,j}^l = f(X_{i,j}^l) \tag{6.15}$$

We assume error or loss mechanisms such as mean-squared-error E is used to measure the difference between the predictions and the actual labels. The errors have to be propagated back and need to update the weights of the filter and the inputs received at the layer l. Thus, in the backpropagation process, we are interested in the gradient of the error E with respect to (w.r.t.) the input $(\frac{\partial E}{\partial X})$ and the filter weights $(\frac{\partial E}{\partial W})$.

6.3.1 Gradient with Respect to Weights $\frac{\partial E}{\partial \mathbf{W}}$

We will first consider the impact of single pixel (m', n') of the kernel, given by $W_{m',n'}$ on the error E, using the chain rule:

$$\frac{\partial E}{\partial W^l_{m',n'}} = \sum_{i=0}^{h-k_1} \sum_{j=0}^{w-k_2} \frac{\partial E}{\partial X^l_{i,j}} \frac{\partial X^l_{i,j}}{\partial W^l_{m',n'}} \tag{6.16}$$

If we consider $(\delta^l_{i,j})$ as the gradient error for the layer l w.r.t input, the above equation can be rewritten as:

$$\frac{\partial E}{\partial W^l_{m',n'}} = \sum_{i=0}^{h-k_1} \sum_{j=0}^{w-k_2} \delta^l_{i,j} \frac{\partial X^l_{i,j}}{\partial W^l_{m',n'}} \tag{6.17}$$

Writing the change in output in terms of the inputs (the previous layer) we get:

$$\frac{\partial X^l_{i,j}}{\partial W^l_{m',n'}} = \frac{\partial}{\partial W^l_{m',n'}} \left(\sum_m \sum_n W^l_{m,n} O^{l-1}_{i+m,j+n} + b^l \right) \tag{6.18}$$

When we take partial derivatives with respect to $W_{m',n'}$, all values become zero except for the components mapping to $m = m'$ and $n = n'$.

$$\frac{\partial X^l_{i,j}}{\partial W^l_{m',n'}} = \frac{\partial}{\partial W^l_{m',n'}} \left(W^l_{0,0} O^{l-1}_{i+0,j+0} + \cdots + W^l_{m',n'} O^{l-1}_{i+m',j+n'} + \cdots + b^l \right) \tag{6.19}$$

$$\frac{\partial X^l_{i,j}}{\partial W^l_{m',n'}} = \frac{\partial}{\partial W^l_{m',n'}} \left(W^l_{m',n'} O^{l-1}_{i+m',j+n'} \right) \tag{6.20}$$

$$\frac{\partial X^l_{i,j}}{\partial W^l_{m',n'}} = O^{l-1}_{i+m',j+n'} \tag{6.21}$$

Substituting the result back, one gets:

$$\frac{\partial E}{\partial W^l_{m',n'}} = \sum_{i=0}^{h-k_1} \sum_{j=0}^{w-k_2} \delta^l_{i,j} O^{l-1}_{i+m',j+n'} \tag{6.22}$$

The whole process can be summarized as the convolution of the gradients with rotation of 180 degrees $\delta^l_{i,j}$ of layer l with the outputs of layer $l-1$. i.e. $O^{l-1}_{i+m',j+n'}$. Thus, the new weights to be updated can be computed very similarly to the forward pass.

$$\frac{\partial E}{\partial W^l_{m',n'}} = \text{rotate}_{180}\{\delta^l_{i,j}\} \times O^{l-1}_{m',n'} \tag{6.23}$$

6.3.2 Gradient with Respect to the Inputs $\frac{\partial E}{\partial \mathbf{X}}$

Next, we are interested in how a change to single input, given by $X_{i',j'}$ affects the error E. Borrowing from computer vision, the input pixel $X_{i',j'}$ after convolution affects a region bounded by top left $(i'+k_1-1, j'+k_2-1)$ and bottom right (i',j').

So, we obtain:

$$\frac{\partial E}{\partial X^l_{i',j'}} = \sum_{m=0}^{k_1-1} \sum_{n=0}^{k_2-1} \frac{\partial E}{\partial X^{l+1}_{i'-m,j'-n}} \frac{\partial X^{l+1}_{i'-m,j'-n}}{\partial X^l_{i',j'}} \tag{6.24}$$

$$\frac{\partial E}{\partial X^l_{i',j'}} = \sum_{m=0}^{k_1-1} \sum_{n=0}^{k_2-1} \delta^{l+1}_{i'-m,j'-n} \frac{\partial X^{l+1}_{i'-m,j'-n}}{\partial X^l_{i',j'}} \tag{6.25}$$

Expanding just the rate of change of the inputs, one obtains:

$$\frac{\partial X^{l+1}_{i'-m,j'-n}}{\partial X^l_{i',j'}} = \frac{\partial}{\partial X^l_{i',j'}} \left(\sum_{m'} \sum_{n'} W^{l+1}_{m',n'} O^l_{i'-m+m',j'-n+n'} + b^{l+1} \right) \tag{6.26}$$

Writing this in terms of input layer l, we get:

$$\frac{\partial X^{l+1}_{i'-m,j'-n}}{\partial X^l_{i',j'}} = \frac{\partial}{\partial X^l_{i',j'}} \left(\sum_{m'} \sum_{n'} W^{l+1}_{m',n'} f(X^l_{i'-m+m',j'-n+n'}) + b^{l+1} \right) \tag{6.27}$$

All partial derivatives result into zero except where $m' = m$ and $n' = n$, $f(X^l_{i'-m'+m,j'-n'+n}) = f(X^l_{i',j'})$, and $W^{l+1}_{m',n'} = W^{l+1}_{m,n}$ in the relevant output regions.

$$\frac{\partial X^{l+1}_{i'-m,j'-n}}{\partial X^l_{i',j'}} = \frac{\partial}{\partial X^l_{i',j'}} \left(W^{l+1}_{m',n'} f\left(X^l_{0-m+m',0-n+n'}\right) + \cdots + W^{l+1}_{m,n} f\left(X^l_{i',j'}\right) \right.$$

$$\left. + \cdots + b^{l+1} \right) \tag{6.28}$$

$$\frac{\partial X^{l+1}_{i'-m,j'-n}}{\partial X^l_{i',j'}} = \frac{\partial}{\partial X^l_{i',j'}} \left(W^{l+1}_{m,n} f\left(X^l_{i',j'}\right) \right) \tag{6.29}$$

$$\frac{\partial X^{l+1}_{i'-m,j'-n}}{\partial X^l_{i',j'}} = W^{l+1}_{m,n} f'\left(X^l_{i',j'}\right) \tag{6.30}$$

Substituting, one obtains:

$$\frac{\partial E}{\partial X^l_{i',j'}} = \sum_{m=0}^{k_1-1} \sum_{n=0}^{k_2-1} \delta^{l+1}_{i'-m,j'-n} W^{l+1}_{m,n} f'\left(X^l_{i',j'}\right) \tag{6.31}$$

The term

$\sum_{m=0}^{k_1-1} \sum_{n=0}^{k_2-1} \delta^{l+1}_{i'-m,j'-n} W^{l+1}_{m,n}$ can be seen as the flipped filter, or the filter rotated by 180 degrees performing convolution with the δ matrix.

Thus,

$$\frac{\partial E}{\partial X^l_{i',j'}} = \left(\delta^{l+1}_{i',j'} \times \text{rotate}_{180}\left\{W^{l+1}_{m,n}\right\} \right) f'\left(X^l_{i',j'}\right) \tag{6.32}$$

6.3.3 Max Pooling Layer

As we saw in the pooling section, the pooling layer does not have any weights but just reduces the size of the input based on the spatial operations performed. In the forward pass, the pooling operation leads to conversion of a matrix or a vector to a single scalar value.

In max pooling, the *winner* neuron or cell is remembered; when the backpropagation needs to be done, the entire error is passed on to that winner neuron, as others have not contributed. In average pooling, an $n \times n$ pooling block during backpropagation divides the total scalar value by $\frac{1}{n \times n}$ and distributes it equally across the block.

6.4 Text Inputs and CNNs

Various NLP and NLU tasks in text mining use a CNN as the feature engineering block. We will start with basic text classification to highlight some important analogies between images in computer vision and mappings in text and the necessary changes in the CNN process.

6.4.1 Word Embeddings and CNN

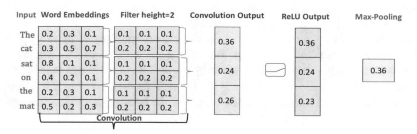

Fig. 6.8: A simple text and CNN mapping is shown

Let us assume all training data are in the form of sentences with labels and of given maximum length s. The first transformation is to convert the sentences into a vector representation. One way is to perform a lookup function for each word in the sentence for its fixed-dimensional representation, such as Word Embeddings. Let us assume that the lookup for word representation in a fixed vocabulary size V yields a vector of fixed dimension and let that be d; thus, each vector can be mapped to $\mathbb{R}^d$. The rows of the matrix represent words of the sentences, and the columns can be the fixed-length vector corresponding the representation. The sentence is a real matrix $\mathbf{X} \in \mathbb{R}^{s \times d}$.

The general hypothesis, especially in classification tasks, is that the words which are local in the sequence, similar to n-grams, form complex higher-level features when combined. This combination of local words in proximity is analogous to computer vision, where local pixels can be combined to form features such as lines, edges, and real objects. In computer vision with image representations, the convolution layer had filters smaller in size than the inputs performing convolution operations via *sliding* across an image in patches. In text mining, the first layer of convolution generally has filters of the same dimension d as the input but has varying height h, normally referred to as the filter size.

Figure 6.8 illustrates this on the sentence "The cat sat on the mat" which is tokenized into $s = 6$ words $\{The, cat, sat, on, the, mat\}$. A lookup operation obtains a 3-dimensional word embeddings ($d = 3$) for each word. A single convolution filter with height or size $h = 2$ starts producing the feature map. The output goes through a non-linear activation of ReLU with 0.0 threshold that then feeds into a 1-max pooling. Figure 6.8 illustrates the end state obtained by using the same shared kernel across all the inputs, producing the single output at the end of the 1-max pooling.

Figure 6.9 gives a generalized methodology from sentences to output in a simple CNN framework. In the real world, many representations of the input words can exist similar to images having color channels in the computer vision field. Different word vectors can map to the channels that may be **static** (i.e., pre-trained) using a

well-known corpus and do not change. They can also be **dynamic**, where even if
they were pre-trained, backpropagation can fine-tune it. There are various applications that use not only word embeddings for the representation but also POS tags
for the words or the position in the sequence. The application and the NLP task at
hand determine the particular representations.

Generally, outputs result in **regions**, and there are multiple of these varying in
size, generally from 2 to 10, i.e., sliding over 2–10 words. For each region size,
there can be multiple learned filters, given by n_f. Similar to the image representation
calculations derived above, there are $\frac{s-h}{stride} + 1$ regions, where stride is the number
of words filters slide across. Thus, the output of the convolution layers are vectors
of dimensions $\mathbb{R}^{s-h+1}$ for a stride of 1. Formally,

$$o_i = \mathbf{W} \cdot \mathbf{X}[i : i+h-1,:] \tag{6.33}$$

This equation represents how a filter matrix $\mathbf{W}$ of height h slides over the region
matrix given by $\mathbf{X}[i : i+h-1,:]$ with unit stride. For each filter, the weights are
shared as in image-based frameworks giving the local feature extraction. The output
then flows into a non-linear activation function $f(a)$, most commonly ReLU. The
output generates a feature map c_i with bias b for each filter or region as in:

$$c_i = f(o_i) + b \tag{6.34}$$

The output vector of each of the feature maps goes through a pooling layer, as discussed before. The pooling layer provides the downsampling and invariance property, as mentioned above. The pooling layer also helps in addressing the variable
length of words in sentences by yielding a reduced dimension $\hat{\mathbf{c}}$ for the entire vector. Since vectors represent words or sequences over time, max pooling in the text
domain is referred to as **max pooling over time**. How two different sizes of texts
get reduced to same-dimensional representation through max pooling over time is
shown in Fig. 6.10.

In short-text classification, 1-max pooling has been effective. In document
classification, k-max pooling has been a better choice.

The output of the pooling layers is concatenated and passed on to the softmax
layer, which performs classification based on the number of categories or labels.
There are many hyperparameter choices when representing text and performing
CNN operations on text, such as type of word representation, choice of how many
word representations, strides, paddings, filter width, number of filters, activation
function, pooling size, pooling type, and number of CNN blocks before the final
softmax layer, to name a few.

The total number of hyperparameters for one simple block of CNN with output
for text processing can be given by:

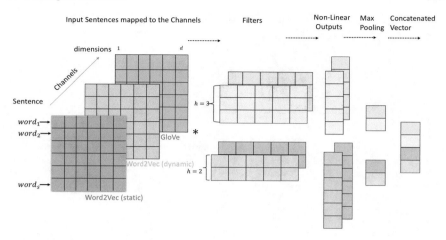

Fig. 6.9: A simple text and CNN mapping. Sentence to word representations, mapping to the embeddings with different channels act as an input. Different filters for each height, i.e., 2 filters each of sizes 2 and 3, shown in two shades, capture different features that are then passed to non-linear processing to generate outputs. These go through max pooling operation, which selects the maximum values from each filter. The values are concatenated to form a final output vector

$$\text{parameters} = \underbrace{(V+1) \times d}_{\text{WordEmbeddings(static)}} + \underbrace{((h \times d) \times n_f) + n_f +}_{\text{Filters}} \underbrace{n_f + 1}_{\text{SoftmaxOutput}} \quad (6.35)$$

Now, we will provide a formal treatment from input to output for a simple CNN in the text domain with word representations. For a variable-length sentence from training data, with words of maximum length s having a similar lookup for word embeddings of vector size d, we obtain an input vector in $\mathbb{R}^d$. For all other sentences which have fewer than s words, we can use padding with 0 or random values in the inputs. As in the previous section, there can be many representations of these words, such as static or dynamic, different embeddings for the words as word2vec, GloVe, etc., and even different embeddings such as positional, tag-based (POS-tag), etc., all forming the input channels with the only constraint that they are all of the same dimension.

The convolution filter can be seen as weights in 1d of the same length as the word embedding but sliding over words of size h. This can create different word windows $W_{1:h}, W_{2:h}..W_{n-h+1:n}$. These feature vectors are represented by $[o_1, o_2, \cdots, o_{n-h+1}]$ in $\mathbb{R}^{n-h+1}$, which go through non-linear activation, such as ReLU. The output of non-linear activation has the same dimension as the input and can be represented as $[c_1, c_2, \cdots, c_{n-h+1}]$ in $\mathbb{R}^{n-h+1}$. Finally, the outputs of the non-linear activation

Fig. 6.10: Two simple sentences with padding and 3-dimensional word embeddings, going through same single filter of size or height 2, a ReLU with a threshold of 0, and max pooling over time, result in the similar output value corresponding the 2-gram of "The cat"

function are further passed to a max pooling layer which finds a single scalar value $\hat{c}$ in $\mathbb{R}$ for each filter. In practice, there are multiple filters for each size h, and there are multiple-sized filters for h varying from 2 to 10 in the general case. The output layer connects all the max pooling layer outputs $\hat{c}$ into a single vector and uses the softmax function for classification. The considerations of padding sentences at the beginning and the end, stride lengths, narrow or wide convolutions, and other hyperparameters are as in the general convolution mapping process.

6.4.2 Character-Based Representation and CNN

In many classification-related tasks, the vocabulary size grows large, and taking into account unseen words in training even with embeddings results in suboptimal performance. Work by Zhang et al. [ZZL15] uses character-level embeddings instead of word-level embeddings in the training input to overcome such issues. The researchers show that character embeddings result in open vocabulary, a way to handle misspelled words, to name a few benefits. Figure 6.11 shows the designed CNN to have many convolution blocks of 1d convolution layer, non-linear activations, and k-max pooling, all stacked together to form a deep convolution layer. The representation uses the well-known set of 70 alphanumeric characters with all lowercase letters, digits, and other characters using one-hot encoding. Combination of characters with fixed size $l = 1024$ as a chunk of the input at a given time forms the input for variable-length text. A total of 6 blocks of 2 sets of CNNs, each of different size, the larger of 1024 dimensions and the smaller of 256 dimensions are used. Two layers have filter height of 7, and the rest have height of 3, with 3-pooling in the first few and strides of 1. The last 3 layers are fully connected. The final layer is a soft-max layer for classification.

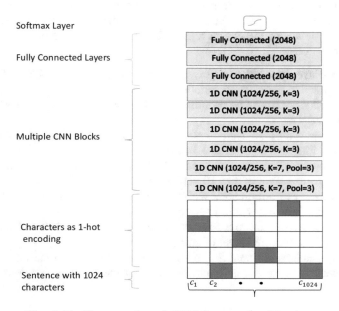

Fig. 6.11: Character-based CNN for text classification

6.5 Classic CNN Architectures

In this section, we will visit some of the standard CNN architectures. We will discuss their structure in detail and will give some historical context for each. Though many of them have been popular in the computer vision domain, they are still applicable with variations in the text and speech domains, as well.

6.5.1 LeNet-5

LeCun et al. presented LeNet-5 as one of the first implementations of CNNs and showed impressive results on the handwritten digit recognition problem [LeC+98]. Figure 6.12 shows the complete design of LeNet-5. LeCun demonstrated the concept of decreasing the height and width via convolutions, increasing the filter/channel size, and having fully connected layers with a cost function to propagate the errors, which are now the backbone of all CNN frameworks. LeNet-5 used the MNIST dataset of 60K training data for training and learning the weights. In all of our discussions from now on, the representation of a layer will be given by $n_w \times n_h \times n_c$, where n_w, n_h, n_c are the width, height, and the number of channels/filters. Next, we will give the details of the design regarding inputs, outputs, number of filters, and pooling operations.

- The input layer uses only the grayscale pixel values and is $32 \times 32 \times 1$ in size. It is normalized to mean 0 and variance 1.
- A filter of $5 \times 5 \times 6$ with no padding and stride $s = 1$ is used to create a layer with size $28 \times 28 \times 6$.
- This is followed by an average pooling layer with filter width $f = 2$ and stride $s = 2$, resulting in a layer of size $14 \times 14 \times 6$.
- Another convolution layer of $5 \times 5 \times 16$ is applied with no padding and stride $s = 1$ to create a layer of size $10 \times 10 \times 16$.
- Another average pooling layer with filter width $f = 2$ and stride $s = 2$, resulting in reduced height and width, yields a layer of size $5 \times 5 \times 16$.
- This is then connected to a fully connected layer of size 120 and followed by another fully connected layer of size 84.
- These 84 features are then fed to an output function which uses the Euclidean radial basis function for determining which of the 10 digits are represented by these features.

1. LeNet-5 used tanh function for non-linearity instead of ReLU, which is more popular in today's CNN frameworks.
2. Sigmoid non-linearity was applied after the pooling layer.
3. LeNet-5 used the Euclidean radial basis function instead of softmax, which is more popular today.
4. The number of parameters/weights in LeNet-5 was approximately 60K with approximately 341K multiplications and accumulations (MACS).
5. The concept of no padding, which results in lowering the size, was used back then, but it is not very popular these days.

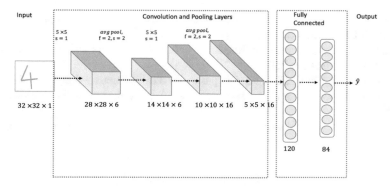

Fig. 6.12: LeNet-5

6.5.2 AlexNet

AlexNet, designed by Krizhevsky et al. [KSH12a], was the first deep learning architecture that won the ImageNet challenge in 2012 by a large margin (around 11.3%). AlexNet was responsible in many ways for focusing attention on deep learning research [KSH12a]. Its design is very similar to LeNet-5, but with more layers and filters resulting in a larger network and more parameters to learn. The work in [KSH12a] showed that with deep learning frameworks, features can be learned instead of manually generated with deep domain understanding. The details of the AlexNet designs are listed below (Fig. 6.13):

- Unlike LeNet-5, AlexNet used all the three channels of inputs. The size $227 \times 227 \times 3$ is convolved with $11 \times 11 \times 96$ filter, stride of $s = 4$, and with nonlinearity performed by ReLU giving an output of $55 \times 55 \times 96$.
- This goes through a max pooling layer with size 3×3 and stride $s = 2$ and reduces the output to $27 \times 27 \times 96$.
- This layer goes through a local response normalization (LRN) which effectively normalizes the values across the depth of the channels and then another convolution of size $5 \times 5 \times 256$, stride $s = 1$, padding $f = 2$, with ReLU applied to get an output of $27 \times 27 \times 256$.
- This goes through a max pooling layer with size 3×3 and stride $s = 2$, reducing the output to $13 \times 13 \times 256$.
- This is followed by LRN, another convolution of size $3 \times 3 \times 384$, stride $s = 1$, padding $f = 1$, with ReLU applied to get an output of $13 \times 13 \times 384$.
- This is followed by another convolution of size $3 \times 3 \times 384$, stride $s = 1$, padding $f = 1$, with ReLU applied to get an output of $13 \times 13 \times 384$.

- This is followed by convolution of size $3 \times 3 \times 256$, stride $s = 1$, padding $f = 1$, with ReLU applied to get an output of $13 \times 13 \times 256$.
- This goes through a max pooling layer with size 3×3 and stride $s = 2$, reducing the output to $6 \times 6 \times 256$.
- This output $6 \times 6 \times 256 = 9216$ is passed to a fully connected layer of size 9216, followed by a dropout of 0.5 applied to two fully connected layers with ReLU of size 4096.
- The output layer is a softmax layer with 100 classes or categories of images to learn.

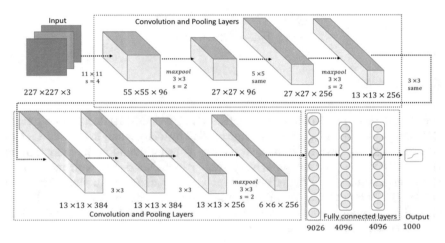

Fig. 6.13: AlexNet

1. AlexNet used ReLU and showed it to be a very effective non-linear activation function. ReLU showed six times performance improvement over sigmoid on a CIFAR dataset, which was the reason why the researchers chose ReLU.
2. The number of parameters/weights in was approximately 63.2 million, and approximately 1.1 billion computations which is significantly higher than LeNet-5.
3. Speedup was obtained via two GPUs. The layers were split, so that each GPU worked in parallel with its output to the next layer collocated in its

6.5.3 VGG-16

VGG-16, also known as VGGNet, by Simonyan and Zisserman, is known for its uniform design and has been very successful in many domains [SZ14]. The uniformity in having all the convolutions being 3×3 with stride $s = 1$ and max pooling with 2×2 along with the channels increasing from 64 to 512 in multiples of 2 makes it very appealing and easy to set up. It has been shown that stacking convolutions of 3×3 with stride $s = 1$ in three layers is equivalent to 7×7 convolutions and with a significantly reduced number of computations. VGG-16 has two fully connected layers at the end with a softmax layer for classification. VGG-16's only disadvantage is the huge network with approximately 140 million parameters. Even with a GPU setup, a long time would be required for training the model (Fig. 6.14).

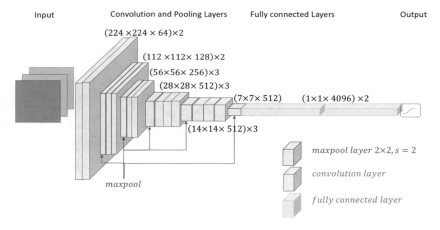

Fig. 6.14: VGG-16 CNN

6.6 Modern CNN Architectures

We will discuss changes that have led to modern CNN architectures in different domains, including in text mining.

6.6.1 Stacked or Hierarchical CNN

The basic CNN mapping to sentences with a convolution filter of size k is shown to be analogous to the $n$gram token detector in classic NLP settings. The idea of a stacked or hierarchical CNN is to extend the principle by adding more layers. In doing so, the receptive field size is increased, and larger windows of words or contexts will be captured as features, as shown in Fig. 6.15.

If we consider $(\mathbf{W}, \mathbf{b})$ as the parameters corresponding to weights and biases, $\oplus$ as the concatenation operation done on sentence of length n with sequence of the word embeddings $\mathbf{e}_{1:n}$ each of d dimension, and k as the window size of the convolution, the output is given by: $\mathbf{c}_{1:m}$

$$\mathbf{c}_{1:m} = CONV^k_{(\mathbf{W},\mathbf{b})}(\mathbf{e}_{1:n}) \tag{6.36}$$

$$\mathbf{c}_i = f(\oplus(\mathbf{w}_{i:i+k-1}) \cdot \mathbf{W} + \mathbf{b}) \tag{6.37}$$

where $m = n - k + 1$ for narrow convolutions, and $m = n + k + 1$ for wide convolutions.

If we consider p layers, where the convolution of one feeds into another, then we can write

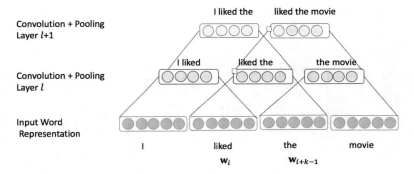

Fig. 6.15: A hierarchical CNN

$$\mathbf{c}^1_{1:m_1} = CONV^{k_1}_{(\mathbf{W}^1,\mathbf{b}^1)}(\mathbf{w}_{1:n})$$

$$\mathbf{c}^2_{1:m_2} = CONV^{k_2}_{(\mathbf{W}^2,\mathbf{b}^2)}(\mathbf{c}^1_{1:m_1})$$

$$\cdots \tag{6.38}$$

$$\mathbf{c}^p_{1:m_p} = CONV^{k_p}_{(\mathbf{W}^p,\mathbf{b}^p)}(\mathbf{c}^{p-1}_{1:m_{p-1}})$$

As the layers feed into each other, the effective window size or the receptive field to capture the signal increases. For instance, if in sentiment classification there is a sentence—*The movie is not a very good one*, a convolution filter with size 2 will not capture the sequence "not a very good one" but a stacked CNN with the same size 2 will capture this in the higher layers.

6.6.2 Dilated CNN

In the stacked CNN, we assumed strides of size 1, but if we generalize the stride to size s, we can then write the convolution operation as:

$$\mathbf{c}_{1:m} = \mathrm{CONV}^{k,s}_{(\mathbf{W},\mathbf{b})}(\mathbf{w}_{1:n}) \tag{6.39}$$

$$\mathbf{c}_i = f(\oplus(\mathbf{w}_{1+(i-1)s:(s+k)i}) \cdot \mathbf{W} + \mathbf{b}) \tag{6.40}$$

A dilated CNN can be seen as a special version of the stacked CNN. One way is to have the stride size of each layer be $k-1$ when the kernel size is k.

$$\mathbf{c}_{1:m} = \mathrm{CONV}^{k,k-1}_{(\mathbf{W},\mathbf{b})}(\mathbf{w}_{1:n}) \tag{6.41}$$

Convolutions of size $k \times k$ on the l layered CNN without pooling result in receptive fields of size $l \times (k-1)+k$, which is linear with the number of layers l. A dilated CNN helps increase the receptive field exponentially with respect to the number of layers. Another approach is to keep the stride size constant, as in $s = 1$, but perform length shortening at each layer using local pooling by using maximum or average as values. Figure 6.16 shows how by progressively increasing the dilations in the layers, the receptive fields can be exponentially increased to cover a larger field in every layer [YK15].

1. A dilated CNN helps in capturing the structure of sentences over longer spans of text and is effective in capturing the context.
2. A dilated CNN can have fewer parameters and so increase the speed of training while capturing more context.

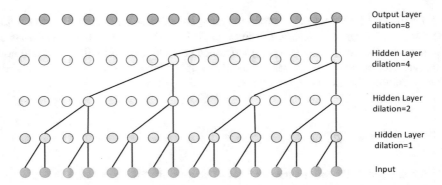

Fig. 6.16: A dilated CNN showing the increase in the receptive fields with dilations changing to 1, 2, *and* 4

6.6.3 Inception Networks

Inception networks by Szegedy et al. [Sze+17] are currently one of the best-performing CNNs, especially in computer vision. The core of an inception network is a repetitive inception block. An inception block uses many filters, such as 1×1, 3×3, and 5×5, as well as max pooling without the need to choose any one of them. The central idea behind using 1×1 filters is to reduce the volume and hence the computation before feeding it to a larger filter such as 3×3.

Figure 6.17 shows a sample $28 \times 28 \times 192$ output from a previous layer convolved with a $3 \times 3 \times 128$ filter to give an output of $28 \times 28 \times 128$; i.e., 128 filters at the output result in about 174 million MACs. By having $1 \times 1 \times 96$ filters to reduce the volume and then convolve with $3 \times 3 \times 128$ filters, the total computations reduce to approximately 100 million MACs, almost a saving of 60%. Similarly, by using $1 \times 1 \times 16$ filters preceding $5 \times 5 \times 32$ filters, the total computations can be reduced from approximately 120 to 12 million, a reduction by a factor of ten. The 1×1 filter that reduces the volume is also called a bottleneck layer.

Figure 6.18 gives a pictorial view of a single inception block with a sample 2d input with width and height of 28×28 and depth of 192, producing an output of width and height of 28×28 and depth of 256. The 1×1 filter plays a dual role in reducing the volume for other filters, as well as producing the output. The $1 \times 1 \times 64$ filter is used to produce an output of $28 \times 28 \times 64$. The input, when passed through the $1 \times 1 \times 96$ filter and convolved with $3 \times 3 \times 128$, generates a $28 \times 28 \times 128$ output. The input, when passed through the $1 \times 1 \times 16$ filter and then through the $5 \times 5 \times 32$ filter, produces a $28 \times 28 \times 32$ output. Max pooling with stride 1 and padding is used to produce an output of size $28 \times 28 \times 192$. Since the max pooling output has many channels, each channel passes through another $1 \times 1 \times 32$ filter for reducing the volume. Thus, each filtering operation can occur in parallel and generate an output that is concatenated together for a final size of $28 \times 28 \times 256$.

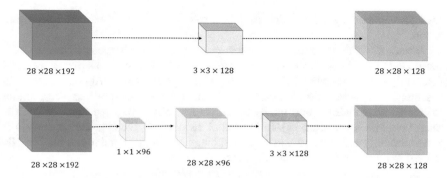

Fig. 6.17: Computational cost savings with having a 1×1 filter precede a 3×3 filter

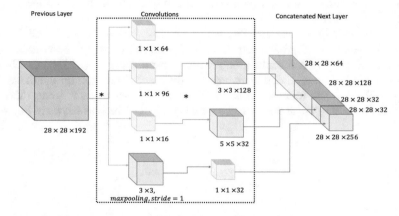

Fig. 6.18: An inception block with multiple filters of size 1×1, 3×3, 5×5, producing a concatenated output

A 1×1 convolution block plays an important role in reducing the volume for a larger filter size convolution. An inception block allows multiple filter weights to be learned, thus removing the need to select one of the filters. Parallel operations of convolutions across different filters and concatenation give further improved performance.

6.6.4 Other CNN Structures

In many NLP tasks, such as in sentiment analysis, features using syntactic elements and other structural information in the language have yielded improved performance. Figure 6.19 shows a syntactic and semantic representation connected through dependency arcs, which map to a tree-based or a graph-based representation.

A basic CNN does not capture such dependencies in the language. Hence, structured CNN have been proposed to overcome this shortcoming. Dependency CNN (DCNN), as shown in Fig. 6.20, is one way of representing word dependencies in a sentence via word embeddings, then performing tree convolutions very similar to matrix convolutions, and finally employing a fully connected layer before the output classification [Ma+15]. In Mou et al. [Mou+14], the application domain is programming languages instead of natural languages, and a similar tree-based CNN is used to learn the feature representation.

The syntactic and dependencies can be captured with a graph representation $G = (V, E)$, where words act as nodes or vertices V, and relationships between them are modeled as edges E. Convolution operations can be then performed on these graph structures [Li+15].

As seen in Chap. 5, each embeddings framework such as word2vec, GloVe, etc. capture different distributional semantics, and each can be of different dimensions. Zhang et al. proposed a multi-group norm constraint CNN (MGNC-CNN) that can combine different embeddings, each of different dimension on the same sentence [ZRW16]. The regularization can be applied either to each group (MGNC-CNN) or can be applied at the concatenation layer (MG-CNN) as shown in Fig. 6.21.

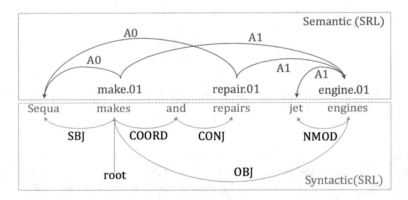

Fig. 6.19: A sentence with syntactic and semantic dependencies is shown

In many applications, such as machine translation, text entailment, question answering, and others, there is often a need to compare two inputs for similarity. Most of these frameworks have some form of **Siamese** structure with two parallel frameworks for each sentence, combining convolution layers, nonlinear transformations, pooling, and stacking, till features are combined at the end. Work of Bromley et al. for signature comparison has been the general inspiration behind many of such frameworks [Bro+94].

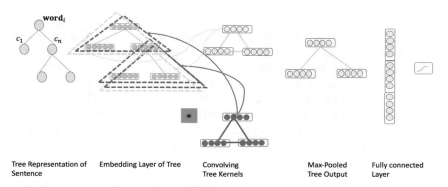

Tree Representation of   Embedding Layer of Tree   Convolving   Max-Pooled   Fully connected
Sentence   Tree Kernels   Tree Output   Layer

Fig. 6.20: A tree-based structured CNN that can capture syntactic and semantic dependencies

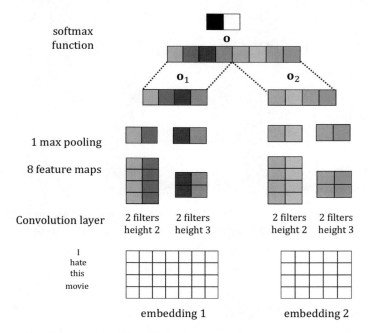

Fig. 6.21: MG-CNN and MGNC-CNN showing different embeddings of different dimensions used for classification. MG-CNN will have norm constraints applied at layer $\mathbf{o}$ while MGNC-CNN will have norm constraints applied at layers $\mathbf{o}_1$ and $\mathbf{o}_2$, respectively

We will illustrate one such recent framework by Wenpeng et al [Yin+16a], who refers to it as a basic bi-CNN, as shown in Fig. 6.22. The framework is further modified to have a shared attention layer and performs very well on diverse applications, such as answer selection, paraphrase identification, and text entailment. The twin networks, each consisting of CNN and pooling blocks, process one of the two sentences and the final layer solves the sentence pair task. The Input layer has word embeddings from word2vec concatenated for the words in each sentence. Each block uses wide convolutions so that all words in the sentence can provide signals as compared to a narrow convolution. The tanh activation function $\tanh(\mathbf{W}\mathbf{x_i} + b)$ is used as the nonlinear transformation. Next, an average pooling operation is performed on each. Finally, the output of both average pooling layers is concatenated and passed to a logistic regression function for binary classification.

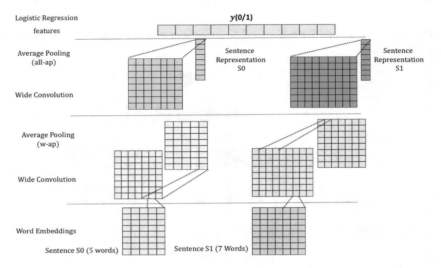

Fig. 6.22: A basic bi-CNN by [Yin+16a] for sentence pair tasks using wide convolutions, average pooling, and logistic regression for binary classification

6.7 Applications of CNN in NLP

In this section, we will discuss some of the applications of CNN in various text mining tasks. Our goal is to summarize this research and provide insights into popular CNNs and modern designs. CNNs with various modifications and even combinations with other frameworks, such as LSTMs, have indeed been used in different NLP tasks. Since a CNN by itself can capture local features and combinations of these through further combinations, CNNs have been primarily used in text/document classification, text categorization, and sentiment classification tasks. CNNs lose the order of the sequences, though they have been used in sequence-based tasks, such as POS tagging, NER, chunking, and more. To be truly effective in such settings, they either need to be combined with other frameworks or have positional features encoded.

6.7.1 Text Classification and Categorization

Many text classification tasks which employ *n*-grams of words to capture local interactions and features have seen lots of success using CNN frameworks in the last few years. CNN-based frameworks can easily capture temporal and hierarchical features in variable-length text sequences. Word or character embeddings are generally the first layers in these frameworks. Based on the data and type, either pre-trained or static embeddings are used to obtain a vector representation of the words in sentences.

Collobert et al. [CW08c], [Col+11] uses a one-layer convolution block for modeling sentences to perform many NLP tasks. Yu et al. [Yu+14] also use a one-layer CNN to model a classifier to select question answer mappings. Kalchbrenner et al. extend the idea to form a dynamic CNN by stacking CNN and using dynamic *k*-max pooling operations over long sentences [KGB14b]. This research significantly improved over existing approaches at the time in much short text and multiclass classification. Kim extends the single-block CNN by adding multiple channels in the input and multiple kernels of various lengths to give higher-order combinations of *n*grams [Kim14b]. This work also performed various impact analysis of static vs. dynamic channels, the importance of max pooling, and more to detect which basic blocks yielded lower error rates. Yin et al. extend Kim's multi-channel, variable kernel framework to use hierarchical CNN, obtaining further improvements [YS16]. Santos and Gatti use character-to-sentence level representation with CNN frameworks for effective sentiment classification [SG14]. Johnson and Zhang explore the usage of region embeddings for effective short text categorization, due to the ability to capture contexts over a larger span where word embeddings fail [JZ15]. Wang and others perform semantic clustering using density peaks on pre-trained word embeddings forming a representation they call semantic cliques, with such semantic units used further with convolutions for short-text mining [Wan+15a].

Zhang et al. explore the use of character-level representations for a CNN instead of word-level embeddings [ZZL15]. On a large dataset, character-level modeling of sentences performs very well when compared to traditional sparse representations or even deep learning frameworks, such as word embedding CNNs or RNNs. Conneau et al. design a very deep CNN along with modifications such as shortcuts to learn more complex features [Con+16]. Xiao and Cho's research further extends the character level encoding for entire document classification task when combined with RNNs with a lower number of parameters [XC16].

6.7.2 Text Clustering and Topic Mining

In their work, Xu et al. use a CNN for short-text clustering in a completely unsupervised manner [Xu+17]. The original keyword features from the text are used to generate compact binary code with locality-preserving constraints. Deep feature representation is obtained using word embeddings with the dynamic CNN in [KGB14b]. The outputs of the CNN are made to fit the binary codes during the training process. The deep features thus evolved from the CNN layers are passed to normal clustering algorithms, such as k-means, to give the final clustering of the data. Experiments show that this method does significantly better than traditional feature-based methods on various datasets.

Lau et al. jointly learn the topics and language models using CNN-based frameworks that result in better coherent, effective, and interpretable topic models [LBC17]. Document context is captured using a CNN framework, and the resulting document vectors are combined with topic vectors to give an effective document-topic representation. The language model is composed of the same document vectors from above and LSTMs.

6.7.3 Syntactic Parsing

In seminal work by Collobert et al., many NLP tasks, such as POS Tagging, Chunking, Named Entity Resolution, and Semantic Role Labeling are performed for the first time using word embeddings and CNN blocks [Col+11]. The research shows the strength of using CNNs in finding features in an automated way rather than hand-crafted task-specific features used for similar tasks. Zheng et al. show that character-based representations, CNN-based frameworks, and dynamic programming can be very effective in performing syntactic parsing without any task-specific feature engineering in languages as complex as Chinese [Zhe+15]. Santos and Zadrozny show that using character-level embeddings jointly with word-level embeddings and deep CNNs can further improve POS Tagging tasks [DSZ14] in English and Portuguese. Zhu et al. propose a recursive CNN (RCNN) to capture complex structure in the dependency trees. An RCNN has a basic k-ary tree as a unit that can capture the parsing tree with the relationship between nodes and children. This structure can be applied recursively to map the representation for the entire dependency tree. The RCNN is shown to be very effective as a re-ranking model in dependency parsers [Zhu+15].

6.7.4 Information Extraction

As discussed in Chap. 3, Information Extraction (IE) is a general category which has various sub-categories, such as entity extraction, event extraction, relation extraction, coreference resolution, and entity linking to name a few. In the work of Chen et al., instead of using hand-coded features, the researchers employ word-based representations to capture lexical and sentence level features that work with a modified CNN for superior multiple event extractions in text [Che+15]. The researchers use word context, positions, and event type in their representations and embeddings flowing into a CNN with multiple feature maps. Instead of a CNN with max pooling, which can miss multiple events happening in a sentence, the researchers employ dynamic multi pooling. In dynamic multi pooling, the feature maps are split into three parts, finding a maximum for each part.

As discussed above in the section on NLP, Zheng et. al [Zhe+15] and Santos et. al [DSZ14] employ CNNs for relation classification without any hand-coded features. In the work of Vu et al., relation classification uses a combination of CNNs and RNNs. In a sentence which has entities and relations between them, the researchers perform a split between left and middle, and middle and right part of the sentence, flowing into two different word embeddings and CNN layers with max pooling. This design gives special attention to the middle part, which is an important aspect in relation classification as compared to previous research. Bi-directional RNNs with an additional hidden layer are introduced to capture relation arguments from succeeding words. The combined approach shows significant improvements over traditional feature-based and even independently used CNNs and RNNs.

Nguyen and Grisham use a CNN-based framework for relation extraction [NG15b]. They use word embeddings and position embeddings concatenated as the input representation of sentences with entities having relations. They employ a CNN with multiple filter sizes and max pooling. It is interesting to see that the performance of their framework is better than all handcrafted feature engineering-based machine learning systems that use many morphological and lexical features.

In the work of Adel et al., the researchers compare many techniques from a traditional feature-based machine learning to CNN-based deep learning for relation classification in the context of slot filling [AS17a]. Similar to the above work, they break the sentences into three parts for capturing the contexts and use a CNN with k-max pooling.

6.7.5 Machine Translation

Hu et al. highlight how CNNs can be used to encode both semantic similarities and contexts in translation pairs and thus yield a more effective translations [Hu+15]. Another interesting aspect of this research is its employment of a curriculum training, where the training data is categorized from easy to difficult, and uses phrase to sentence for contexts encoding for effective translations.

Meng et al. build a CNN-based framework for guiding signals from both source and target during machine translation [Men+15]. Using CNNs with gating provides guidance on which parts of source text have influence on the target words. Fusing them with entire source sentence for context yields a better joint model.

Gehring et al. use a CNN with an attention module, thus showing not only a fast performing and parallelizable implementation, but also a more effective model compared to LSTM-based ones. Using word and positional embeddings in the input representation, stacking the filters on top of each other for hierarchical mappings, gated linear units, and multi-step attention module gives the researchers a real edge on English-French and English-German translation [Geh+17b].

6.7.6 Summarizations

Denil et. al show that by having a hierarchy of word embeddings that compose sentence embeddings, which in turn compose document embeddings, and using dynamic CNNs gives useful document summarizations, as well as effective visualizations [Den+14]. The research also highlights that the composition can capture from low-level lexical features to high-level semantic concepts in various tasks, including summarization, classification, and visualization.

Cheng and Lapata develop a neural framework combining CNN for hierarchical document encoding and attention extractor for effective document summarizations [CL16]. Mapping the representations very close to the actual data, where there is the composition of words into sentences, sentences to paragraphs, paragraphs to entire document using CNN, and max pooling gives the researchers a clear advantage in capturing both local and global sentential information.

6.7.7 Question and Answers

Dong et al. use multi-column CNNs for analyzing questions from various aspects such as answer path, answer type, and answer contexts [Don+15b]. The embedding of answer layers using entities and relations using low-dimensional embedding space is utilized along with a scoring layer on top to rank candidate answers. The work shows that without hand-coded or engineered features, the design provides a very effective question answering system.

Severyn and Moschitti show that using relational information given by the matches between the words used in the question and answers with a CNN-based framework gives very effective of question-answering system [SM15]. If we can map the question to finding facts from the database, Yin et al. show in their work that a two-stage approach using a CNN-based framework can yield excellent results [Yin+16b]. The facts in the answers are mapped to subject, predicate, and object. Then entity linking from the mention in the question to the subject employing a character-level CNN is the first stage of the pipeline. Matching the predicate in the fact with the question using a word-level CNN with attentive max pooling is the second stage of the pipeline.

6.8 Fast Algorithms for Convolutions

CNNs, in general, are more parallel in nature as compared to other deep learning architectures. However, as training data size has increased, need for faster predictions in near real-time, and GPU based hardware for parallelizing operations are becoming more widespread, convolution operations in CNN have gone through many enhancements. In this section, we will discuss some fast algorithms for CNN and give insights into how convolutions can be made faster with fewer floating point operations [LG16].

6.8.1 Convolution Theorem and Fast Fourier Transform

This theorem states that convolution in the time domain (any discrete input such as image or text) is equivalent to pointwise multiplications in the frequency domain. We can represent this transformation as taking fast Fourier transform (FFT) of the input and the kernel, multiplying it and taking inverse FFT.

$$(\mathbf{f} \times \mathbf{g})(t) = \mathcal{F}^{-1}(\mathcal{F}(\mathbf{f}) \cdot \mathcal{F}(\mathbf{g})) \tag{6.42}$$

Convolution operations is an n^2 algorithm, whereas it has been shown that FFT is $n \log(n)$. Thus, for a larger sequence even with 2 operations of FFT and inverse FFT, it can be shown that $n + 2n \log(n) < n^2$, thus giving a significant speedup.

6.8.2 Fast Filtering Algorithm

Winograd algorithms use computational tricks to reduce the multiplications in convolution operations. For example, if a 1d input data of size n needs to be convolved with a filter of size r to give an m-size output, then it will take $m \times r$ multiplications in normal. The minimal filtering algorithms $F(m,r)$ can be shown to need only $\mu(F(m,r)) = m+r-1$ multiplications. Let us consider a simple example with input $[d_0,d_1,d_2,d_3]$ of size 4 convolved with filter $[g_0,g_1,g_2]$ of size 3 to give size 2 outputs $[m_1,m_2]$. In traditional convolution, we would need 6 multiplications but by arranging the inputs as shown:

$$\begin{bmatrix} d_0 & d_1 & d_2 \\ d_1 & d_2 & d_3 \end{bmatrix} \begin{bmatrix} g_0 \\ g_1 \\ g_2 \end{bmatrix} = \begin{bmatrix} m_1+m_2+m_3 \\ m_2-m_3-m_4 \end{bmatrix} \tag{6.43}$$

where

$$m_1 = (d_0-d_2)g_0 \tag{6.44}$$

$$m_2 = (d_1+d_2)\frac{(g_0+g_1+g_2)}{2} \tag{6.45}$$

$$m_3 = (d_2-d_1)\frac{(g_0-g_1+g_2)}{2} \tag{6.46}$$

$$m_4 = (d_1-d_3)g_2 \tag{6.47}$$

Thus, multiplications are reduced to only $m+r-1$, i.e., 4, and savings are only $\frac{6}{4} = 1.5$. The additions $\frac{(g_0+g_1+g_2)}{2}$ and $\frac{(g_0-g_1+g_2)}{2}$ can be precomputed from the filters, giving additional performance benefits. These fast filtering algorithms can be written in matrix form as:

$$Y = \mathbf{A}^\mathsf{T}[(\mathbf{Gg}) \circ (\mathbf{B}^\mathsf{T}\mathbf{d})] \tag{6.48}$$

where $\circ$ is element-wise multiplication. For the 1d example, the matrices are

$$\mathbf{B}^\mathsf{T} = \begin{bmatrix} 1 & 0 & -1 & 0 \\ 0 & 1 & 1 & 0 \\ 0 & -1 & 1 & 0 \\ 0 & 1 & 0 & -1 \end{bmatrix} \quad \mathbf{G} = \begin{bmatrix} 1 & 0 & 0 \\ \frac{1}{2} & \frac{1}{2} & \frac{1}{2} \\ \frac{1}{2} & \frac{-1}{2} & \frac{1}{2} \\ 0 & 0 & 1 \end{bmatrix} \tag{6.49}$$

$$\mathbf{A}^\mathsf{T} = \begin{bmatrix} 1 & 1 & 1 & 0 \\ 0 & 1 & -1 & -1 \end{bmatrix} \tag{6.50}$$

$$\mathbf{g} = \begin{bmatrix} g_0 & g_1 & g_2 \end{bmatrix}^\mathsf{T} \tag{6.51}$$

$$\mathbf{d} = \begin{bmatrix} d_0 & d_1 & d_2 & d_3 \end{bmatrix}^\mathsf{T} \tag{6.52}$$

The 2d minimal algorithms can be expressed in terms of nested 1d algorithms. For example, $F(m,r)$ and $F(n,s)$ can be used to compute $m \times n$ outputs for a filter of size $r \times s$. Thus

$$\mu(F(m \times n, r \times s)) = \mu(F(m,r))\mu(F(n,s)) = (m+r-1)(n+s-1) \qquad (6.53)$$

Similarly, the matrix form for a 2d algorithm can be written as:

$$Y = \mathbf{A}^\mathsf{T}[(\mathbf{G}^\mathsf{T}\mathbf{g}\mathbf{G}) \circ (\mathbf{B}^\mathsf{T}\mathbf{d}\mathbf{B})]\mathbf{A}^\mathsf{T} \qquad (6.54)$$

A filter g is now of size $r \times r$ and each input can be considered to be a tile of dimension $(m+r-1) \times (m+r-1)$. Generalization to a non-square matrix can be done by nesting $F(m,r)$ and $F(n,s)$ as above. Thus, for $F(2 \times 2, 3 \times 3)$, a normal convolution will use $4 \times 9 = 36$ multiplications, whereas fast filter algorithms need only $(2+3-1) \times (2+3-1) = 16$, giving a savings of $\frac{36}{16} = 2.25$.

Here are some practical tips in regards to CNNs especially for classification tasks.

- For the classification task, it is always good to start with Yoon Kim et al. proposed CNN with word representation [Kim14b].
- Using pre-trained embeddings with word2vec or GloVe as compared to 1-hot vector representation as a single static channel for mapping sentences should be done before fine-tuning or introducing multiple channels.
- Having multiple filters such as $[3,4,5]$, number of feature maps ranging from 60 to 500, and ReLU as the activation function often gives a good performance [ZW17].
- 1-max pooling as compared to average pooling and k-max pooling gives better results [ZW17]
- Choice of regularization technique, i.e., $L1$ or $L2$ or dropout, etc. depends on the dataset and always good to try without the regularization and with regularization and compare the validation metrics.
- Learning curves and variances in them across multiple cross-validation gives an interesting idea of "robustness" of the algorithm. Flatter the curves and smaller the variances, highly likely the validation metric estimates are accurate.
- Understand the predictions from the model on the validation data to look for the patterns in false positives and false negatives. Is it because of spelling mistakes? Is it because of dependencies and orders? Is it because of lack of training data to cover the cases?
- Character-based embeddings can be useful provided there is enough training data.
- Adding other structures such as LSTM, hierarchical, attention-based should be done incrementally to see the impact of each combination.

6.9 Case Study

To get hands-on experience on a real-world data analysis problem that involves many of the techniques and frameworks described in this chapter, we will use sentiment classification from text. In particular, we will utilize on the public U.S. airline sentiment dataset, scraped from Twitter for classifying tweets as positive, negative, or neutral. Negative tweets can be further classified for their reason.

We will evaluate the effectiveness of different deep learning techniques involving CNNs with various input representations for sentiment classification. In this case study, the classification will be based on the text of the tweet only, and not on any tweet metadata. We will explore various representations of text data, such as word embeddings trained from the data, pre-trained word embeddings, and character embeddings. We have not done a lot of hyperparameter optimization for each method to show the best it can produce without further fine-tuning. Readers are welcome to use the notebook and code to explore fine-tuning themselves.

6.9.1 Software Tools and Libraries

First, we need to describe the main open source tools and libraries we will use for our case study.

- **Keras** (www.keras.io) is a high-level deep learning API written in Python which gives a common interface to various deep learning backends, such as TensorFlow, CNTK, and Theano. The code can run seamlessly on CPUs and GPUs. All experiments with CNN are done using Keras API.
- **TensorFlow** (https://www.tensorflow.org/) is a popular open source machine learning and deep learning library. We use TensorFlow as our deep learning library but Keras API as the basic API for experimenting.
- **Pandas** (https://pandas.pydata.org/) is a popular open source implementation for data structures and data analysis. We will use it for data exploration and some basic processing.
- **scikit-learn** (http://scikit-learn.org/) is a popular open source for various machine learning algorithms and evaluations. We will use it only for sampling and creating datasets for estimations in our case study.
- **Matplotlib** (https://matplotlib.org/) is a popular open source for visualization. We will use it to visualize performance.

Now we are ready to focus on the four following sub-tasks.

- Exploratory data analysis
- Data preprocessing and data splits
- CNN model experiments and analysis
- Understanding and improving models

6.9.2 *Exploratory Data Analysis*

The total data has 14,640 labeled data, 15 features/attributes of which only the attribute text will be used for learning. The classes are in three categories of positive, negative, and neutral. We will take 15% of the total data for testing from the whole dataset in a stratified way and similarly 10% from training data for validation. Normally cross-validation (CV) is used for both model selection and parameter tuning, and we will use the validation set to reduce the time to run. We did compare CV estimates with separate validation set and both looked comparable.

Figure 6.23 shows the class distribution. We see that there is a skew in class distribution towards negative sentiments as compared to positive and neutral.

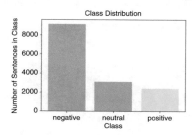

Fig. 6.23: Number of instances across different classes

One interesting step in EDA is to plot the word cloud for positive and negative sentiment data from the entire dataset to understand some of the most frequent words used and maybe correlating with that sentiment. Tokens that are over-represented in the cloud are mostly adjectives such as "thanks," "great," "good," "appreciate," etc. in the positive tweets while the negative sentiment word cloud in has reasons "luggage," "canceled flight," "website," etc. as shown in Fig. 6.24.

6.9.3 *Data Preprocessing and Data Splits*

We perform some further basic data processing to remove stop words and mentions from the text as they are basically not useful in our classification task. The listing below shows the basic data cleanup code.

```
# remove stop words with exceptions
def remove_stopwords(input_text):
    stopwords_list = stopwords.words('english')
    # Some words which might indicate a certain sentiment are kept
    whitelist = ["n't", "not", "no"]
    words = input_text.split()
    clean_words = [word for word in words if (
```

Fig. 6.24: Word Cloud for negative sentiments

```
8          word not in stopwords_list or word in whitelist) and
     len(word) > 1]
9     return " ".join(clean_words)
10 # remove mentions
11
12
13 def remove_mentions(input_text):
14     return re.sub(r'@\w+', '', input_text)
15
16
17 tweets = tweets[[TEXT_COLUMN_NAME, LABEL_COLUMN_NAME]]
18 tweets[TEXT_COLUMN_NAME] = tweets[TEXT_COLUMN_NAME].apply(
19     remove_stopwords).apply(remove_mentions)
20 tweets.head()
```

Next, we will split the entire data into training and testing with 85% for training and
15% for testing. We will build various models using same training data and evaluate
with respect to same test data to get a clear comparison.

```
1 X_train, X_test, y_train, y_test = train_test_split(
2     tweets[TEXT_COLUMN_NAME], tweets[LABEL_COLUMN_NAME],
      test_size=0.15, random_state=37)
```

We then perform tokenization, splits of training data into training and validation and
sequence mapping with fixed size. We use the maximum text length in our corpus
to determine the sequence length and do padding.

```
1 # tokenization with max words defined and filters to remove
       characters
2 tk = Tokenizer(num_words=NB_WORDS,
3                filters='!"#$%&()*+,-./:;<=>?@[\\]^_`{|}~\t\n',
4                lower=True,
5                split=" ")
6 tk.fit_on_texts(X_train)
```

```
7
8  # understand the sequence distribution
9  seq_lengths = X_train.apply(lambda x: len(x.split(' ')))
```

```
1  # convert train and test to sequence using the tokenizer
       trained on the training data
2  X_train_total = tk.texts_to_sequences(X_train)
3  X_test_total = tk.texts_to_sequences(X_test)
4
5  # pad the sequences to a maximum length
6  X_train_seq = pad_sequences(X_train_total, maxlen=MAX_LEN)
7  X_test_seq = pad_sequences(X_test_total, maxlen=MAX_LEN)
8
9  # perform encoding of
10 le = LabelEncoder()
11 y_train_le = le.fit_transform(y_train)
12 y_test_le = le.transform(y_test)
13 y_train_one_hot = to_categorical(y_train_le)
14 y_test_one_hot = to_categorical(y_test_le)
```

6.9.4 CNN Model Experiments

Once we have preprocessed and created training, validation and test sets, we will
perform various modeling analysis on the data. We will first do some basic analysis
using simple CNN and then proceed to run various configurations and modifications
of CNN discussed in the chapter.

Next, we show a basic code of CNN which input layer with maximum length of
sentence 24, which outputs a 100-dimensional vectors which are convoluted with
64 filters each of height or size 3, equivalent to 3-gram, going through a ReLU non-
linear activation function, then a max pooling layer that gets flattened so that it is
input of a fully connected layer which outputs to soft max layer with 3 outputs for
3 classes.

```
1  # basic CNN Model to understand how it works
2  def base_cnn_model():
3      # Embedding
4      # Layer->Convolution1D->MaxPooling1D->Flatten ->
       FullyConnected->Classifier
5      model = Sequential(
6          [Embedding(input_dim=10000, output_dim=100,
       input_length=24),
7              Convolution1D(filters=64, kernel_size=3, padding='
       same', activation='relu'),
8              MaxPooling1D(),
9              Flatten(),
10             Dense(100, activation='relu'),
11             Dense(3, activation='softmax')])
12     model.summary()
13     return model
14
```

```
15
16 #train and validate
17 base_cnn_model = base_cnn_model()
18 base_history = train_model(
19     base_cnn_model,
20     X_train_seq,
21     y_train,
22     X_valid_seq,
23     y_valid)
```

A single channel using pre-trained embeddings and Yoon Kim's CNN model with multiple filters is shown here

```
1  # create embedding matrix for the experiment
2  emb_matrix = create_embedding_matrix(tk, 100, embeddings)
3
4  # single channel CNN with multiple filters
5
6
7  def single_channel_kim_cnn():
8      text_seq_input = Input(shape=(MAX_LEN,), dtype='int32')
9      text_embedding = Embedding(NB_WORDS + 1,
10                                 EMBEDDING_DIM,
11                                 weights=[emb_matrix],
12                                 trainable=True,
13                                 input_length=MAX_LEN)(
14     text_seq_input)
15
16     filter_sizes = [3, 4, 5]
17     convs = []
18     # parallel layers for each filter size with conv1d and max
19        pooling
18     for filter_size in filter_sizes:
19         l_conv = Convolution1D(
20             filters=128,
21             kernel_size=filter_size,
22             padding='same',
23             activation='relu')(text_embedding)
24         l_pool = MaxPooling1D(filter_size)(l_conv)
25         convs.append(l_pool)
26     # concatenate outputs from all cnn blocks
27     merge = concatenate(convs, axis=1)
28     convol = Convolution1D(128, 5, activation='relu')(merge)
29     pool1 = GlobalMaxPooling1D()(convol)
30     dense = Dense(128, activation='relu', name='Dense')(pool1)
31     # classification layer
32     out = Dense(3, activation='softmax')(dense)
33     model = Model(
34         inputs=[text_seq_input],
35         outputs=out,
36         name="KimSingleChannelCNN")
37     model.summary()
38     return model
39
```

```
40
41  single_channel_kim_model = single_channel_kim_cnn()
42  single_channel_kim_model_history = train_model(
43      single_channel_kim_model, X_train_seq, y_train,
        X_valid_seq, y_valid)
```

We will list different experiments with its name and their purpose before highlighting the results from each.

1. **Base CNN**. A basic single block of CNN with convolution with filter of size 3, max pooling and a softmax layer.
2. **Base CNN + Dropout**. To see the impact of dropout on the base CNN.
3. **Base CNN + Regularization**. To see the impact of $L2$ regularization on the base CNN.
4. **Multi-filters**. To see the impact of adding more filters [2,3,4,5] to CNN.
5. **Multi-filters + Increased Maps**. To see the impact of increasing the filter maps from 64 to 128.
6. **Multi-filters + Static Pre-trained Embeddings**. To see the impact of using pre-trained word embeddings in CNN.
7. **Multi-filters + Dynamic Pre-trained Embeddings**. To see the impact of using pre-trained word embeddings in CNN that are trained on the training set.
8. **Yoon Kim's Single Channel**. Single channel CNN using widely known architecture [Kim14b].
9. **Yoon Kim's Multiple Channel**. Multiple channel CNN using widely known architecture [Kim14b] to see the impact on increasing the channels. Static and dynamic embeddings are used as two different channels.
10. **Kalchbrenner et al. Dynamic CNN**. Kalchbrenner et al. based dynamic CNN with K-max pooling [KGB14b].
11. **Multichannel Variable MVCNN**. We use two embedding layers with static and dynamic channels [YS16].
12. **Multigroup MG-CNN**. We use three different channels (two inputs with embedding layers using GloVe and one using fastText) with different dimensions (100 and 300) [ZRW16].
13. **Word-level Dilated CNN**. Exploring the concept of dilations with the reduced parameters and larger coverage using word-level inputs [YK15].
14. **Character-level CNN**. We explore the character-level embeddings instead of the word-level embeddings [ZZL15].
15. **Very Deep Character-level CNN**. Impact of a very deep level CNN with multiple layers [Con+16].
16. **Character-level Dilated CNN**. Exploring the concept of dilations with the reduced parameters and larger coverage using character-level inputs [YK15].
17. **C-LSTM**. Exploring C-LSTM to verify how CNN can be used to capture the local features of phrases and RNN to capture global and temporal sentence semantics [Zho+15].
18. **AC-BiLSTM** Exploring the bi-directional LSTM with CNN [LZ16].

We use some practical deep learning aspects while training the models as high-lighted below:

```
# use the validation loss to detect the best weights to be
    saved
checkpoints.append(ModelCheckpoint(checkpoint_file, monitor='
    val_loss', verbose=0, save_best_only=True,
    save_weights_only=True, mode='auto', period=1))
# output to TensorBoard
checkpoints.append(TensorBoard(log_dir='./logs', write_graph=
    True, write_images=False))
# if no improvements in 10 epochs, then quit
checkpoints.append(EarlyStopping(monitor='val_loss', patience
    =10))
```

In the Table 6.1 we will highlight the results of running different CNN architectures given above and the results we track with accuracy and average precision.

Table 6.1: CNN test results summary

Experiments	Accuracy %	Average precision %
Base CNN	77.77	82
Base CNN + dropout	70.85	78
Base CNN + regularization	78.32	83
Multi-filters	80.55	86
Multi-filters + increased maps	79.18	85
Multi-filters + static pre-trained embeddings	77.41	84
Multi-filters + dynamic pre-trained embeddings	78.96	85
Yoon Kim's Single Channel	79.50	85
Yoon Kim's Multiple Channel	80.05	86
Kalchbrenner et al. dynamic CNN	78.68	85
Multichannel variable MVCNN	79.91	85
Multigroup CNN MG-CNN	**81.96**	**87**
Word-level dilated CNN	77.81	84
Character-level CNN	73.36	81
Very deep character-level CNN	67.89	73
Character-level dilated CNN	74.18	78
C-LSTM	79.14	85
AC-BiLSTM	79.46	86

Bold indicates best result or accuracy amongst all the experiments

We will list some high level analysis and observations from Table 6.1 and our analysis from various runs below:

- Basic CNN with L2 regularization seems to improve on overfitting from both the angles of reducing the loss and cutting the max loss. Dropout seems to hurt the performance of basic CNN.

- Multiple layers and multiple filters seem to improve both accuracy and average precision by more than 2%.
- Using pre-trained embeddings which get trained in the data gives one of the best performances and is very much in line with many research.
- Multigroup Norm constraint MG-CNN show the best results in both accuracy and average precision in word-based representation. Using three embedding channels with two different embeddings with different sizes seems to give the edge.
- Yoon Kim's model with two channels has second best performance and confirms that it should be always a model to try in classification problems. The performance of dual channel along with MG-CNN confirms that increasing number of channels helps the model in general.
- Increasing the depth and complexity of CNN and hence the parameters has not much effect on generalization, again can be accounted for small training data size.
- Character-based representation shows relatively poor performance and that is well in line with most research because of limited corpus and training size.
- Introducing complexity by combining CNN and LSTM does not improve the performance and again can be attributed to the task complexity and the size of training data.

6.9.5 Understanding and Improving the Models

In this section, we will give some practical tips and tricks to help the researchers gain insights into model behaviors and improve them further.

One way of understanding the model behavior is to look at the predictions at various layers using some form of dimensionality reduction techniques and visualization techniques. To explore the behavior of last but one layer before the classification layer, we first create a clone of the model by removing the last layer and using the test set to generate high dimensional outputs from this layer. We then use PCA to get 30 components from 128-dimensional outputs and finally project this using TSNE. As shown in Fig. 6.25, we see a clear reason why **Yoon Kim's Single Channel CNN** performs better than the **Basic CNN**.

Next, we will analyze the false positives and false negatives to understand the patterns and the causes, to improve the models further.

Table 6.2 highlights some of the text and probable cause. Having words such as **late flight**, **overhead**, etc. are so overrepresented in the negatives that it

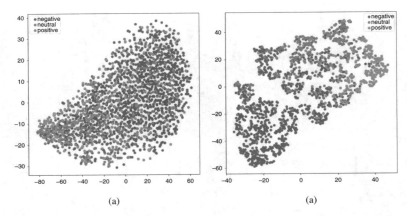

(a)   (a)

Fig. 6.25: A layer before last with 128 dimensions is used to visualize the test data with PCA and TSNE. (**a**) Hidden layer from Basic CNN. (**b**) Hidden layer from Yoon Kim's Single Channel CNN

causes even sentences which have these to be classified negatives. Adding more positives with similar language and using average pooling might help. Adding support for emojis and even embeddings that have been trained on them can improve further on examples that use them.

Table 6.2: False negatives

Text	Prediction	Probable cause
Kudos ticket agents making passengers check bags big fit **overhead**	Negative	Keyword overrepresented
Thankful united ground staff put last seat **last flight** out home **late flight** still home	Negative	Keyword overrepresented
Emoji love flying	Neutral	Emojis

Table 6.3 highlights some of the text and probable causes. Having words such as **awesome**, **thanks**, etc. are so overrepresented in the positives that it causes even sentences which have these to be classified positives. Adding more negatives with similar language and using average pooling might help. Having sarcasm based datasets, training embeddings on them and using it as input channel can improve the performance.

Using Lime for model explanations, especially false positives and false negatives gives insights into the reason with weights associated with the keywords as shown in Fig. 6.26.

Table 6.3: False positives

Text	Prediction	Probable cause
Forget reservations **thank great company** i've flighted flight once again **thank you**	Positive	Keyword overrepresented and sarcasm
Thanks finally made it and missed meetings now	Positive	Keyword overrepresented and sarcasm
My flight cancelled led mess please **thank awesome** out	Positive	Keyword overrepresented and sarcasm

```
def keras_wrapper(texts):
    _seq = tk.texts_to_sequences(texts)
    _text_data = pad_sequences(_seq, maxlen=MAX_LEN)
    return single_channel_kim_model.predict(_text_data)

exp = explainer.explain_instance('forget reservations thank
    great company i have       cancelled flighted flight once
    again thank you',
keras_wrapper,
num_features=10,
labels=[0, 2]
)
```

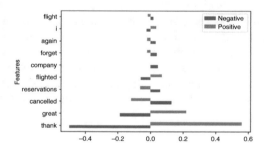

Fig. 6.26: Lime outputs weights for words for both positive and negative class

6.9.6 Exercises for Readers and Practitioners

Some other interesting problems and research questions that readers and practitioners can further attempt are listed below:

1. What is the measurable impact of preprocessing, such as removing stop words, mentions, stems, and others on the CNN performance?
2. Does the embedding dimension have an impact on the CNN; for example, a 100-dimensional embedding vs. a 300-dimensional embedding?
3. Does the type of embeddings such as word2vec, GloVE, and other word embeddings change the performance significantly across CNN frameworks?
4. Is there an impact in performance when multiple embeddings such as word, POS Tags, positional are used for sentence representation with CNN frameworks?
5. Does hyperparameter tuning done more robustly across various parameters improve validation and hence the test results significantly?
6. Many researchers use the ensemble of models, with different parameters as well as with different model types. Does that improve performance?
7. Do pre-trained character embeddings improve performance as compared to one that we tune on the limited training data?
8. Using some of the standard CNN frameworks such as AlexNet, VGG-16, and others with modifications for text processing and doing a survey of these on the dataset seems like further interesting research.

6.10 Discussion

Geoffery Hinton in his talk "What is wrong with convolutional neural nets?" given at MIT highlights some of the issues with CNNs, especially around max pooling. The talk explains how max pooling can "ignore" some of the important signals because of the bias it has towards finding the "key" features. Another issue highlighted was around how the filters can capture different features and build higher level features but fail to capture the "relationship" between these features to some extent. An example being, presence of features detecting eye, ears, mouth in face recognition using CNNs with different filters and layers cannot separate out an image with presence of these but not in the right place. Capsule Networks, using capsules as a basic building block is seen as an alternative design to overcome these issues [SFH17].

References

[AS17a]   Heike Adel and Hinrich Schütze. "Global Normalization of Convolutional Neural Networks for Joint Entity and Relation Classification". In: *EMNLP*. Association for Computational Linguistics, 2017, pp. 1723–1729.

[Bro+94]   Jane Bromley et al. "Signature Verification using a "Siamese" Time Delay Neural Network". In: *Advances in Neural Information Pro-*

cessing Systems 6. Ed. by J. D. Cowan, G. Tesauro, and J. Alspector. Morgan-Kaufmann, 1994, pp. 737–744.

[Che+15]   Yubo Chen et al. "Event Extraction via Dynamic Multi-Pooling Convolutional Neural Networks". In: *ACL (1)*. The Association for Computer Linguistics, 2015, pp. 167–176.

[CL16]   Jianpeng Cheng and Mirella Lapata. "Neural Summarization by Extracting Sentences and Words". In: *Proceedings of the 54th Annual Meeting of the Association for Computational Linguistics*. Association for Computational Linguistics, 2016, pp. 484–494.

[Col+11]   R. Collobert et al. "Natural Language Processing (Almost) from Scratch". In: *Journal of Machine Learning Research* 12 (2011), pp. 2493–2537.

[CW08b]   Ronan Collobert and Jason Weston. "A Unified Architecture for Natural Language Processing: Deep Neural Networks with Multitask Learning". In: *Proceedings of the 25th International Conference on Machine Learning*. ICML '08. 2008.

[Con+16]   Alexis Conneau et al. "Very Deep Convolutional Networks for Natural Language Processing". In: *CoRR* abs/1606.01781 (2016).

[Den+14]   Misha Denil et al. "Modelling, Visualising and Summarising Documents with a Single Convolutional Neural Network." In: *CoRR* abs/1406.3830 (2014).

[Don+15b]   Li Dong et al. "Question Answering over Freebase with Multi-Column Convolutional Neural Networks". In: *Proceedings of the International Joint Conference on Natural Language Processing*. Association for Computational Linguistics, 2015, pp. 260–269.

[DSZ14]   Cícero Nogueira Dos Santos and Bianca Zadrozny. "Learning Character-level Representations for Part-of-speech Tagging". In: *Proceedings of the 31st International Conference on International Conference on Machine Learning - Volume 32*. ICML'14. 2014, pp. II–1818–II–1826.

[Geh+17b]   Jonas Gehring et al. "Convolutional Sequence to Sequence Learning". In: *Proceedings of the 34th International Conference on Machine Learning*. Ed. by Doina Precup and Yee Whye Teh. Vol. 70. Proceedings of Machine Learning Research. 2017, pp. 1243–1252.

[Hu+15]   Baotian Hu et al. "Context-Dependent Translation Selection Using Convolutional Neural Network". In: *ACL (2)*. The Association for Computer Linguistics, 2015, pp. 536–541.

[JZ15]   Rie Johnson and Tong Zhang. "Semi-supervised Convolutional Neural Networks for Text Categorization via Region Embedding". In: *Advances in Neural Information Processing Systems 28*. Ed. by C. Cortes et al. 2015, pp. 919–927.

[KGB14b]   Nal Kalchbrenner, Edward Grefenstette, and Phil Blunsom. "A Convolutional Neural Network for Modelling Sentences". In: *CoRR* abs/1404.2188 (2014).

[Kim14b]   Yoon Kim. "Convolutional Neural Networks for Sentence Classifica-
            tion". In: *CoRR* abs/1408.5882 (2014).

[KSH12a]   Alex Krizhevsky, I Sutskever, and G. E Hinton. "ImageNet Classifi-
            cation with Deep Convolutional Neural Networks". In: *Advances in
            Neural Information Processing Systems (NIPS 2012)*. 2012, p. 4.

[LBC17]   Jey Han Lau, Timothy Baldwin, and Trevor Cohn. "Topically Driven
            Neural Language Model". In: *ACL (1)*. Association for Computa-
            tional Linguistics, 2017, pp. 355–365.

[LG16]   Andrew Lavin and Scott Gray. "Fast Algorithms for Convolu-
            tional Neural Networks". In: *CVPR*. IEEE Computer Society, 2016,
            pp. 4013–4021.

[LB95]   Y. LeCun and Y. Bengio. "Convolutional Networks for Images,
            Speech, and Time-Series". In: *The Handbook of Brain Theory and
            Neural Networks*. 1995.

[LeC+98]   Yann LeCun et al. "Gradient-Based Learning Applied to Document
            Recognition". In: *Proceedings of the IEEE*. Vol. 86. 1998, pp. 2278–
            2324.

[Li+15]   Yujia Li et al. "Gated Graph Sequence Neural Networks". In:
            *CoRR*abs/1511.05493 (2015).

[LZ16]   Depeng Liang and Yongdong Zhang. "AC-BLSTM: Asymmetric
            Convolutional Bidirectional LSTM Networks for Text Classifica-
            tion". In: *CoRR* abs/1611.01884 (2016).

[Ma+15]   Mingbo Ma et al. "Tree-based Convolution for Sentence Modeling".
            In: *CoRR* abs/1507.01839 (2015).

[Men+15]   Fandong Meng et al. "Encoding Source Language with Convolu-
            tional Neural Network for Machine Translation". In: *ACL (1)*. The
            Association for Computer Linguistics, 2015, pp. 20–30.

[Mou+14]   Lili Mou et al. "TBCNN: A Tree-Based Convolutional Neu-
            ral Network for Programming Language Processing". In: *CoRR*
            abs/1409.5718 (2014).

[NG15b]   Thien Huu Nguyen and Ralph Grishman. "Relation Extraction: Per-
            spective from Convolutional Neural Networks". In: *Proceedings
            of the 1st Workshop on Vector Space Modeling for Natural Lan-
            guage Processing*. Association for Computational Linguistics, 2015,
            pp. 39–48.

[RSA15]   Oren Rippel, Jasper Snoek, and Ryan P. Adams. "Spectral Repre-
            sentations for Convolutional Neural Networks". In: *Proceedings of
            the 28th International Conference on Neural Information Processing
            Systems - Volume 2*. NIPS'15. 2015, pp. 2449–2457.

[SFH17]   Sara Sabour, Nicholas Frosst, and Geoffrey E Hinton. "Dynamic
            Routing Between Capsules". In: 2017, pp. 3856–3866.

[SG14]   Cicero dos Santos and Maira Gatti. "Deep Convolutional Neural Net-
            works for Sentiment Analysis of Short Texts". In: *Proceedings of
            COLING 2014, the 25th International Conference on Computational
            Linguistics: Technical Papers*. 2014.

[SM15]   Aliaksei Severyn and Alessandro Moschitti. "Learning to Rank Short Text Pairs with Convolutional Deep Neural Networks". In: *Proceedings of the 38th International ACM SIGIR Conference on Research and Development in Information Retrieval*. SIGIR '15. 2015, pp. 373–382.

[SZ14]   Karen Simonyan and Andrew Zisserman. "Very Deep Convolutional Networks for Large-Scale Image Recognition". In: 2014.

[Sze+17]   Christian Szegedy et al. "Inception-v4, Inception-ResNet and the Impact of Residual Connections on Learning". In: *AAAI*. AAAI Press, 2017, pp. 4278–4284.

[Wan+15a]   Peng Wang et al. "Semantic Clustering and Convolutional Neural Network for Short Text Categorization". In: *Proceedings the 7th International Joint Conference on Natural Language Processing*. 2015.

[XC16]   Yijun Xiao and Kyunghyun Cho. "Efficient Character-level Document Classification by Combining Convolution and Recurrent Layers". In: *CoRR* abs/1602.00367 (2016).

[Xu+17]   Jiaming Xu et al. "Self-Taught convolutional neural networks for short text clustering". In: *Neural Networks* 88 (2017), pp. 22–31.

[YS16]   Wenpeng Yin and Hinrich Schütze. "Multichannel Variable-Size Convolution for Sentence Classification". In: *CoRR* abs/1603.04513 (2016).

[Yin+16a]   Wenpeng Yin et al. "ABCNN: Attention-Based Convolutional Neural Network for Modeling Sentence Pairs". In: *Transactions of the Association for Computational Linguistics* 4 (2016), pp. 259–272.

[Yin+16b]   Wenpeng Yin et al. "Simple Question Answering by Attentive Convolutional Neural Network". In: *Proceedings of COLING 2016, the 26th International Conference on Computational Linguistics: Technical Papers*. The COLING 2016 Organizing Committee, 2016, pp. 1746–1756.

[YK15]   Fisher Yu and Vladlen Koltun. "Multi-Scale Context Aggregation by Dilated Convolutions". In: *CoRR* abs/1511.07122 (2015).

[Yu+14]   Lei Yu et al. "Deep Learning for Answer Sentence Selection". In: *CoRR* abs/1412.1632 (2014).

[ZF13b]   Matthew D. Zeiler and Rob Fergus. "Stochastic Pooling for Regularization of Deep Convolutional Neural Networks". In: *CoRR* abs/1301.3557 (2013).

[ZZL15]   Xiang Zhang, Junbo Jake Zhao, and Yann LeCun. "Character level Convolutional Networks for Text Classification". In: *CoRR* abs/1509.01626 (2015).

[ZRW16]   Ye Zhang, Stephen Roller, and Byron C. Wallace. "MGNC-CNN: A Simple Approach to Exploiting Multiple Word Embeddings for Sentence Classification". In: *Proceedings of the 2016 Conference of the North American Chapter of the Association for Computational Linguistics: Human Language Technologies*. Association for Computational Linguistics, 2016, pp. 1522–1527.

[ZW17]   Ye Zhang and Byron Wallace. "A Sensitivity Analysis of (and Practitioners' Guide to) Convolutional Neural Networks for Sentence Classification". In: *Proceedings of the Eighth International Joint Conference on Natural Language Processing (Volume 1: Long Papers)*. Asian Federation of Natural Language Processing, 2017, pp. 253–263.

[Zhe+15]   Xiaoqing Zheng et al. "Character-based Parsing with Convolutional Neural Network". In: *Proceedings of the 24th International Conference on Artificial Intelligence*. IJCAI'15. 2015, pp. 1054–1060.

[Zho+15]   Chunting Zhou et al. In: *CoRR* abs/1511.08630 (2015).

[Zhu+15]   Chenxi Zhu et al. "A Re-ranking Model for Dependency Parser with Recursive Convolutional Neural Network". In: *Proceedings of International Joint Conference on Natural Language Processing*. 2015, pp. 1159–1168.

Chapter 7
Recurrent Neural Networks

7.1 Introduction

In the previous chapter, CNNs provided a way for neural networks to learn a hierarchy of weights, resembling that of n-gram classification on the text. This approach proved to be very effective for sentiment analysis, or more broadly text classification. One of the disadvantages of CNNs, however, is their inability to model contextual information over long sequences.[1] In many situations in NLP, it is desirable to capture long-term dependencies and maintain the contextual order between words to resolve the overall meaning of a text. In this chapter, we introduce recurrent neural networks (RNNs) that extend deep learning to sequences.

Sequential information and long-term dependencies in NLP traditionally relied on HMMs to compute context information, for example, in dependency parsing. One of the limitations of using a Markov chain for sequence focused tasks is that the generation of each prediction is limited to a fixed number of previous states. RNNs, however, relax this constraint, accumulating information from each time step into a "hidden state." This allows sequential information to be "summarized" and predictions can be made based on the entire history of the sequence.

Another advantage of RNNs is their ability to learn representations for variable length sequences, such as sentences, documents, and speech samples. This allows two samples of differing lengths to be mapped into the same feature space, allowing them to be comparable. In the context of language translation, for example, an input sentence may have more words than its translation, requiring a variable number of computational steps. Thus, it is highly beneficial to have knowledge of the entire length of the sentence before predicting the translation. We will study this example more at the end of this chapter.

In this chapter, we begin by describing the basic building blocks of RNNs and how they retain memory. We then describe the training process for RNNs and discuss the vanishing gradient problem, regularization, and RNN variants. Next we

---

[1] This statement is made in a basic context of CNNs and RNNs. The CNN vs. RNN superiority debate in sequential contexts is an active area of research.

© Springer Nature Switzerland AG 2019
U. Kamath et al., *Deep Learning for NLP and Speech Recognition*,
https://doi.org/10.1007/978-3-030-14596-5_7

show how to incorporate text input in recurrent architectures, leveraging word and character representations. We then introduce some traditional RNN architectures in NLP, and then move towards more modern architectures. The chapter is concluded with a case study on neural machine translation and a discussion about the future directions of RNNs.

7.2 Basic Building Blocks of RNNs

An RNN is a standard feed-forward neural network applied to vector inputs in a sequence. However, in order to incorporate sequential context into the next time step's prediction, a "memory" of the previous time steps in the sequence must be preserved.

7.2.1 Recurrence and Memory

First we will look at the idea of recurrence conceptually. Let us define a T length input sequence as X, where $X = \{\mathbf{x}_1, \mathbf{x}_2, \ldots, \mathbf{x}_T\}$, such that $\mathbf{x}_t \in \mathbb{R}^N$ is a vector input at time t. We then define our memory or history up to and including time t as $\mathbf{h}_t$.[2] Thus, we can define our output $\mathbf{o}_t$ as:

$$\mathbf{o}_t = f(\mathbf{x}_t, \mathbf{h}_{t-1}) \tag{7.1}$$

where the function f maps memory and input to an output. The memory from the previous time step is $\mathbf{h}_{t-1}$, and the input is $\mathbf{x}_t$. For the initial case $\mathbf{x}_1$, $\mathbf{h}_0$ is the zero vector $\mathbf{0}$.

Abstractly, the output $\mathbf{o}_t$ is considered to have summarized the information from the current input $\mathbf{x}_t$ and the previous history from $\mathbf{h}_{t-1}$. Therefore, $\mathbf{o}_t$ can be considered the history vector for the entire sequence up to and including time t. This yields the equation:

$$\mathbf{h}_t = \mathbf{o}_t = f(\mathbf{x}_t, \mathbf{h}_{t-1}) \tag{7.2}$$

Here we see where the term "recurrence" comes from: the application of the same function for each instance, wherein the output is directly dependent on the previous result.

More formally, we can extend this concept to neural networks by redefining the transformation function f as follows:

$$\mathbf{h}_t = f(\mathbf{U}\mathbf{x}_t + \mathbf{W}\mathbf{h}_{t-1}) \tag{7.3}$$

---

[2] This history vector will be called the hidden state later on for obvious reasons.

where **W** and **U** are weight matrices $\mathbf{W}, \mathbf{U} \in \mathbb{R}^{(N \times N)}$, and f is a non-linear function, such as tanh, σ, or ReLU. Figure 7.1 shows a diagram of the simple RNN that we have described here.

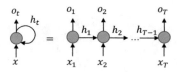

Fig. 7.1: Diagram of a recurrent neural network

7.2.2 PyTorch Example

The code snippet below illustrates a PyTorch implementation of the simple RNN previously described. It illustrates the recurrent computation in a modern framework.

```
# PyTorch RNN Definition
import torch.nn as nn
from torch.autograd import Variable
import torch.optim as optim

class RNN(nn.Module):

    def __init__(self, input_size):
        super(RNN, self).__init__()

        self.input_size = input_size
        self.hidden_size = input_size
        self.output_size = input_size

        self.U = nn.Linear(input_size, self.hidden_size)
        self.W = nn.Linear(self.hidden_size, self.output_size)

    def forward(self, input, hidden):
        Ux = self.U(input)
        Wh = self.W(hidden)
        output = Ux + Wh
        return output

rnn = RNN(input_size)

# Training the network
optimizer = optim.Adam(rnn.parameters(), lr=learning_rate,
    weight_decay=1e-5)
```

```
29  for epoch in range(n_epoch):
30      for data, target in train_loader:
31          # Get samples
32          input = Variable(data)
33
34          # Forward Propagation
35          hidden = Variable(torch.zeros(1, rnn.hidden_size))
36          for i in range(input.size()[0]):
37              output = rnn(input[i], hidden)
38              hidden = output
39
40          # Error Computation
41          loss = F.nll_loss(output, target)
42
43          # Clear gradients
44          optimizer.zero_grad()
45
46          # Backpropagation
47          loss.backward()
48
49          # Parameter Update
50          optimizer.step()
```

In this snippet, we perform a classification (and subsequently an error computation) at every time step. Instead of performing the computation as the outputs are computed, the error is computed after the forward propagation has completed for each time step. The error with respect to each time step is being backpropagated. This snippet by itself is incomplete, because our input size, output size, and hidden size will normally differ depending on the problem, as we will see in the upcoming sections.

7.3 RNNs and Properties

Let us now focus on a typical implementation of an RNN, how it is trained, and some of the difficulties that are introduced in training them.

7.3.1 Forward and Backpropagation in RNNs

RNNs are trained through backpropagation and gradient descent similar to feed-forward networks we have seen previously: forward propagating an example, calculating the error for a prediction, computing the gradients for each set of weights via backpropagation, and updating the weights according to the gradient descent optimization method.

The forward propagation equations for $\mathbf{h}_t$ and the output prediction $\hat{\mathbf{y}}_t$ are:

$$\mathbf{h}_t = \tanh(\mathbf{U}\mathbf{x}_t + \mathbf{W}\mathbf{h}_{t-1})$$
$$\hat{\mathbf{y}}_t = \text{softmax}(\mathbf{V}\mathbf{h}_t) \tag{7.4}$$

where the learnable parameters are $\mathbf{U}$, $\mathbf{W}$, and $\mathbf{V}$.[3] $\mathbf{U}$ incorporates the information from $\mathbf{x}_t$, $\mathbf{W}$ incorporates the recurrent state, and $\mathbf{V}$ learns a transformation to the output size and classification. A diagram of this RNN is shown in Fig. 7.2.

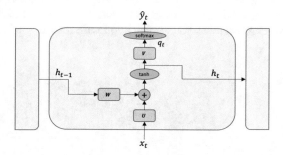

Fig. 7.2: Forward propagation of a simple RNN

We compute the error using cross-entropy loss at each time step t, where $\mathbf{y}_t$ is the target.

$$E_t = -\mathbf{y}_t \log \hat{\mathbf{y}}_t. \tag{7.5}$$

This gives us the overall loss with the following:

$$L(\mathbf{y}, \hat{\mathbf{y}}) = -\frac{1}{N} \sum_t \mathbf{y}_t \log \hat{\mathbf{y}}_t. \tag{7.6}$$

The gradients are computed by evaluating every path that contributed to the prediction $\hat{\mathbf{y}}_t$. This process is called backpropagation through time (BPTT). This process is illustrated in Fig. 7.3.

The parameters of our RNN are $\mathbf{U}, \mathbf{V}$, and $\mathbf{W}$, so we must compute the gradient of our loss function with respect to these matrices. Figure 7.4 shows backpropagation through a step of the RNN.

---

[3] It is common to split the single weight matrix $\mathbf{W}$ of an RNN in Eq. (7.3) into two separate weight matrices, here $\mathbf{U}$ and $\mathbf{W}$. Doing this allows for a lower computational cost and forces separation between the hidden state and input in the early stages of training.

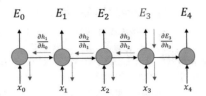

Fig. 7.3: Backpropagation through time shown with respect to the error at $t = 3$. The error E_3 is comprised of input from each previous time step and the inputs to those time steps. This figure excludes backpropagation with respect to $E_1, E_2,$ and E_4

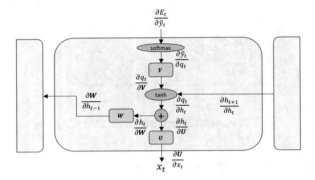

Fig. 7.4: Backpropagation through a single time step of a simple RNN

7.3.1.1 Output Weights (V)

The weight matrix $\mathbf{V}$ controls the output dimensionality of $\hat{\mathbf{y}}$, and does not contribute to the recurrent connection. Therefore, computing the gradient is the same as a linear layer.

For convenience, let

$$\mathbf{q}_t = \mathbf{V}\mathbf{h}_t. \tag{7.7}$$

Then,

$$\frac{\partial E_t}{\partial V_{i,j}} = \frac{\partial E_t}{\partial \hat{y}_{t_k}} \frac{\partial \hat{y}_{t_k}}{\partial q_{t_l}} \frac{\partial q_{t_l}}{\partial V_{i,j}}. \tag{7.8}$$

From our definition of E_t (7.5), we have that:

$$\frac{\partial E_t}{\partial \hat{y}_{t_k}} = -\frac{y_{t_k}}{\hat{y}_{t_k}}. \tag{7.9}$$

Backpropagation through the softmax function can be computed as:

$$\frac{\partial \hat{y}_{t_k}}{\partial q_{t_l}} = \begin{cases} -\hat{y}_{t_k}\hat{y}_{t_l}, & k \neq l \\ \hat{y}_{t_k}\left(1 - \hat{y}_{t_k}\right), & k = l \end{cases}. \tag{7.10}$$

If we combine (7.9) and (7.10) we obtain the sum over all values of k to produce $\frac{\partial E_t}{\partial q_{t_l}}$:

$$-\frac{y_{t_l}}{\hat{y}_{t_l}}\hat{y}_{t_l}\left(1 - \hat{y}_{t_l}\right) + \sum_{k \neq l}\left(-\frac{y_{t_k}}{\hat{y}_{t_k}}\right)\left(-\hat{y}_{t_k}\hat{y}_{t_l}\right) = -y_{t_l} + y_{t_l}\hat{y}_{t_l} + \sum_{k \neq l} y_{t_k}\hat{y}_{t_l} \tag{7.11a}$$

$$= -y_{t_l} + \hat{y}_{t_l}\sum_k y_{t_k}. \tag{7.11b}$$

Recall that all $\mathbf{y}_t$ are one-hot vectors, meaning that all values are in the vector are zero except for one indicating the class. Thus, the sum is 1, so

$$\frac{\partial E_t}{\partial q_{t_l}} = \hat{y}_{t_l} - y_{t_l} \tag{7.12}$$

Lastly, $\mathbf{q}_t = \mathbf{V}\mathbf{h}_t$, so $q_{t_l} = V_{l,m}h_{t_m}$. Therefore,

$$\frac{\partial q_{t_l}}{\partial V_{i,j}} = \frac{\partial}{\partial V_{i,j}}\left(V_{l,m}h_{t_m}\right) \tag{7.13a}$$

$$= \delta_{il}\delta_{jm}h_{t_m} \tag{7.13b}$$

$$= \delta_{il}h_{t_j}. \tag{7.13c}$$

Now we combine (7.12) and (7.13c) to obtain:

$$\frac{\partial E_t}{\partial V_{i,j}} = \left(\hat{y}_{t_i} - y_{t_i}\right)h_{t_j}, \tag{7.14}$$

which is recognizable as the outer product. Hence,

$$\frac{\partial E_t}{\partial \mathbf{V}} = \left(\hat{\mathbf{y}}_t - \mathbf{y}_t\right) \otimes \mathbf{h}_t, \tag{7.15}$$

where $\otimes$ is the outer product.

7.3.1.2 Recurrent Weights (W)

The parameter $\mathbf{W}$ appears in the argument for $\mathbf{h}_t$, so we will have to check the gradient in both $\mathbf{h}_t$ and $\hat{\mathbf{y}}_t$. We must also make note that $\hat{\mathbf{y}}_t$ depends on $\mathbf{W}$ both directly and indirectly (through $\mathbf{h}_{t-1}$). Let $\mathbf{z}_t = \mathbf{U}\mathbf{x}_t + \mathbf{W}\mathbf{h}_{t-1}$. Then $\mathbf{h}_t = \tanh(\mathbf{z}_t)$. At first it seems that by the chain rule we have:

$$\frac{\partial E_t}{\partial W_{i,j}} = \frac{\partial E_t}{\partial \hat{y}_{t_k}}\frac{\partial \hat{y}_{t_k}}{\partial q_{t_l}}\frac{\partial q_{t_l}}{\partial h_{t_m}}\frac{\partial h_{t_m}}{\partial W_{i,j}} \tag{7.16}$$

Note that of these four terms, we have already calculated the first two, and the third is simple:

$$\frac{\partial q_{t_l}}{\partial h_{t_m}} = \frac{\partial}{\partial h_{t_m}} \left(V_{l,b} h_{t_b} \right) \tag{7.17a}$$

$$= V_{l,b} \delta_{b,m} \tag{7.17b}$$

$$= V_{l,m} \tag{7.17c}$$

The final term, however, requires an implicit dependence of $\mathbf{h}_t$ on $\mathbf{W}_{i,j}$ through $\mathbf{h}_{t-1}$ as well as a direct dependence. Hence, we have:

$$\frac{\partial h_{t_m}}{\partial W_{i,j}} \rightarrow \frac{\partial h_{t_m}}{\partial W_{i,j}} + \frac{\partial h_{t_m}}{\partial h_{t-1_n}} \frac{\partial h_{t-1_n}}{\partial W_{i,j}}. \tag{7.18}$$

But we can just apply this again to yield:

$$\frac{\partial h_{t_m}}{\partial W_{i,j}} \rightarrow \frac{\partial h_{t_m}}{\partial W_{i,j}} + \frac{\partial h_{t_m}}{\partial h_{t-1_n}} \frac{\partial h_{t-1_n}}{\partial W_{i,j}} + \frac{\partial h_{t_m}}{\partial h_{t-1_n}} \frac{\partial h_{t-1_n}}{\partial h_{t-2_p}} \frac{\partial h_{t-2_p}}{\partial W_{i,j}}. \tag{7.19}$$

This process continues until we reach $h_{(-1)}$, which was initialized to a vector of zeros ($\mathbf{0}$). Notice that the last term in (7.19) collapses to $\frac{\partial h_{t_m}}{\partial h_{t-2_n}} \frac{\partial h_{t-2_n}}{\partial W_{i,j}}$ and we can turn the first term into $\frac{\partial h_{t_m}}{\partial h_{t_n}} \frac{\partial h_{t_n}}{\partial W_{i,j}}$. Then, we arrive at the compact form:

$$\frac{\partial h_{t_m}}{\partial W_{i,j}} = \frac{\partial h_{t_m}}{\partial h_{r_n}} \frac{\partial h_{r_n}}{\partial W_{i,j}}, \tag{7.20}$$

where we sum over all values of r less than t in addition to the standard dummy index n. More clearly, this is written as:

$$\frac{\partial h_{t_m}}{\partial W_{i,j}} = \sum_{r=0}^{t} \frac{\partial h_{t_m}}{\partial h_{r_n}} \frac{\partial h_{r_n}}{\partial W_{i,j}}, . \tag{7.21}$$

This term is responsible for the vanishing/exploding gradient problem: the gradient exponentially shrinking to 0 (vanishing) or exponentially growing larger (exploding). The multiplication of the term $\frac{\partial h_{t_m}}{\partial h_{r_n}}$ by the term $\frac{\partial h_{r_n}}{\partial W_{i,j}}$ means that the product will be smaller if both terms are less than 1 or larger if the terms are greater than 1. We will address this problem in more detail momentarily.
Combining all of these yields:

$$\frac{\partial E_t}{\partial W_{i,j}} = \left(\hat{y}_{t_l} - y_{t_l} \right) V_{l,m} \sum_{r=0}^{t} \frac{\partial h_{t_m}}{\partial h_{r_n}} \frac{\partial h_{r_n}}{\partial W_{i,j}}. \tag{7.22}$$

7.3.1.3 Input Weights (U)

Taking the gradient of $\mathbf{U}$ is similar to doing it for $\mathbf{W}$ since they both require taking sequential derivatives of the $\mathbf{h}_t$ vector. We have:

$$\frac{\partial E_t}{\partial U_{i,j}} = \frac{\partial E_t}{\partial \hat{y}_{t_k}} \frac{\partial \hat{y}_{t_k}}{\partial q_{t_l}} \frac{\partial q_{t_l}}{\partial h_{t_m}} \frac{\partial h_{t_m}}{\partial U_{i,j}}. \tag{7.23}$$

Note that we only need to calculate the last term now. Following the same procedure as for W, we find that:

$$\frac{\partial h_{t_m}}{\partial U_{i,j}} = \sum_{r=0}^{t} \frac{\partial h_{t_m}}{\partial h_{r_n}} \frac{\partial h_{r_n}}{\partial U_{i,j}}, \tag{7.24}$$

and thus we have:

$$\frac{\partial E_t}{\partial U_{i,j}} = \left(\hat{y}_{t_l} - y_{t_l}\right) V_{l,m} \sum_{r=0}^{t} \frac{\partial h_{t_m}}{\partial h_{r_n}} \frac{\partial h_{r_n}}{\partial U_{i,j}}. \tag{7.25}$$

The difference between $\mathbf{U}$ and $\mathbf{W}$ appears in the actual implementation since the values of $\frac{\partial h_{r_n}}{\partial U_{i,j}}$ and $\frac{\partial h_{r_n}}{\partial W_{i,j}}$ differ.

7.3.1.4 Aggregate Gradient

The error for all time steps is the summation of E_t's according to our loss function (7.6). Therefore, we can sum the gradients for each of the weights in our network ($\mathbf{U}$, $\mathbf{V}$, and $\mathbf{W}$) and update then with the accumulated gradients.

7.3.2 Vanishing Gradient Problem and Regularization

One of the most difficult parts of training RNNs is the vanishing/exploding gradient problem (often referred to as just the vanishing gradient problem).[4] During back-propagation, the gradients are multiplied by the weight's contribution to the error at each time step, shown in Eq (7.21). The impact of this multiplication at each time step dramatically reduces or increases the gradient propagated to the previous time step which will in turn be multiplied again. The recurrent multiplication in the backpropagation step causes an exponential effect for any irregularity.

- If the weights are small, the gradients will shrink exponentially.
- If the weights are large, the gradients will grow exponentially.

In the case when the contribution is very small, the weight update may be a negligible change, potentially causing the network to stop training. Practically, this usually

---

[4] The tanh activation function bounds the gradient between 0 and 1. This has the effect of shrinking the gradient in these circumstances.

leads to underflow or overflow errors when not considered. One way to alleviate this issue is to use the second-order derivatives to predict the presence of vanishing/exploding gradients by using Hessian-free optimization techniques. Another approach is to initialize the weights of the network carefully. However, even with careful initialization, it can still be challenging to deal with long-range dependencies.

A common initialization for RNNs is to initialize the initial hidden state to $\mathbf{0}$. The performance can typically be improved by allowing this hidden state to be learned [KB14].

Adaptive learning rate methods, such as Adam [KB14], can be useful in recurrent networks, as they cater to the dynamics of individual weights, which can vary significantly in RNNs.

There are many methods used to combat the vanishing gradient problem, many of them focusing on careful initialization or controlling the size of the gradient being propagated. The most commonly used method to combat vanishing gradients is the addition of gates to RNNs. We will focus more on this approach in the next section. RNN sequences can be very long. For example, if an RNN used for speech recognition samples 20 ms windows with a stride of 10 ms will produce an output sequence length of 999 time steps for a 10-s clip (assuming no padding). Thus, the gradients can vanish/explode very easily [BSF94b].

7.3.2.1 Long Short-Term Memory

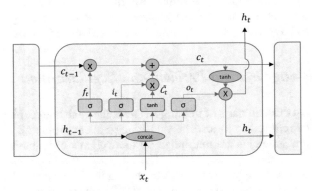

Fig. 7.5: Diagram of an LSTM cell

Long short-term memory (LSTM) utilizes gates to control the gradient propagation in the recurrent network's memory [HS97b]. These gates (referred to as the input, output, and forget gates) are used to guard a memory cell that is carrying the hidden state to the next time step. The gating mechanisms are themselves neural network layers. This allows the network to learn the conditions for when to forget, ignore, or keep information in the memory cell. Figure 7.5 shows a diagram of an LSTM.

The LSTM cell is formally defined as:

$$
\begin{aligned}
\mathbf{i}_t &= \sigma(\mathbf{W}_i \mathbf{x}_t + \mathbf{U}_i \mathbf{h}_{t-1} + \mathbf{b}_i) \\
\mathbf{f}_t &= \sigma(\mathbf{W}_f \mathbf{x}_t + \mathbf{U}_f \mathbf{h}_{t-1} + \mathbf{b}_f) \\
\mathbf{o}_t &= \sigma(\mathbf{W}_o \mathbf{x}_t + \mathbf{U}_o \mathbf{h}_{t-1} + \mathbf{b}_o) \\
\tilde{\mathbf{c}}_t &= \tanh(\mathbf{W}_c \mathbf{x}_t + \mathbf{U}_c \mathbf{h}_{t-1}) \\
\mathbf{c}_t &= \mathbf{f}_t \circ \mathbf{c}_{t-1} + \mathbf{i}_t \circ \tilde{\mathbf{c}}_t \\
\mathbf{h}_t &= \mathbf{o}_t \circ \tanh(\mathbf{c}_t)
\end{aligned}
\tag{7.26}
$$

The forget gate controls how much is remembered from step to step. Some recommend initializing the bias of the forget gate to 1 in order for it to remember more initially [Haf17].

7.3.2.2 Gated Recurrent Unit

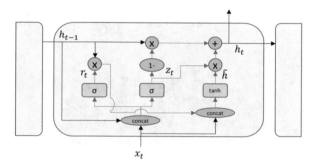

Fig. 7.6: Diagram of a GRU

The gated recurrent unit (GRU) is another popular gating structure for RNNs [Cho+14]. The GRU combines the gates in the LSTM to create a simpler update rule with one less learned layer, lowering the complexity and increasing efficiency. The choice between using LSTM or GRU is largely decided empirically. Despite a number of attempts to compare the two methods, no generalizable conclusion has been reached [Chu+14]. The GRU uses fewer parameters, so it is usually chosen when performance is equal between the LSTM and GRU architectures. The GRU is shown in Fig. 7.6. The equations for the update rules are shown below:

$$
\begin{aligned}
\mathbf{z}_t &= \sigma(\mathbf{W}_z \mathbf{x}_t + \mathbf{U}_z \mathbf{h}_{t-1}) \\
\mathbf{r}_t &= \sigma(\mathbf{W}_r \mathbf{x}_t + \mathbf{U}_r \mathbf{h}_{t-1}) \\
\tilde{\mathbf{h}}_t &= \tanh(\mathbf{W}_h \mathbf{x}_t + \mathbf{U}_h \mathbf{h}_{t-1} \circ \mathbf{r}_t) \\
\mathbf{h}_t &= (1 - \mathbf{z}_t) \circ \tilde{\mathbf{h}}_t + \mathbf{z}_t * \mathbf{h}_{t-1}
\end{aligned}
\tag{7.27}
$$

In the GRU, the new candidate state, $\tilde{\mathbf{h}}_t$, is combined with the previous state, with $\mathbf{z}_t$ determining how much of the history is carried forward or how much the new candidate replaces the history. Similar to setting the LSTM's forget gate bias for improved memory in the early stages, the GRU's reset gate biases can be set to -1 to achieve a similar effect [Haf17].

7.3.2.3 Gradient Clipping

A simple way to limit gradient explosion is to force the gradients to a specific range. Limiting the gradient's range can solve a number of problems, specifically preventing overflow errors when training. It is typically good practice to track the gradient norm to understand its characteristics, and then reduce the gradient when it exceeds the normal operating range. This concept is commonly referred to as gradient clipping.

The two most common ways to clip gradients are:

- L_2 norm clipping with a threshold t.

$$\nabla_{\text{new}} = \nabla_{\text{current}} \circ \frac{t}{L_2(\nabla)} \tag{7.28}$$

- Fixed range

$$\nabla_{\text{new}} = \begin{cases} t_{\min} & \text{if } \nabla < t_{\min} \\ \nabla & \\ t_{\max} & \text{if } \nabla > t_{\max} \end{cases} \tag{7.29}$$

With a maximum threshold $t_{\max}$ and minimum threshold $t_{\min}$.

7.3.2.4 BPTT Sequence Length

The computation involved in recurrent network training depends heavily on the number of time steps in the input. One way to fix/limit the amount of computation in the training process is to set a maximum sequence length for the training procedure.

Common ways to set the sequence length are:

- Pad training data to the longest desired length
- Truncate the number of steps backpropagated during training.

In the early stages of training, overlapping sequences with truncated backpropagation can help the network converge quicker. Increasing the truncation length as training progresses can also help convergence in the early stages of learning, particularly for complex sequences, or when the maximum sequence length in a dataset is quite long.

Setting a maximum sequence length can be useful in a variety of situations. In particular, when:

- static computational graph requires a fixed size input,
- the model is memory constrained, or
- gradients are very large at the beginning of training.

7.3.2.5 Recurrent Dropout

Recurrent networks, like other deep learning networks, are prone to overfitting. Dropout, being a common regularization technique, is an intuitive choice to apply to RNNs as well, however, the original form must be modified. If the original form of dropout is applied at each step, then the combination of masks can cause little signal to be passed over longer sequences. Instead, we can reuse the same mask at each step [SSB16] to prevent loss of information between time steps.

Additional techniques such as variational dropout [GG16] and zoneout [Kru+16] have similar aims, by dropping out input or output gates in LSTMs or GRUs.

7.4 Deep RNN Architectures

As with the entire field of deep learning, many of the architectures and techniques are an area of active research. In this section, we describe a few architectural variants to illustrate the expressive power and extensions of the basic RNN concepts that have been introduced so far.

7.4.1 Deep RNNs

Just as we have stacked multiple fully connected and convolutional layers, we can also stack layers of recurrent networks [EHB96]. The hidden state in a stacked RNN composed of l vanilla RNN layers can be defined as follows:

$$\mathbf{h}_t^{(l)} = f\left(\mathbf{W}\left[\mathbf{h}_{t-1}^{(l)}; \mathbf{h}_t^{(l-1)}\right]\right) \tag{7.30}$$

where $\mathbf{h}_t^{(l-1)}$ is the output of the previous RNN layer at time t. This is illustrated in Fig. 7.7. Anecdotally, when convolutional layers were stacked, the network was learning a hierarchy of spatially correlated features. Similarly, when recurrent networks are stacked it allows longer ranges of dependencies and more complex sequences to be learned [Pas+13].

Because the weights in RNNs are quadratic in size, it can also be more efficient to have multiple smaller layers rather than larger ones. Another benefit is computational optimization for fused RNN layers [AKB16].

A common problem with stacking RNNs, however, is the vanishing gradient problem due to the depth and number of time steps. However, RNNs have been

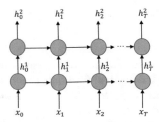

Fig. 7.7: Diagram of a stacked RNN with $l = 2$

able to gain inspiration from other areas of deep learning, incorporating residual connections and highway networks seen in deep convolutional networks.

7.4.2 Residual LSTM

In Prakash et al. [Pra+16], the authors used residual connections between layers of the LSTM to provide a stronger gradient to lower layers for the purpose of paraphrase generation. Residual layers, typically applied in convolutional networks, allow "residuals" of lower level information to pass on to later layers of the network. This provides lower level information to higher layers and also allows a larger gradient to be passed to the earlier layers, because there is a more direct connection to the output. In Kim et al. [KEL17], the authors used residual connections to improve word error rates on a deep speech network and concluded that the lack of accumulation on the highway path, while using a projection matrix to scale the LSTM output.

In the LSTM definition in Eq. (7.26), h_t is changed to:

$$\mathbf{h}_t = \mathbf{o}_t \cdot (\mathbf{W}_p \cdot \tanh(\mathbf{c}_t) + \mathbf{W}_h \mathbf{x}_t) \tag{7.31}$$

where $\mathbf{W}_p$ is the projection matrix and $\mathbf{W}_h$ is an identity matrix that matches the sizes of $\mathbf{x}_t$ to $\mathbf{h}_t$. When the dimensions of $\mathbf{x}_t$ and $\mathbf{h}_t$ are the same, this equation becomes:

$$\mathbf{h}_t = \mathbf{o}_t \cdot (\mathbf{W}_p \cdot \tanh(\mathbf{c}_t) + \mathbf{x}_t). \tag{7.32}$$

Note that the output gate is applied after the addition of the input $\mathbf{x}_t$.

7.4.3 Recurrent Highway Networks

Recurrent highway networks (RHN) [Zil+16] offer an approach to gate the gradient propagation between recurrent layers in multilayer RNN architectures. The authors present an extension of the LSTM architecture that allows for gated connections

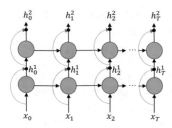

Fig. 7.8: A two layer residual LSTM

between the recurrent layers, allowing an increase in the number of layers that can be stacked for deep RNNs.

For an RHN with L layers and output $\mathbf{s}^{(L)}$, the networks is is described as:

$$
\begin{aligned}
\mathbf{s}_t^{(l)} &= \mathbf{h}_t^{(l)} \cdot \mathbf{t}_t^{(l)} + \mathbf{s}_t^{(l-1)} \cdot \mathbf{c}_t^{(l)} \\
\mathbf{h}_t^{(l)} &= \tanh\left(\mathbf{W}_H \mathbf{x}_t \mathbb{1}_{\{l=1\}} + \mathbf{R}_{H^l} \mathbf{s}_t^{(l-1)} + \mathbf{b}_{H^l} \right) \\
\mathbf{t}_t^{(l)} &= \sigma\left(\mathbf{W}_T \mathbf{x}_t \mathbb{1}_{\{l=1\}} + \mathbf{R}_{T^l} \mathbf{s}_t^{(l-1)} + \mathbf{b}_{T^l} \right) \\
\mathbf{c}_t^{(l)} &= \sigma\left(\mathbf{W}_C \mathbf{x}_t \mathbb{1}_{\{l=1\}} + \mathbf{R}_{C^l} \mathbf{s}_t^{(l-1)} + \mathbf{b}_{C^l} \right)
\end{aligned}
\tag{7.33}
$$

with $\mathbb{1}$ denoting the indicator function.

A number of useful properties are gained from RHNs, specifically that the Jacobian eigenvalue is regulated across time steps, facilitating more stable training. The authors reported impressive results on a language modeling task using a 10 layer deep RHN.

7.4.4 Bidirectional RNNs

So far we have only considered the accumulation of a memory context in the forward direction. In many situations it is desirable to know what will be encountered in future time steps to inform the prediction at time t. Bidirectional RNNs [SP97] allow for both the forward context and "backward" context to be incorporated into a prediction. This is accomplished by running two RNNs over a sequence, one in

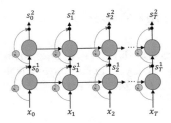

Fig. 7.9: A diagram of a two layer highway LSTM. Not that the highway connection uses a learned gate along the connection to the next layer

the forward direction and one in the backward direction. For an input sequence $X = \{\mathbf{x}_1, \mathbf{x}_2, \ldots, \mathbf{x}_T\}$, the forward context RNN receives the inputs in forward order $t = \{1, 2, \ldots, T\}$, and the backward context RNN receives the inputs in reverse order $t = \{T, T - 1, \ldots, 1\}$. These two RNNs together constitute a single bidirectional layer. Figure 7.10 shows a diagram of a bidirectional RNN.

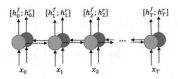

Fig. 7.10: Diagram of a bidirectional RNN. Here the outputs are concatenated to form a single output vector holding the forward and backward context

The output of the two RNNs, $\mathbf{h}^f$ and $\mathbf{h}^r$, is often joined to form a single output vector either by summing the two vectors, concatenating, averaging, or another method.

In NLP, there are many uses for this type of structure. For example, this has proven very useful for the task of phoneme classification in speech recognition, where knowledge of the future context can better inform the predictions at any forward time step. Bidirectional networks typically outperform forward-only RNNs in most tasks. Furthermore, this approach can be extended to other forms of recurrent networks such as bidirectional LSTMs (BiLSTM). These techniques follow logically, with one LSTM network operating over the inputs in the forward direction and another with the inputs in the reverse direction, combining the outputs (concatenation, addition, or another method).

One limitation of bidirectional RNNs is that full input sequence must be known before prediction, because the reverse RNN requires $\mathbf{x}_T$ for the first computation. Thus bidirectional RNNs cannot be used for real-time applications. However, depending on the requirements of the application, having a fixed buffer for the input can alleviate this restriction.

7.4.5 SRU and Quasi-RNN

The recurrent connections restrict the amount of computation that can be parallelized, because the information must be processed sequentially. Thus, the computational cost of RNNs is high compared to CNNs. Two techniques introduced to speed up computation involve eliminating some of the sequential dependencies. These techniques allow networks to become much deeper for a lower computational cost. The first technique introduces a semi-recurrent unit (SRU) [LZA17]. The approach processes the input at each time step simultaneously and applies a light-weight recurrent computation afterward. The SRU incorporates skip and highway connections to improve the feature propagation in the network. The SRU is defined as:

$$
\begin{aligned}
\tilde{\mathbf{x}}_t &= \mathbf{W}\mathbf{x}_t \\
\mathbf{f}_t &= \sigma(\mathbf{W}_f \mathbf{x}_t + \mathbf{b}_f) \\
\mathbf{r}_t &= \sigma(\mathbf{W}_r \mathbf{x}_t \mathbf{b}_r) \\
\mathbf{c}_t &= \mathbf{f}_t \circ \mathbf{c}_{t-1} + (1 - \mathbf{f}_t) \circ \tilde{\mathbf{x}}_t \\
\mathbf{h}_t &= \mathbf{r}_t \circ g(\mathbf{c}_t) + (1 - \mathbf{r}_t) \circ \mathbf{x}_t
\end{aligned}
\tag{7.34}
$$

where $\mathbf{f}$ is the forget gate, $\mathbf{r}$ is the reset gate, and $\mathbf{c}$ is the memory cell.

This approach was applied to text classification, question answering, language modeling, machine translation, and speech recognition, achieving competitive results with a reduction in training times by up to $10\times$ over the LSTM counterpart.

The quasi-recurrent neural network (QRNN) [Bra+16] is a different approach with the same goal. The QRNN applies convolutional layers to parallelize the input computation that is being supplied to the reduced recurrent component. This network was applied to the task of sentiment analysis and also achieved competitive results with a significant reduction in training and prediction time.

7.4.6 Recursive Neural Networks

Recursive neural networks (RecNN) are a generalized form of recurrent neural networks that allow effective manipulation of graphical structures. A recursive neural network can learn information related to labeled directed acyclic graphs, while recurrent networks only process ordered sequences [GK96]. In NLP, the main application for recursive neural networks is dependency parsing [SMN10] and learning morphological word vectors [LSM13b].

The data is structured as a tree with parent nodes at the top and children nodes stemming from them. The aim is to learn the appropriate graphical structure of the data by predicting a tree and reducing the error with respect to the target tree structure.

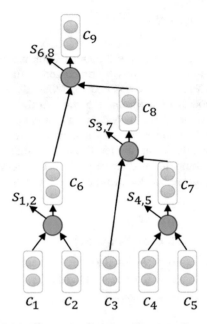

Fig. 7.11: Diagram of a recursive neural network

For simplicity, we consider a branching factor of 2 (2 children for each parent). The structure prediction, a recursive neural network aims to achieve two outputs:

- A semantic vector representation, $p(\mathbf{x}_i, \mathbf{x}_j)$, merging the children nodes $\mathbf{c}_i$ and $\mathbf{c}_j$
- A score s indicating how likely the children nodes are to be merged.

The network can be described as follows:

$$s_{ij} = \mathbf{U}\dot{p}(\mathbf{c}_i, \mathbf{c}_j)$$
$$p(\mathbf{c}_i, \mathbf{c}_j) = f(W[\mathbf{c}_i; \mathbf{c}_j] + \mathbf{b}) \tag{7.35}$$

where $\mathbf{W}$ is the weight matrix for the shared layer and $\mathbf{U}$ is the weight matrix for the score computation.

The score of a tree is the sum of the scores at each node:

$$S = \sum_{n \in \text{nodes}} s_n \tag{7.36}$$

The error computation for recursive neural networks uses max-margin parsing:

$$E = \sum_i s(x_i, y_i) - \max_{y \in A(x_i)} \left(s(x_i, y) + \Delta(y, y_i) \right) \tag{7.37}$$

The quantity $\Delta(y, y_i)$ computes the loss for all incorrect decisions

Backpropagation through structure (BPTS) similar to BPTT computes the derivatives at each node in the graph. The derivatives are split at each node and passed on to the children nodes. In addition to the gradient with respect to the predicted node, we also compute the gradient with respect to the score values as well.

LSTM and GRU units have also been applied to recursive networks to combat the vanishing gradient problem [TSM15]. Recursive networks have been used in areas, such as relation classification [Soc+12], sentiment analysis [Soc+13], and phrase similarity [TSM15].

Recursive neural networks demonstrate powerful extensions to sequence-based neural architectures. Although their use has decreased in popularity with the introduction of attention-based architectures, the concepts they present for improved computational efficiency are useful.

7.5 Extensions of Recurrent Networks

Recurrent neural networks can be used to accomplish many types of sequence problems. Until now, we have focused on a many-to-many example with a one-to-one mapping from an input to an output with the same number of time steps. However, RNNs can be used for many types of sequence oriented problems by modifying where the error is computed. Figure 7.12 shows the types of sequence problems that can be solved with RNNs.

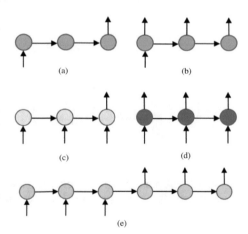

Fig. 7.12: Recurrent neural networks can address a variety of sequence-based problems. (**a**) shows a one-to-one sequence (this would be equivalent to deep neural networks with shared weights). (**b**) illustrates a one-to-many sequence task, generating a series of outputs given one input. (**c**) is a many-to-one task, which could represent a text classification task, predicting a single classification after the entire text has been seen. (**d**) shows a many-to-many sequence task with a one-to-one alignment between the number of input and output time steps. This structure is common in language modeling. (**e**) shows a many-to-many without a specific alignment between the input and output. The number of inputs and outputs steps can also be different lengths. This technique is commonly referred to as sequence-to-sequence and is commonly seen in neural machine translation

RNNs have tremendous flexibility and can be extended to address a wide range of sequence tasks. The limitations of feed-forward neural networks remain: a tendency to overfit without proper regularization, need for large datasets, and computational requirements. Additionally, sequence models introduce other considerations, such as vanishing gradients with longer sequence lengths and "forgetting" of earlier context. These difficulties have led to various extensions, best practices, and techniques to alleviate these issues.

7.5.1 Sequence-to-Sequence

Many NLP and speech tasks are sequence oriented. One of the most common architectural approaches for these tasks is sequence-to-sequence, often abbreviated seq-to-seq or seq2seq. The seq-to-seq approach resembles an autoencoder, having a recurrent encoder and a recurrent decoder, shown in Fig. 7.13. The final hidden state of the encoder functioning as the "encoding" passed to the decoder; however, it is typically trained in a supervised fashion with a specific output sequence. The seq-to-seq approach was born out of neural machine translation, having an input sentence in one language and a corresponding output sentence in a separate language. The aim is to summarize the input with encoder and decode to the new domain with the decoder.

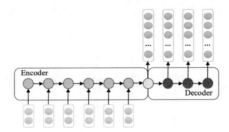

Fig. 7.13: Seq-to-seq model with an RNN-based encoder and decoder. The first hidden state of the decoder is the last hidden state of the encoder (shown in yellow)

One difficulty of this approach is that the hidden state tends to reflect the most recent information, losing memory of earlier content. This forces a limitation for long sequences, where all information about that sequence must be summarized into a single encoding.

7.5.2 *Attention*

Forcing a single vector to summarize all the information from the previous time steps is a drawback. In most applications, the information from a generated sequence from the decoder will have some correlation with the input sequence. For example, in machine translation, the beginning of the output sentence likely depends on the beginning of the input sentence and less so on the end of the sentence, which has been seen more recently. In many situations, it would be helpful to have not only the summarized knowledge, but also the ability to focus on different parts of the input to better inform the output at a particular time step.

Attention [BCB14a] has been one of the most popular techniques to address this issue by paying specific attention to parts of the sequence for each word in the output sequence. Not only does this allow us to improve the quality of our predictions, it also allows insight into the network by viewing what inputs were relied upon for the prediction.

If s_i is the attention augmented hidden state at time i, it takes three inputs:

- the previous hidden state of the decoder $\mathbf{s}_{i-1}$,
- the prediction from the previous time step $\mathbf{y}_{i-1}$, and
- a context vector $\mathbf{c}_i$ which weighs the appropriate hidden states for the given time step.

$$\mathbf{s}_i = f(\mathbf{s}_{i-1}, \mathbf{s}_i, \mathbf{c}_i) \tag{7.38}$$

The context vector, c_i, is defined as:

$$\max - \mathrm{marginparsing}_i = \sum_{j=1}^{T_x} \alpha_{ij} \mathbf{h}_j. \tag{7.39}$$

where the attention weights are:

$$\alpha_{ij} = \frac{\exp(e_{ij})}{\sum_{k=1}^{T_x} \exp(e_{ik})} \tag{7.40}$$

and

$$e_{ij} = a(\mathbf{s}_{i-1}, \mathbf{h}_j). \tag{7.41}$$

The function $a(\mathbf{s}, \mathbf{h})$ is referred to as the *alignment model*. This function scores how influential input $\mathbf{h}_j$ should be on the output at position i.

It is fully differentiable and deterministic because it is considering all time steps that have contributed to the output. A drawback of using all of the previous time steps is that it requires a large amount of computation for long sequences. Other techniques relax this dependency by being selective about the number of states that inform the context vector. Doing this creates a non-differentiable loss, however, and training requires Monte Carlo sampling for the estimation of the gradient during backpropagation.

An additional benefit of attention is that it provides a score for each time step, identifying what inputs were most useful for the prediction. This can be very useful for interpretability when inspecting the quality of a network or gaining intuition about what the model is learning, as shown in Fig. 7.15. Attention mechanisms are covered in more detail in Chap. 9.

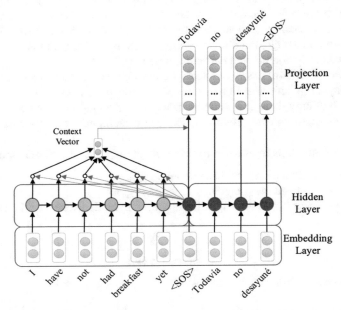

Fig. 7.14: Attention is applied to the first step of decoding for a neural machine translation model. A similarity score is computed for the hidden state at each time step in the encoder and the current hidden state of the decoder. These scores are used to weigh the contribution of that time step. These weights are used to produce the context vector that is supplied to the decoder

7.5.3 Pointer Networks

Pointer networks [VFJ15] are an application of attention-based, sequence-to-sequence models. Contrary to other attention-based models, it selects words (points) to be used as the output instead of accumulating the input sequence into a context vector. The output dictionary in this scenario must grow with the length of the input sequence. To accommodate this, an attention mechanism is used as a pointer, rather than mixing the information for decoding.

$$\mathbf{u}_j^i = \mathbf{v}^\mathsf{T} \tanh(\mathbf{W}_1 \mathbf{e}_j + \mathbf{W}_2 \mathbf{d}_i)$$
$$p(C_i | C_1, \dots, C_{i-1}, P) = \mathrm{softmax}(\mathbf{u}^i) \tag{7.42}$$

where $\mathbf{e}_j$ is the output of the encoder at time $j \in \{1, \dots, n\}$, $\mathbf{d}_i$ is the decoder output at time step i, and C_i is the index at time i, and $\mathbf{v}$, $\mathbf{W}_1$, and $\mathbf{W}_2$ are learnable parameters.

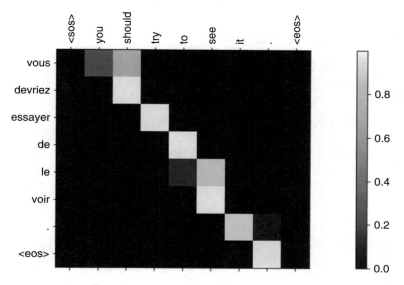

Fig. 7.15: Attention weights on an English-to-French machine translation task. Notice how the attended area of the network is correlated with the output sequence

This model showed success finding planar convex hulls, computing Delaunay triangulations, and producing solutions to the traveling salesman problem.

7.5.4 Transformer Networks

The success of attention on seq-to-seq tasks prompts the question of whether it can be directly applied to the input, reducing or even eliminating the need for recurrent connections in the network. Transformer networks [Vas+17b] applied this attention directly to the input with great success, beating both recurrent and convolutional models in machine translation. Instead of relying on RNNs to accumulate a memory of previous states as in sequence-to-sequence models, the transformer uses "multi-headed" attention directly on the input embeddings. This alleviates the sequential dependencies of the network allowing much of the computation to be performed in parallel.

Attention is applied directly to the input sequence, as well as the output sequence as it is being predicted. The encoder and decoder portions are combined, using another attention mechanism before predicting a probability distribution over the output dictionary.

Multi-head attention, shown in Fig. 7.16, is defined by three input matrices: $\mathbf{Q}$ the set of queries packed into a matrix, keys $\mathbf{K}$, and values $\mathbf{V}$.

$$\text{Attention}(\mathbf{Q}, \mathbf{K}, \mathbf{V}) = \text{softmax}\left(\frac{\mathbf{Q}\mathbf{K}^{\mathsf{T}}}{\sqrt{d_k}}\right)\mathbf{V} \tag{7.43}$$

Multi-head attention is then defined as:

$$\text{MultiHead}(\mathbf{Q}, \mathbf{K}, \mathbf{V}) = \text{Concat}(\text{head}_1, \ldots, \text{head}_h)\mathbf{W}^O \tag{7.44}$$

where

$$\text{head}_i(\mathbf{Q}, \mathbf{K}, \mathbf{V}) = \text{Attention}\left(\mathbf{Q}\mathbf{W}_i^Q, \mathbf{K}\mathbf{W}_i^K, \mathbf{V}\mathbf{V}_i^Q\right). \tag{7.45}$$

The parameters of all $\mathbf{W}$ matrices are projection matrices.

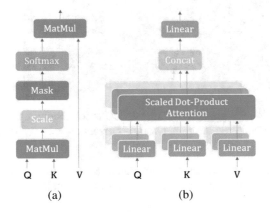

Fig. 7.16: Illustration of scaled dot-product attention referred to as attention in the text and multi-head attention. (**a**) Scaled dot-product attention, (**b**) multi-head attention

The encoder and decoder apply multiple layers of multi-head attention with residual connections and additional fully connected layers. Because much of the computation is happening in parallel, a masking technique and offsetting are used to ensure that the network only uses information that is available up to time $t - 1$ when predicting for time t. The transformer network reduces the number of steps required for prediction that significantly improve the computation time, while achieving state-of-the-art results on the translation task.

7.6 Applications of RNNs in NLP

Incorporating text into recurrent networks is a straight-forward process, resembling the CNN classification in the previous chapter. The words of a sentence are converted into word embeddings and passed as a time series into our network. In this case we do not have to worry about a minimum length to our sequence, because the word context is learned in the RNN's memory rather than as a combination of the inputs.

In Yin et al. [Yin+17], the authors do a wide comparison of CNN and RNN architectures for a variety of NLP tasks such as text classification, entailment, answer selection, and POS tagging. In this work the authors train basic CNN and RNN architectures, showing that RNNs perform well on most tasks, with CNNs proving superior only in certain matching cases where the main features are essentially key phrases. Overall, CNNs and RNNs have different methods of modeling sentences. CNNs tend to learn features similar to *n*-grams, while RNNs aim to maintain long-range dependencies for defining context.

7.6.1 Text Classification

Figure 7.17 shows the structure of a simple text classification task for an input sentence. With a recurrent network we are able to sequentially encode the word embeddings at each time step. Once the entire sequence is encoded, we use the last hidden state to predict the class. The network is trained using BPTT and learns to sequentially weigh the words for the classification task.

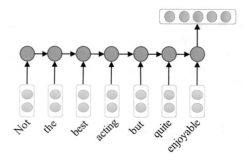

Fig. 7.17: Simple RNN-based text classifier for sentiment classification

In Lee and Dernoncourt [LD16], the authors compared CNN and RNN architectures for short-text classification. The addition of sequential information via CNN and RNN architectures significantly improved the results on dialog act characterization.

In sentiment classification, Wang et al. [Wan+15b] encoded tweets using an LSTM network to predict sentiment. In their work they showed the robustness of RNNs to capture complexities contained within the structure of the tweets, particularly the effect of negation phrases, such as when the word *not* negated a phrase. In Lowe et al. [Low+15], the authors introduced an architecture called dual-LSTM for semantic matching. This architecture encodes questions and answers and uses the inner product of the question and answer vector to rank the candidate responses.

7.6.2 Part-of-Speech Tagging and Named Entity Recognition

In Huang et al. [HXY15], word features and embeddings were applied to POS, NER, and chunking tasks with a bidirectional LSTM with CRF to boost performance. In Ma and Hovy [MH16] a bidirectional LSTM was used for end-to-end classification for POS on WSJ. Ma and Hovy [MH16] used an end-to-end method to improve on these results. Their method does not rely on context features that were applied in other works, such as POS, lexicon features, and task-dependent preprocessing. In Lample et al. [Lam+16b], a bidirectional LSTM was used in conjunction with CRF to achieve state-of-the-art performance on NER in four languages on the CoNLL-2003 dataset. This work also extended the base RNN-CRF architecture to stack-LSTM (LSTM units used to mimic a stack data structure with pushing and pulling capabilities). Character embeddings are often incorporated in addition to word embeddings to capture additional information about a word's semantic structure as well as to inform predictions on OOV words.

7.6.3 Dependency Parsing

In Dyer et al. [Dye+15], stack-LSTMs that allow for pushing and pulling operations were used to predict dependency parsing for variable-length text by predicting the dependency tree transitions. Kiperwasser and Goldberg [KG16] simplified the architecture by removing the need for the stack-LSTM, relying on bidirectional LSTMs to predict the dependency tree transitions.

7.6.4 Topic Modeling and Summarization

In Ghosh et al. [Gho+16], the contextual LSTM (C-LSTM) was introduced for word prediction, sentence selection, and topic prediction. The C-LSTM concatenates a topic embedding with the word embedding at each time step in the training of the network. This work functions similarly to language model training in that the goal is to predict the next word; however, it is extended to include the topic context into the prediction as well. Thus, the aim is to predict the next word and the topic of the sentence so far.

7.6.5 Question Answering

In Tan et al. [Tan+15], the authors train a question RNN and an answer RNN to yield a respective embedding for each. The two networks are then trained simultaneously by using a hinge loss objective to enforce a cosine similarity between the two most probable pairs. Another approach, dynamic memory networks [XMS16], incorporates a series of components to make up a question answering system. This system used a combination of recurrent networks and attention mechanisms to construct input, question, and answer modules that utilize an episodic memory to condition the predictions.

7.6.6 Multi-Modal

The effectiveness of deep learning in other applications such as images and video has led to a variety of multi-modal applications. These applications require generating language based on an input medium. These applications include image and video captioning, visual question answering, and visual speech recognition.

Image captioning was one of the first ways that deep convolutional networks for images were combined with text. In Vinyals et al. [Vin+15b], the authors utilized a pre-trained convolutional network for image classification to generate an image embedding for the initial state of an LSTM network. The LSTM network was trained to predict each word of the caption. The initial approach led to advancements in RNN architectures [Wan+16a].

Video captioning showed a similar development, with [Ven+14] utilizing a pre-trained CNN model to extract image features for each video frame to be used as input into a recurrent network for text generation. Pan et al. [Pan+15a] extended this method by striding over the output frames of the earlier recurrent layers to create a "hierarchical recurrent neural encoder" to reduce the number of time steps considered in the output layer of the stacked RNN.

In visual question answering, language generation is used to generate an answer for a textual question related to a visual input. An end-to-end approach was shown with the neural-image-QA network in Malinowski et al. [MRF15], where the input image and question conditioned the LSTM network to generate a textual answer.

7.6.7 Language Models

In the previous chapters we have briefly discussed language models. Recall that a language model provides a way to determine the probability of a sequence of words. For example, an n-gram language model determines the probability of a sequence of words $P(w_1, \ldots, w_m)$ by looking at the probability of each word given its n preceding words:

$$P(w_1, \ldots, w_m) \approx \prod_{i=1}^{m} P(w_i | w_{i-(n-1)}, \ldots, w_{i-1}). \tag{7.46}$$

Language models are particularly interesting in NLP as they can provide additional contextual information to situations where a prediction might be semantically similar, yet syntactically different. In the case of speech recognition, two words that sound the same such as "to" and "two" have different meanings. But the phrase "*set a timer for to minutes*" doesn't make sense, whereas "*set a timer for two minutes*" does.

Language models are often used when language is being generated and for domain adaptation (where there may be large amounts of unlabeled text data and limited labeled data). The concept of *n*-gram language models can also be implemented with RNNs, which benefit from not having to set a hard cutoff for the number of grams considered. Additionally, similar to word vectors, these models can be trained in an unsupervised manner over a large corpus of data.

The language model is trained to predict the next word in the sequence given the previous context, i.e., the hidden state. This allows the language model to target learning:

$$P(w_1, \ldots, w_m) = \prod_{i=1}^{m} P(w_i | w_1, \ldots, w_{i-1}). \tag{7.47}$$

An example of an RNN-based language model is shown in Fig. 7.18.

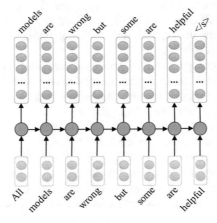

Fig. 7.18: An RNN language model trained to predict the next word in the sequence given the entire history of the sequence. Note that each time step is focused on classification, therefore the target outputs are the size of the vocabulary, not the size of the input word embeddings

In language modeling, a good practice is to have a single embedding matrix for both the input and the output sequence, allowing parameters to be shared, reducing the total number of parameters that need to be learned. Additionally, introducing a "down-projection" layer to reduce the state of a large RNN is typically useful when the output contains a large number of elements. This projection layer reduces the size of the final linear projection, as is often the case in language modeling [MDB17].

7.6.7.1 Perplexity

Perplexity is a measure of how well a model can represent the domain, shown by its ability to predicting a sample. For language models, perplexity can quantify the language model's ability to predict the validation or test data. The language model performs well if it produces a high probability for a sentence in the test set. Perplexity is the inverse probability normalized by the number of words.

We can define the perplexity measure for a test set of sentences $(s_1, \ldots, s_m)$ with:

$$PP(s_1, \ldots, s_m) = 2^{-\frac{1}{M} \sum_{i=1}^{m} \log_2 p(s_i)} \tag{7.48}$$

where M is the vocabulary size of the test set. Because perplexity gives the inverse probability of the dataset, a lower perplexity implies a better result.

7.6.7.2 Recurrent Variational Autoencoder

Recurrent variational autoencoders (RVAE) are an extension of recurrent language models [KW13, RM15]. The goal of a RVAE is to incorporate variational inference in the training process of the autoencoder to capture global features in latent variables. In Bowman et al. [Bow+15], the authors utilized a VAE architecture to generate sentences from a language model.

7.6.8 Neural Machine Translation

Machine translation has been one of the largest benefactors of the success of recurrent neural networks. Traditional approaches were based around statistical models that were computationally expensive and required heavy domain expertise to tune them. Machine translation is a natural fit for RNNs because input sentences may differ in length and order from the desired output. Early architectures for neural machine translation (NMT) relied on a recurrent encoder–decoder architecture. A very simple illustration of this is shown in Fig. 7.19.

NMT takes an input sequence of words $X = (x_1, \ldots, x_m)$ and maps them to an output sequence $Y = (y_1, \ldots, y_n)$. Note n is not necessarily the same as m. Using an embedding space, input X is mapped to a vector representation that is utilized by a recurrent network to encode the sequence. A decoder then uses the final RNN hidden state (the encoded input) to predict the translated sequence of words, Y (some have also shown success with subword translation [DN17]).

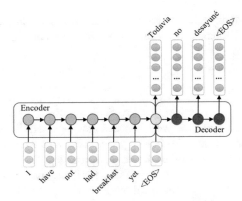

Fig. 7.19: Diagram of a one hidden layer encoder–decoder neural machine translation architecture. Note how the input and output sequences can have different lengths, and are truncated when the end of sentence (<EOS>)tag is reached

It is often beneficial to reinforce the sequence as it is being predicted by the decoder network. Passing the predicted output sequence as input, as shown in Fig. 7.20, can improve predictions. During training, the ground truth can be passed as the input into the next time step at some frequency. This referred to as "teacher forcing," because it is using the true predictions to help when training. The alternative is to use the decoder's predicted output, which can cause difficulty in converging in the early stages of training. Teacher forcing is phased out as training continues, allowing the model to learn the appropriate dependencies. Scheduled sampling is a way to combat this issue by switching between predicting with the targets and the network output.

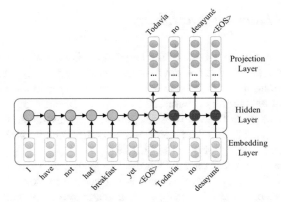

Fig. 7.20: A one hidden layer encoder–decoder neural machine translation architecture with the previous word of the prediction being used as the input at the following time step. Note that the embedding matrix has entries for both languages

In practice, the encoder and decoder do not need to be more than 2–4 layers deep, and bidirectional encoders usually outperform unidirectional ones [Bri+17].

7.6.8.1 BLEU

The most common metric used to evaluate machine translation is BLEU. BLEU (bilingual evaluation understudy) is a quality evaluation metric for machine translation, designed to align with human evaluations of natural language. It allows a translation to be compared with a set of target translations to evaluate the quality.

The score is bound between 0 and 1 with higher values indicating better performance. Often in literature, the score will be multiplied by 100 to approximate a percentage correlation. At its core the BLEU score is a precision measurement. It computes the precision for reference n-grams in the targets.

A perfect match would look like the following:

```
from nltk.translate.bleu_score import sentence_bleu
targets = [['i', 'had', 'a', 'cup', 'of', 'black', 'coffee', '
    at', 'the', 'cafe']]
prediction = ['i', 'had', 'a', 'cup', 'of', 'black', 'coffee',
    'at', 'the', 'cafe']
score = sentence_bleu(targets, prediction) * 100
print(score)

> 100.0
```

Alternatively, if none of the reference words are present in the prediction, then we get a score of 0.

```
from nltk.translate.bleu_score import sentence_bleu
targets = [['i', 'had', 'a', 'cup', 'of', 'black', 'coffee', '
    at', 'the', 'cafe']]
prediction = ['what', 'are', 'we', 'doing']
score = sentence_bleu(targets, prediction) * 100
print(score)

> 0
```

If we change one or two words in the transcript, then we see a drop in the score.

```
from nltk.translate.bleu_score import sentence_bleu
targets = [['i', 'had', 'a', 'cup', 'of', 'black', 'coffee', '
    at', 'the', 'cafe']]
prediction = ['i', 'had', 'a', 'cup', 'of', 'black', 'tea', '
    at', 'the', 'cafe']
score = sentence_bleu(targets, prediction) * 100
print(score)

> 65.8037

targets = [['i', 'had', 'a', 'cup', 'of', 'black', 'coffee', '
    at', 'the', 'cafe']]
```

```
10  prediction = ['i', 'had', 'a', 'cup', 'of', 'black', 'tea', '
        at', 'the', 'house']
11  score = sentence_bleu(targets, prediction) * 100
12  print(score)
13
14  > 58.1430
```

In these examples, BLEU-1 score is presented; however, higher n-grams would be given a better indicator of the quality. BLEU-4 is commonly seen in NMT, giving the correlation when considering the 4-gram precision between the hypothesis and the target translation.

7.6.9 Prediction/Sampling Output

There are a variety of ways to evaluate the output of a language model.

7.6.9.1 Greedy Search

If we predict the most likely word at each step, we may not yield the best sequence probability over all. The best decision early in the process may not maximize the overall probability of the sequence. In fact there is a decision tree of possibilities to be decoded for the best possible outcome. Because of the tree-like structure in language model outputs, there are a variety of methods to parse them.

7.6.9.2 Random Sampling and Temperature Sampling

Another way we can parse the output of our model is by using a random search. In a random search, the next word in the sentence is chosen according to the probability distribution of the next state. The random sampling technique can help achieve diversity in the results. However, sometimes the predictions for language models can be very confident, making the output results look similar to the greedy search results. A common way to improve the diversity of the predictions is to use a concept called temperature. Temperature is a method that exponentially transforms the probabilities and renormalizes to redistribute the highest probabilities among the top classes.

One method of sampling from the language model is to use "temperature sampling." This method selects an output prediction by applying a freezing function, defined by:

$$f_\tau(p)_i = \frac{p_i^{\frac{1}{\tau}}}{\sum_j p_j^{\frac{1}{\tau}}} \tag{7.49}$$

where $\tau \in [0, 1]$ is the temperature parameter that controls how "warm" the predictions are. The lower the temperature, the less diverse the results.

Another desirable quality for NLP is language generation. In Sutskever et al. [SVL14b], a deep RNN-based encoder–decoder architecture is used to generate unique sentences. The networks encode a "source" word sequence into a fixed length "encoding," a vector via an RNN. The decoder uses the "encoding" as the initial hidden state, and produces the response.

7.6.9.3 Optimizing Output: Beam Search Decoding

Greedy search makes an independence assumption between each time step for the decoding. We are relying on our RNNs to correctly inform the dependency between each time step. We can provide a prior to our predictions to ensure that we avoid simple errors (such as conjugation). We can do this by biasing our prediction on a scoring mechanism that informs whether a particular sequence is more probable than another.

When using our trained model to predict on new data, we are relying on the model to produce the correct output given the most confident prediction. However, in many situations it is desirable to impose a prior on the output, biasing it towards a particular domain. For example, in speech recognition an acoustic model's performance can be greatly improved by incorporating a language model as the bias for the output predictions.

The output that we are obtaining from our machine translation model, for example, is a probability distribution over the vocabulary at each time step, thus creating a tree of possibilities for how we could parse the output.

Often it is too computationally expensive to explore the entire tree of possibilities, so the most common search method is the beam search. Beam search is a searching algorithm that keeps a fixed maximum on the number of possible states in memory. This provides a flexible approach to optimizing the output of a network after it is trained, balancing speed and quality.

If we consider the output sequence of our network $(y_1, \ldots, y_m)$ where y_t is a softmax output over our vocabulary, then we can compute the probability of the overall sequence with the product of the probabilities at each time step :

$$P(y_1, \ldots, y_m) = \prod_{i=1}^{m} p_i(y_i) \qquad (7.50)$$

We can decode it by conditioning our output on the probability of transitioning from one word to the next.

If we have a language model that gives us the probability of a sequence of words, we can use this model to bias the prediction of our output by computing the probability of the different paths that can be taken through the tree of possible transitions.

Let $\mathbf{y}$ be a sequence of words and $P(\mathbf{y})$ be the probability of that sequence according to our language model. We will use a beam search to explore multiple hypotheses of sequences at time t, $\mathcal{H}_{t-1}$, with a beam size of k.

$$\mathcal{H}_t := \left\{ \left(w_1^1,\ldots,w_t^1\right),\ldots,\left(w_1^k,\ldots,w_t^k\right)\right\}$$
$$\mathcal{H}_3 := \{(\text{cup of tea}),(\text{cup of coffee})\}$$

With beam search we keep track of our top k hypotheses, and choose the path that maximizes $P(\mathbf{y})$. We will collect the probability of each hypothesis $P(h_t)$ in $\mathcal{P}_t$. The index order of $\mathcal{H}_t$ and $\mathcal{P}_t$ should be tied to keep them in sequence when sorting. We begin each hypothesis with the $<$SOS$>$ token and end the hypothesis once the $<$EOS$>$ token is reached. The hypothesis with the highest score is the one that is selected.

Algorithm 1: Beam Search

Data: $\hat{\mathbf{y}}$, beamWidth
Result: $\mathbf{y}$ with highest $p(\mathbf{y})$
begin
    $\mathcal{H}_0 = \{(< SOS >)\}$
    $\mathcal{P}_0 = \{0\}$
    for t *in* 1 *to* T **do**
        for h *in* $\mathcal{H}_{t-1}$ **do**
            for $\hat{y} \in \mathcal{Y}$ **do**
                $\hat{\mathbf{y}} = (y_1^h,\ldots,y_{t-1}^h,\hat{y})$
                $\mathcal{H}_t\mathrel{+}= \hat{\mathbf{y}}$
                $\mathcal{P}_t\mathrel{+}= P(\hat{\mathbf{y}})$
        $\mathcal{H}_t = \text{sort}(\mathcal{H}_t)$ according to highest $\mathcal{P}_t$
        $\mathcal{H}_t = \mathcal{H}_t[1,\ldots,\text{beamWidth}]$

7.7 Case Study

Here, we apply the concepts of recurrent neural networks for neural machine translation. Specifically, basic RNN, LSTM, GRU, and transformer sequence-to-sequence architectures are explored with an English-to-French translation task. We begin with an exploratory process of generating a dataset for the task. Next we explore sequence-to-sequence architectures, comparing the effects of various hyperparameters and architecture designs on quality.

The dataset we use is a large set of English sentences with French translations from the Tatoeba website. The original data is a raw set of paired examples with no designated train, val, and test splits, so we create these during the EDA process.

7.7.1 Software Tools and Libraries

The popularity and diversity of problems sequence-to-sequence models can solve have led to many high-performance implementations. In this case study, we focus on the PyTorch-based Fairseq(-py) repository [Geh+17a], produced by Facebook AI Research (FAIR). This library holds implementations of many of the common seq-to-seq patterns with optimized dataloaders and batch support.

Additionally, we use the PyTorch text package and spaCy [HM17] to perform EDA and data preparation. These packages provide many useful functions for text processing and dataset creation, specifically with a focus on deep learning data loaders (although we do not use them here).

7.7.2 Exploratory Data Analysis

The raw format of the text contained in the Tatoeba dataset is a tab-separated English sentence followed by the French translation, with one pair per line. Counting the number of lines gives us a total of $135,842$ English–French pairs. By selecting a few random samples, as in Fig. 7.21, we can see that it contains punctuation, capitalization, as well as unicode characters. Unicode should come as no surprise when considering a translation task; however, it must be considered when dealing with any computational representation due to variations in libraries and their support for unicode characters.

```
Cheers! Santé !
I want to join you.    Je veux me joindre à vous.
I was busy cooking.    J'étais occupée à faire la cuisine.
```

Fig. 7.21: Examples from the English–French dataset

7.7.2.1 Sequence Length Filtering

First, we inspect the sequence lengths in the dataset. We use spaCy for both English and French tokenization. The tokenizers can be applied by torchtext fields when reading in the data, automatically applying the tokenizer. Fields in torchtext are generic data types for a dataset. In our example, there are two types of fields, a source field represented as "SRC" which will contain details on how the English sentences should be processed, while a second field called the "TRG" contains the target French data and its type handling. We can attach a tokenizer to each as follows.

```
1  def tokenize_fr(text):
2      """
3      Tokenizes French text from a string into a list of strings
4      """
5      return [tok.text for tok in spacy_fr.tokenizer(text)]
6
7  def tokenize_en(text):
8      """
9      Tokenizes English text from a string into a list of strings
10     """
11     return [tok.text for tok in spacy_en.tokenizer(text)]
12
13 SRC = Field(tokenize=tokenize_en, init_token='<sos>',
        eos_token='<eos>', lower=True)
14 TRG = Field(tokenize=tokenize_fr, init_token='<sos>',
        eos_token='<eos>', lower=True)
15
16 SRC.build_vocab(train_data, min_freq=0)
17 TRG.build_vocab(train_data, min_freq=0)
```

Torchtext can take a tokenizer of any type, as it is just a function that operates on the text that is passed in. The tokenizers in spaCy are useful as they have stop words, token exceptions, and various types of punctuation handling.

Another consideration when training sequence-based models is the length of the examples. We plot a histogram of the sequence lengths in Fig. 7.22. The longer sen-

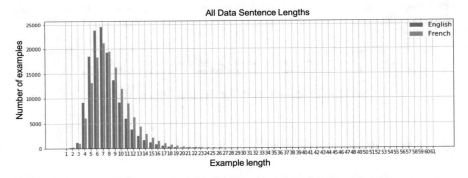

Fig. 7.22: Histogram of sentence lengths for both English and French. Notice that a majority of the sentences are short, and there are very few long sentences

tences are likely to hold a more complex structure, and likely have longer range dependencies. We would not expect to learn these examples, as they are under represented in the dataset. If we desire to learn translation for longer sentences, we would have to collect more data, or intelligently break long examples into shorter ones, where we have more data. Additionally, long examples can lead to memory concerns with mini-batch training, as the batch size can be larger with shorter examples.

For this case study, we remove longer examples by setting a threshold on the length of our examples. We select a limit of 20 time steps on input or output sequence, which will allow for a maximum of 18 actual words in the sequences after incorporating the $<$ sos $>$ and $<$ eos $>$ tokens. This restriction means our max length incorporates all of the sequence lengths that have significant data. The resulting length distribution is shown in Fig. 7.23.

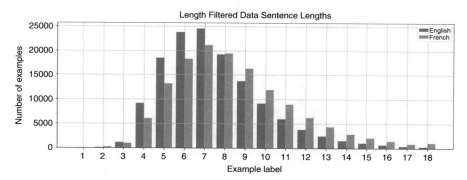

Fig. 7.23: Histogram of sentence lengths for both English and French after filtering the max length to 18 (20 if we include $<$ sos $>$ and $<$ eos $>$ tokens)

After filtering the longer examples we create our training, validation, and testing splits, without replacement using a shuffling index technique shown below.

```
n_examples = len(all_data)
idx_array = list(range(n_examples))
random.shuffle(idx_array)
train_indexs = idx_array[:int(0.8*n_examples)] # 80% training
    data
val_indexs = idx_array[int(0.8*n_examples):int(0.9*n_examples)
    ] # 10% validation data
test_indexs = idx_array[int(0.9*n_examples):] # 10% testing
    data
```

This technique should provide each split of the datasets with similar characteristic. The final dataset allocates 80% for training, 10% for validation, and 10% for testing. We save the data into files so that they can be used in other experiments if desired, without having to repeat all the preprocessing. Inspecting the resulting data splits shows a similar length distribution for each, as depicted in Fig. 7.24.

7.7.2.2 Vocabulary Inspection

The vocabulary object offers many common NLP functions, such as indexed access to the terms, simplified embedding creation, and frequency filtering.

We now load and tokenize the data splits. The overall vocabulary size is:

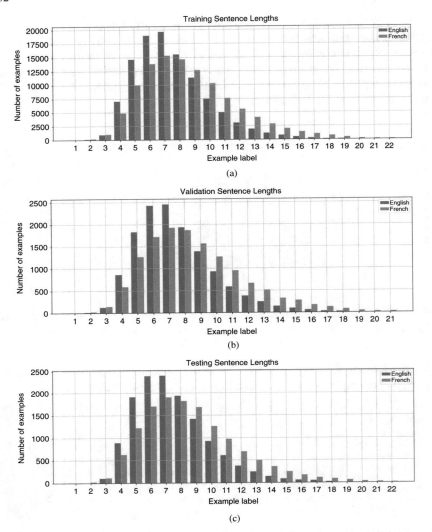

Fig. 7.24: Histogram of sentence lengths for the (**a**) training data, (**b**) validation data, and (**c**) testing data

```
train_data, valid_data, test_data = FrenchTatoeba.splits(path=
    data_dir,
exts=('.en', '.fr'),
fields=(SRC, TRG))

SRC.build_vocab(train_data, min_freq=0)
TRG.build_vocab(train_data, min_freq=0)
```

```
 8 print("English vocabulary size:", len(SRC.vocab))
 9 print("French vocabulary size:", len(TRG.vocab))
10
11 > English vocabulary size: 12227
12 > French vocabulary size: 20876
```

The vocabulary frequencies are shown in Fig. 7.25. The distribution displays a "long tail" effect, where a small subset of tokens have high counts, for example "." that occurs in almost all sentences, and other tokens that are seen just once, for example "stitch." In the most extreme case of a word only appearing once, training relies solely on that single example to inform the model, likely leading to overfitting. Additionally, the distribution for the softmax will assign some probability to

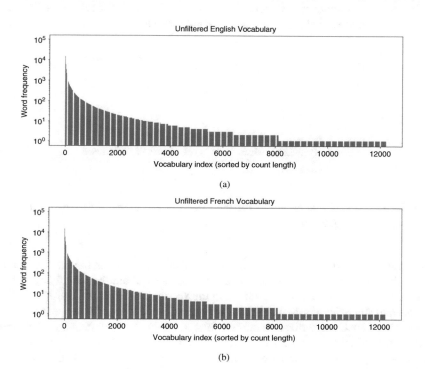

(a)

(b)

Fig. 7.25: Unfiltered word frequency for (**a**) English and (**b**) French. The counts were sorted and placed on a log scale to capture the severity of the word representations in this dataset. As we can see, there are many words that are used rarely, while a small subset is used frequently

these terms. As the infrequent words occupy the majority of the vocabulary, much of the probability mass will be assigned to these terms in the early stages, slowing learning. A common approach is to map infrequent words to the unknown token, $< unk >$. This allows the model to ignore a likely invalid representation of an underrepresented set of terms. We can enforce a minimum frequency by setting it as an argument when building the vocabulary.

The training dataset is used to create the vocabulary (using the validation data is considered data snooping). We set a minimum frequency of 5 during vocabulary creation. Evaluating the effects of this parameter is left as an exercise.

```
1 SRC.build_vocab(train_data, min_freq=5)
2 TRG.build_vocab(train_data, min_freq=5)
```

When investigating the final vocabulary, we still notice that there is a long tail distribution for the frequency of the words, shown in Fig. 7.26. This shouldn't be too much of a surprise, given that we chose a minimum frequency of 5. If the threshold is too high, removing many words, then the model becomes too restricted in its learning, with many values mapping to the unknown token.

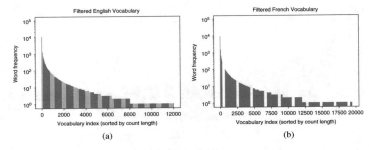

Fig. 7.26: Term frequency graph for the filtered vocabulary of the (**a**) English and (**b**) French training data

Figure 7.27 shows the top 50 terms for English and French in the training set. An analysis of the list leads to some interesting questions about the data. For example, one of the most common words shown in the vocabulary list is the word "n't." This seems odd since there is no word "n't," in the English language. A deeper inspection reveals that spaCy tokenization splits contractions in this way, leaving an isolated token "n't" whenever a contraction such as "don't" or "can't" appear. The same situation occurs when the contraction "I'm" is processed. This illustrates the importance of iterative improvements on data, as preprocessing is a fundamental component of the feature generation, as well as post-processing if results are computed on the final output.

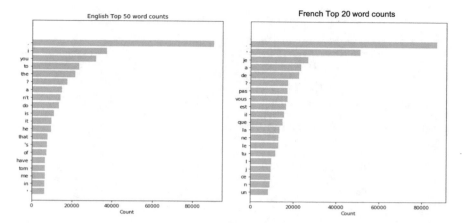

Fig. 7.27: Frequency counts for the top 20 terms from the training set for (**a**) English and (**b**) French

The final counts of our data splits are shown below.

```
1  Training set size: 107885
2  Validation set size: 13486
3  Testing set size: 13486
4  Size of English vocabulary: 4755
5  Size of French vocabulary: 6450
```

7.7.3 Model Training

Now that the dataset is ready, we investigate models and their performance on the training and validation sets. Specifically, we focus on various simple RNNs, LSTMs, and GRUs. Each of these architectures is investigated with respect to learning rate, depth, and bidirectionality. Each technique involves optimization of multiple hyperparameters to regularize the network, while also changing the training dynamics of the network. To alleviate a full grid search over all possible hyperparameters, we only tune learning rate to the introduced architecture. This does not completely alleviate the need to tune other parameters, but it makes the problem tractable.

Each model we train utilizes the script shown in Fig. 7.28. Note GRU and RNN configurations are not implemented in fairseq. We added these to the library for the purposes of this comparison.

Each model is trained for a maximum of 100 epochs. We reduce the learning rate when validation performance plateaus, and stop when learning plateaus. The embedding dimension is fixed at 256 and dropout for the input and output are set to 0.2. For simplicity, we fix the hidden size to 512 for all experiments (except for bidirectional architectures). A bidirectional provides two hidden states to the decoder, and therefore the decoder size must double. Some may argue that com-

```
1  python train.py datasets/en-fr \
2          --arch {rnn_type} \
3          --encoder-dropout-out 0.2 \
4          --encoder-layers {n_layers} \
5          --encoder-hidden-size 512 \
6          --encoder-embed-dim 256 \
7          --decoder-layers {n_layers} \
8          --decoder-embed-dim 256 \
9          --decoder-hidden-size 512 \
10         --decoder-attention False \
11         --decoder-dropout-out 0.2 \
12         --optimizer adam --lr {lr} \
13         --lr-shrink 0.5 --max-epoch 100 \
14         --seed 1 --log-format json \
15         --num-workers 4 \
16         --batch-size 512 \
17         --weight-decay 0
```

Fig. 7.28: Base training configuration for our fairseq model training. The rnn type, number of layers, and learning rate (lr) can be controlled by inserting parameter appropriately

parability between models would only be achieved if the models have the same number of parameters. For example, LSTMs have roughly $4\times$ the number of parameters as standard RNNs; however, for simplicity and clarity, we maintain a fixed hidden representation. In the following figures, each model name takes the form, {rnn_type}_{lr}_{num_layers}_{metric}.

7.7.3.1 RNN Baseline

First, we investigate the performance of a single layer, unidirectional RNN as a baseline for our experiments. We perform a manual grid search on the learning rate to find a reasonable starting value. The resulting validation curves for these selections are shown in Fig. 7.29.

The validation curves show how much the learning rate impacts the capacity of the model for RNNs, giving drastically different learning curves.

We also compute our testing results for this model to be used as a comparison at the end. Note that the test result is not used in any way to tune or improve our models. All tuning is done using the validation set. Any tuning should be done on the validation set. The testing result for our best RNN model is:

```
1  Translated 13486 sentences:
2  Generate test with beam=1: BLEU4 = 15.46
```

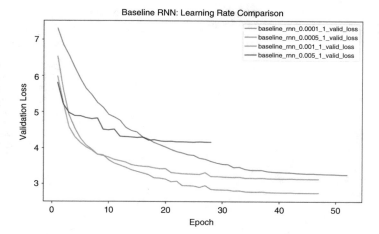

Fig. 7.29: Validation loss for a single layer RNN with different learning rates on English–French translation

7.7.3.2 RNN, LSTM, and GRU Comparison

Next, we compare RNN, LSTM, and GRU architectures. We vary the learning rate for each, as the dynamics are likely different for each architecture. The validation results are shown below in Fig. 7.30.

Upon inspection, we notice that some configurations take much longer to converge than others. In particular, with a learning rate of 0.0001, both the GRU and LSTM architectures reach the maximum 100 epochs. Secondly, we see that the LSTM and GRU architectures converge to lower losses, much faster, and with higher learning rates than RNN architectures. The GRU appears to be the best performing model here, but both the LSTM and GRU show similar convergence.

7.7.3.3 RNN, LSTM, and GRU Layer Depth Comparison

We now compare the effect of depth on each architecture. Here, we vary the depth configuration in addition to the learning rate for each architecture. The depths explored are 1, 2, and 4-layers deep. Results are displayed in Fig. 7.31.

Now that we have many models, it becomes more difficult to draw general conclusions about their properties. If we compare the RNN models, we notice that many of the configurations converge to a much higher validation loss than either the GRU or LSTM architectures. We also observe that the deeper architectures tend to perform well with lower learning rates than their shallower counterparts. Additionally,

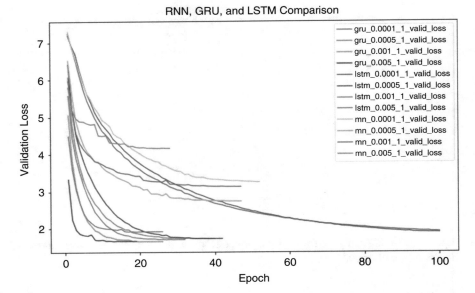

Fig. 7.30: Comparison of single layer RNN, GRU, and LSTM networks on English–French translation

both the LSTM and GRU architectures achieve their best models with a depth of 2 layers and a learning rate of 0.001.

7.7.3.4 Bidirectional RNN, LSTM, and GRU Comparison

Next, we look at the effects of bidirectional models. Many of the models perform similarly. Figure 7.32b shows the perplexity of the models predictions (ppl) instead of the validation loss. This value is 2^{loss}, exaggerating the effects in the graph, which can be useful when visibly inspecting curves.

Once again we see that the LSTM and GRU architectures outperform the RNN architectures, with the GRU architecture performing slightly better.

7.7.3.5 Deep Bidirectional Comparison

So far, the best performing models have been the 2-layer LSTM and GRU models and the single layer bidirectional LSTM and GRU models. Here, we combine the two components to see if the benefits are complimentary. In this set of experiments we remove the under-performing RNN models for clarity. The results are shown in Fig. 7.33.

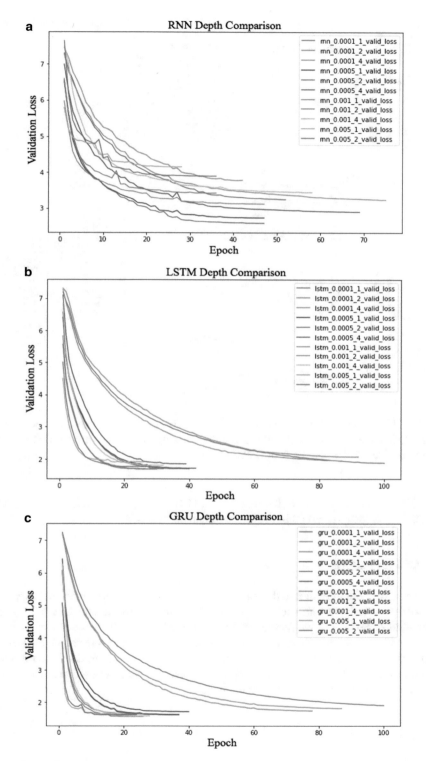

Fig. 7.31: Depth comparison for (**a**) RNN, (**b**) LSTM, and (**c**) GRU architectures

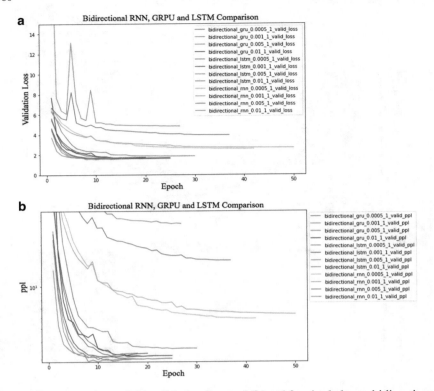

Fig. 7.32: Comparison of (**a**) validation loss and (**b**) ppl for single layer, bidirectional RNN, GRU, and LSTM networks. Note that although the colors are similar, the top two lines are RNN models (not GRU models)

This set of results shows that the 2-layer GRU architecture, with a learning rate of 0.001, is the best model in the bidirectional comparison.

7.7.3.6 Transformer Network

We now turn our attention to the transformer architecture, where attention is applied directly to the input sequence without incorporating recurrent networks. Similar to previous experiments, we fix the input and output dimensionality to 256, set 4 attention heads in both the encoder and decoder, and fix the fully connected layers size to 512. We explore a small selection of depths and vary the learning rates accordingly. The results are shown in Fig. 7.34, with the 4-layer transformer architecture using a learning rate of 0.0005 performing the best.

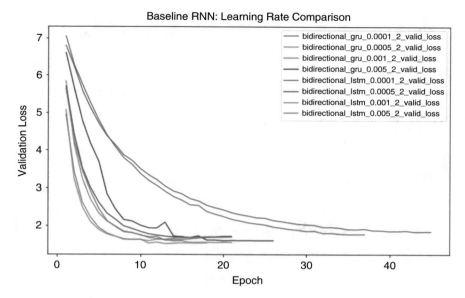

Fig. 7.33: Comparison of 2-layer bidirectional GRU and LSTM architectures

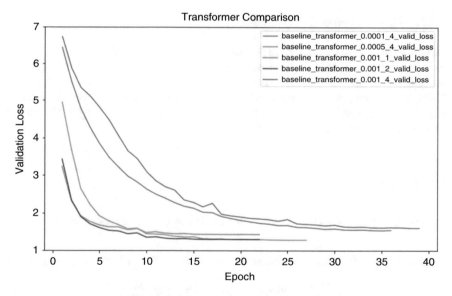

Fig. 7.34: Comparison of transformer architectures with different learning rates and depths. Note the depth is the same for both the encoder and decoder

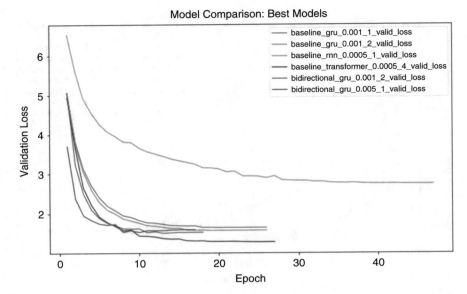

Fig. 7.35: Comparison of best NMT models from previous trials

7.7.3.7 Comparison of Experiments

Having explored many types of architectures for machine translation, we now compare the outputs of each experiment. This set includes the best performing RNN from the baseline experiments, the single-layer unidirectional and bidirectional GRU, the 2-layer unidirectional and bidirectional GRU, and the 4-layer transformer network. Comparing loss of these models on the validation set (Fig. 7.35) we see that the 4-layer transformer network is our best performer.

7.7.4 Results

We now compare the models from each experiment on the test set (Table 7.1).

Table 7.1: NMT network performance on the test set. The best result is highlighted

Network type	Learning rate	BLEU4
Baseline RNN (1 layer)	0.0005	15.46
GRU, 1-layer	0.001	36.17
GRU, 2-layer	0.001	38.53
GRU, 1-layer, bidirectional	0.005	40.63
GRU, 2-layer, bidirectional	0.001	40.60
Transformer, 4-layer	0.0005	**44.07**

When we sample outputs from the model (Fig. 7.36) we see that the results look pretty good. Notice how the model may produce reasonable translations even though it may not predict the target exactly.

```
Input: are you surprised ?
Target: êtes - vous surpris ?
Hypothesis: êtes - vous surprises ?

Input: i have evidence .
Target: j' ai des preuves .
Hypothesis: je dispose de preuves .

Input: i do n't know how many more times i 'll be able to do this .
Target: j' ignore combien de fois je serai encore capable de faire ça .
Hypothesis: je ne sais pas combien de fois je serai capable de faire ça .
```

Fig. 7.36: Output from the best performing NMT model

In conclusion, we have shown that for our task it is almost always preferable to use GRU or LSTM architectures over base RNNs. Additionally, we have shown that the initial learning rate has a significant impact on a model's quality, even when using adaptive learning rate methods. Furthermore, the learning rate needs to be tuned for each configuration of the model given the dynamic nature of the deep networks. Lastly, deeper networks are not always better. On this dataset, the 2-layer recurrent architectures outperformed 4-layer counterparts. And a single-layer, bidirectional GRU showed marginal improvements over the 2-layer counterpart on the final testing set, even though it performed slightly worse on the validation loss comparison. These results show the importance of tuning hyperparameters for not only the application, but also the dataset. In real-world applications, it is recommended to tune as many hyperparameters as possible to achieve the best result.

7.7.5 Exercises for Readers and Practitioners

Other interesting problems for readers and practitioners include:

1. Add L2 regularization to the training and see if it improves generalization on the testing set.
2. Prune the vocabulary, to remove more infrequent terms (for example, words that appear fewer than 20 times). What effect does this have on the training (performance, quality)?
3. Tune the beam search parameter to the validation dataset. What effect does this have on the test data? What is the effect on prediction time?
4. Experiment with tuning other hyperparameters in encoder and decoder.
5. What would need to be changed to modify the architecture for the question answering task?
6. Initialize the network with pre-trained embeddings

7.8 Discussion

The results from RNNs on many NLP tasks are quite impressive, achieving state-of-the-art results in almost every area. Their effectiveness is remarkable given their simplicity. However, in practice, real-world settings require additional considerations, such as small datasets, lack of diversity in data, and generalization. Following is a short discussion focusing on these concerns and common debates that arise.

7.8.1 Memorization or Generalization

All of the deep learning techniques that have been discussed so far come with the risk of overfitting. Additionally, many of the academic tasks for various NLP tasks are heavily focused on a particular problem with ample data that may not represent a real-world task.[5] The correlations between training and testing data allow some level of overfitting to be advantageous to both the validation and testing sets; however, it is arguable whether or not these correlations are just representative of the domain itself. The difficulty is knowing whether or not the network is memorizing certain sequences that are significant to lower the overall cost or learning correlations of underlying semantic structure of the problem. Some of the symptoms of memorization are illustrated in the need for decoding algorithms such as beam search and random selection with temperature to produce variety in the output sequences.

In Ref. [Gre16], Grefenstette explored the question of whether or not recurrent networks are capable of learning push-down automate, which is arguably the simplest form of computation required for natural language. This work cites some of the limitations of "simple RNNs" as:

- Non-adaptive capacity
- Target sequence modeling dominates training
- Gradient-starved encoder.

The suggestion, focused specifically on simple RNNs, was that RNNs are arguably only capable of learning finite state machines.

In Liska et al. [LKB18], the authors studied the ability of RNNs to learn a composition structure, which would show an RNN's ability to transfer learning from one task to another. A small number of the RNNs in the experiment showed that it was possible to learn compositional solutions without architectural constraints, although many of the RNN attempts were not successful. The results achieved show that gradient descent and evolutionary strategies may be a compelling direction for learning compositional structures.

---

[5] This is not to say that academic benchmarks are not relevant, but rather to point out the importance of domain and technological understanding for domain adaptation.

7.8.2 Future of RNNs

One suggestion from Grefenstette's presentation [Gre16] was to treat recurrence as an API. We have seen indications of this suggestion in this chapter already with LSTM and GRU cells. In those examples the recurrence API only needs to satisfy the interaction: given an input and previous state produce an output and updated state. This abstraction paves the way for a variety of memory-based architectures such as dynamic memory networks [XMS16] and the stack-LSTM [Dye+15]. Future directions point towards adding stacks and queues to have a more interactive memory model similar to RAM with architectures such as neural Turing machines [GWD14a].

References

[AKB16]   Jeremy Appleyard, Tomas Kocisky, and Phil Blunsom. "Optimizing performance of recurrent neural networks on GPUs". In: *arXiv preprint arXiv:1604.01946* (2016).

[BCB14a]   Dzmitry Bahdanau, Kyunghyun Cho, and Yoshua Bengio. "Neural machine translation by jointly learning to align and translate". In: *arXiv preprint arXiv:1409.0473* (2014).

[BSF94b]   Yoshua Bengio, Patrice Simard, and Paolo Frasconi. "Learning long-term dependencies with gradient descent is difficult". In: *IEEE transactions on neural networks* 5.2 (1994), pp. 157–166.

[Bow+15]   Samuel R. Bowman et al. "Generating Sentences from a Continuous Space". In: *CoRR* abs/1511.06349 (2015).

[Bra+16]   James Bradbury et al. "Quasi-Recurrent Neural Networks". In: *CoRR* abs/1611.01576 (2016).

[Bri+17]   Denny Britz et al. "Massive exploration of neural machine translation architectures". In: *arXiv preprint arXiv:1703.03906* (2017).

[Cho+14]   Kyunghyun Cho et al. "Learning phrase representations using RNN encoder-decoder for statistical machine translation". In: *arXiv preprint arXiv:1406.1078* (2014).

[Chu+14]   Junyoung Chung et al. "Empirical evaluation of gated recurrent neural networks on sequence modeling". In: *arXiv preprint arXiv:1412.3555* (2014).

[DN17]   Michael Denkowski and Graham Neubig. "Stronger baselines for trustable results in neural machine translation". In: *arXiv preprint arXiv:1706.09733* (2017).

[Dye+15]   Chris Dyer et al. "Transition-Based Dependency Parsing with Stack Long Short-Term Memory". In: *CoRR* abs/1505.08075 (2015).

[EHB96]   Salah El Hihi and Yoshua Bengio. "Hierarchical recurrent neural networks for long-term dependencies". In: *Advances in neural information processing systems*. 1996, pp. 493–499.

[GG16]   Yarin Gal and Zoubin Ghahramani. "A theoretically grounded appli-
            cation of dropout in recurrent neural networks". In: *Advances in neu-
            ral information processing systems.* 2016, pp. 1019–1027.

[Geh+17a]   Jonas Gehring et al. "Convolutional Sequence to Sequence Learning".
            In: *Proc. of ICML.* 2017.

[Gho+16]   Shalini Ghosh et al. "Contextual lstm (clstm) models for large scale
            nlp tasks". In: *arXiv preprint arXiv:1602.06291* (2016).

[GK96]   Christoph Goller and Andreas Kuchler. "Learning task-dependent dis-
            tributed representations by backpropagation through structure". In:
            Neural Networks, 1996., IEEE International Conference on. Vol. 1.
            IEEE. 1996, pp. 347–352.

[GWD14a]   Alex Graves, Greg Wayne, and Ivo Danihelka. "Neural turing ma-
            chines". In: *arXiv preprint arXiv:1410.5401* (2014).

[Gre16]   Ed Grefenstette. *Beyond Seq2Seq with Augmented RNNs.* 2016.

[Haf17]   Danijar   Hafner.   "Tips   for   Training   Recurrent   Neu-
            ral   Networks".   In: (2017).   URL: https://danijar.com/
            tips-for-training-recurrent-neural-networks/

[HS97b]   Sepp Hochreiter and Jürgen Schmidhuber. "Long short-term mem-
            ory". In: *Neural computation* 9.8 (1997), pp. 1735–1780.

[HM17]   Matthew Honnibal and Ines Montani. "spaCy 2: Natural language un-
            derstanding with Bloom embeddings, convolutional neural networks
            and incremental parsing". In: *To appear* (2017).

[HXY15]   Zhiheng Huang, Wei Xu, and Kai Yu. "Bidirectional LSTM-CRF
            models for sequence tagging". In: *arXiv preprint arXiv:1508.01991*
            (2015).

[KEL17]   Jaeyoung Kim, Mostafa El-Khamy, and Jungwon Lee. "Residual
            LSTM: Design of a Deep Recurrent Architecture for Distant Speech
            Recognition". In: *CoRR* abs/1701.03360 (2017).

[KB14]   Diederik P Kingma and Jimmy Ba. "Adam: A method for stochastic
            optimization". In: *arXiv preprint arXiv:1412.6980* (2014).

[KW13]   Diederik P Kingma and Max Welling. "Auto-encoding variational
            Bayes". In: *arXiv preprint arXiv:1312.6114* (2013).

[KG16]   Eliyahu Kiperwasser and Yoav Goldberg. "Simple and accurate de-
            pendency parsing using bidirectional LSTM feature representations".
            In: *arXiv preprint arXiv:1603.04351* (2016).

[Kru+16]   David Krueger et al. "Zoneout: Regularizing rnns by randomly pre-
            serving hidden activations". In: *arXiv preprint arXiv:1606.01305*
            (2016).

[Lam+16b]   Guillaume Lample et al. "Neural architectures for named entity recog-
            nition". In: *arXiv preprint arXiv:1603.01360* (2016).

[LD16]   Ji Young Lee and Franck Dernoncourt. "Sequential short-text classi-
            fication with recurrent and convolutional neural networks". In: *arXiv
            preprint arXiv:1603.03827* (2016).

[LZA17]   Tao Lei, Yu Zhang, and Yoav Artzi. "Training RNNs as Fast as
            CNNs". In: *CoRR* abs/1709.02755 (2017).

[LKB18]   Adam Liska, Germán Kruszewski, and Marco Baroni. "Memorize or generalize? Searching for a compositional RNN in a haystack". In: *CoRR* abs/1802.06467 (2018).

[Low+15]   Ryan Lowe et al. "The Ubuntu dialogue corpus: A large dataset for research in unstructured multi-turn dialogue systems". In: *arXiv preprint arXiv:1506.08909* (2015).

[LSM13b]   Thang Luong, Richard Socher, and Christopher Manning. "Better word representations with recursive neural networks for morphology". In: *Proceedings of the Seventeenth Conference on Computational Natural Language Learning.* 2013, pp. 104–113.

[MH16]   Xuezhe Ma and Eduard Hovy. "End-to-end sequence labeling via bi-directional lstm-cnns-crf". In: *arXiv preprint arXiv:1603.01354* (2016).

[MRF15]   Mateusz Malinowski, Marcus Rohrbach, and Mario Fritz. "Ask your neurons: A neural-based approach to answering questions about images". In: *Proceedings of the IEEE international conference on computer vision.* 2015, pp. 1–9.

[MDB17]   Gábor Melis, Chris Dyer, and Phil Blunsom. "On the state of the art of evaluation in neural language models". In: *arXiv preprint arXiv:1707.05589* (2017).

[Pan+15a]   Pingbo Pan et al. "Hierarchical Recurrent Neural Encoder for Video Representation with Application to Captioning". In: *CoRR* abs/1511.03476 (2015).

[Pas+13]   Razvan Pascanu et al. "How to construct deep recurrent neural networks.". In: *arXiv preprint arXiv:1312.6026* (2013).

[Pra+16]   Aaditya Prakash et al. "Neural Paraphrase Generation with Stacked Residual LSTM Networks". In: *CoRR* abs/1610.03098 (2016).

[RM15]   Danilo Jimenez Rezende and Shakir Mohamed. "Variational inference with normalizing flows". In: *arXiv preprint arXiv:1505.05770* (2015).

[SP97]   Mike Schuster and Kuldip K Paliwal. "Bidirectional recurrent neural networks". In: *IEEE Transactions on Signal Processing* 45.11 (1997), pp. 2673–2681.

[SSB16]   Stanislau Semeniuta, Aliaksei Severyn, and Erhardt Barth. "Recurrent Dropout without Memory Loss". In: *CoRR* abs/1603.05118 (2016).

[SMN10]   Richard Socher, Christopher D Manning, and Andrew Y Ng. "Learning continuous phrase representations and syntactic parsing with recursive neural networks". In: *Proceedings of the NIPS-2010 Deep Learning and Unsupervised Feature Learning Workshop.* Vol. 2010. 2010, pp. 1–9.

[Soc+12]   Richard Socher et al. "Semantic compositionality through recursive matrix-vector spaces". In: *Proceedings of the 2012 joint conference on empirical methods in natural language processing and computational natural language learning.* Association for Computational Linguistics. 2012, pp. 1201–1211.

[Soc+13]   Richard Socher et al. "Reasoning with neural tensor networks for knowledge base completion". In: *Advances in neural information processing systems.* 2013, pp. 926–934.

[SVL14b]   Ilya Sutskever, Oriol Vinyals, and Quoc V Le. "Sequence to sequence learning with neural networks". In: *Advances in neural information processing systems.* 2014, pp. 3104–3112.

[TSM15]   Kai Sheng Tai, Richard Socher, and Christopher D Manning. "Improved semantic representations from tree-structured long short-term memory networks". In: *arXiv preprint arXiv:1503.00075* (2015).

[Tan+15]   Ming Tan et al. "LSTM-based deep learning models for non-factoid answer selection". In: *arXiv preprint arXiv:1511.04108* (2015).

[Vas+17b]   Ashish Vaswani et al. "Attention is all you need". In: *Advances in Neural Information Processing Systems.* 2017, pp. 5998–6008.

[Ven+14]   Subhashini Venugopalan et al. "Translating videos to natural language using deep recurrent neural networks". In: *arXiv preprint arXiv:1412.4729* (2014).

[VFJ15]   Oriol Vinyals, Meire Fortunato, and Navdeep Jaitly. "Pointer networks". In: *Advances in Neural Information Processing Systems.* 2015, pp. 2692–2700.

[Vin+15b]   Oriol Vinyals et al. "Show and tell: A neural image caption generator". In: *Proceedings of the IEEE conference on computer vision and pattern recognition.* 2015, pp. 3156–3164.

[Wan+16a]   Cheng Wang et al. "Image captioning with deep bidirectional LSTMs". In: *Proceedings of the 2016 ACM on Multimedia Conference.* ACM. 2016, pp. 988–997.

[Wan+15b]   Xin Wang et al. "Predicting polarities of tweets by composing word embeddings with long short-term memory". In: *Proceedings of the 53rd Annual Meeting of the Association for Computational Linguistics and the 7th International Joint Conference on Natural Language Processing (Volume 1: Long Papers).* Vol. 1. 2015, pp. 1343–1353.

[XMS16]   Caiming Xiong, Stephen Merity, and Richard Socher. "Dynamic memory networks for visual and textual question answering". In: *International conference on machine learning.* 2016, pp. 2397–2406.

[Yin+17]   Wenpeng Yin et al. "Comparative study of CNN and RNN for natural language processing". In: *arXiv preprint arXiv:1702.01923* (2017).

[Zil+16]   Julian G. Zilly et al. "Recurrent Highway Networks". In: *CoRR*abs/1607.03474 (2016).

Chapter 8
Automatic Speech Recognition

8.1 Introduction

Automatic speech recognition (ASR) has grown tremendously in recent years, with deep learning playing a key role. Simply put, ASR is the task of converting spoken language into computer readable text (Fig. 8.1). It has quickly become ubiquitous today as a useful way to interact with technology, significantly bridging in the gap in human–computer interaction, making it more natural. Historically, ASR is tightly coupled with computational linguistics, given its close connection with natural language, and phonetics, given the variety of speech sounds that can be produced by humans. This chapter introduces the fundamental concepts of speech recognition with a focus on HMM-based methods.

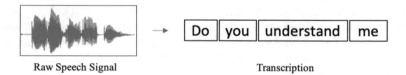

Raw Speech Signal   Transcription

Fig. 8.1: The focus of ASR is to convert a digitized speech signal into computer readable text, referred to as the transcript

Simply put, ASR can be described as follows: given an input of audio samples X from a recorded speech signal, apply a function f to map it to a sequence of words W that represent the transcript of what was said.

$$W = f(X) \tag{8.1}$$

However, finding such a function is quite difficult, and requires consecutive modeling tasks to produce the sequence of words.

© Springer Nature Switzerland AG 2019
U. Kamath et al., *Deep Learning for NLP and Speech Recognition*,
https://doi.org/10.1007/978-3-030-14596-5_8

These models must be robust to variations in speakers, acoustic environments, and context. For example, human speech can have any combination of time variation (speaker speed), articulation, pronunciation, speaker volume, and vocal variations (raspy or nasally speech) and still result in the same transcript.

Linguistically, additional variables are encountered such as prosody (rising intonation when asking a question), mannerisms, spontaneous speech, also known as filler words ("um"s or "uh"s), all can imply different emotions or implications, even though the same words are spoken. Combining these variables with any number of environmental scenarios such as audio quality, microphone distance, background noise, reverberation, and echoes exponentially increases the complexity of the recognition task.

The topic of speech recognition can include many tasks such as keyword spotting, voice commands, and speaker verification (security). In the interest of concision, we focus mainly on the task of speech-to-text (STT), specifically, large vocabulary continuous speech recognition (LVCSR) in this chapter. We begin by discussing error metrics commonly used for ASR systems. Next, we discuss acoustic features and processing, as well as phonetic units used for speech recognition. These concepts are combined as we introduce statistical speech recognition, the classical approach to ASR. We then introduce the DNN/HMM hybrid model, showing how the classical ASR pipeline incorporates deep learning. At the end of the chapter, a case study compares two common ASR frameworks.

8.2 Acoustic Features

The selection of acoustic features for ASR is a crucial step. Features extracted from the acoustic signal are the fundamental components for any model building as well as the most informative component for the artifacts in the acoustic signal. Thus, the acoustic features must be descriptive enough to provide useful information about the signal, as well as resilient enough to the many perturbations that can arise in the acoustic environment.

8.2.1 Speech Production

Let us first begin with a quick overview of how humans produce speech. While a full study of the anatomy of the human vocal system is beyond the scope of this book, some knowledge of human speech production can be helpful. The physical production of speech consists of changes in air pressure that produces compression waves that our ears interpret in conjunction with our brain. Human speech is created from the vocal tract and modulated with the tongue, teeth, and lips (often referred to as articulators):

- Air is pushed up from the lungs and vibrates the vocal cords (producing quasi-periodic sounds).
- The air flows into the pharynx, nasal, and oral cavities.
- Various articulators modulate the waves of air.
- Air escapes through the mouth and nose.

Human speech is usually limited to the range 85 Hz–8 kHz, while human hearing is in the range 20 Hz–20 kHz.

8.2.2 Raw Waveform

The waves of air pressure produced are converted into a voltage via a microphone and sampled with an analog-to-digital converter. The output of the recording process is a 1-dimensional array of numbers representing the discrete samples from the digital conversion. The digitized signal has three main properties: sample rate, number of channels, and precision (sometimes referred to as bit depth). The **sample rate** is the frequency at which the analog signal is sampled (in Hertz). The number of **channels** refers to audio capture with multiple microphone sources. Single-channel audio is referred to as monophonic or mono audio, while stereo refers to two-channel audio. Additional channels such as stereo and multi-channel audio can be useful for signal filtering in challenging acoustic environments [BW13]. The **precision** or **bit depth** is the number of bits per sample, corresponding to the resolution of the information.

Standard telephone audio has a sampling rate of 8 kHz and 16-bit precision. CD quality is 44.1 kHz, 16-bit precision, while contemporary speech processing focuses on 16 kHz or higher.

Sometimes **bit rate** is used to measure the overall quality of audio computed by:

$$\text{bit rate} = \text{sample rate} \times \text{precision} \times \text{number of channels}. \qquad (8.2)$$

The raw speech signal is high dimensional and difficult to model. Most ASR systems rely on features extracted from the audio signal to reduce the dimensionality and filter unwanted signals. Many of these features come from some form of spectral analysis that converts the audio signal to a set of features that strengthen signals that mimic the human ear. Many of these methods depend on computing a short time Fourier transform (STFT) on the audio signal using FFT, filter banks, or some combination of the two [PVZ13].

8.2.3 MFCC

Mel frequency cepstral coefficients (MFCC) [DM90] are the most commonly used features for ASR. Their success relies upon their ability to perform similar types of filtering that correlates to the human auditory system and their low dimensionality.

There are seven steps to computing the MFCC features [MBE10]. The overall process is shown in Fig. 8.2. These steps are similar for most feature generation techniques, with some variability in the types of filters that are used and the filter banks applied. We discuss each step individually:

1. Pre-emphasis
2. Framing
3. Hamming windowing
4. Fast Fourier transform
5. Mel filter bank processing
6. Discrete cosine transform (DCT)
7. Delta energy and delta spectrum.

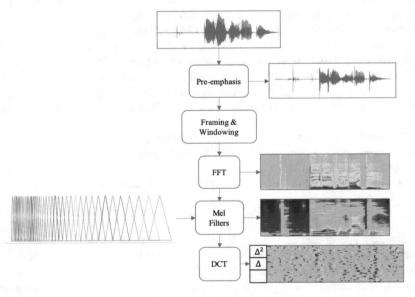

Fig. 8.2: Diagram of MFCC processing with a visual representation for various parts of the process. All spectrograms and features are shown in log-space

8.2.3.1 Pre-emphasis

Pre-emphasis is the first step in MFCC feature generation. In speech production (and signal processing in general), the energy of higher frequency signals tends to be lower. Pre-emphasis processing applies a filter to the input signal that emphasizes the amplitudes of higher frequencies and lowers the amplitudes of lower frequency bands. For example:

$$y_t = x_t - \alpha x_{t-1} \tag{8.3}$$

would make the output less dependent on a strong signal from the previous time steps.

8.2.3.2 Framing

The acoustic signal is perpetually changing in speech. Modeling this changing signal is done by treating small segments sampled from the audio as stationary. Framing is the process of separating the samples from the raw audio into fixed length segments referred to as **frames**. These segments are converted to the frequency domain with an FFT, yielding a representation of the strength of frequencies during each frame. The segments signify the boundaries between the phonetic representations of speech. The phonetic sounds associated with speech tend to be in the range of 5–100 ms, so the length of frames is usually chosen to account for this. Typically, frames are in the range of 20 ms for most ASR systems, with a 10 ms overlap, yielding a resolution of 10 ms for our frames.

8.2.3.3 Windowing

Windowing multiplies the samples by a scaling function. The purpose of this function is to smooth the potentially abrupt effects of framing that can cause sharp differences at the edges of frames. Applying windowing functions to the samples therefore tapers the changes to the segment to dampen signals near the edges of the frame that may have harsh effects after the application of the FFT.

Many windowing functions can be applied to a signal. The most commonly used for ASR are Hann windowing and Hamming windowing.

Hann window:

$$w(n) = 0.5 \left(1 - \cos\left(\frac{2\pi n}{N-1} \right) \right) = \sin^2\left(\frac{\pi n}{N-1} \right) \tag{8.4}$$

Hamming window:

$$w(n) = 0.54 - 0.46 \cos\left(\frac{2\pi n}{N-1} \right) \tag{8.5}$$

where N is the window length and $0 \leq n \leq N - 1$.

8.2.3.4 Fast Fourier Transform

A short-time Fourier transform (STFT) converts the 1-dimensional signal from the time domain into the frequency domain by using the frames and applying a discrete Fourier transform (DFT) to each. An illustration of the DFT conversion is shown in Fig. 8.3. The fast Fourier transforms (FFT) is an efficient algorithm to compute the DFT under suitable circumstances and is common for ASR.

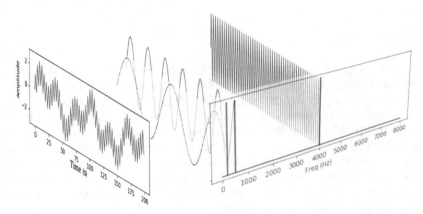

Fig. 8.3: The desired effect of an FFT on an input signal (shown on the left) and the normalized FFT output in the frequency domain (shown on the right)

The **spectrogram** is a 3-dimensional visual FFT transformation of the acoustic signal and is often a valuable set of features itself. The STFT representation can be advantageous because it makes the fewest assumptions about the speech signal (aside from the raw waveform). For some end-to-end systems, the spectrogram is used as input, because it provides a higher resolution frequency description. The plot itself, shown in Fig. 8.4, has time along the x-axis, the frequency bins on the y-axis, and the intensity of that frequency in the z-axis, which is usually represented by the color.

The magnitude spectrogram can be computed by:

$$S_m = |\text{FFT}(x_i)|^2 \tag{8.6}$$

The power spectrogram is sometimes more helpful because it normalizes the magnitude by number of points considered

$$S_p = \frac{|\text{FFT}(x_i)|^2}{N} \tag{8.7}$$

where N is the number of points considered for the FFT computation (typically 256 or 512).

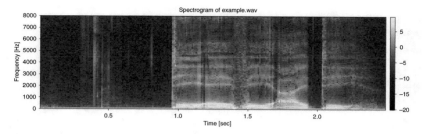

Fig. 8.4: Log spectrogram of an audio file

Most of the significant frequencies are in the lower portion of the frequency spectrum, so the spectrogram is typically mapped into the log scale.

8.2.3.5 Mel Filter Bank

The features created from the STFT transformation of the audio aim to simulate conversions made by the human auditory system processes. The Mel filter bank is a set of bandpass filters that mimic the human auditory system. Rather than follow a linear scale, these triangular filters act logarithmic at higher frequencies and linear at lower frequencies, which is typical in speech signals. Figure 8.5 shows the Mel filter bank. The filter bank usually has 40 filters.

The conversion between the Mel (m) and Hertz (f) domains can be accomplished by:

$$m = 2595 \log_{10}\left(1 + \frac{f}{700}\right)$$
$$f = 700\left(10^{\frac{m}{2595}} - 1\right)$$

(8.8)

Each of the filters produces an output that is the weighted sum of the spectral frequencies that correspond to each filter. These values map the input frequencies into the Mel scale.

8.2.3.6 Discrete Cosine Transform

The discrete cosine transform (DCT) maps the Mel scale features into the time domain. The DCT function is similar to a Fourier transform but uses only real numbers (a Fourier transform produces complex numbers). It compresses the input data into a set of cosine coefficients that describe the oscillations in the function. The output of this conversion is referred to as the MFCC.

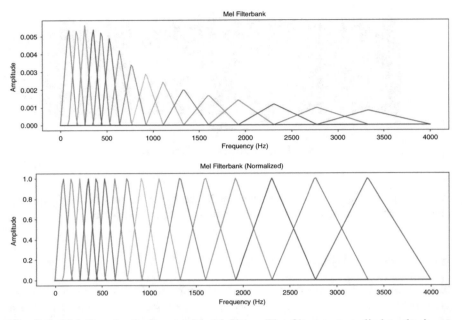

Fig. 8.5: Mel filter bank shown with 16 filters. The filters are applied to the input signal to produce the Mel-scale output

8.2.3.7 Delta Energy and Delta Spectrum

The delta energy (delta) and delta spectrum (also known as "delta delta" or "double delta") features provide information about the slope of the transition between frames. The delta energy features are the difference between consecutive frames' coefficients (the current and previous frames). The delta spectrum features are the difference between consecutive delta energy features (the current and previous delta energy features). The equations for computing the delta energy and delta spectrum features are:

$$d_t = \frac{\sum_{n=1}^{N} n(c_{t+n} - c_{t-n})}{2\sum_{n=1}^{N} n^2} \tag{8.9}$$

$$dd_t = \frac{\sum_{n=1}^{N} n(d_{t+n} - d_{t-n})}{2\sum_{n=1}^{N} n^2} \tag{8.10}$$

8.2.4 Other Feature Types

Many acoustic features have been proposed over the years, applying different filters and transforms to highlight various aspects of the acoustic spectrum. Many of these approaches relied on hand engineered features such as MFCCs, gammatone features [Sch+07], or perceptual linear predictive coefficients [Her90]; however, MFCCs remain the most popular.

One of the downsides of MFCC features (or any manually engineered feature set) is the sensitivity to noise due to its dependence on the spectral form. Low dimensionality of the feature space was highly beneficial with earlier machine learning techniques, but with deep learning approaches, such as convolutional neural networks, higher resolution features can be used or even learned.

Overall MFCC features are efficient to compute, apply useful filters for ASR, and decorrelate the features. They are sometimes combined with additional speaker-specific features (typically i-vectors) to improve the robustness of the model.

8.2.4.1 Automatically Learned

Various attempts have been tried to learn the feature representations directly, rather than relying on engineered features, which may not be best for the overall task of reducing WER. Some of the approaches include: supervised learning of features with DNNs [Tüs+14], CNNs on raw speech for phone classification [PCD13], combined CNN-DNN features [HWW15], or even unsupervised learning with RBMs [JH11].

Automatically learned features improve quality in specific scenarios but can also be limiting across domains. Features produced with supervised training learn to distinguish between the examples in the dataset and may be limited in unobserved environments. With the introduction of end-to-end models for ASR, these features are tuned during the end-to-end task alleviating the two-stage training process.

8.3 Phones

Following from NLP, the most logical linguistic representation for transforming speech into a transcript may seem to be words, ultimately because a word-level transcript is the desired output and there is meaning attached at the word-level. Practically speaking, however, speech datasets tend to have few transcribed examples per word, making word-level modeling difficult. A shared representation for words is desirable, to obtain sufficient training data for the variety of words that are possible. For example, **phonemes** can be used to phonetically discretize words in a particular language. Swapping one phoneme with another changes the meaning of the word (although this may not be the case for the same phonemes in another language). For example, if the third phone in the word *sweet* [swit] is changed from [i] to [ɛ], the meaning of the whole word changes: *sweat* [swɛt].

Phonemes, themselves, tend to be too strict to use practically due to the attachment of meaning. Instead **phones** are used as a phonetic representation for the linguistic units (with potentially multiple phones mapping to a single phoneme).

Phones do not map to any specific language, but rather, are absolute to speech itself, distinguishing sounds that signify speech. Figure 8.6 shows the phone set for English.

AA	AY	EH	HH	L	OY	T	W
AE	B	ER	IH	M	P	TH	Y
AH	CH	EY	IY	N	R	UH	Z
AO	D	F	JH	NG	S	UW	ZH
AW	DH	G	K	OW	SH	V	

Fig. 8.6: English phone set, based on the ARPAbet symbols for ASR as used in the CMU Sphinx framework. The phone set is made up of 39 phones

With phones, words are mapped to their phonetic counterpart by using a phonetic dictionary similar to the one shown in Fig. 8.7. A phonetic entry should be present for each word in the vocabulary (sometimes more than one entry if there are multiple ways to pronounce a word). By using phones to represent words, the shared representations can be learned from many examples across words, rather than modeling the full words.

Word	Phone Representation
a	AH
aardvark	AA R D V AA R K
aaron	EH R AH N
aarti	AA R T IY
...	...
zygote	Z AY G OW T

Fig. 8.7: Phonetic dictionary for supported words in an ASR system. Note: the stress of the syllable is sometimes included adding an additional features to the phone representations

If every word were pronounced with the same phones, then a mapping from the audio to the set of phones to words would be a relatively straight-forward transformation. However, audio exists as a continuous stream, and a speech signal does not necessarily have defined boundaries between the phone units or even words. The signal can take many forms in the audio stream and still map to the same interpretable output. For example, the speaker's pace, accent, cadence, and environment can all play significant roles in how to map the audio stream into an output sequence. The words spoken depend not only on the phone at any given moment, but also on the states that have come before and after the context. This natural dynamic in speech places a strong emphasis on the dependency of the surrounding context and phones.

Combining phone states is a common strategy to improve quality, rather than relying on their canonical representations. Specifically, the transitions between words can be more informative than single phone states. In order to model this,

diphones—parts of two consecutive phones, **triphones**, or extended to **senones** (triphone context-dependent units) can be used as the linguistic representation or intermediary rather than phones themselves. Many methods exist for combining the phone representations with additional context, modeling them directly or by learning a statistical hierarchy of the state combinations, and most traditional approaches rely on these techniques.

Although ASR focuses on *recognition* rather than *interpretation* (e.g., the accuracy on recognizing spoken words rather than context-dependent word sequence modeling), the contextual understanding is an important aspect. In the case of homophones, two words with the same phonetic representation and different spellings, predicting the correct word relies entirely on the surrounding context. In this case, some of the issues can be overcome with a language model, discussed later. Incorrect phonetic substitutions further complicate matters. For example, in English, the representations of *pin* [P IH N] and *pen* [P EH N] are distinct. However, although these words do have different phonetic representations, they are commonly mistakenly said interchangeably or pronounced similarly, requiring the correct selection to depend on the context more so than the phones themselves. With the inclusion of accents, phonetic representations can contain even more conflicts, requiring alternative methods to determine speaker-specific features. These types of scenarios are crucial in ASR, for there are many times that humans may say the wrong word, and yet the context and intent can still be interpreted. All of these real-world factors of spoken language contribute the complexity of automatic speech recognition in practice.

8.4 Statistical Speech Recognition

Statistical ASR focuses on predicting the most probable word sequence given a speech signal, via an audio file or input stream. Early approaches did not use a probabilistic focus, aiming to optimize the output word sequence by applying templates for reserved words to the input acoustic features (this was historically used for recognizing spoken digits). Dynamic time warping (DTW) was an early way to expand this templating strategy by finding the "lowest constrained path" for the templates. This approach allowed for variations in the input time sequence and output sequence; however, it was difficult to come up with appropriate constraints, such as distance metrics, how to choose templates, and the lack of a statistical, probabilistic foundation. These drawbacks made the DTW-templating approach challenging to optimize.

A probabilistic approach was soon formed to map the acoustic signal to a word sequence. Statistical sequence recognition introduced a focus on maximum posterior probability estimation. Formally, this approach is a mapping from a sequence of acoustic, speech features, X, to a sequence of words, W. The acoustic features are a sequence of feature vectors of length T: $X = \{\mathbf{x}_t \in \mathbb{R}^D | t = 1, \ldots, T\}$, and the word sequence is defined as $W = \{w_n \in V | n = 1, \ldots, N\}$, having a length N, where V is the vocabulary. The most probable word sequence W^* can be estimated by maximizing $P(W|X)$ for all possible word sequences, V^*. Probabilistically this can be written as:

$$W^* = \underset{W \in V^*}{\text{argmax}}\, P(W|X) \tag{8.11}$$

Solving this quantity is the center of ASR. Traditional approaches factorize this quantity, optimizing models to solve each component, whereas more recent end-to-end deep learning methods focus on optimizing for this quantity directly.

Using Bayes' theorem, statistical speech recognition is defined as:

$$P(W|X) = \frac{P(X|W)P(W)}{P(X)} \tag{8.12}$$

The quantity $P(W)$ represents the language model (the probability of a given word sequence) and $P(X|W)$ represents the acoustic model. Because this equation drives the maximization of the numerator to achieve the most likely word sequence, the goal does not depend on $P(X)$, and it can be removed:

$$W^* = \underset{W \in V^*}{\text{argmax}}\, P(X|W)P(W) \tag{8.13}$$

An overview of statistical ASR is illustrated in Fig. 8.8.

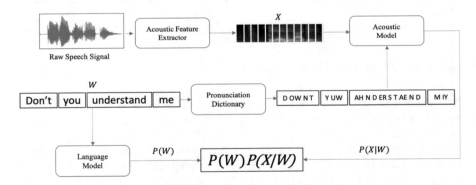

Fig. 8.8: Diagram of statistical speech recognition

Often, one of the most challenging components of speech recognition is the significant difference between the number of steps in the input sequence compared to the output sequence ($T \gg N$). For example, extracted acoustic features may represent a 10 ms frame from the audio signal. A typical ten-word utterance could have a duration of 3-s utterance, leading to an input sequence length of 300 and a target output sequence of 10 [You96]. Thus, a single word can spread many frames and take a variety of forms, as shown in Fig. 8.9. It is, therefore, sometimes beneficial to split a word into sub-components that span fewer frames.

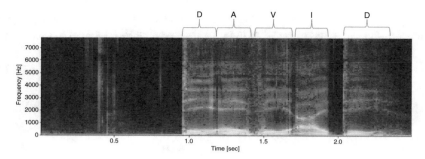

Fig. 8.9: Spectrogram of a 16 kHz speech utterance, reciting the letters "D A V I D." The spectrogram has been created with 20 ms frames with a 10 ms overlap, yielding an spectrogram size of 249×161. The output sequence has a length of 5 corresponding to each of the characters in the vocabulary

8.4.1 Acoustic Model: $P(X|W)$

The statistical definition in Eq. (8.13) can be augmented to incorporate the mapping acoustic features to phones and then from phones to words:

$$
\begin{aligned}
W^* &= \underset{W}{\operatorname{argmax}} P(X|W)P(W) \\
&= \underset{W}{\operatorname{argmax}} \sum_S P(X,S|W)P(W) \\
&\approx \underset{W,S}{\operatorname{argmax}} P(X|S)P(S|W)P(W)
\end{aligned}
\tag{8.14}
$$

where $P(X|S)$ maps the acoustic features to phone states and $P(S|W)$ maps phones to words (commonly referred to as the pronunciation model).

Equation (8.13) showed two factors $P(X|W)$ and $P(W)$. Each of these factors are considered models and therefore have learnable parameters, Θ_A and Θ_L, for the acoustic model and language model, respectively.

$$
W^* = \underset{W \in V^*}{\operatorname{argmax}} P(X|W,\Theta_A)P(W,\Theta_L)
\tag{8.15}
$$

This model now depends on predicting the likelihood of observations X, with the factor $P(X|W,\Theta_A)$. Solving this quantity requires a state-based modeling approach (HMMs). If a discrete-state model is assumed, the probability of an observation can be defined by introducing a state sequence S, where $S = \{s_t \in \{s^{(i)}, \ldots s^{(Q)}\} | t = 1, \ldots, T\}$ into $P(X|W)$.

$$
P(X|W) = \sum_S P(X|S)P(S|W)
\tag{8.16}
$$

Equation 8.16 can be factorized further using the chain rule of probability to produce the framewise likelihood. For notational convenience, let $\mathbf{x}_{1:n} = \mathbf{x}_1, \mathbf{x}_2, \ldots, \mathbf{x}_n$.

$$P(X|S) = \prod_{t=1}^{T} P(\mathbf{x}_t|\mathbf{x}_{1:t-1}, S) \tag{8.17}$$

Using the conditional independence assumption, this quantity can be reduced to:

$$P(X|S) \approx \prod_{t=1}^{T} P(\mathbf{x}_t|\mathbf{s}_t) \tag{8.18}$$

The conditional independence assumption limits the context that is considered for prediction. We assume that any observation $\mathbf{x}_t$ is only dependent on the current state, s_t, and not on the history of observations $\mathbf{x}_{1:t-1}$, as shown in Fig. 8.10. This assumption reduces the computational complexity of the problem; however, it limits the contextual information included in any decision. The conditional independence assumption is often one of the biggest hurdles in ASR, due to the contextual nature of speech. Thus, a variety of techniques are centered around providing "context features" to improve quality.

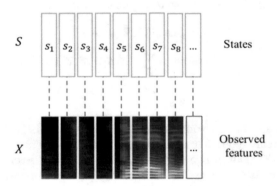

Fig. 8.10: State alignment with feature observations

The conditional independence assumption allows us to compute the probability of an observation by summing over all possible state sequences S because the actual state sequence that produced X is never known. The set of states Q can vary depending on the modeling approach of the ASR system. In a simple system, the target states are sub-word units (such as English phones).

The transition alignments between frames are not known beforehand. We use an HMM, allowing us to learn the temporal dilation, and train it using the expectation maximization (EM) algorithm. In general, the EM algorithm estimates the state occupation probabilities with the current HMM parameters and then re-estimates the HMM parameters based on the estimation.

An HMM is composed of two stochastic processes: a *hidden* part that is a Markov chain and an *observable* process that is probabilistically dependent on the Markov chain. The aim is to model probability distributions of the states that produce the observable events, which are acoustic features. Formally, the HMM is defined by:

1 A set of Q states $S = \{s^{(1)}, \ldots, s^Q\}$. The Markov chain can only be in one state at a time. In a simple ASR model, the state set S could be the set of phones for the language.
2 The *initial state* probability distribution, $\pi = \{P(s^{(i)}|t = 0)\}$, where t is the time index.
3 A probability distribution that defines the transitions between states: $a_{ij} = P(s_t^{(j)}|s_{t-1}^{(i)})$. The *transition probabilities* a_{ij} are independent of time t.
4 Observations X from our feature space F. In our case, this feature space can be all continuous acoustic features that are input into our model. These features are given to us by the acoustic signal.
5 A set of probability distributions, *emission* probabilities (sometimes referred to as *output* probabilities). This set of distributions describe the properties of the observations yielded by each state, i.e., $b_x = \{b_i(x) = P(x|s^{(i)})\}$

- Emission distributions: $b_x = P(x|s)$
- Transition probabilities: $a_{ij} = P(s_t|s_{t-1})$
- Initial state probabilities: $\pi = P(s_1)$.

The transitions between states s_t only depend on the previous state s_{t-1}. The lexicon model (discussed in the next section) provides the initial transition state probabilities. These transitions can be self-loops, allowing the time dilation that is necessary to allow elasticity in the frame-based prediction.

The HMM is optimized, learning π, a, and $b(x)$ by training on the acoustic observations X and the target sequence of phone states Y. An initial estimate for $P(s^{(j)}|s^{(i)})$ can be obtained from the lexicon model, $P(S|W)$. The forward-recursion algorithm is used to score the current model parameters $a, b(x)$, yielding parameters to obtain $P(X|S)$. The Viterbi algorithm is used to avoid computing the sum of all paths and serves as an approximation of the forward algorithm:

$$
\begin{aligned}
P(X|S) &= \sum_{\{\text{path}_l\}} P(X, S|\lambda) \quad \text{Baum-Welch} \\
&\simeq \max_{\text{path}_l} P(X, S|\lambda) \qquad \text{Viterbi}
\end{aligned}
\tag{8.19}
$$

Training is typically accomplished by the forward–backward (or Baum–Welch) and Viterbi algorithms [Rab89b]. In our case, the emission probabilities target maximizing the probability of the sample given the model. Because of this, the Viterbi algorithm focuses only on the most likely path in the set of possible state sequences (Fig. 8.11). Modeling the emission probability density function is usually accomplished using a Gaussian or mixture of Gaussians.

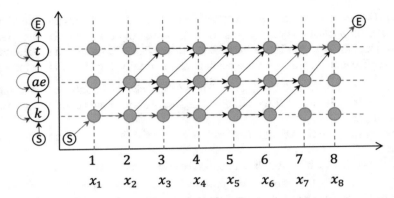

Fig. 8.11: All possible state transitions to produce the 3-phone word, "cat" for an 8-frame utterance. The Viterbi path applied to the possible state transitions is shown in red

8.4.1.1 *LexiconModel* : $P(S|W)$

A model for $P(S|W)$ can be constructed by representing the probability of a state sequence given a word sequence. This model is commonly referred to as the pronunciation or lexicon model. We factorize this using the probabilistic chain rule to obtain:

$$P(S|W) = \prod_{t=1}^{T} P(s_t|s_{1:t-1}, W) \qquad (8.20)$$

Once again, using the conditional independence assumption, this quantity is approximated by:

$$P(s_t|s_{1:t-1}, W) = P(s_t|s_{t-1}, W) \qquad (8.21)$$

The introduction of the conditional independence assumption is also the first-order Markovian assumption, allowing us to implement the model as a first-order HMM. The states of the model, s_t, are not directly observable; therefore, we are not able to observe the transition from one phonetic unit to another; however, the observations x_t do depend on the current state s_t. The HMM allows us to infer information about the state sequence from the observations.

First, the word vocabulary V is converted into the state representations for each term to create a word model.

The lexicon model can be used to determine the initial probability of each state $P(s_1)$ by counting the occurrence rate for the beginning of each word. The transition probabilities accumulate over the lexical version of the transcript targets for the acoustic data.

A state-based word-sequence model can be created for each word in the vocabulary, as shown in Fig. 8.12.

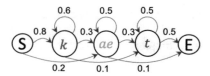

Fig. 8.12: Phone state model for the 3-phone word "cat" with transition probabilities

8.4.2 *LanguageModel* : $P(W)$

The language model $P(W)$ is typically an n-gram language model leveraging the probabilistic chain rule. It is factorized by using the conditional independence assumption except with an $(m-1)$-th order Markov assumption, where m is the number of grams to be considered. It can be described as:

$$
\begin{aligned}
P(W) &= \prod_{n=1}^{N} P(w_n | w_{w_1:w_{n-1}}) \\
&\approx \prod_{n=1}^{N} P(w_n | w_{n-m-1:n-1})
\end{aligned}
\tag{8.22}
$$

HMMs are robust models for training and decoding sequences. When training the HMM models, we focused on training individual models for the state alignments, and then combine them into a single HMM for continuous speech recognition. HMMs also allow the word sequence to be incorporated, creating a state sequence based on words and apply the word sequence priors as well. Furthermore, HMMs support compositionality; therefore, the time dilation, pronunciation, and word sequences (grammar) are handled in the same model by composing the individual components:

$$
\begin{aligned}
P(q|M) &= P(q, \phi, w | M) \\
&= P(q|\phi) \cdot P(\phi|w) \cdot P(w_n | w_{w_1:w_{n-1}})
\end{aligned}
\tag{8.23}
$$

Unfortunately, many of the assumptions that are needed to optimize HMMs limit their functionality due to [BM12]:

- HMM and DNN models are trained independently of each other and yet depend on each other.
- A priori choices of statistical distributions rely on linguistic information from handcrafted pronunciation dictionaries. These are subject to human error.
- The first-order Markov assumption often referred to as the conditional independence assumption (states are only dependent on their previous state) forces strict

limitations on the number of context states considered for an individual prediction.

- The decoding process is complex.

8.4.3 HMM Decoding

The decoding process for an HMM-based ASR model finds the optimal word sequence, combining the various models. The process decodes a state sequence from the acoustic features initially and then decodes to the optimal word sequence from the state sequence. Phonetic decoding has traditionally relied on interpreting the HMMs probability lattice constructed for each word from the phonetic lexicons according to the acoustic features. Decoding can be done using the Viterbi algorithm on the HMM output lattice, but this is expensive for large vocabulary tasks. Viterbi decoding performs an exact search efficiently, making it infeasible for a large vocabulary task. Beam search is often used instead to reduce the computation. The decoding process uses backtracking to keep track of the word sequence produced.

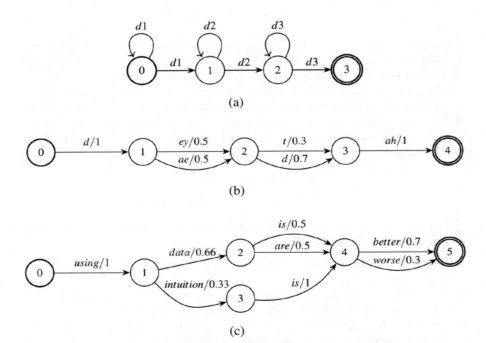

Fig. 8.13: (**a**) HMM state representation, (**b**) phone state transitions for the word "data," and (**c**) grammar state model. This figure is adapted from [MPR08]

During prediction, decoding the HMM typically relies on using weighted automata and transducers. In a simple case, weighted finite state acceptors (WFSA), the automata are composed of a set of states (initial, intermediate, and final), a set of transitions between states with a label and weight, and final weights for each final state. The weights express the probability, or cost, of each transition. You can express HMMs in the form of finite state automata. In this approach, a transition connects each state. WFSA accept or deny possible decoding paths depending on the states and the possible transitions. The topology could represent a word, the possible word pronunciation(s), or the probabilities of the states in the path to result in this word, (Fig. 8.13). Decoding, therefore, depends on combining the state models from the HMM with the pronunciation, dictionary, and n-gram language models that must be combined in some way.

Usually, weighted finite state transducers (WFST) are used to represent the different levels of state transition in the decoding phase [MPR08]. WFSTs transduce an input sequence to an output sequence. WFSTs add an output label, which can be used to tie different levels of the decoding relationships together, such as phones and words. A WFSA is a WFST without the output label. The WFST representation allows models to be combined and optimized jointly via its structural properties with efficient algorithms: compositionality, determinism, and minimization. The composition property allows for different types of WFSTs to be constructed independently and composed together, such as combining a lexicon (phones to words) WFST and a probabilistic grammar. Determinism forces unique initial states, where no two transitions leaving a state share the same input label. Minimization combines redundant states and can be thought of as suffix sharing. Thus, the whole decoding algorithm for a DNN-HMM hybrid model can be represented by WFSTs via four transducers:

- HMM: mapping HMM states to CD phones
- Context-dependency: mapping CD phones to phones
- Pronunciation lexicon: mapping phones to words
- Word-level grammar: mapping words to words.

In Kaldi, for example, these transducers are referred to as H, C, L, and G, respectively. Compositionality allows a composition between L and G into a single transducer, L G, that maps phone sequences to a word sequence. Practically, the composition of these transducers may grow too large, so the conversion usually takes the form: $HCLG$, where

$$HCLG = \min(\det(H \circ \min(\det(C \circ \min(\det(L \circ G)))))). \tag{8.24}$$

8.5 Error Metrics

The most commonly used metric for speech recognition is word error rate (WER). WER measures the edit distance between the prediction and the target by considering the number of insertions, deletions, and substitutions, using the Levenshtein distance measure.

Word error rate is defined as:

$$WER = 100 \times \frac{I+D+S}{N} \qquad (8.25)$$

where

- I is the number of word insertions,
- D is the number of word deletions,
- S is the number of word substitutions, and
- N is the total number of words in the target.

For character-based models and character-based languages, error metrics focus on CER (character error rate), sometimes referred to as LER (letter error rate). Character-based models will be explored more in Chap. 9:

$$CER = 100 \times \frac{I+D+S}{N} \qquad (8.26)$$

where

- I is the number of character insertions,
- D is the number of character deletions,
- S is the number of character substitutions, and
- N is the total number of characters in the target.

CER and WER are used to identify how closely a prediction resembles its target, giving a measurement of the overall system. They are straight-forward to compute and give a straight-forward summary of the recognition system's quality. Figure 8.14 shows the scripts to compute WER and CER. A few examples of WER and CER are shown in Figs. 8.15, 8.16, 8.17.

One of the drawbacks to edit distance metrics, however, is that they do not give any indication of what the errors might be. Measuring specific types of errors, therefore, would require additional investigation for improving models, such as SWER (salient word error rate) or looking at concept accuracy. In [MMG04], the authors suggested improvements to the WER metric in the form of MER (match error rate) and WIR (word information loss). These metrics can be useful when the information communicated is more important than the edit cost, with the added benefit of providing probabilistic interpretations (as WER can be greater than 100).

8.6 DNN/HMM Hybrid Model

GMMs were a popular choice because they are capable of modeling $P(\mathbf{x}_t|s_t)$ directly. Additionally, they provide a probabilistic interpretation of the input, modeling the distribution under each state. However, the Gaussian distribution at each state is a strong assumption itself. In practice, the features may be strongly non-Gaussian. DNNs showed significant improvements over GMMs with their ability

```
1  import Levenshtein as Lev
2
3  def wer(s1, s2):
4      """
5      Computes the Word Error Rate, defined as the edit distance
       between the
6      two provided sentences after tokenizing to words.
7      Arguments:
8      s1 (string): space-separated sentence
9      s2 (string): space-separated sentence
10     """
11
12     # build mapping of words to integers
13     b = set(s1.split() + s2.split())
14     word2char = dict(zip(b, range(len(b))))
15
16     # map the words to a char array (Levenshtein packages only
       accepts
17     # strings)
18     w1 = [chr(word2char[w]) for w in s1.split()]
19     w2 = [chr(word2char[w]) for w in s2.split()]
20     wer_lev = Lev.distance(''.join(w1), ''.join(w2))
21     wer_inst = float(wer_lev)/len(s1.split()) * 100
22     return 'WER: {0:.2f}'.format(wer_inst)
23
24 def cer(s1, s2):
25     """
26     Computes the Character Error Rate, defined as the edit
       distance.
27     Arguments:
28     s1 (string): space-separated sentence
29     s2 (string): space-separated sentence
30     """
31     s1, s2, = s1.replace(' ', ''), s2.replace(' ', '')
32     cer_inst = float(Lev.distance(s1, s2)) / len(s1) * 100
33     return 'CER: {0:.2f}'.format(cer_inst)
34
```

Fig. 8.14: Python functions to compute WER and CER

to learn non-linear functions. The DNN cannot provide the conditional likelihood directly. The framewise posterior distribution is used to turn the probabilistic model of $P(\mathbf{x}_t|s_t)$ into a classification problem $P(s_t|\mathbf{x}_t)$ using the pseudo-likelihood trick as an approximation of the joint probability. The application of the pseudo-likelihood is referred to as the "hybrid-approach."

$$\prod_{t=1}^{T} P(\mathbf{x}_t|s_t) \propto \prod_{t=1}^{T} \frac{P(s_t|\mathbf{x}_t)}{p(s_t)} \tag{8.27}$$

```
1  prediction = 'the cat sat on the mat'
2  target = 'the cat sat on the mat'
3  print('Prediction: ' + prediction, '\nTarget: ' + target)
4  print(wer(prediction, target))
5  print(cer(prediction, target))
6
7  > Prediction: the cat sat on the mat
8  > Target: the cat sat on the mat
9  > WER: 0.00
10 > CER: 0.00
```

Fig. 8.15: An exact match between the prediction and the target yields a WER and CER of 0

```
1  prediction = 'the cat sat on the mat'
2  target = 'the cat sat on the hat'
3  print('Prediction: ' + prediction, '\nTarget: ' + target)
4  print(wer(prediction, target))
5  print(cer(prediction, target))
6
7  > Prediction: the cat sat on the mat
8  > Target: the cat sat on the hat
9  > WER: 16.67
10 > CER: 5.88
```

Fig. 8.16: Changing one character of the predicted word yields a larger increase in WER because the entire word is wrong, albeit phonetically similar. The change in CER is much smaller by comparison because there are more characters than words; thus a single character change has less effect

```
1  prediction = 'cat mat'
2  target = 'the cat sat on the mat'
3  print('Prediction: ' + prediction, '\nTarget: ' + target)
4  print(wer(prediction, target))
5  print(cer(prediction, target))
6
7  > Prediction: cat mat
8  > Target: the cat sat on the hat
9  > WER: 200.00
10 > CER: 183.33
```

Fig. 8.17: WER and CER are typically not treated as percentages, because they can exceed 100%. The loss or insertion of large sections can greatly increase the WER and CER

The numerator is a DNN classifier, trained with a set of input features as the input $\mathbf{x}_t$ and target state s_t. In a simple case, if we consider 1-state per phone, then the number of classifier categories will be $len(q)$. The denominator $P(s_t)$ is the prior probability of the state s_t. Note that training the framewise model requires framewise alignments with $\mathbf{x}_t$ as the input and s_t as the target, as shown in Fig. 8.18. This alignment is usually created by leveraging a weaker HMM/GMM alignment system or through human-created labels. The quality and quantity of the alignment labels are typically the most significant limitations with the hybrid-approach.

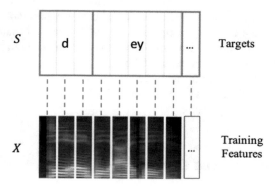

Fig. 8.18: In order to apply the DNN classifier, a framewise target must exist. A constrained alignment is computed using an existing classifier to align the acoustic features and the known sequence of states

Classifier construction requires the selection of target states (words, phones, triphone states). Selection of the states, Q, can make a significant difference in the quality and complexity of the task. First, it must support the recognition task to obtain the alignments. Second, it must be practical for the classification task. For example, phones may be easier to train a classifier for, however, getting framewise labels for the training data and a decoding scheme may be much more difficult to obtain. Alternatively, word-based states are straight-forward to create but are harder to get framewise alignments and train a classifier.

8.7 Case Study

In this case study, the main focus is on training ASR systems using open source frameworks. We begin by training a traditional ASR engine and then move towards the more advanced models in the frameworks, ending with a TDNN model.

8.7.1 Dataset: Common Voice

In this case study we focus on building ASR models for the Common Voice[1] dataset released by Mozilla. Common Voice is a 500 h speech corpus of recorded speech from text. It is composed of crowdsourced speakers recoding one utterance per example. These recordings are then peer-reviewed to assess the quality of the transcript-recording pair. Depending on the number of positive and negative votes that each utterance receives, it is labeled as either valid, invalid, or other. The valid category contains samples that have had at minimum two reviews, and the majority confirms the audio matches the text. The invalid category similarly has had at least two reviews, with the majority confirming the audio does not match the text. The other category contains all the files with less than two votes or with no majority consensus. Each of the sub-groups, valid and other, is further split into train, test, and dev (validation). The "cv-valid-train" dataset contains a total of 239.81 h of audio in total. Overall, the dataset is complex, containing a variety of accents, recording environments, ages, and genders.

8.7.2 Software Tools and Libraries

Kaldi is one of the most widely used toolkits for ASR, developed mainly for researchers and professional use. It is developed primarily by Johns Hopkins University and built entirely in C++ with shell scripts tying various components of the library together. The design focuses on providing a flexible toolkit that can be modified and extended to the task. Collections of scripts referred to as "recipes" are used to connect the components to perform training and inference.

CMU Sphinx is an ASR toolkit that developed at Carnegie Mellon University. It also relies on HMM-based speech recognition and n-gram language models for ASR. There have been various releases of the Sphinx toolkit with Sphinx 4 being the most current. A version of Sphinx called PocketSphinx is more suitable for embedded systems (typically it comes with some losses in quality due to the restrictions of the hardware).

8.7.3 Sphinx

In this section, we train a Sphinx ASR model on the Common Voice dataset. This framework relies on a variety of packages, mainly based on C++. In light of this, much of the work in this ASR introduction chapter focuses on the scripts and associated concepts, namely the data preparation. Like many frameworks, once the data is appropriately formatted, the framework is relatively straight-forward.

---

[1] https://voice.mozilla.org/en/data.

8.7.3.1 Data Preparation

The data preparation required for Sphinx is the most important step. Sphinx is configured to look in specific places for certain things and expects consistency between files and file names. The conventional structure is a top level directory with the same name as the dataset. This name is used as the file name for the subsequent files. Inside this directory, there are two directories "wav" and "etc." The "wav" directory contains all training and testing audio files in wav form. The "etc" directory contains all configuration and transcript files. Figure 8.19 shows the file structure.

```
1   /common_voice
2           etc/
3                   common_voice.dic
4                   common_voice.filler
5                   common_voice.idngram
6                   common_voice.lm
7                   common_voice.lm.bin
8                   common_voice.phone
9                   common_voice.vocab
10                  common_voice_test.fileids
11                  common_voice_test.transcription
12                  common_voice_train.fileids
13                  common_voice_train.transcription
14          wav/
15                  train_sample000000.wav
16                  test_sample000000.wav
17                  train_sample000001.wav
18                  test_sample000001.wav
19                  ...
20
```

Fig. 8.19: Files that are created for Sphinx

Common Voice initially comes packaged with "mp3" files. These will be converted to "wav" by using the "SoX" tool.[2] The processing script is shown in Fig. 8.20.

After we create the "wav" files, we create a list of all files that should be used for training, and separately, validation (referred to as testing for the Sphinx framework). The file list contains the ".fileids" files. The ".fileids" contain a single file name per line without the file extension. This is illustrated in Fig. 8.21.

Next, we create the transcript files. The transcript files have one utterance transcript per line with the "fileid" specified at the end. A sample of a transcript file is shown in Fig. 8.22.

---

[2] http://sox.sourceforge.net/.

```
1  def convert_to_wav(x):
2      file_path, wav_path = x
3      file_name = os.path.splitext(os.path.basename(file_path))[0]
4      cmd = "sox {} -r {} -b 16 -c 1 {}".format(
5          file_path,
6          args.sample_rate,
7          wav_path)
8      subprocess.call([cmd], shell=True)
9
10 with ThreadPool(10) as pool:
11     pool.map(convert_to_wav, train_wav_files)
12
```

Fig. 8.20: Converts mp3 files to wav files, using the sox library. The function is parallelized to increase the speed of the conversion

```
1  train_sample000000
2  train_sample000001
3  train_sample000002
4  ...
5
```

Fig. 8.21: Sample from the "common_voice_train.fileids" file

```
1  <s> learn to recognize omens and follow them the old king had
       said </s> (train_sample -000000)
2  <s> everything in the universe evolved he said </s> (train_sample
       -000001)
3  <s> you came so that you could learn about your dreams said the
       old woman </s> (train_sample -000002)
4  ...
5
```

Fig. 8.22: Sample from the "common_voice_train.transcript" file

Once the transcript files are created, we turn our attention to the phonetic units used. In this example, the same phones are used as illustrated in Fig. 8.6, with one additional phone *<SIL>* to symbolize the silent token.

The next step is to create the phonetic dictionary. We create a list of all words in the training dataset transcripts. The script in Fig. 8.23 shows a simple way of doing this.

Next, we create a phonetic dictionary (lexicon) using the word and phone lists. Creating the lexicon model typically requires linguistic expertise or existing models to create mappings for these words, as the phonetic representation should match

```
1  import collections
2  import os
3
4  counter = collections.Counter()
5  with open(csv_file) as csvfile:
6      reader = csv.DictReader(csvfile)
7      for row in reader:
8          trans = row['text']
9          counter += collections.Counter(trans.split())
10
11 with open(os.path.join(etc_dir,'common_voice.words'), 'w') as f:
12     for item in counter:
13         f.write(item.lower() + '\n')
14
```

Fig. 8.23: This script creates a file, "common_voice.words" that contains one word per line from the training data. Note: each word is only represented once in this file

the pronunciation. To ease this dependency, CMU Lextool[3] is used to create our phonetic dictionary, and saved as "common_voice.dic". Note: there is some extra processing required here to ensure that there are no additional phones added to the representation than those specified in our ".phone" file. Additionally, in this example, all phones and transcripts are represented as lower case. The phonetic dictionary needs to match as well. A sample from the phonetic dictionary is shown in Fig. 8.24

```
1  a        ah
2  a(2)     ey
3  monk     m ah ng k
4  dressed  d r eh s t
5  in       ih n
6  black    b l ae k
7  came     k ey m
8  to       t uw
9  ...
10
```

Fig. 8.24: Sample from the "common_voice_train.dic" file

Our final step is to create a language model. Most language models follow the ARPA format, representing the n-grams and its associated probabilities once per line, with section delimiters for increases in the number of grams. We create a 3-gram language model using the CMUCLMTK, a language modeling toolkit from CMU. The script counts the different n-grams and computes the probabilities of

---

[3] http://www.speech.cs.cmu.edu/tools/lextool.html.

each. The script is shown in Fig. 8.25 and a sample of the language model is shown in Fig. 8.26.

```
1  # Create vocab file
2  text2wfreq < etc/common_voice_train.transcription | wfreq2vocab >
       etc/common_voice.vocab
3
4  # Create n-gram count from training transcript file
5  text2idngram -vocab etc/common_voice.vocab -idngram etc/
       common_voice.idngram < etc/common_voice_train.transcription
6
7  # Create language model from n-grams
8  idngram2lm -vocab_type 0 -idngram etc/common_voice.idngram -vocab
       etc/common_voice.vocab -arpa etc/common_voice.lm
9
10 # Convert language model to binary (compression)
11 sphinx_lm_convert -i etc/common_voice.lm -o etc/common_voice.lm.
       DMP
12
```

Fig. 8.25: CMUCLMTK creating the language modeling file

With the preprocessing complete, we are ready to train the ASR models.

8.7.3.2 Model Training

The model training process for Sphinx is straight-forward, with everything set up to follow the configuration file. To generate the configuration file, we run:

```
sphinxtrain -t common_voice setup
```

With the setup config, the Sphinx model can be trained by running:

```
sphinxtrain run
```

The training function runs a series of scripts that check the configuration and setup. It then performs a series of transforms on the data to produce the features, and then trains a series of models. The Sphinx framework achieves a WER of 39.824 and CER of 24.828 on the Common Voice validation set.

8.7.4 Kaldi

In this section, we train a series of Kaldi models to train a high-quality ASR model on the Common Voice dataset. Training a high-quality model requires intermediate models to align the acoustic feature frames to the phonetic states. The code and explanations are adapted from the Kaldi tutorial.

```
 1  \data\
 2  ngram  1=8005
 3  ngram  2=31528
 4  ngram  3=49969
 5
 6  \1−grams:
 7  −6.8775  </s>        0.0000
 8  −0.9757  <s>         −4.8721
 9  −1.6598  a           −4.5631
10  −5.0370  aaron       −1.2761
11  −4.5116  abandon  −1.7707
12  −3.9910  abandoned          −2.2851
13  ...
14
15  \2−grams:
16
17  −1.9149  <s> a  −3.2624
18  −4.1178  <s> abigail  0.0280
19  −2.8197  <s> about  −2.4474
20  −4.0634  <s> abraham  0.0483
21  −2.9228  <s> absolutely  −1.8134
22  ...
23
24  \3−grams:
25  −0.9673  <s> a boy
26  −1.6977  <s> a breeze
27  −2.6800  <s> a bunch
28  −1.5866  <s> a card
29  −2.2998  <s> a citation
30
```

Fig. 8.26: Sample from the "common_voice.lm" file in the standard ARPA format

8.7.4.1 Data Preparation

The data preparation in Kaldi is similar to the Sphinx preparation, requiring transcription and audio ID files, shown in Fig. 8.27. Kaldi has a set of scripts in place to automate the construction of these files, to reduce the manual work required in the Sphinx setup.

Prepare a mapping from the ".wav" files to the audio path. We create an utterance ID for each of the files. This utterance ID is used to tie the file to the different representations in the training pipeline. In Kaldi, this is treated as a simple text file, with the ".scp" extension.

Secondly, a file mapping the utterance ID to the utterance transcript is created (Fig. 8.28).[4] This file will be used to create the utterance.

A corpus file contains all the utterances from the dataset. It is used to compute the word-level decoding graphs for the system.

---

[4] If there are additional labels like speaker and gender, these can also be used in the process. Common Voice does not have these labels, so each utterance is treated independently.

```
1    # spk2gender  [<speaker-id> <gender>]
2    # wav.scp     [<uterranceID> <full_path_to_audio_file>]
3    # text        [<uterranceID> <text_transcription>]
4    # utt2spk     [<uterranceID> <speakerID>]
5    # corpus.txt  [<text_transcription>]
6
```

Fig. 8.27: Files that need to be created for Kaldi

```
1  dad_4_4_2  four  four  two
2  july_1_2_5  one  two  five
3  july_6_8_3  six  eight  three
4  # and so on...
```

Fig. 8.28: Sample from the transcript file

These files allow the rest of the required files to be generated, such as the "lexicon.txt" file, which contains all the words from the dictionary with the phone transcript. Additionally, there are non-silence and silence phone files that provide ways to handle non-speech audio (Fig. 8.29).

```
1    !SIL  sil
2    <UNK>  spn
3    eight  ey  t
4    five  f  ay  v
5    four  f  ao  r
6    nine  n  ay  n
7    one  hh  w  ah  n
8    one  w  ah  n
9    seven  s  eh  v  ah  n
10   six  s  ih  k  s
11   three  th  r  iy
12   two  t  uw
13   zero  z  ih  r  ow
14   zero  z  iy  r  ow
15
```

Fig. 8.29: Sample from the "lexicon.txt" file

Once these files are prepared, Kaldi scripts can be used to create the ARPA language model and vocabulary. With Sphinx, all the terms in the dictionary needed a lexical entry to be entered manually (we leveraged premade dictionaries and additional inference dictionaries to accomplish this). The CMU dictionary is also used in this case, except in Kaldi, a pre-trained model is used to estimate the pronunciations of OOV words.[5] Once the dictionary is prepared, the lexical model is built, with the phonetic representation of each word in the dataset. An FST is then constructed from the transcripts and lexicon model and used to train the model.

The next step in the data preprocessing is to produce MFCCs for all of the training data. A file for each utterance is saved individually to reduce feature generation for various experiments. During this process, we also create two smaller datasets: a dataset of the 10 k shortest utterances and a dataset of the 20 k shortest utterances. These are used to build the earlier models. Once features are extracted, we can train our models.

8.7.4.2 Model Training

Much of the model training is scripted in Kaldi and is straight-forward after the data preparation is completed. The first model that is trained is an HMM-GMM model. It is trained for 40 epochs through the 10 k shortest utterances. This model is then used to align the 20 k utterances. After each model is trained, we rebuild the decoding graph and apply it to the test set. This model yields a WER of 52.06 on the Validation (Fig. 8.30).

```
1  steps/train_mono.sh —boost−silence 1.25 —nj 20 —cmd "run.pl —
       mem 8G" \
2        data/train_10kshort data/lang exp/mono || exit 1;
3  (
4        utils/mkgraph.sh data/lang_test exp/mono exp/mono/graph
5        for testset in valid_dev; do
6            steps/decode.sh —nj 20 —cmd "run.pl —mem 8G" exp/mono/
       graph \
7                data/$testset exp/mono/decode_$testset
8        done
9  )&
10
```

Fig. 8.30: Train the monophone "mono" model with alignments on the 10 k shortest utterances subset from the training data

Next, we use the 20 k alignments to train a new model incorporating the delta and double delta features. This model will also leverage triphones in the training process. The previous process will be performed again, with a separate training script, producing a model that achieves a WER of 25.06 (Fig. 8.31).

The third model that is trained is an LDA+MLLT model. This model will be used to compute better alignments using the learnings of the previous model on the 20 k dataset. So far, we have been using the 13-dimensional MFCC features. In this

---

[5] Note: It is possible to add specific words to the lexicon by exiting the *lexicon-iv.txt* file.

```
1  steps / align_si . sh ——boost—silence  1.25  ——nj  10  ——cmd "run . pl ——
     mem  8G" \
2      data / train_20k  data / lang  exp / mono  exp / mono_ali_train_20k
3
4  steps / train_deltas . sh ——boost—silence  1.25  ——cmd  "$train_cmd" \
5      2000  10000  data / train_20k  data / lang  exp / mono_ali_train_20k
     exp / tri1
6
7  # decode  using  the  tri1  model
8  (
9      utils / mkgraph . sh  data / lang_test  exp / tri1  exp / tri1 / graph
10     for  testset  in  valid_dev ;  do
11         steps / decode . sh  ——nj  20  ——cmd  "$decode_cmd"  exp / tri1 /
     graph \
12             data / $testset  exp / tri1 / decode_$testset
13     done
14 )&
15
```

Fig. 8.31: Train another monophone "tri1" model with alignments from the "mono" model on the 20 k training subset

model, multiple frames are considered in a single input t to provide more context at each state. The added input dimensionality increases the computational requirements of the classifier, so we use linear discriminant analysis (LDA) to reduce the dimensionality of the features. Additionally, a maximum likelihood linear transform (MLLT) to further decorrelate the features and make them "orthogonal" to be better modeled by diagonal-covariance Gaussians [Rat+13]. The resulting model yields a WER of 21.69 (Fig. 8.32).

The next model that is trained is the speaker adapted model referred to as "LDA+MLLT+SAT". The 20 k dataset is aligned again using the previous model, and using the same architecture as the previous model with the additional speaker adapted features. Because our data doesn't include speaker tags, we would not expect to get gains in this area, and we do not see any. The resulting model yields a WER of 22.25 (Fig. 8.33).

We now apply the alignments computed with the previous model to the entire training dataset. We train another LDA+MLLT+SAT model on the new alignments. The resulting model gives a WER of 17.85 (Fig. 8.34).

The final model is a TDNN model [PPK15]. The TDNN model is an 8 layer DNN, with batch normalization. This model requires a GPU to train due to the depth and need for parallelization (Fig. 8.35).

The final model, after integrating the 8-layer DNN, achieves a WER of 4.82 on the validation set.

```
1  steps/align_si.sh —nj 10 —cmd "$train_cmd" \
2      data/train_20k data/lang exp/tri1 exp/tri1_ali_train_20k
3
4  steps/train_lda_mllt.sh —cmd "$train_cmd" \
5  —splice—opts "——left—context=3 ——right—context=3" 2500 15000 \
6      data/train_20k data/lang exp/tri1_ali_train_20k exp/tri2b
7
8  # decode using the LDA+MLLT model
9  utils/mkgraph.sh data/lang_test exp/tri2b exp/tri2b/graph
10 (
11      for testset in valid_dev; do
12          steps/decode.sh —nj 20 —cmd "$decode_cmd" exp/tri2b/
    graph \
13              data/$testset exp/tri2b/decode_$testset
14      done
15 )&
16
```

Fig. 8.32: Train the LDA+MLLT "tri2b" model with alignments from the "tri1" model

```
1  # Align utts using the tri2b model
2  steps/align_si.sh —nj 10 —cmd "$train_cmd" —use—graphs true \
3      data/train_20k data/lang exp/tri2b exp/tri2b_ali_train_20k
4
5  steps/train_sat.sh —cmd "$train_cmd" 2500 15000 \
6      data/train_20k data/lang exp/tri2b_ali_train_20k exp/tri3b
7
8  # decode using the tri3b model
9  (
10      utils/mkgraph.sh data/lang_test exp/tri3b exp/tri3b/graph
11      for testset in valid_dev; do
12          steps/decode_fmllr.sh —nj 10 —cmd "$decode_cmd" \
13              exp/tri3b/graph data/$testset exp/tri3b/
    decode_$testset
14      done
15 )&
16
```

Fig. 8.33: Train the LDA+MLLT+SAT "tri3b" model with alignments from the "tri2b" model

8.7.5 Results

The results achieved throughout this case study are summarized in Table 8.1. The best Kaldi and Sphinx model are then evaluated on the test set. These results are shown in Table 8.2. We see that the addition of deep learning in the final Kaldi TDNN model shows significant quality improvements over the traditional learning algorithms for the acoustic model.

```
1  # Align utts in the full training set using the tri3b model
2  steps/align_fmllr.sh —nj 20 —cmd "$train_cmd" \
3      data/valid_train data/lang \
4      exp/tri3b exp/tri3b_ali_valid_train
5
6  # train another LDA+MLLT+SAT system on the entire training set
7  steps/train_sat.sh   —cmd "$train_cmd" 4200 40000 \
8      data/valid_train data/lang \
9      exp/tri3b_ali_valid_train exp/tri4b
10
11 # decode using the tri4b model
12 (
13     utils/mkgraph.sh data/lang_test exp/tri4b exp/tri4b/graph
14     for testset in valid_dev; do
15         steps/decode_fmllr.sh —nj 20 —cmd "$decode_cmd" \
16             exp/tri4b/graph data/$testset \
17             exp/tri4b/decode_$testset
18     done
19 )&
20
```

Fig. 8.34: Train the LDA+MLLT+SAT "tri4b" model with alignments from the "tri3b" model

```
1  local/chain/run_tdnn.sh —stage 0
2
```

Fig. 8.35: Script: Train the TDNN model using the "tri4b" model

8.7.6 Exercises for Readers and Practitioners

Some other interesting problems readers and practitioners can try on their own include:

1. How are additional words added to the vocabulary?
2. Evaluate the real-time factor (RTF) for this system.

Table 8.1: Speech recognition performance on Common Voice validation set. Best result is shaded

Approach	WER
Sphinx	39.82
Kaldi monophone (10 k sample)	52.06
Kaldi triphone (with delta and double delta, 20 k sample)	25.06
Kaldi LDA+MLLT (20 k sample)	21.69
Kaldi LDA+MLLT+SAT (20 k sample)	22.25
Kaldi LDA+MLLT+SAT (all data)	17.85
Kaldi TDNN (all data)	4.82

Table 8.2: Speech recognition performance on Common Voice test set. Best result is shaded

Approach	WER
Sphinx	53.85
Kaldi TDNN	4.44

3. What are some ways to improve quality on accented speech?
4. How many states are in the set for a diphone model? How many for a triphone model?

References

[BM12]   Herve A Bourlard and Nelson Morgan. *Connectionist speech recognition: a hybrid approach*. Vol. 247. Springer Science & Business Media, 2012.

[BW13]   Michael Brandstein and Darren Ward. *Microphone arrays: signal processing techniques and applications*. Springer Science & Business Media, 2013.

[DM90]   Steven B Davis and Paul Mermelstein. "Comparison of parametric representations for monosyllabic word recognition in continuously spoken sentences". In: *Readings in speech recognition*. Elsevier, 1990, pp. 65–74.

[Her90]   Hynek Hermansky. "Perceptual linear predictive (PLP) analysis of speech". In: *the Journal of the Acoustical Society of America* 87.4 (1990), pp. 1738–1752.

[HWW15]   Yedid Hoshen, Ron J Weiss, and Kevin W Wilson. "Speech acoustic modeling from raw multichannel waveforms". In: *Acoustics, Speech and Signal Processing (ICASSP), 2015 IEEE International Conference on*. IEEE. 2015, pp. 4624–4628.

[JH11]   Navdeep Jaitly and Geoffrey Hinton. "Learning a better representation of speech soundwaves using restricted Boltzmann machines". In: *Acoustics, Speech and Signal Processing (ICASSP), 2011 IEEE International Conference on*. IEEE. 2011, pp. 5884–5887.

[MPR08]   Mehryar Mohri, Fernando Pereira, and Michael Riley. "Speech recognition with weighted finite-state transducers". In: *Springer Handbook of Speech Processing*. Springer, 2008, pp. 559–584.

[MMG04]   Andrew Cameron Morris, Viktoria Maier, and Phil Green. "From WER and RIL to MER and WIL: improved evaluation measures for connected speech recognition". In: *Eighth International Conference on Spoken Language Processing*. 2004.

[MBE10]   Lindasalwa Muda, Mumtaj Begam, and Irraivan Elamvazuthi. "Voice recognition algorithms using Mel frequency cepstral coefficient

(MFCC) and dynamic time warping (DTW) techniques". In: *arXiv preprint arXiv:1003.4083* (2010).

[PCD13]   Dimitri Palaz, Ronan Collobert, and Mathew Magimai Doss. "Estimating phoneme class conditional probabilities from raw speech signal using convolutional neural networks". In: *arXiv preprint arXiv:1304.1018* (2013).

[PVZ13]   Venkata Neelima Parinam, Chandra Sekhar Vootkuri, and Stephen A Zahorian. "Comparison of spectral analysis methods for automatic speech recognition." In: *INTERSPEECH*. 2013, pp. 3356–3360.

[PPK15]   Vijayaditya Peddinti, Daniel Povey, and Sanjeev Khudanpur. "A time delay neural network architecture for efficient modeling of long temporal contexts". In: *Sixteenth Annual Conference of the International Speech Communication Association*. 2015.

[Rab89b]   Lawrence R Rabiner. "A tutorial on hidden Markov models and selected applications in speech recognition". In: *Proceedings of the IEEE* 77.2 (1989), pp. 257–286.

[Rat+13]   Shakti P Rath et al. "Improved feature processing for deep neural networks." In: *Interspeech*. 2013, pp. 109–113.

[Sch+07]   Ralf Schluter et al. "Gammatone features and feature combination for large vocabulary speech recognition". In: *Acoustics, Speech and Signal Processing, 2007. ICASSP 2007. IEEE International Conference on*. Vol. 4. IEEE. 2007, pp. IV–649.

[Tüs+14]   Zoltán Tüske et al. "Acoustic modeling with deep neural networks using raw time signal for LVCSR". In: *Fifteenth Annual Conference of the International Speech Communication Association*. 2014.

[You96]   Steve Young. "A review of large-vocabulary continuous-speech". In: *IEEE signal processing magazine* 13.5 (1996), p. 45.

Part III
Advanced Deep Learning Techniques for Text and Speech

Chapter 9
Attention and Memory Augmented Networks

9.1 Introduction

In deep learning networks, as we have seen in the previous chapters, there are good architectures for handling spatial and temporal data using various forms of convolutional and recurrent networks, respectively. When the data has certain dependencies such as out-of-order access, long-term dependencies, unordered access, most standard architectures discussed are not suitable. Let us consider a specific example from the bAbI dataset where there are stories/facts presented, a question is asked, and the answer needs to be inferred from the stories. As shown in Fig. 9.1, it requires out of order access and long-term dependencies to find the right answer.

Deep learning architectures have made significant progress in the last decade in capturing implicit knowledge as features in various NLP tasks. Many tasks such as question answering or summarization require storage of explicit knowledge so that it can be used in the tasks. For example, in the bAbI dataset, information about Mary, Sandra, John, football, and its location is captured for answering the question such as "where is the football?" Recurrent networks such as LSTM and GRU cannot capture such information over very long sequences. Attention mechanisms, memory augmented networks, and some combinations of the two are currently the top techniques which address many of the issues discussed above. In this chapter, we will discuss in detail many popular techniques of attention mechanisms and memory networks that have been successfully employed in speech and text.

Though the attention mechanism has become very popular in NLP and speech in recent times after Bahdanau et al. proposed their research, it had been previously introduced in some forms in neural architectures. Larochelle and Hinton highlight the usefulness of "fixation points" to improve performance in the image recognition tasks [LH10]. Denil et al. also proposed a similar attention model for object tracking

© Springer Nature Switzerland AG 2019
U. Kamath et al., *Deep Learning for NLP and Speech Recognition*,
https://doi.org/10.1007/978-3-030-14596-5_9

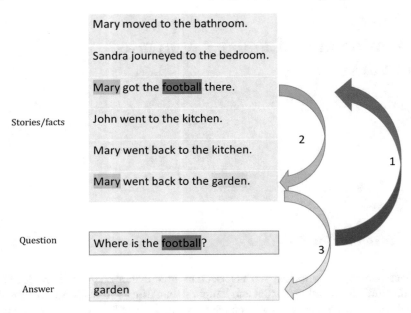

Fig. 9.1: Question and answer task

and recognition inspired by neuroscience models [Den+12]. Weston et al. pioneered the modern-day memory augmented networks but the origins trace back to the early 1960s by Steinbuch and Piske [SP63]. Das et al. used neural network pushdown automaton (NNPDA) with external stack memory addressing the issues of recurrent networks in learning context-free grammars [DGS92]. Mozer in his work addressing the complex time series had two separate parts in the architectures: (a) short-term memory to capture past events and (b) associator to use the short-term memory for classification or prediction [Moz94].

9.2 Attention Mechanism

In a much more general way, **attention** is a concept very well known in human psychology, where humans who are limited by processing bottlenecks have a selective focus on certain part of the information and ignore the rest of the visible information. Mapping the same concept of human psychology to sequence data such as text streams or audio streams, when we focus on certain parts of sequences or regions and blur the remaining ones during the learning, the process is called an **attention mechanism**. Attention was introduced in Chap. 7 while introducing recurrent networks and sequence modeling. Since many techniques using attention are either

related to or are used in memory augmented networks, we will cover some of the
modern techniques that have broad application.

9.2.1 The Need for Attention Mechanism

Let us consider a translation from English to French for a sentence "I like coffee"
which in French is 'J'aime le café." We will use a machine translation use case
with the sequence-to-sequence models to highlight the need for the attention mechanism. Let us consider a simple RNN with encoder–decoder network as shown in
Fig. 9.2. We see in the above neural machine translation that the entire sentence

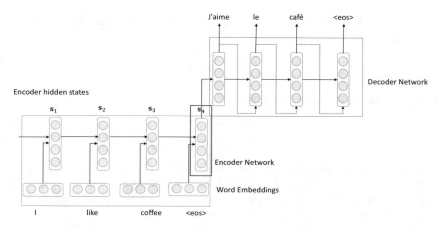

Fig. 9.2: Encoder–decoder using RNN for neural machine translation

is compressed into a single representation given by the hidden vector s_4, which is
a representation of the entire sentence, and is used by the decoder sequence as an
input for the translation. As the sequence length of the input increases, encoding
the entire information in that single vector becomes infeasible. The input sequence
in text normally has a complex phrase structure and long-distance relationships between words, which all seem to be cramped in the single vector at the end. Also, in
reality, all of the hidden values from the encoder network carry information that can
influence the decoder output at any timestep. By not using all of the hidden outputs
but just the single one, their influence may be diluted in the process. Finally, each
output from the decoder may be influenced differently by each of the inputs, and it
may not happen in the same order as in the input sequence.

9.2.2 Soft Attention

In this section, we will introduce the attention mechanism as a way to overcome the issues with recurrent networks. We will start with the attention mechanism by Luong et al. which is more general and then describe how it differs from the original Bahdanau et al. attention-based paper [LPM15, BCB14b].

The attention mechanism has the following in the encoder and the decoder: (Fig. 9.3)

- The source sequence which is of length n given by $\mathbf{x} = \{x_1, x_2, \ldots, x_n\}$.
- The target sequence which is of length m given by $\mathbf{y} = \{y_1, y_2, \ldots, y_m\}$.
- Encoder hidden states $\mathbf{s}_1, \mathbf{s}_2, \ldots, \mathbf{s}_n$.
- The decoder sequence has the hidden state given by $\mathbf{h}_i$ for the output at $i = 1, 2, \ldots, m$.
- The source-side context vector $\mathbf{c}_i$ at position i is the weighted average of previous states and alignment vector $\mathbf{a}_i$:

$$\mathbf{c}_i = \sum_j a_{i,j} \mathbf{s}_j \qquad (9.1)$$

- The alignment scores are given by:

$$\mathbf{a}_i = align(\mathbf{h}_i, \mathbf{s}_j) \qquad (9.2)$$
$$= \text{softmax}(score(\mathbf{h}_i, \mathbf{s}_j)) \qquad (9.3)$$

  a_{ij} are called the **alignment weights**. The equation above captures how every input element can influence the output element at a given position. The predefined function *score* is called the attention score function and there are many variants of these that will be defined in the next section.
- The source side context vector $\mathbf{c}_i$ and the hidden state $\mathbf{h}_i$ are combined using concatenation $[\mathbf{c}_i; \mathbf{h}_i]$ and the non-linear tanh operation to give the attention hidden vector $\tilde{\mathbf{h}}_i$:

$$\tilde{\mathbf{h}}_i = \tanh(\mathbf{W}_c[\mathbf{c}_i; \mathbf{h}_i]) \qquad (9.4)$$

  where the weights $\mathbf{W}_c$ are learned in the training process.
- The attention hidden vector $\tilde{\mathbf{h}}_i$ is passed through a softmax function to generate the probability distribution given by:

$$P(y_i | y < i, x) = \text{softmax}(\mathbf{W}_s \tilde{\mathbf{h}}_i) \qquad (9.5)$$

- Bahdanau et al. use bidirectional LSTM layers in the encoders, and concatenate the hidden states.
- Bahdanau et al. use the previous state, i.e., $\mathbf{h}_{i-1}$, and the computational path is $\mathbf{h}_{i-1} \longrightarrow \mathbf{a}_i \longrightarrow \mathbf{c}_i \longrightarrow \tilde{\mathbf{h}}_i$ as compared to Luong et al. which has $\mathbf{h}_i \longrightarrow \mathbf{a}_i \longrightarrow \mathbf{c}_i \longrightarrow \tilde{\mathbf{h}}_i$.

- Bahdanau et al. use the linear combination of previous state and encoder states in the scoring function given by $\text{score}(\mathbf{s}_j, \mathbf{h}_i) = \mathbf{v}_a^{\mathsf{T}} \tanh(\mathbf{W}_a \mathbf{s}_j + \mathbf{U}_a \mathbf{h}_i)$.
- Luong et al. have the **input-feeding** mechanism where the attentional hidden vectors $\tilde{\mathbf{h}}_i$ are concatenated with the target input.

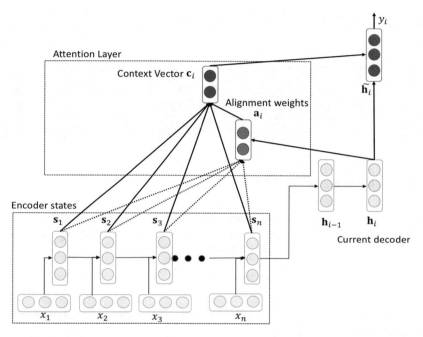

Fig. 9.3: Step-by-step computation process for soft attention in the encoder–decoder network

9.2.3 Scores-Based Attention

Table 9.1 gives different ways attention score functions can be computed to give different flavors of attention.

- The multiplicative and additive score functions generally give similar results, but multiplicative score functions are faster in both computation and space-efficiency using efficient matrix multiplication techniques.

Table 9.1: Attention score summary

Score name	Score description	Parameters	References
Concat (additive)	$\text{score}(\mathbf{s}_j, \mathbf{h}_i) = \mathbf{v}_a^\mathsf{T} \tanh(\mathbf{W}_a[\mathbf{s}_j; \mathbf{h}_i])$	$\mathbf{v}_a$ and $\mathbf{W}_a$ trainable	[LPM15]
Linear (additive)	$\text{score}(\mathbf{s}_j, \mathbf{h}_i) = \mathbf{v}_a^\mathsf{T} \tanh(\mathbf{W}_a \mathbf{s}_j + \mathbf{U}_a \mathbf{h}_i)$	$\mathbf{v}_a$, $\mathbf{U}_a$, and $\mathbf{W}_a$ trainable	[BCB14b]
Bilinear (multiplicative)	$\text{score}(\mathbf{s}_j, \mathbf{h}_i) = \mathbf{h}_i^\mathsf{T} \mathbf{W}_a \mathbf{s}_j$	$\mathbf{W}_a$ trainable	[LPM15]
Dot (multiplicative)	$\text{score}(\mathbf{s}_j, \mathbf{h}_i) = \mathbf{h}_i^\mathsf{T} \mathbf{s}_j$	No parameters	[LPM15]
Scaled dot (multiplicative)	$\text{score}(\mathbf{s}_j, \mathbf{h}_i) = \frac{\mathbf{h}_i^\mathsf{T} \mathbf{s}_j}{\sqrt{n}}$	No parameters	[Vas+17c]
Location-based	$\text{score}(\mathbf{s}_j, \mathbf{h}_i) = \text{softmax}(\mathbf{W}_a \mathbf{h}_i^\mathsf{T})$	$\mathbf{W}_a$ trainable	[LPM15]

- Additive attention performs much better when the input dimension is large. The scaled dot-product method defined above has been used to mitigate that issue in the general dot product.

9.2.4 Soft vs. Hard Attention

The only difference between **soft attention** and **hard attention** is that in hard attention it picks one of the encoder states rather than a weighted average over all the inputs as in soft attention. The hard attention is given by:

$$\mathbf{c}_i = \underset{a_{i,j}}{\arg\max}\{\mathbf{s}_1, \mathbf{s}_2, \dots, \mathbf{s}_n\} \tag{9.6}$$

Thus the difference between hard attention and soft attention is based on the search when the context is computed.

Hard attention uses the $\arg\max$ function which is not a continuous function, not differentiable and hence cannot be used in standard backpropagation methods. Techniques such as reinforcement learning to select the discrete part and Monte Carlo based sampling are often used. Another technique is to use the Gaussian trick given in the next section.

9.2.5 Local vs. Global Attention

The soft attention methods such as the Bahdanau's research are also referred to as **global attention**, as each decoder state takes "all" of the encoder inputs while computing the context vector. The process of iterating over all the inputs can be both computationally expensive and many times impractical when the sequence length is large.

Luong et al. introduced **local attention** which is a combination of the soft attention and the hard attention to overcome these issues [Luo+15]. One of the ways local attention can be achieved is to have a small window of the encoder hidden states used for computing context. This is called **predictive alignment** and it restores differentiability.

At any decoder state for time i, the network generates an aligned position p_i, and a window of size D around either side of the hidden state of the position, i.e., $[p_i - D, p_i + D]$ is used to compute the context vector $\mathbf{c}$. The position p_i is a scalar computed using a sigmoid function on the current decoder hidden state $\mathbf{h}_i$ and using the sentence length S as given by:

$$p_t = S \cdot \text{sigmoid}(\mathbf{v}_p^\mathsf{T} \tanh(\mathbf{W}_p \mathbf{h}_i)) \tag{9.7}$$

where $\mathbf{W}_p$ and $\mathbf{v}_p$ are the model parameters to be learned to predict the position and S is the length of the sequence and $p_i \in [0, S]$. The difficulty is in how to focus around the location p_i without using the non-differentiable $\arg \max$. One way to focus the alignment near p_i, a Gaussian distribution is centered around p_i with standard deviation $\sigma = \frac{D}{2}$ given by:

$$\mathbf{a}_i = align(\mathbf{s}_j, \mathbf{h}_i) \exp\left(-\frac{(s - p_t)^2}{2\sigma^2}\right) \tag{9.8}$$

The schematic is shown in Fig. 9.4.

9.2.6 Self-Attention

Lin et al. introduced the concept of self-attention or intra-attention where the premise is that by allowing a sentence to attend to itself many relevant aspects can be extracted [Lin+17]. Additive attention is used to compute the score for each hidden state $\mathbf{h}_i$:

$$score(\mathbf{h}_i) = \mathbf{v}_a^\mathsf{T} \tanh(\mathbf{W}_a \mathbf{h}_i) \tag{9.9}$$

Then, using all the hidden states $\mathbf{H} = \{\mathbf{h}_1, \ldots, \mathbf{h}_n\}$ attention vector $\mathbf{a}$:

$$\mathbf{a} = \text{softmax}(\mathbf{v}_a \tanh(\mathbf{W}_a \mathbf{H}^\mathsf{T})) \tag{9.10}$$

where $\mathbf{W}_a$ and $\mathbf{v}_a$ are weight matrices and vectors learned on the training data. The final sentence vector $\mathbf{c}$ is computed by:

$$\mathbf{c} = \mathbf{H}\mathbf{a}^\mathsf{T} \tag{9.11}$$

Instead of just using the single vector $\mathbf{v}_a$, several hops of attention are performed by using a matrix $\mathbf{V}$ which captures multiple relationships existing in the sentences and allows us to extract an attention matrix $\mathbf{A}$ as:

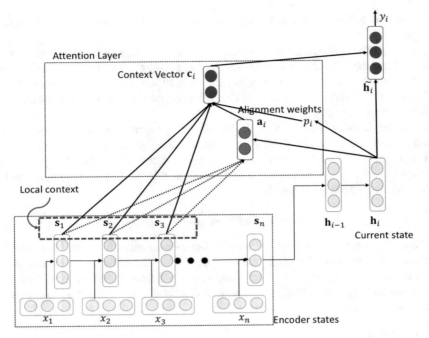

Fig. 9.4: Step-by-step computation process for local attention in the encoder-decoder network

$$\mathbf{A} = softmax(\mathbf{V}_a tanh(\mathbf{W}_a \mathbf{H}^\mathsf{T})) \qquad (9.12)$$

$$\mathbf{C} = \mathbf{A}\mathbf{H} \qquad (9.13)$$

To encourage diversity and penalize redundancy in the attention vectors, we use the following orthogonality constraint as a regularization technique:

$$\Omega = |(\mathbf{A}\mathbf{A}^\mathsf{T} - \mathbf{I})|_F^2 \qquad (9.14)$$

9.2.7 Key-Value Attention

Key-value attention by Daniluk et al. is another variant which splits the hidden layer into key-value where the keys are used for attention distribution and the values for context representation [Dan+17]. The hidden vector $\mathbf{h}_j$ is split into a key $\mathbf{k}_j$ and a value $\mathbf{v}_j$: $[\mathbf{k}_j; \mathbf{v}_j] = \mathbf{h}_j$. The attention vector $\mathbf{a}_i$ of length L is given by:

$$\mathbf{a}_i = \mathrm{softmax}(\mathbf{v}_a \tanh(\mathbf{W}_1[\mathbf{k}_{i-L}; \cdots ; \mathbf{k}_{i-1}] + \mathbf{W}_2 \mathbf{1}^\mathsf{T})) \qquad (9.15)$$

where $\mathbf{v}_a, \mathbf{W}_1, \mathbf{W}_2$ are the parameters. The context is then represented as:

$$\mathbf{c}_i = [\mathbf{v}_{i-L}; \cdots ; \mathbf{v}_{i-1}]\mathbf{a}^\mathsf{T} \qquad (9.16)$$

9.2.8 Multi-Head Self-Attention

Vaswani et al. in their work propose **transformer** network using multi-head self-attention without any recurrent networks to achieve state-of-the-art results in machine translation [Vas+17c]. We will describe the multi-head self-attention in a step-by-step manner in this section and as illustrated in Fig. 9.5. The source

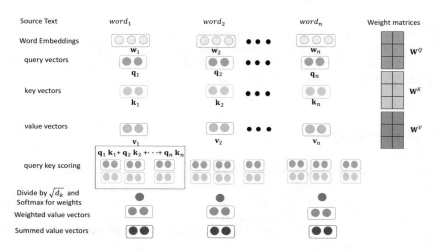

Fig. 9.5: Multi-head self-attention step-by-step computation

input with $word_1, word_2, \ldots, word_n$ words is first mapped to an embedding layer to get vectors for the words $\mathbf{x}_1, \mathbf{x}_2, \ldots, \mathbf{x}_n$. There are three matrices $\mathbf{W}^Q, \mathbf{W}^K$, and $\mathbf{W}^V$ called the query, key, and value weight matrices that are trained during the training process. The word embedding vectors are multiplied by the matrices $\mathbf{W}^Q, \mathbf{W}^K$, and $\mathbf{W}^V$ to get the query, key, and value vectors, respectively, for each word given by $\mathbf{q}, \mathbf{k}$, and $\mathbf{v}$. Next is to calculate a score for each word for every other word in the sentence by using the dot product of that query vector $\mathbf{q}$ and the key vectors $\mathbf{k}$ for every word. This scoring captures a single interaction of the word with every other word. For instance, for the first word:

$$score_1 = \mathbf{q}_1 \mathbf{k}_1 + \mathbf{q}_2 \mathbf{k}_2 + \cdots + \mathbf{q}_n \mathbf{k}_n$$

This is then divided by the length of the key vector $\sqrt{d_k}$ and a softmax is computed to get weights between 0 and 1. This score is then multiplied with all the value vectors to get the weighted value vectors. This gives attention or focus to specific words in the sentence rather than every word. Then the value vectors are summed to compute the output attention vector given for that word given by:

$$\mathbf{z}_1 = score_1\mathbf{v}_1 + score_1\mathbf{v}_2 + \cdots + score_1\mathbf{v}_n$$

Now, instead of this step-by-step computation, the whole thing can be computed by taking the sentence representation as embedding matrix of all word vectors $\mathbf{X}$ and multiplying it with the respective weight matrices $\mathbf{W}^Q, \mathbf{W}^K$, and $\mathbf{W}^V$ to get the matrices for all the words as $\mathbf{Q}, \mathbf{K}$, and $\mathbf{V}$ and then using the equation to compute the attention:

$$attention(\mathbf{Q},\mathbf{K},\mathbf{V}) = \mathbf{Z} = \text{softmax}\left(\frac{(\mathbf{Q}\mathbf{K}^\mathsf{T})}{\sqrt{d_k}}\right)\mathbf{V} \qquad (9.17)$$

Instead of using just one attention as computed above, they use **multi-head attention** where there are many such attention matrices computed for the input and can be represented by $\mathbf{Z}_0, \mathbf{Z}_1, \ldots, \mathbf{Z}_m$. These matrices are concatenated and multiplied by another weight matrix $\mathbf{W}^Z$ to get the final attention $\mathbf{Z}$.

9.2.9 Hierarchical Attention

Yang et al. used hierarchical attention for document classification tasks showing the advantage of having attention mechanisms at sentence level for the context and the word level for importance [Yan+16]. As shown in Fig. 9.6, the overall idea is to have word level encoding using bidirectional GRUs, word level attention, sentence level encoding, and sentence level attention hierarchically. We will briefly explain each of these components.

Let us consider input as a set of documents, each document has a maximum L sentences, and each sentence has a maximum T words such that w_{it} represents a tth word in the ith sentence in a document. The sentences with all the words go through an embedding matrix $\mathbf{W}_e$ that converts them to a vector $\mathbf{x}_{ij} = \mathbf{W}_e\mathbf{w}_{ij}$. It then goes through bidirectional GRU as:

$$\mathbf{x}_{it} = \mathbf{W}_e\mathbf{w}_{it}, t \in [1, T] \qquad (9.18)$$

$$\mathbf{h}_{it}^F = \text{GRU}^F(\mathbf{x}_{it}), t \in [1, T] \qquad (9.19)$$

$$\mathbf{h}_{it}^R = \text{GRU}^R(\mathbf{x}_{it}), t \in [T, 1] \qquad (9.20)$$

The hidden state for the word $\mathbf{w}_{it}$ is obtained by concatenating the two vectors from above $\mathbf{h}_{it} = [\mathbf{h}_{it}^F; \mathbf{h}_{it}^R]$, thus summarizing all the information around it.

The word annotation $\mathbf{h}_{it}$ gets fed to a one-layer MLP first to get the hidden representation $\mathbf{u}_{it}$, which is then used to measure importance with a word level context vector $\dot{\mathbf{u}}_w$, get a normalized importance through a softmax, and use that to compute the sentence vector $\mathbf{s}_i$ with weighted sum of annotations and weights. The context vector $\mathbf{u}_w$ is initialized randomly and then learned in the training process. The intuition behind the context vector $\mathbf{u}_w$ according to the authors is that it captures a fixed query like "what is the informative word" in the sentence.

$$\mathbf{u}_{it} = \tanh(\mathbf{W}_w \mathbf{h}_{it} + b_w) \tag{9.21}$$

where $\mathbf{W}_w, b_w$ are the parameters learned from the training process.

$$\alpha_{it} = \frac{\exp(\mathbf{u}_{it}^\mathsf{T} \mathbf{u}_w)}{\sum_t \exp(\mathbf{u}_{it}^\mathsf{T} \mathbf{u}_w)} \tag{9.22}$$

$$\mathbf{s}_i = \sum_t \alpha_{it} \mathbf{h}_{it} \tag{9.23}$$

Given L of the $\mathbf{s}_i$ sentence vector, the document hidden vectors are computed similar to word vectors using bidirectional GRUs.

$$\mathbf{h}_i^F = \mathrm{GRU}^F(\mathbf{s}_i), i \in [1, L] \tag{9.24}$$

$$\mathbf{h}_i^R = \mathrm{GRU}^R(\mathbf{s}_i), i \in [L, 1] \tag{9.25}$$

Similar to word annotations, concatenating both the vectors captures all the summarizations across sentences from both directions given by $\mathbf{h}_i = [\mathbf{h}_i^F; \mathbf{h}_i^R]$. The sentence context vector $\mathbf{u}_s$ is used in a similar way to the word context vector $\mathbf{u}_w$ to obtain attention among the sentences to get a document vector $\mathbf{v}$:

$$\mathbf{u}_i = \tanh(\mathbf{W}_s \mathbf{h}_i + b_s) \tag{9.26}$$

where $\mathbf{W}_s, b_s$ are the parameters learned from the training process.

$$\alpha_i = \frac{\exp(\mathbf{u}_i^\mathsf{T} \mathbf{u}_s)}{\sum_i \exp(\mathbf{u}_i^T \mathbf{u}_s)} \tag{9.27}$$

$$\mathbf{v} = \sum_i \alpha_i \mathbf{h}_i \tag{9.28}$$

The document vector $\mathbf{v}$ goes through the softmax for classification and negative log-likelihood of label to prediction is used for training.

In practice, if there is a document classification task, hierarchical attention becomes a good choice compared to other attention mechanisms or even other classification techniques. It helps to find both important keywords in the sentences and important sentences in the document through the learning process.

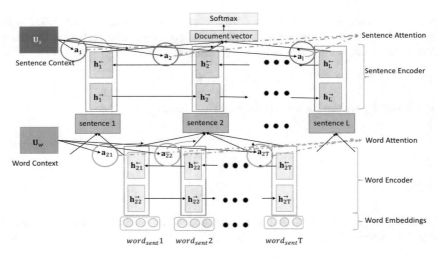

Fig. 9.6: Hierarchical attention used in document classification

9.2.10 Applications of Attention Mechanism in Text and Speech

Many NLP and NLU research have used attention mechanisms for tasks such as sentence embedding, language modeling, machine translation, syntactic constituency parsing, document classification, sentiment classification, summarization, and dialog systems to name a few. Lin et al.'s self-attention using LSTM for sentence embedding showed significant improvements over other embeddings on a variety of tasks such as sentiment classification and textual entailment [Lin+17]. Daniluk et al. applied attention mechanisms to language modeling and showed comparable results to memory augmented networks [Dan+17]. Neural machine translation implementations discussed in the chapter has achieved the state-of-the-art results [BCB14b, LPM15, Vas+17c]. Vinyals et al. showed that attention mechanisms for syntactic constituency parsing could not only attain state-of-the-art results but also improve on speed [Vin+15a]. Yang et al.'s research showed that using hierarchical attention can outperform many CNN and LSTM based networks by a large margin [Yan+16]. Wang et al. showed that attention-based LSTM could achieve state-of-the-art results in the aspect-level sentiment classification [Wan+16b]. Rush et al. showed how local attention methods could give significant improvements in the text summarization task [RCW15].

Chorowski et al. introduced how attention mechanisms can achieve better normalization for smoother alignments and using previous alignments for generating features in speech recognition [Cho+15b]. Bahdanau et al. used end-to-end attention-based networks for large vocabulary speech recognition problems [Bah+16b]. Listen, attend, and spell (LAS), an attention-based model, has been shown to outperform sequence-to-sequence approach [Cha+16a]. Zhang et al. in their research showed how using attention mechanisms with convolutional networks can achieve state-of-the-art results in the speech emotion recognition problem [Zha+18].

9.3 Memory Augmented Networks

Next, we will describe some well-known memory augmented networks that have been very effective in NLP and speech research.

9.3.1 Memory Networks

Memory networks (MemNN) by Weston et al. were motivated by the ability to store information coming from stories or knowledge base facts so that various questions pertaining to these can be easily answered [WCB14]. Memory networks have been extended in many ways for various other applications but performing question and answer on the stories or facts can be considered its basic application that we will focus on our narrative.

Memory networks consists of memory $\mathbf{m}$ indexed by $\mathbf{m}_i$ and has four components as shown in Fig. 9.7

1. Input Feature Map $\mathbf{I}$: This component converts the incoming data to the internal feature representation. This component can do any task-specific preprocessing such as converting the text to embeddings or POS representations, to name a few. Given an input x, the goal is to convert to an internal feature representation $I(x)$.

2. Generalization $\mathbf{G}$: This component uses the representation of the input from above and updates the memory by using any transformation if necessary. The transformation can be as simple as using the representation as is or making a coreference resolution, to complex reasoning based on the tasks. This transformation is given by:

$$\mathbf{m}_{H(x)} = I(x) \qquad (9.29)$$

In general the updating of the memories $\mathbf{m}_i$ for the new input is given by $H(\cdot)$, which is a general function that can be used to do various things from simplest such as finding a slot index in the memory to the complex part of finding the slot if full or forget certain memory slot. Once the slot index is detected, the G stores the input $I(x)$ in that location:

$$\mathbf{m}_i = G(\mathbf{m}_i, I(x), \mathbf{m}) \forall i \qquad (9.30)$$

3. Output **O**: This is the "read" part from the memory where necessary inference to deduce relevant parts of memories for generating the response happens. This can be represented as computing the output features given the new input and memory as $o = O(I(x), \mathbf{m})$.
4. Response **R**: This component converts the output from memory into a representation that the outside world can understand. The decoding of the output features to give the final response can be given as $r = R(o)$.

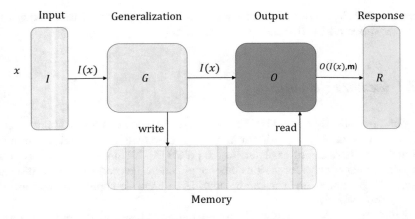

Fig. 9.7: Memory networks

In the paper the input component stores the sentence as is, for both the stories and questions. The memory write or generalization is also basic writing to the next slot, i.e., $\mathbf{m}_N = x$, $N = N + 1$. Most of the work is done in the output O, R part of the network.

The output module finds the closest match for the input using k memories that support the fact and a scoring function

$$o_k = \arg\max_{i=1,\dots,n} s_O(x, \mathbf{m}_i) \qquad (9.31)$$

The s_O is the scoring function that matches the input question or the input fact/story sentence to all the existing memory slots for the best match. In the simplest case, they choose $k = 2$ for the output inferencing. This can be represented as:

$$o_1 = O_1(x, \mathbf{m}) = \arg\max_{i=1,\dots,N} s_O(x, \mathbf{m}_i) \qquad (9.32)$$

$$o_2 = O_2(x, \mathbf{m}) = \arg\max_{i=1,\dots,N} s_O([x, \mathbf{m}_{O1}], \mathbf{m}_i) \qquad (9.33)$$

The input $[x, \mathbf{m}_{O1}, \mathbf{m}_{O2}]$ is given to the response component which generates a single word with highest ranking given by:

$$r = \arg\max_{w \in W} s_R([x, \mathbf{m}_{O1}, \mathbf{m}_{O2}], w) \tag{9.34}$$

where W is the set of all words in the vocabulary and s_R is the scoring function to match words to the inputs. In the paper the scoring functions s_O and s_R have same form and can be written as:

$$s(x, y) = \phi_x(x)^T U^T U \phi_y(y) \tag{9.35}$$

The matrix U is a $n \times D$ dimensional, where n is the embedding size and D is the number of features. The matrices ϕ_x, ϕ_y represent a mapping of the original text to the D-dimensional feature space. The feature space chosen in the paper was the bag of the words over the vocabulary W and $D = 3|W|$ for both s_O, and s_R, i.e., every word has three representations one for $\phi_y(\cdot)$ and two for $\phi_x(\cdot)$ based on whether the word is in the input or the supporting memories and can be modeled separately. The parameters of U in both o and r are separate and trained using the marginal loss function given by:

$$\sum_{\bar{f} \neq \mathbf{m}_{O1}} \max(0, \gamma - s_O(x, \mathbf{m}_{O1}) + s_O(x, \bar{f})) +$$

$$\sum_{\bar{f}' \neq \mathbf{m}_{O2}} \max(0, \gamma - s_O([x, \mathbf{m}_{O1}], \mathbf{m}_{O2}) + s_O([x, \mathbf{m}_{O1}], \bar{f}')) +$$

$$\sum_{\bar{r} \neq r} \max(0, \gamma - s_R([x, \mathbf{m}_{O1}, \mathbf{m}_{O2}], r) + s_R([x, \mathbf{m}_{O1}, \mathbf{m}_{O2}], \bar{r})) \tag{9.36}$$

where $\bar{f}, \bar{f}', \bar{r}$ are other choices apart from the true label, i.e., it adds a margin loss if the score of the wrong choices is greater than the ground truth minus γ.

The scoring function o_1 and o_2 given in Eqs. 9.32 and 9.33 can be computationally expensive when the memory storage is large. The paper uses a couple of tricks such as hashing the words and clustering word embeddings in a cluster k. The clustering approach gives a nice trade-off between speed and accuracy with the cluster size k choice.

Let us take a simple example with two supporting facts dataset from the bAbI already in the memory slots given in the table below. When the question "Where is the

memory slot ($\mathbf{m}_i$)	sentence
1	Mary moved to the bathroom.
2	Sandra journeyed to the bedroom.
3	John went to the kitchen.
4	Mary got the football there.
5	Mary went back to the kitchen.
6	Mary went back to the garden.

football?" is asked, the input after going through $k = 2$, $x =$ "Where is the football?" is matched to everything in memory and the slot $\mathbf{m}_{O1} =$ "Mary got the football there" and using this, i.e., $[x, \mathbf{m}_{O1}]$, it will perform another similarity search and find $\mathbf{m}_{O2} =$ "Mary went back to the garden" giving rise to the new output $[x, \mathbf{m}_{O1}, \mathbf{m}_{O2}]$. The R component uses $[x, \mathbf{m}_{O1}, \mathbf{m}_{O2}]$ input to generate an output response $r =$ "garden."

9.3.2 End-to-End Memory Networks

To overcome the issues of memory networks such as the need for each of the component to be trained in a supervised manner, issues of training hard attention, to name a few, Sukhbaatar et al. proposed end-to-end memory networks or MemN2N. MemN2N overcomes many of the disadvantages of MemNN by having soft attention while reading from the memory, performing multiple lookups or hops on memory, and training end-to-end with backpropagation with minimal supervision [Suk+15].

9.3.2.1 Single Layer MemN2N

The MemN2N takes three inputs; (a) the story/facts/sentences $\mathbf{x}_1, \mathbf{x}_2, \ldots, \mathbf{x}_n$, (b) the query/question $\mathbf{q}$, and (c) the answer/label a. We will walk through different components and interactions of MemN2N architecture next considering only one layer of memory and the controller.

9.3.2.2 Input and Query

The input sentences, for instance, x_i is the ith sentence with words w_{ij} as given by $x_i = x_{i1}, x_{i2}, \ldots, x_{in}$, are converted into a memory representation $\mathbf{m}_1, \mathbf{m}_2, \ldots, \mathbf{m}_n$ of dimension d using embedding matrix $\mathbf{A}$ of dimension $d \times |V|$, where $|V|$ is the size of the vocabulary. The operation is given by:

$$\mathbf{m}_i = \sum_j \mathbf{A} x_{ij} \tag{9.37}$$

The paper discusses different ways of combining word embeddings for all the words in the sentence, for example, by performing a *sum* operation on all the word embeddings to get a sentence embedding. Similarly, the query or the question sentence is mapped to a vector of dimension d using embedding matrix $\mathbf{B}$ of dimension $d \times |V|$.

9.3.2.3 Controller and Memory

The query representation $\mathbf{u}$ from the embedding matrix B for the controller is then matched with every memory index $\mathbf{m}_i$ using the dot product for similarity and softmax for choosing the state. The operation can be given by:

$$p_i = \frac{\exp(\mathbf{u}^T \mathbf{m}_i)}{\sum_j \exp(\mathbf{u}^T \mathbf{m}_j)} \tag{9.38}$$

9.3.2.4 Controller and Output

Each input sentences $\mathbf{x}_i$ are also mapped to the controller as vectors $\mathbf{c}_i$ of dimension d using a third embedding matrix $\mathbf{C}$ of dimension $d \times |V|$. The output is then combined using the softmax outputs p_i and the vector $\mathbf{c}_i$ as:

$$\mathbf{o} = \sum_i p_i \mathbf{c}_i \tag{9.39}$$

9.3.2.5 Final Prediction and Learning

The output vector $\mathbf{o}$ and input query with embeddings $\mathbf{u}$ are combined and then passed through a final weight matrix $\mathbf{W}$ and a softmax to produce the label:

$$\hat{a} = \frac{\exp(\mathbf{W}(\mathbf{o} + \mathbf{u}))}{\sum_j \exp(\mathbf{W}(\mathbf{o} + \mathbf{u}))} \tag{9.40}$$

The true label a and the predicted label $\hat{a}$ are used to then train the networks including the embeddings A, B, C, and W using cross-entropy loss and stochastic gradient descent. A single layer MemN2N with complete flow from input sentences, query, and answer is shown in Fig. 9.8.

9.3.2.6 Multiple Layers

The single-layered MemN2N is then extended to multiple layers as shown in Fig. 9.9 in the following way:

- Each layer has its own memory embedding matrix A for input and controller/output embedding matrix C.
- Each layer input $K + 1$ combines the output of current layer $\mathbf{o}^k$ and its input $\mathbf{u}^k$ using:

$$\mathbf{u}^{k+1} = \mathbf{o}^k + \mathbf{u}^k \tag{9.41}$$

- The top layer uses the output with softmax function in a similar way to generate a label $\hat{a}$.

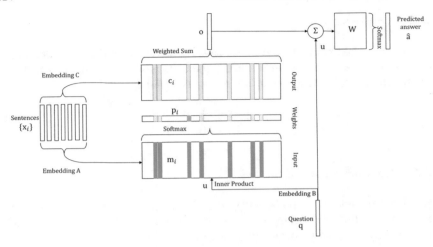

Fig. 9.8: Single-layered MemN2N

- The final output $\hat{a}$ is similarly compared to the actual label a, and the entire network is trained using cross-entropy loss and stochastic gradient descent.

Since many of the tasks such as QA require temporal context, i.e., an entity was at some place before going to another place, the paper modifies the memory vector to encode the temporal context using a temporal matrix. For example, the input memory mapping can be written as:

$$\mathbf{m}_i = \sum_j \mathbf{A} x_{ij} + \mathbf{T}_A(i) \tag{9.42}$$

where $\mathbf{T}(i)$ is the ith row of a temporal matrix $\mathbf{T}$.

9.3.3 Neural Turing Machines

Graves et al. proposed a memory augmented network called neural Turing machines (NTM) for performing complex tasks that were repetitive and needing information over longer periods [GWD14b]. As shown in Fig. 9.10, it has a neural network component called **controller** which interacts with the outside world and the inner **mem-**

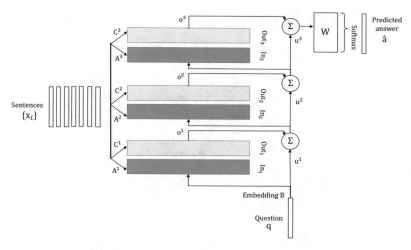

Fig. 9.9: Multiple layered MemN2N

ory for all its operations. Drawing inspiration from Turing machines, the controller interacts with the memory using the **read heads** and the **write heads**. Since memory read or write can be seen as discrete and non-continuous operations, it cannot be differentiated, and thus most gradient-based algorithms cannot be used as is. One of the most important concepts introduced in the research was to overcome this issue by using **blurry** operations in both read and write that interact with all the memory elements with varying degrees. By using these blurry operations, all reading and writing can be continuous, differentiable, and learned effectively using gradient-based algorithms such as stochastic gradient descent.

Let us consider memory $\mathbf{M}$ to be a two-dimensional matrix $(N \times M)$ with N rows corresponding to the memory and M columns for each row where values get stored. Next, we will discuss different operations in NTM.

9.3.3.1 Read Operations

Attention is used to move the read (and write) heads in NTM. The attention mechanism can be written as a length-N normalized **weight vector** $\mathbf{w}_t$, reading contents from the memory $\mathbf{M}_t$ at a given time t. Individual elements of this weight vector will be referred to as $w_t(i)$.

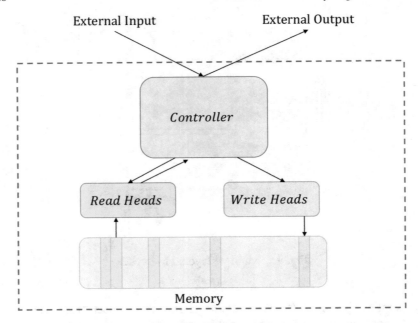

External Input   External Output

Fig. 9.10: Neural Turing machines

The constraints on weight vectors are:

$$\forall i \in \{1 \ldots N\} 0 \leq w_t(i) \leq 1 \tag{9.43}$$

$$\sum_{i=1}^{N} w_t(i) = 1 \tag{9.44}$$

The read head will return a length-M **read vector** r_t which is a linear combination of the memory's rows scaled by the weight vector as given by:

$$\mathbf{r}_t \leftarrow \sum_{i=1}^{M} w_t(i) \mathbf{M}_t(i) \tag{9.45}$$

As the above equation is differentiable, the whole read operation is differentiable.

9.3.3.2 Write Operations

Writing in NTM can be seen as two distinct steps: erasing the memory content and then adding new content. The erasing operation is done through a length-M **erase vector** $\mathbf{e}_t$ in addition to the weight vector $\mathbf{w}_t$ to specify which elements in the row should be completely erased, left unchanged, or some changes carried out. Thus the weight vector $\mathbf{w}_t$ gives us a row to attend and the erase vector $\mathbf{e}_t$ erases the elements in that row giving the update:

$$\mathbf{M}_t^{erased}(i) \leftarrow \mathbf{M}_{t-1}(i)[\mathbf{1} - w_t(i)\mathbf{e}_t] \tag{9.46}$$

After the erase state, i.e., $\mathbf{M}_{t-1}$ converted to $\mathbf{M}_t^{erased}$, the write head uses a length-M **add vector** $\mathbf{a}_t$ to complete the writing as given by:

$$\mathbf{M}_t(i) \leftarrow \mathbf{M}_t^{erased}(i) + w_t(i)\mathbf{a}_t \tag{9.47}$$

Since both erase and write operations are differentiable, entire write operation is differentiable.

9.3.3.3 Addressing Mechanism

The weights used in reading and writing are computed based on two addressing mechanisms: (a) **content-based addressing** and (b) **location-based addressing**. The idea behind content-based addressing is to take information generated from the controller, even if it is partial, and find an exact match in the memory. In certain tasks, especially variable-based operations, it is imperative to find the location of the variables for tasks such as iterations and jumps. In such cases, location-based addressing is very useful.

The weights are computed in different stages and passed on to the next stage.

We will walk through every step in the process of computing the weights as given in Fig. 9.11. The first stage known as content addressing takes the two inputs: a **key**

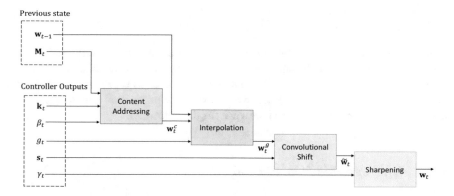

Fig. 9.11: NTM addressing steps

vector $\mathbf{k}_t$ of length-M and a scalar **key strength** β_t. The key vector $\mathbf{k}_t$ is compared to every vector $\mathbf{M}_t(i)$ using a similarity measure $K[\cdot,\cdot]$. The key strength, β_t, acts

puts focus on certain terms or deemphasizes them. The content-based addressing produces the output $\mathbf{w}_t^c$ as given by:

$$\mathbf{w}_t^c = \frac{\exp(\beta_t K[\mathbf{k}_t, \mathbf{M}_t(i)])}{\sum_j \exp(\beta_t K[\mathbf{k}_t, \mathbf{M}_t(j)])} \tag{9.48}$$

Location-based addressing is performed in the next three stages. The second stage, called **interpolation** $g_t \in (0,1)$, takes a scalar parameter from the controller head which is used to combine the content weight from the previous step $\mathbf{w}_t^c$ and previous time step's weight vector $\mathbf{w}_{t-1}$ to generate the **gated weighting** w_t^g given by:

$$\mathbf{w}_t^g \leftarrow g_t \mathbf{w}_t^c + (1 - g_t) \mathbf{w}_{t-1} \tag{9.49}$$

The next stage is a convolutional shift which works to shift attention to other rows. It takes a shift vector $\mathbf{s}_t$ from the controller head as input and the previous interpolated output $\mathbf{w}_t^g$. The shift vector can have various values such as $+1$ to shift forward one row, 0 to stay as is, and -1 to shift backward one row. The operation is a shift modulo N so that attention shift of the bottom moves the head to the top and vice versa. The convolution shift is given by $\tilde{w}_t$ and the operation is:

$$\tilde{w}_t(i) \leftarrow \sum_{j=0}^{N-1} w_t(j)^g s_t(i - j) \tag{9.50}$$

The final stage is sharpening, which prevents the previous convolution shifted weights from blurring using another parameter $\gamma \geq 1$ from the controller head. The final output of the weight vector $\mathbf{w}_t$ is given by:

$$\mathbf{w}_t(i) \leftarrow \frac{\tilde{w}_t(i)^{\gamma_t}}{\sum_j \tilde{w}_t(j)^{\gamma_t}} \tag{9.51}$$

Thus the address of reading and writing is computed by the above operations and all of the parts are differentiable and hence can be learned by gradient-based algorithms. The controller network has many choices such as type of neural network, number of read heads, number of write heads, etc. The paper used both feed-forward and LSTM based recurrent neural network for the controller.

9.3.4 Differentiable Neural Computer

Graves et al. proposed a differentiable neural computer (DNC) as an extension and improvement over the neural Turing machines [Gra+16]. It follows the same high-level architecture of controller with multiple read heads and single write head affecting the memory as given in Fig. 9.12. We will describe the changes that DNC makes to NTM in this section.

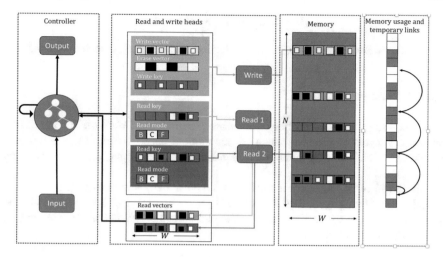

Fig. 9.12: DNC addressing scheme

9.3.4.1 Input and Outputs

The controller network receives input vector $\mathbf{x}_t \in \mathbb{R}^X$ at every time step and generates an output $\mathbf{y}_t \in \mathbb{R}^Y$. It also receives as input rather R read vectors the previous time step as $\mathbf{r}_{t-1}^1, \ldots, \mathbf{r}_{t-1}^R$ from the memory matrix $\mathbf{M}_{t-1} \in \mathbb{R}^{N \times W}$ via the read heads. The input and the read vectors are concatenated as single controller input $\mathbf{x}_{con_t} = [\mathbf{x}_t; \mathbf{r}_{t-1}^1, \ldots, \mathbf{r}_{t-1}^R]$. The controller uses a neural network such as LSTM.

9.3.4.2 Memory Reads and Writes

Location selection happens using weight vectors that are non-negative and sum to 1. The complete set of "allowed" weighting over N locations in the memory is given by a non-negative orthant and constraints as:

$$\Delta_N = \left\{ \alpha \in \mathbb{R}^N \alpha_i \in [0,1], \sum_{i=1}^N \alpha_i \leq 1 \right\} \tag{9.52}$$

The read operation is carried out using R read weights $\{\mathbf{w}_t^{r,1}, \ldots, \mathbf{w}_t^{r,R}\} \in \Delta_N$, thus giving read vectors $\{\mathbf{r}_t^1, \ldots, \mathbf{r}_t^R\}$ by equation:

$$\mathbf{r}_t^i = \mathbf{M}_t^\mathsf{T} \mathbf{w}_t^{r,i} \tag{9.53}$$

The read vectors get appended to the controller input at the next time step.

The write operation is carried out by write weighting $\mathbf{w}_t^w \in \mathbb{R}^N$ together with write vector $\mathbf{v}_t \in \mathbb{R}^W$ and the erase vector $\mathbf{e}_t \in [0,1]^W$ both emitted by the controller to modify the memory as:

$$\mathbf{M}_t = \mathbf{M}_{t-1} \circ (\mathbf{E} - \mathbf{w}_t^w \mathbf{e}_t^\mathsf{T}) + \mathbf{w}_t^w \mathbf{v}_t^\mathsf{T} \tag{9.54}$$

where $\circ$ represents element-wise multiplication and $\mathbf{E}$ is $N \times M$ matrix of ones.

9.3.4.3 Selective Attention

The weightings from controller outputs are parameterized over the memory rows with three forms of attention mechanisms: content-based, memory allocation, and temporal order. The controller interpolates among these three mechanisms using scalar gates.

Similar to NTM, selective attention uses a partial key vector $\mathbf{k}_t$ of length-W and a scalar **key strength** β_t. The key vector $\mathbf{k}_t$ is compared to every vector $\mathbf{M}_t[i]$ using a similarity measure $K[\cdot, \cdot]$ to find the closest to the key normally using cosine similarity as given by:

$$C(\mathbf{M}, \mathbf{k}, \beta)[i] = \frac{\exp(\beta_t K[\mathbf{k}_t, \mathbf{M}_t[i]])}{\sum_j \exp(\beta_t K[\mathbf{k}_t, \mathbf{M}_t[j]])} \tag{9.55}$$

The $C(\mathbf{M}, \mathbf{k}, \beta)$ NTM's drawback of allocating only contiguous blocks of memory is also overcome in DNC. DNC defines the concept of a differentiable **free list** for tracking the **usage** ($\mathbf{u}_t$) of every memory location. Usage is increased after each write ($\mathbf{w}_t^w$) and optionally decreased after each read ($\mathbf{w}_t^{r,i}$) by **free gates** ($\mathbf{f}_t^i$) given by:

$$\mathbf{u}_t = (\mathbf{u}_{t-1} + \mathbf{w}_{t-1}^w - \mathbf{u}_{t-1} \circ \mathbf{w}_{t-1}^w) \circ \prod_{i=1}^{R} (\mathbf{1} - \mathbf{f}_t^i \mathbf{w}_t^{r,i}) \tag{9.56}$$

The controller uses an allocation gate ($g_t^a \in [0,1]$) to interpolate between writing to the newly allocated location in the memory ($\mathbf{a}_t$) or an existing location found by content ($\mathbf{c}_t^w$) with $g_t^w \in [0,1]$ being the write gate:

$$\mathbf{w}_t^w = g_t^w [(g_t^a \mathbf{a}_t + (1 - g_t^a) \mathbf{c}_t^w)] \tag{9.57}$$

Another drawback of NTM was the inability to retrieve memories preserving temporal order which is very important in many tasks. DNC overcomes this by having an ability to iterate through memories in the order they were written. A **precedence weighting** ($\mathbf{p}_t$) keeps track of which memory locations were written to most recently using:

$$\mathbf{p}_t = \left(1 - \sum_i \mathbf{w}_t^w[i]\right) \mathbf{p}_{t-1} + \mathbf{w}_t^w \tag{9.58}$$

A **temporal link matrix** ($\mathbf{L}_t[i,j] \in \mathbb{R}^{N \times N}$) represents the degree to which location i was the location after location j. The matrix gets updated using the precedence weight vector $\mathbf{p}_t$ as given by:

$$\mathbf{L}_t[i,j] = (1 - \mathbf{w}_t^w[i] - \mathbf{w}_t^w[j]) \mathbf{L}_{t-1}[i,j] + \mathbf{w}_t^w[i] \mathbf{p}_{t-1}[j] \tag{9.59}$$

The controller can use the temporal link matrix to retrieve the write before ($\mathbf{b}_t^i$) or after ($\mathbf{f}_t^i$) the last read location ($\mathbf{w}_{t-1}^{r,i}$) allowing the forward and backward movement in time given by the following equations:

$$\mathbf{b}_t^i = \mathbf{L}_t^\mathsf{T} \mathbf{w}_{t-1}^{r,i} \tag{9.60}$$

$$\mathbf{f}_t^i = \mathbf{L}_t \mathbf{w}_{t-1}^{r,i} \tag{9.61}$$

In the paper, the temporal link matrix is $N \times N$ and thus the operation related to memory and computation is of the order $\mathcal{O}(N^2)$. Since the matrix is sparse, the authors have approximated it using a fixed length K to approximate the vectors $\hat{\mathbf{w}}_\mathbf{t}^W$, $\hat{\mathbf{p}}_{t-1}$ for write weight and the precedence weighting. This is further used to compute the approximate temporal link matrix $\hat{\mathbf{L}}_t$ and thus the new forward and backward movement $\hat{\mathbf{f}}_t^i$ and $\hat{\mathbf{b}}_t^i$, respectively. They saw faster performance without any noticeable degradation to the effectiveness using an approximate method.

The read head i computes the content weight vector $\mathbf{c}_t^{r,i}$ using the read key $\mathbf{k}_t^{r,i}$ using:

$$\mathbf{c}_t^{r,i} = C(\mathbf{M}_t, \mathbf{k}_t^{r,i}, \beta_t^{r,i}) \tag{9.62}$$

The read head gets the inputs from three-way gates (π_t^i) and uses it to interpolate among iterating forward, backward, or by content given by:

$$\mathbf{w}_t^{r,i} = \pi_t^i[1]\mathbf{b}_t^i + \pi_t^i[2]\mathbf{c}_t^{r,i} + \pi_t^i[2]\mathbf{f}_t^i \tag{9.63}$$

9.3.5 Dynamic Memory Networks

Kumar et al. proposed dynamic memory networks (DMN), where many tasks in NLP can be formulated as a triplet of facts–question–answer and end-to-end learning can happen effectively [Kum+16]. We will describe the components of DMN as shown in Fig. 9.13. We will use the small example given in Fig. 9.13 to explain each step.

9.3.5.1 Input Module

The input module takes the stories/facts, etc. as sentences in raw form, transforms them into a distributed representation using embeddings such as GloVe from the memory module, and encodes it using a recurrent network such as GRU. The input can be a single sentence or list of sentences concatenated together—say of T_I words given by $w_1, \ldots, w_{T_I}$. Every sentence is converted by adding an end-of-sentence token and then the words are concatenated. Each end-of-sentence generates a hidden state corresponding to that sentence like $\mathbf{h}_t = GRU(\mathbf{L}(w_t), \mathbf{h}_{t-1})$, where w_t is the word w_t index at time t and $\mathbf{L}$ is the embedding matrix. The outputs of this input

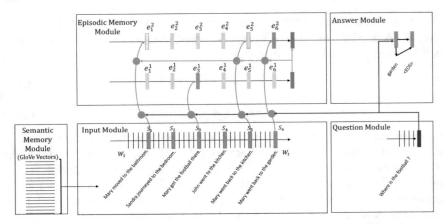

Fig. 9.13: Dynamic memory networks (DMN)

module are length T_c fact sequences of hidden state of each sentence and $\mathbf{c}_t$ is the fact at step t. In the simplest case where each sentence output is encoded as a fact T_c is equal to the number of sentences.

9.3.5.2 Question Module

The question module is similar to the input module, where the question sentence of T_Q words is converted to an embedding vector, and given to the recurrent network. The GRU based recurrent network is used to model it given by $\mathbf{q}_t = GRU(\mathbf{L}(w_t^Q), \mathbf{q}_{t-1})$, where $\mathbf{L}$ is the embedding matrix. The hidden state, which is the final state at the end of the question, was given by $\mathbf{q} = \mathbf{q}_{T_Q}$. The word embedding matrix $\mathbf{L}$ is shared across both the input and the question module.

9.3.5.3 Episodic Memory Module

The hidden states of the input module across all the sentences and the question modules output are the inputs for the episodic memory module. Episodic memory has an attention mechanism to focus on the states from the input and a recurrent network which updates its episodic memory. The episodic memory updates are part of an iterative process. In each iteration the attention mechanism attends over the input module's hidden states mapping to the fact representation $\mathbf{c}$, question $\mathbf{q}$, and the past memory $\mathbf{m}^{i-1}$ to produce an episode $\mathbf{e}^i$. The episode is then used along with previous memory $\mathbf{m}^{i-1}$ to update episodic memory $\mathbf{m}^i = GRU(\mathbf{e}^i, \mathbf{m}^{i-1})$. The GRU is initialized with a question as the state, i.e., $\mathbf{m}^0 = \mathbf{q}$. The iterative nature of the episodic memory helps in focusing on different sections of the inputs and thus has the transitive nature required for inferencing. The number of passes T_M is a hyperparameter and the final episodic memory $\mathbf{m}^{T_M}$ is given to the answer module.

The attention mechanism has a feature generation part and a scoring part using the gating mechanism. The gating happens through a function G that takes as input a candidate fact $\mathbf{c}_t$, previous memory $\mathbf{m}^{i-1}$, and the question $\mathbf{q}$ to compute a scalar g_t which acts as a gate:

$$g_t^i = G(\mathbf{c}_t, \mathbf{m}^{i-1}, \mathbf{q}) \tag{9.64}$$

The feature vector $\mathbf{z}(\mathbf{c}, \mathbf{m}, \mathbf{q})$ which feeds into the scoring function G above using different similarities between the input facts, previous memory, and the question as given by:

$$\mathbf{z}(\mathbf{c}, \mathbf{m}, \mathbf{q}) = [\mathbf{c} \circ \mathbf{m}; \mathbf{c} \circ \mathbf{q}; |\mathbf{c} - \mathbf{m}|; \mathbf{c} - \mathbf{m}] \tag{9.65}$$

where $\circ$ is the element-wise product between the vectors. The scoring function G is the standard two-layer feed-forward network where

$$G(\mathbf{c}, \mathbf{m}, \mathbf{q}) = \sigma(\mathbf{W}^{(2)} \tanh(\mathbf{W}^{(1)} \mathbf{z}(\mathbf{c}, \mathbf{m}, \mathbf{q}) + b^1) + b^2) \tag{9.66}$$

where weights $\mathbf{W}^{(1)}, \mathbf{W}^{(2)}$ and biases b^1, b^2 are learned through the training process. The episode at iteration i uses the GRU with sequences $\mathbf{c}_1, \ldots, \mathbf{c}_{T_C}$ weighted by the gate g^i and the final hidden state is used to update as given by:

$$\mathbf{h}_t^i = g_t^i GRU(\mathbf{c}_t, \mathbf{h}_{t-1}^i,) + (1 - g_t^i)\mathbf{h}_{t-1}^i \tag{9.67}$$

$$\mathbf{e}^i = \mathbf{h}_{T_C}^i \tag{9.68}$$

Either maximum iteration is set or a supervised symbol to mark the end of phase token is passed to stop the iteration.

9.3.5.4 Answer Module

The answer module can either be triggered at the end of every episodic memory iteration or the final one based on the task. It is again modeled as a GRU with an input question, last hidden state $\mathbf{a}_{t-1}$, and previous prediction y_{t-1}. The initial state a_0 is initialized to the last memory as $\mathbf{a}_0 = \mathbf{m}^{T_M}$. Thus the updates can be written as:

$$y_t = softmax(\mathbf{W}^a \mathbf{a}_t) \tag{9.69}$$

$$a_t = GRU([y_{t-1}, \mathbf{q}], \mathbf{a}_{t-1}) \tag{9.70}$$

9.3.5.5 Training

The end-to-end training is done in a supervised manner where the answer generated by the answer module is compared to the real labeled answer, and a cross-entropy loss is propagated back using stochastic gradient descent. To give a concrete example, let us consider the story with sentences $s_1, \ldots, s_6$ as inputs to the input module and question q *Where is the football?* to be passed to the question module as shown in Fig. 9.13. In the first pass of the episodic memory let us assume it will try to attend to the word *football* from the question, all the facts c coming as hidden states from the input modules and will score all the facts from input where *football* appears and give maximum to facts such as *Mary got the football there*. In the next iteration, it will take the output from this episodic state and try to focus on the next part *Mary* and thus select all statements such as *Mary moved to the bathroom*, *Mary got the football there*, *Mary went back to the kitchen*, and *Mary went back to the garden*. From these let us assume it will select the last sentence *Mary went back to the garden*. The selection of the right sentences to focus happens in an end-to-end manner using backpropagation, where the actual label from the answer module *garden* compares the generated output to propagate the errors back.

9.3.6 Neural Stack, Queues, and Deques

Grefenstette et al. explore learning interactions between the controller and memory using the traditional data structures such as stacks, queues, and deques. They produce superior generalization when compared with RNNs [Gre+15]. In the next few sections, we will explore the basic working of neural stack architecture and then generalize it to others.

9.3.6.1 Neural Stack

The neural stack is a differentiable structure which allows storing the vectors through push operations, and retrieving the vectors through pop operations analogous to the stack data structure as shown in Fig. 9.14.

The entire stack content at a given time t is denoted by matrix $\mathbf{V}_t$, each row corresponding to memory address i contains a vector $\mathbf{v}_t$ of size m such that it is in the space $\mathbb{R}^m$. Associated with each index in the matrix is a strength vector giving the weight associated with that index of content and is given by $\mathbf{s}_t$. The push signal is given by a scalar $d_t \in (0, 1)$ and the pop signal is given by a scalar $u_t \in (0, 1)$. The value read from the stack is given by $\mathbf{r}_t \in \mathbb{R}^m$.

The necessary operations for the neural stack are given by the following three equations for $\mathbf{V}_t$, $\mathbf{s}_t$, and $\mathbf{r}_t$:

$$\mathbf{V}_t[i] = \begin{cases} \mathbf{V}_{t-1}[i] & \text{if } 1 \leq i < t \\ \mathbf{v}_t & \text{if } i = t, \mathbf{V}_t[i] = \mathbf{v}_t \text{ for all } i \leq t \end{cases} \tag{9.71}$$

Equation 9.71 captures the updates to the stack as an ever-growing list-like structure of a neural stack where every old index gets the value from the previous time step and the new vector is pushed on the top:

$$
\mathbf{s}_t[i] = \begin{cases} \max\left(0, \mathbf{s}_{t-1}[i] - \max\left(0, u_t - \sum_{j=i+1}^{t-1} \mathbf{s}_{t-1}[j]\right)\right) & \text{if } 1 \le i < t \\ d_t & \text{if } i = t \end{cases}
\tag{9.72}
$$

Equation 9.72 captures the weight updates, where the case $i = t$ means that we directly pass the push weight $d_t \in (0,1)$. Removing a entry from stack doesn't remove it physically but sets the strength value at the index 0. Each of the strengths lower down the stacks changes based on the following calculation, subtract the pop signal strength u_t and the relative sum above that index $i+1$ and below index at value $t-1$ and cap it by finding maximum between that value and 0. Then subtract it with the current value at the index s_{t+1} and cap it by finding maximum between that value and 0.

We look at Fig. 9.14 at time $t = 3$ and lowest index $i = 1$ assuming a with previous value of 0.7, it will become $\max(0, 0.7 - \max(0, 0.9 - 0.5)) = 0.3$. Similarly, we can plug the same value for $t = 3$ and next index $i = 2$ with previous value of 0.7, it will become $\max(0, 0.9 - \max(0, 0.9 - 0)) = 0$. Finally, at $t = 3$, the top index $i = 3$ will have the d_t value of 0.9:

$$
\mathbf{r}_t = \sum_{i=1}^{t} \min\left(\mathbf{s}_t[i], \max\left(0, 1 - \sum_{j=i+1}^{t} \mathbf{s}_t[j] \cdot V_t[i]\right)\right)
\tag{9.73}
$$

Equation 9.73 can be seen as the state that the network sees at time t. It is a combination of the index vector and its strength, where the strengths are constrained to sum to 1.

Again when we look at Fig. 9.14 at time $t = 3$, we see that everything is normal combinations except the strength of index 1 is changed from 0.3 to 0.1 because substituting we get $\min(0.3, \max(0, 1 - 0.9)) = 0.1$.

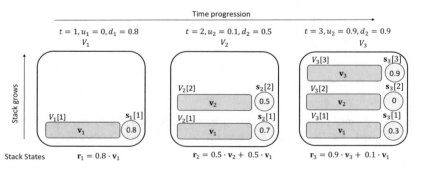

Fig. 9.14: Neural stack states with respect to time and operations of push and pop

9.3.6.2 Recurrent Networks, Controller, and Training

The gradual extension of the neural stack from above as a recurrent network and the
controller actions are shown in Fig. 9.15a, b. The entire architecture marked with
dotted lines is a recurrent network with inputs (a) previous recurrent state H_{t-1} and
(b) current input $\mathbf{i}_t$; and outputs (a) next recurrent state H_t and (b) $\mathbf{o}_t$. The previous
recurrent state H_{t-1} consists of three parts: (a) the previous state vector from RNN
$\mathbf{h}_{t-1}$, (b) the previous stack read $\mathbf{r}_t$, and (c) the state of the stack from previous state
$(\mathbf{V}_{t-1}, \mathbf{s}_t)$. In the implementation, all of the vectors except $\mathbf{h}_0$, which is randomly
initialized, all are set to $\mathbf{0}$ to start with.

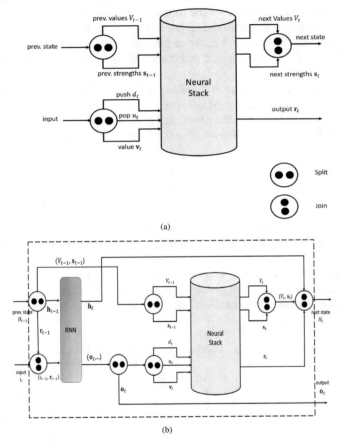

Fig. 9.15: Neural stack with recurrent network and controller. (**a**) Neural stack as
recurrent network. (**b**) Neural stack recurrent network with controller

The current input $\mathbf{i}_t$ is concatenated with previous read of the stack $\mathbf{r}_{t-1}$ to the controller which has its own previous state $\mathbf{h}_{t-1}$ generating next state $\mathbf{h}_t$ and the output $\mathbf{o}'_t$. The output $\mathbf{o}'_t$ results in the push signal scalar d_t, the pop signal scalar u_t, and the value vector $\mathbf{v}_t$ that go as input signals to the neural stack and the output signal $\mathbf{o}_t$ for the whole. The equations are:

$$d_t = \text{sigmoid}(W_d \mathbf{o}'_t + b_d) \tag{9.74}$$

$$u_t = \text{sigmoid}(W_u \mathbf{o}'_t + b_u) \tag{9.75}$$

$$\mathbf{v}_t = \text{sigmoid}(W_v \mathbf{o}'_t + b_v) \tag{9.76}$$

$$\mathbf{o}_t = \text{sigmoid}(W_o \mathbf{o}'_t + b_o) \tag{9.77}$$

The whole structure can be easily adapted to **neural queues** by changing the pop signal to read from the bottom of the list rather than the top and can be written as:

$$\mathbf{s}_t[i] = \begin{cases} \max\left(0, \mathbf{s}_{t-1}[i] - \max\left(0, u_t - \sum_{j=1}^{i-1} \mathbf{s}_{t-1}[j]\right)\right) & \text{if } 1 \leq i < t \\ d_t & \text{if } i = t \end{cases} \tag{9.78}$$

$$\mathbf{r}_t = \sum_{i=1}^{t} \min\left(\mathbf{s}_t[i], \max\left(0, 1 - \sum_{j=1}^{i-1} \mathbf{s}_t[j] \cdot \mathbf{V}_t[i]\right)\right) \tag{9.79}$$

The **Neural DeQue** works similarly to a neural stack, but has the ability to take the input signals of push, pop, and value for both the top and the bottom sides of the list.

9.3.7 Recurrent Entity Networks

Henaff et al. designed a highly parallel architecture with a long dynamic memory which performs well on many NLU tasks known as recurrent entity networks (EntNet) [Hen+16]. The idea is to have **blocks** of memory cells, where each cell can store information about an entity in the sentence so that many entities corresponding to names, locations, and others have information content in the cells. We will discuss the core components of EntNet in Fig. 9.16.

9.3.7.1 Input Encoder

Let us consider specifically a question-answering system with sentences discussing the topic of interest, where the question and answer are both found in the given sentences, though this can be used for many other tasks. Let us consider a setup with training set as $\{(x_i, y_i)_{i=1}^n\}$, x_i is the input sentences, q is the question, and

y_i is the single word answer. The input encoding layer transforms the sequence of words into a fixed length vector. This can be done as the authors describe using BOW representation and end states of RNN. They chose a simple representation given by using a set of vectors $\{\mathbf{f}_1, \ldots, \mathbf{f}_k\}$ with the input embeddings of the words $\{\mathbf{e}_1, \ldots, \mathbf{e}_k\}$ for a given input at given time t:

$$\mathbf{s}_t = \sum_i \mathbf{f}_i \circ \mathbf{e}_i \tag{9.80}$$

where $\circ$ is the Hadamard product or element-wise multiplication. The same set of vectors $\{\mathbf{f}_1, \ldots, \mathbf{f}_k\}$ are used for all the time steps. The embedding matrix $\mathbf{E} \in \mathbb{R}^{|V| \times d}$ transforms each word in the sentence using $E(w) = \mathbf{e} \in \mathbb{R}^d$, where d is the dimension of the embeddings. Like other parameters, the vectors $\{f_1, \ldots, f_k\}$ are learned from the training data jointly with other parameters.

9.3.7.2 Dynamic Memory

As shown in Fig. 9.16b the input encoded sentences flow into blocks of memory cells and the whole network is a form of gated recurrent unit (GRU) with hidden states in these blocks, which concatenated together give the total hidden state of the network. The total blocks $h_1, \ldots, h_m$ are of the order 5–20 and each block h_j has 20–100 units. Each block j is given a hidden state $\mathbf{h}_j \in \mathbb{R}^d$ and a key $\mathbf{w}_j \in \mathbb{R}^d$.

The role of a block is to capture information about an entity with the facts. This is accomplished by associating the weights of the key vectors with the embeddings of entities of interest so that the model learns information about the entities occurring within the text. A generic j block with weight $\mathbf{w}_j$ and hidden state $\mathbf{h}_j$ is given by:

$$\mathbf{g}_j^t \leftarrow \text{sigmoid}(\mathbf{s}_t^\mathsf{T} \mathbf{h}_j^{t-1} + \mathbf{s}_t^\mathsf{T} \mathbf{w}_j^{t-1}) \qquad (gate) \tag{9.81}$$

$$\tilde{\mathbf{h}}_j^t \leftarrow \phi(\mathbf{P} \mathbf{h}_j^{t-1} + \mathbf{Q} \mathbf{w}_j^{t-1} + \mathbf{R} \mathbf{s}_t) \qquad (candidate\ memory) \tag{9.82}$$

$$\mathbf{h}_j^t \leftarrow \mathbf{h}_j^{t-1} + \mathbf{g}_j \circ \tilde{\mathbf{h}}_j^t \qquad (new\ memory) \tag{9.83}$$

$$\mathbf{h}_j^t \leftarrow \frac{\mathbf{h}_j^t}{\|\mathbf{h}_j^t\|} \qquad (reset\ memory) \tag{9.84}$$

where g_j is the gate that decides how much of the memory will be updated, ϕ is the activation function like ReLU, $\mathbf{h}_j^t$ is the new memory that is combining the older timestamp with current, and the normalization in the last step helps in forgetting the previous information. The matrices $\mathbf{P} \in \mathbb{R}^{d \times d}, \mathbf{Q} \in \mathbb{R}^{d \times d}, \mathbf{R} \in \mathbb{R}^{d \times d}$ are shared across all blocks.

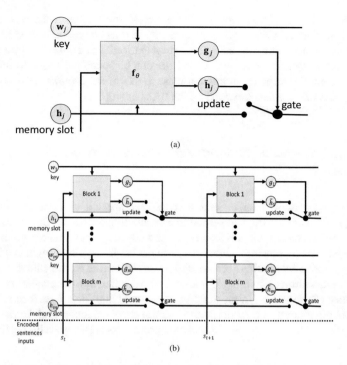

Fig. 9.16: EntNet. (**a**) A single block in the EntNet. (**b**) Recurrent entity networks (EntNet) with multiple blocks

9.3.7.3 Output Module and Training

The output module when presented with question **q** creates a probability distribution over all the hidden states and the entire equations can be written as:

$$p_j = \text{softmax}(\mathbf{q}^\mathsf{T}\mathbf{h}_j) \tag{9.85}$$

$$\mathbf{u} = \sum_j p_j h_j \tag{9.86}$$

$$y = \mathbf{R}\phi(\mathbf{q} + \mathbf{H}\mathbf{u}) \tag{9.87}$$

The matrices $R \in \mathbb{R}^{|V| \times d}$ and $H \in \mathbb{R}^{d \times d}$ are again trained with the rest of the parameters. The function ϕ adds the non-linearity and can be an activation like ReLU. The entire network is trained using backpropagation. The entities can be extracted as part of preprocessing, and the key vectors can be specifically tied to the embeddings of the entities existing in the stories such as {*Mary, Sandra, John, bathroom, bedroom, kitchen, garden, football*} in the bAbI example.

9.3.8 Applications of Memory Augmented Networks in Text and Speech

Most memory networks have been successfully applied to complex NLU tasks such as question answering and semantic role labeling [WCB14, Suk+15, Gra+16, Hen+16]. Sukhbaatar et al. applied end-to-end memory networks to outperform traditional RNNs by increasing memory hops in the language modeling task [Suk+15]. Kumar et al. have interestingly converted most NLP tasks from syntactic to semantic tasks in a question-answering framework and applied dynamic memory networks successfully [Kum+16]. Grefenstette et al. showed significant performance gains, obtained using memory networks such as neural stacks, queues, and deques in transduction tasks such as inversion transduction grammars (ITG) used in machine translation [Gre+15].

9.4 Case Study

In this section, we explore two NLP topics: attention-based NMT and memory networks for question and answering. Each topic follows the same format as used in previous chapters, and provides exercises in the end.

9.4.1 Attention-Based NMT

In this portion of the case study, we compare attention mechanisms on the English-to-French translation task introduced in Chap. 7. The dataset used is composed of translation pairs from the Tatoeba website. This is the same dataset used in the Chap. 7 case study.

9.4.2 Exploratory Data Analysis

For the EDA process, we refer the readers to Sect. 7.7.2 for the steps that were used to create the dataset splits.

The dataset summary is shown below.

```
1  Training set size: 107885
2  Validation set size: 13486
3  Testing set size: 13486
4  Size of English vocabulary: 4755
5  Size of French vocabulary: 6450
```

9.4.2.1 Software Tools and Libraries

When we first explored neural machine translation, we used the fairseq library which leverages PyTorch. To the best of our knowledge, there is not a single library that supports all the different attention mechanisms. Therefore, we combine a collection of libraries to compare the attention approaches. Specifically, we use PyTorch as the deep learning framework, AllenNLP for most attention mechanism implementations, spaCy for tokenization, and torchtext for the data loader. The code contained here extends some of the original work in the PyTorch tutorials with additional functionality and comparisons.

9.4.2.2 Model Training

We compare five different attention mechanisms, training for 100 epochs. For each attention mechanism the model that performs the best on the validation data is chosen to run on the testing data. The models trained are 4-layer bidirectional GRU encoders with a single unidirectional GRU decoder. The encoder and decoder both have a hidden size of 512 with the encoding and decoding embeddings having a size of 256. The models are trained with cross-entropy loss and SGD, with a batch size of 512. The initial learning rate is 0.01 for the encoder and 0.05 for the decoder, and momentum is applied to both with a value of 0.9. A learning rate schedule is used to reduce the learning rate when the validation loss hasn't improved for 5 epochs.

To regularize our model, we add dropout to both the encoder and decoder with a probability of 0.1 and the norms of the gradients are clipped at 10.

We incorporate a batch implementation of the model to leverage the parallel computation of the GPUs. The architecture is the same for each of the models except for the Bahdanau model, which required introducing a weight matrix for the bidirectional output of the encoder in the attention mechanism.

We define the different components of our networks as follows:

```
1  class Encoder(nn.Module):
2      def __init__(self, input_dim, emb_dim, enc_hid_dim,
         dec_hid_dim,
```

```
 3      dropout , num_layers=1, bidirectional=False ):
 4        super ( ) . __init__ ()
 5        self . input_dim = input_dim
 6        self . emb_dim = emb_dim
 7        self . enc_hid_dim = enc_hid_dim
 8        self . dec_hid_dim = dec_hid_dim
 9        self . num_layers = num_layers
10        self . bidirectional = bidirectional
11
12        self . embedding = nn . Embedding ( input_dim , emb_dim )
13        self . rnn = nn . GRU( emb_dim , enc_hid_dim , num_layers=
          num_layers , bidirectional=bidirectional )
14        self . dropout = nn . Dropout ( dropout )
15        if bidirectional :
16            self . fc = nn . Linear ( enc_hid_dim * 2 , dec_hid_dim )
17
18     def forward ( self , src ):
19        embedded = self . dropout ( self . embedding ( src ))
20        outputs , hidden = self . rnn ( embedded )
21
22        if self . bidirectional :
23            hidden = torch . tanh ( self . fc ( torch . cat (( hidden [ −2 ,: ,:] ,
              hidden [ −1 ,: ,:]) , dim=1 )))
24
25        if not self . bidirectional and self . num_layers > 1 :
26            hidden = hidden [ −1 ,: ,:]
27
28        return outputs , hidden
```

```
 1   class Decoder ( nn . Module ):
 2     def __init__ ( self , output_dim , emb_dim , enc_hid_dim ,
         dec_hid_dim , dropout ,
 3       attention , bidirectional_input=False ):
 4        super ( ) . __init__ ()
 5        self . emb_dim = emb_dim
 6        self . enc_hid_dim = enc_hid_dim
 7        self . dec_hid_dim = dec_hid_dim
 8        self . output_dim = output_dim
 9        self . dropout = dropout
10        self . attention = attention
11        self . bidirectional_input = bidirectional_input
12
13        self . embedding = nn . Embedding ( output_dim , emb_dim )
14
15        if bidirectional_input :
16            self . rnn = nn . GRU(( enc_hid_dim * 2 ) + emb_dim ,
              dec_hid_dim )
17            self . out = nn . Linear (( enc_hid_dim * 2 ) + dec_hid_dim +
              emb_dim , output_dim )
18        else :
19            self . rnn = nn . GRU(( enc_hid_dim ) + emb_dim , dec_hid_dim
              )
20            self . out = nn . Linear (( enc_hid_dim ) + dec_hid_dim +
              emb_dim , output_dim )
```

```
21
22      self.dropout = nn.Dropout(dropout)
23
24  def forward(self, input, hidden, encoder_outputs):
25      input = input.unsqueeze(0)
26      embedded = self.dropout(self.embedding(input))
27      hidden = hidden.squeeze(0) if len(hidden.size()) > 2 else
        hidden # batch_size=1 issue
28
29      # Repeat hidden state for attention on bidirectional
        outputs
30      if hidden.size(-1) != encoder_outputs.size(-1):
31          attn = self.attention(hidden.repeat(1, 2),
        encoder_outputs.permute(1, 0, 2))
32      else:
33          attn = self.attention(hidden, encoder_outputs.permute
        (1, 0, 2))
34
35      a = attn.unsqueeze(1)
36
37      encoder_outputs = encoder_outputs.permute(1, 0, 2)
38
39      weighted = torch.bmm(a, encoder_outputs)
40      weighted = weighted.permute(1, 0, 2)
41
42      rnn_input = torch.cat((embedded, weighted), dim=2)
43
44      output, hidden = self.rnn(rnn_input, hidden.unsqueeze(0))
45
46      embedded = embedded.squeeze(0)
47      output = output.squeeze(0)
48      weighted = weighted.squeeze(0)
49
50      output = self.out(torch.cat((output, weighted, embedded),
        dim=1))
51
52      return output, hidden.squeeze(0), attn
```

```
1   class Seq2Seq(nn.Module):
2     def __init__(self, encoder, decoder, device):
3       super().__init__()
4       self.encoder = encoder
5       self.decoder = decoder
6       self.device = device
7
8     def forward(self, src, trg, teacher_forcing_ratio=0.5):
9       batch_size = src.shape[1]
10      max_len = trg.shape[0]
11      trg_vocab_size = self.decoder.output_dim
12
13      outputs = torch.zeros(max_len, batch_size, trg_vocab_size)
        .to(self.device)
14
15      encoder_outputs, hidden = self.encoder(src)
```

```
16    hidden = hidden.squeeze(1)
17
18    output = trg[0,:]  # first input to decoder <sos>
19
20    for t in range(1, max_len):
21      output, hidden, attn = self.decoder(output, hidden,
      encoder_outputs)
22      outputs[t] = output
23      teacher_force = random.random() < teacher_forcing_ratio
24      top1 = output.max(1)[1]
25      output = (trg[t] if teacher_force else top1)
26
27    return outputs
```

We use the attention implementations from AllenNLP for dot product, cosine, and bilinear attention. These functions take the hidden state of the decoder and the output of the encoder and return the attended scores.

```
1  from allennlp.modules.attention import LinearAttention,
2                                         CosineAttention,
3                                         BilinearAttention,
4                                         DotProductAttention
5
6  attn = DotProductAttention() # Changed for each type of model
7  enc = Encoder(INPUT_DIM,
8          ENC_EMB_DIM,
9          ENC_HID_DIM,
10         DEC_HID_DIM,
11         ENC_DROPOUT,
12         num_layers=ENC_NUM_LAYERS,
13         bidirectional=ENC_BIDIRECTIONAL)
14 dec = Decoder(OUTPUT_DIM,
15         DEC_EMB_DIM,
16         ENC_HID_DIM,
17         DEC_HID_DIM,
18         DEC_DROPOUT,
19         attn,
20         bidirectional_input=ENC_BIDIRECTIONAL)
21
22 model = Seq2Seq(enc, dec, device).to(device)
```

In Figs. 9.17 and 9.18 we show the training graphs for the loss and PPL, respectively, for each of the attention models. The three methods that perform the best are Bahdanau, dot product, and bilinear models. Cosine and linear attention struggle to converge. The attention mechanism in linear attention specifically does not correlate with the input sequence at all.

In Figs. 9.19, 9.20, 9.21, 9.22, and 9.23, we give some examples of the decoded attention outputs for three different files, showing what the decoder is attending to during the translation process. In each of the figures, the first two graphs (a) and (b) are inputs with lengths of 10, the maximum seen by the models during training. In most cases, the attention still aligns with the input; however, the predictions are mostly incorrect, typically with high entropy near the time steps close to the maximum training sequence length.

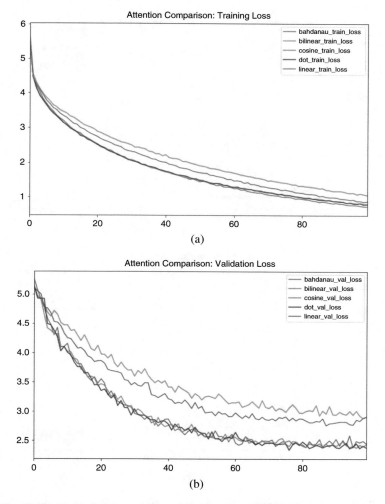

Fig. 9.17: (**a**) Training and (**b**) validation losses for each attention model

9.4.2.3 Bahdanau Attention

The Bahdanau attention employs a fully connected layer to combine the concatenated outputs of the bidirectional layer, rather than duplicating the hidden state. Incorporating this requires slight alterations to accommodate the changes in tensor sizes.

```
class BahdanauEncoder(nn.Module):
    def __init__(self, input_dim, emb_dim, enc_hid_dim,
    dec_hid_dim, dropout):
    super().__init__()
```

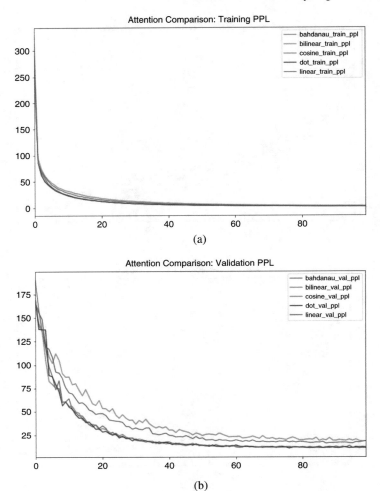

Fig. 9.18: (**a**) Training and (**b**) validation PPL for each attention model

```
 4
 5      self.input_dim = input_dim
 6      self.emb_dim = emb_dim
 7      self.enc_hid_dim = enc_hid_dim
 8      self.dec_hid_dim = dec_hid_dim
 9      self.dropout = dropout
10
11      self.embedding = nn.Embedding(input_dim, emb_dim)
12      self.rnn = nn.GRU(emb_dim, enc_hid_dim, num_layers=4,
        bidirectional=True)
13      self.fc = nn.Linear(enc_hid_dim * 2, dec_hid_dim)
14      self.dropout = nn.Dropout(dropout)
```

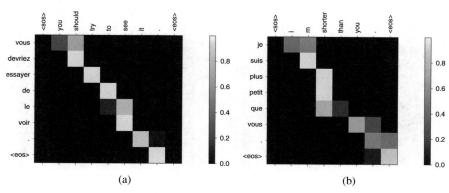

Fig. 9.19: Dot product attention examples

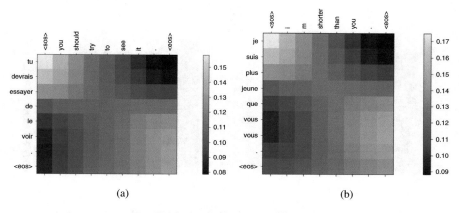

Fig. 9.20: Cosine attention examples

```
15
16   def forward ( self , src ):
17     embedded = self.dropout(self.embedding(src))
18     outputs , hidden = self.rnn(embedded)
19     hidden = torch.tanh(self.fc(torch.cat((hidden[−2,:,:],
       hidden[−1,:,:]), dim=1)))
20     return outputs , hidden
```

Minor alterations are made to the decoder to handle the difference between the hidden size and the encoder output.

```
1   class BahdanauAttention(nn.Module):
2     def __init__(self, enc_hid_dim, dec_hid_dim):
3       super().__init__()
4
5       self.enc_hid_dim = enc_hid_dim
6       self.dec_hid_dim = dec_hid_dim
7
```

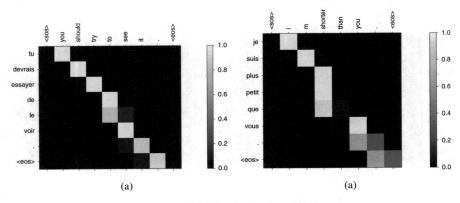

Fig. 9.21: Bilinear attention examples

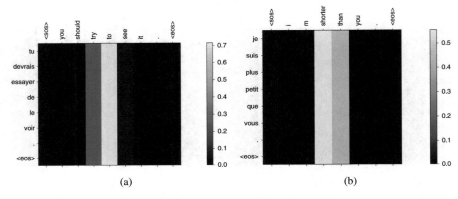

Fig. 9.22: Linear attention examples. Note how the model was unable to learn a useful mapping from the attention mechanism while still being able to translate some examples

```
8    self.attn = nn.Linear((enc_hid_dim * 2) + dec_hid_dim,
     dec_hid_dim)
9    self.v = nn.Parameter(torch.rand(dec_hid_dim))
10
11   def forward(self, hidden, encoder_outputs):
12     batch_size = encoder_outputs.shape[1]
13     src_len = encoder_outputs.shape[0]
14
15     hidden = hidden.unsqueeze(1).repeat(1, src_len, 1)
16
17     encoder_outputs = encoder_outputs.permute(1, 0, 2)
18
19     energy = torch.tanh(self.attn(torch.cat((hidden,
       encoder_outputs), dim=2)))
20     energy = energy.permute(0, 2, 1)
```

```
21
22     v = self.v.repeat(batch_size, 1).unsqueeze(1)
23
24     attention = torch.bmm(v, energy).squeeze(1)
25     return F.softmax(attention, dim=1)
```

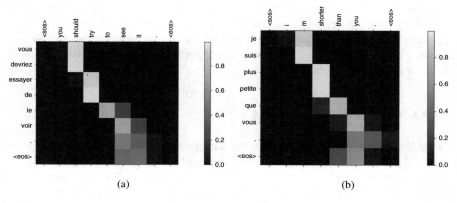

Fig. 9.23: Bahdanau attention examples

9.4.2.4 Results

The best performing model for each of the attention mechanisms was run on the testing set to produce the following results in Table 9.2.

Attention type	Loss	PPL
Dot	17.826	2.881
Bilinear	**13.987**	**2.638**
Cosine	22.098	3.095
Linear	17.918	2.886
Bahdanau	17.580	2.867

Table 9.2: Test results for attention models. Best results are shown in bold

Bilinear attention performed the best in this experiment. We can see from the attention alignments in Fig. 9.21 that the attention output is strongly correlated with the input. Moreover, the strength of attention is highly confident throughout the prediction sequence, ever so slightly losing confidence towards the end of the sequence.

9.4.3 Question and Answering

To help the reader familiarize themselves with the attention and memory networks, we will apply the concepts of this chapter to the question-answering task with the bAbI dataset. The bAbI is a collection of 20 simple QA tasks with limited vocabulary. For each task, there is a set of 1000 training and 1000 stories, test questions and answers as well as an extended training set with 10,000 samples. Despite its simplicity, bAbI effectively captures the complexities of memory and long-range dependencies in question answering. For this case study, we will focus on tasks 1–3, consisting of questions where up to three supporting facts from the stories provide information to support the answer.

9.4.3.1 Software Tools and Libraries

We will implement several architectures with Keras and TensorFlow for this case study. Keras provides a useful example recurrent neural network architecture for the question-answering task that will serve as our baseline. We will contrast performance with several of the memory network-based architectures discussed in this chapter, including a differentiable neural computer model from DeepMind. Rather than providing full coverage of each architecture here, we direct the reader to the notebooks accompanying this chapter for full implementation details.

9.4.3.2 Exploratory Data Analysis

Our first step is to download the bAbI dataset and to extract the training and test sets for our analysis. We will focus on the extended dataset with 10,000 training samples and 1000 test samples. Let's first take a quick look at the samples for tasks QA1, QA2, and QA3:

QA1 Story:	Mary moved to the bathroom. John went to the hallway.
QA1 Query:	Where is Mary?
QA1 Answer:	bathroom
QA2 Story:	Mary moved to the bathroom. Sandra journeyed to the bedroom. Mary got the football there. John went to the kitchen. Mary went back to the kitchen. Mary went back to the garden.
QA2 Query:	Where is the football?
QA2 Answer:	garden
QA3 Story:	Mary moved to the bathroom. Sandra journeyed to the bedroom. Mary got the football there. John went back to the bedroom. Mary journeyed to the office. John journeyed to the office. John took the milk. Daniel went back to the kitchen. John moved to the bedroom. Daniel went back to the hallway. Daniel took the apple. John left the milk there. John travelled to the

> kitchen. Sandra went back to the bathroom. Daniel journeyed to the bathroom. John journeyed to the bathroom. Mary journeyed to the bathroom. Sandra went back to the garden. Sandra went to the office. Daniel went to the garden. Sandra went back to the hallway. Daniel journeyed to the office. Mary dropped the football. John moved to the bedroom.
>
> QA3 Query: Where was the football before the bathroom?
>
> QA3 Answer: office

Analysis of the datasets shows the increasing complexity and long-range memory that is required when progressing from task QA1 to QA3. The distribution of story

Task	train_stories	test_stories	min(story_size)	max(story_size)	query_size	vocab_size
QA1	10,000	1000	12	68	4	21
QA2	10,000	1000	12	552	5	35
QA3	10,000	1000	22	1875	8	36

lengths and question lengths (in terms of the number of tokens) can be seen in Fig. 9.24.

The average length of the stories increases substantially from tasks QA1 to QA3, which makes it significantly more difficult. Remember that for task QA3, there are only three support facts and most of the story is considered "noise." We will see how well different architectures are able to learn to identify the relevant facts from this noise.

9.4.3.3 LSTM Baseline

We use the Keras LSTM architecture example to serve as our baseline. This architecture consists of the following:

1. The tokens of each story and question are mapped to embeddings (that are not shared between them).
2. The stories and questions are encoded using separate LSTMs.
3. The encoded vectors for the story and question are concatenated.
4. These concatenated vectors are used as an input to a DNN whose output is a softmax over the vocabulary.
5. The entire network is trained to minimize the error between the softmax output and the answer.

The Keras model that implements this architecture is:

```
RNN = recurrent.LSTM

sentence = layers.Input(shape=(story_maxlen,), dtype='int32')
encoded_sentence = layers.Embedding(vocab_size,
    EMBED_HIDDEN_SIZE)(sentence)
```

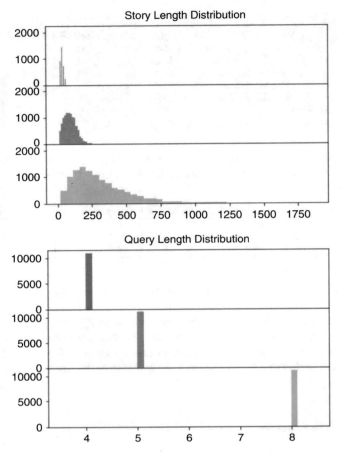

Fig. 9.24: Distributions of story and question lengths in bAbI tasks 1–3

```
 5 encoded_sentence = Dropout(0.3)(encoded_sentence)
 6 encoded_sentence = RNN(SENT_HIDDEN_SIZE,
 7 return_sequences=False)(encoded_sentence)
 8
 9 question = layers.Input(shape=(query_maxlen,), dtype='int32')
10 encoded_question = layers.Embedding(vocab_size,
     EMBED_HIDDEN_SIZE)(question)
11 encoded_question = Dropout(0.3)(encoded_question)
12 encoded_question = RNN(QUERY_HIDDEN_SIZE,
13 return_sequences=False)(encoded_question)
14
15 merged = layers.concatenate([encoded_sentence,encoded_question
     ])
16 merged = Dropout(0.3)(merged)
17 preds = layers.Dense(vocab_size, activation='softmax')(merged)
```

```
18
19  model = Model([sentence, question], preds)
20  model.compile(optimizer='adam',loss='categorical_crossentropy'
       ,metrics=['accuracy'])
```

We train this model using the extended bAbI training sets with 50-dim embeddings, 100-dim encodings, batch size of 32, and the Adam optimizer for 100 epochs. The performance on tasks QA1, QA2, and QA3 is given in Table 9.3. As seen in the

Table 9.3: Baseline LSTM performance

Task	Test set accuracy
QA1	0.51
QA2	0.31
QA3	0.17

results, the longer the stories, the worse the performance of the LSTM model due to the increased "noise" in the data.

9.4.3.4 End-to-End Memory Network

Memory networks offer the opportunity to store long-term information and thereby improve performance, especially on longer sequences such as task QA3. Memory networks are able to store supporting facts as memory vectors which are queried and used for prediction. In the original form by Weston, the memory vectors are learned via direct supervision with hard attention and supervision is required at each layer of the network. This requires significant effort. To overcome this need, end-to-end memory networks as proposed by Sukhbaatar use soft attention in place of supervision that can be learned during training via backpropagation. This end-to-end architecture takes the following steps:

1. Each story sentence and query are mapped to separate embedding representations.
2. The query embedding is compared with the embedding of each sentence in the memory, and a softmax function is used to generate a probability distribution analogous to a soft attention mechanism.
3. These probabilities are used to select the most relevant sentence in memory using a separate set of sentence embeddings.
4. The resulting vector is concatenated with the query embedding and used as input to an LSTM layer followed by a dense layer with a softmax output.
5. The entire network is trained to minimize the error between the softmax output and the answer.

Note that this is termed a 1-hop or single-layered MemN2N, since we query the memory only once. As described earlier, memory layers can be stacked to improve performance, especially where multiple facts are relevant and necessary to predict the answer. The Keras implementation of the architecture is given below.

```
input_sequence = Input ((story_maxlen ,))
input_encoded_m = Embedding (input_dim=vocab_size ,
                            output_dim=EMBED_HIDDEN_SIZE)(
    input_sequence )
input_encoded_m = Dropout (0.3)(input_encoded_m)

input_encoded_c = Embedding (input_dim=vocab_size ,
                            output_dim=query_maxlen )(
    input_sequence )
input_encoded_c = Dropout (0.3)(input_encoded_c)

question = Input ((query_maxlen ,))
question_encoded = Embedding (input_dim=vocab_size ,
                            output_dim=EMBED_HIDDEN_SIZE ,
                            input_length=query_maxlen )(
    question )
question_encoded = Dropout (0.3)(question_encoded)

match = dot ([input_encoded_m , question_encoded ], axes =(2, 2))
match = Activation ('softmax')(match )

response = add ([match , input_encoded_c ])
response = Permute ((2, 1))(response )

answer = concatenate ([response , question_encoded ])
answer = LSTM(BATCH_SIZE)(answer )
answer = Dropout (0.3)(answer )
answer = Dense (vocab_size )(answer )
answer = Activation ('softmax')(answer )

model = Model ([input_sequence , question ], answer )
model.compile (optimizer='adam', loss='
    sparse_categorical_crossentropy ',
                metrics =['accuracy '])
```

We train this single-layered model using the extended bAbI training sets with 50-dim embeddings, batch size of 32, and the adam optimizer for 100 epochs. The performance on tasks QA1, QA2, and QA3 is given in Table 9.4. In comparison to

Table 9.4: End-to-end memory network performance

Task	Accuracy (20 epochs)	Accuracy (100 epochs)
QA1	0.53	0.92
QA2	0.39	0.35
QA3	0.15	0.21

the baseline LSTM, the MemN2N model did significantly better for all three tasks, and especially for QA1.

9.4.4 Dynamic Memory Network

As discussed earlier, dynamic memory networks take memory networks one step further and encode memories using a GRU layer. An episodic memory layer is the key to dynamic memory networks, with its attention mechanisms for feature generation and scoring. Episodic memory is composed of two nested GRUs, where the inner GRU generates the episodes and the outer GRU generates the memory vector from the sequence of episodes. DMNs follow the following steps:

1. The input story sentences and query are encoded using GRUs and passed to the episodic memory module.
2. Episodes are generated by attending over these encodings to form a memory such that sentence encodings with low attention scores are ignored.
3. Episodes along with previous memory states are used to update the episodic memory.
4. The query and memory states serve as inputs to the GRU within the answer module which is used to predict the output.
5. The entire network is trained to minimize the error between the GRU output and answer.

A TensorFlow implementation of the episodic memory module for a dynamic memory network is provided below. Note that EpisodicMemoryModule depends on a soft attention GRU implementation, which is included in the case study code.

```
class EpisodicMemoryModule(Layer):

    # attention network
    self.l_1 = Dense(units=emb_dim, batch_size=batch_size,
activation='tanh')
    self.l_2 = Dense(units=1, batch_size=batch_size,
activation=None)

    # Episode network
    self.episode_GRU = SoftAttnGRU(units=units,
                                   return_sequences=False,
                                   batch_size=batch_size)

    # Memory generating network
    self.memory_net = Dense(units=units, activation='relu')

    for step in range(self.memory_steps):
        attentions = [tf.squeeze(
            compute_attention(fact, question, memory),
axis=1)
            for i, fact in enumerate(fact_list)]
        attentions = tf.stack(attentions)
        attentions = tf.transpose(attentions)
        attentions = tf.nn.softmax(attentions)
        attentions = tf.expand_dims(attentions, axis=-1)
```

```
24        episode = K.concatenate([facts, attentions], axis
      =2)
25        episode = self.episode_GRU(episode)
26
27        memory = self.memory_net(K.concatenate([memory,
      episode, question], axis=1))
28
29        return K.concatenate([memory, question], axis=1)
```

We train a DMN model using the extended bAbI training sets with 50-dim GloVe embeddings, batch size of 50, 100 hidden units, 3 memory steps, and the adam optimizer for just 20 epochs. The performance on tasks QA1, QA2, and QA3 is given in Table 9.5. In comparison with earlier architectures, we can see that dynamic

Table 9.5: Dynamic memory network performance

Task	Test set accuracy
QA1	1.00
QA2	0.47
QA3	0.29

memory networks perform better than MemN2N and LSTM networks for all three tasks, reaching perfect prediction on task QA1.

9.4.4.1 Differentiable Neural Computer

The differentiable neural computer (DNC) is a neural network with an independent memory bank. It is an embedded neural network controller with a collection of pre-set operations for memory storage and management. As an extension of the neural Turing machine architecture, it allows for scaling of memory without having to scale the rest of the network.

The heart of a DNC is a neural network called a controller, which is analogous to a CPU in a computer. This DNC controller can perform several operations on memory concurrently, including reading and writing to multiple memory locations at once and producing output predictions. As before, the memory is a set of locations that can each store a vector of information. The DNC controller can use soft attention to search memory based on the content of each location, or associative temporal links can be traversed forward or backward to recall sequence information in either direction. Queried information can then be used for prediction.

For a given input at each time step, the DNC controller outputs four vectors:

read vector/s: used by the read head/s to address memory locations
erase vector/s: used to selectively erase items from memory
write vector/s: used by the write heads to store information in memory
output vector: used as a feature for output prediction

For this case study, we will apply the TensorFlow-DNC implementation developed by DeepMind to the bAbI extended datasets. The DNC module for this implementation is given by:

```
DNCState = collections.namedtuple('DNCState', ('access_output'
    ,
                                               'access_state',
                                               '
    controller_state'))
class DNC(snt.RNNCore):
    # modules
    self._controller = snt.LSTM(**controller_config)
    self._access = access.MemoryAccess(**access_config)

    # output
    prev_access_output = prev_state.access_output
    prev_access_state = prev_state.access_state
    prev_controller_state = prev_state.controller_state

    batch_flatten = snt.BatchFlatten()
    controller_input = tf.concat([batch_flatten(inputs),
                                  batch_flatten(
    prev_access_output)], 1)

    controller_output, controller_state = self._controller(
    controller_input, prev_controller_state)

    access_output, access_state = self._access(
    controller_output, prev_access_state)

    output = tf.concat([controller_output, batch_flatten(
    access_output)], 1)
    output = snt.Linear(output_size=self._output_size.as_list
    ()[0],
                        name='output_linear')(output)
```

We train a DNC model using the extended bAbI training sets with 50-dim GloVe embeddings, hidden size of 256, memory size of 256×64, 4 read heads, 1 write head, batch size of 1, and the RMSprop optimizer with gradient clipping for 20,000 iterations. The performance on tasks QA1, QA2, and QA3 is given in Table 9.6. It

Table 9.6: Differentiable neural computer performance

Task	Test set accuracy
QA1	1.00
QA2	0.67
QA3	0.55

may not be surprising to see that the DNC model outperforms all previous models, given the increased complexity. The trade-off between accuracy and training time should be carefully weighed when choosing which architecture is most suitable for the task. For simple tasks, a single LSTM implementation may be all that is required. DNCs with their scalable memory are a better choice when complex knowledge is required for task prediction.

9.4.4.2 Recurrent Entity Network

Recurrent entity networks (EntNets) incorporate a fixed bank of dynamic memory cells that allow simultaneous location and content-based updates. Because of this ability, they perform very well and set the state-of-the-art in reasoning tasks such as bAbI. Unlike the DNC which relies on a sophisticated central controller, EntNet is essentially a set of separate, parallel recurrent memories with independent gates for each memory.

The EntNet architecture consists of an input encoder, a dynamic memory, and an output layer. It operates with the following steps:

1. The input story sentences and query are mapped to embedding representations and passed to the dynamic memory layer and output layer, respectively.
2. Key vectors with the embeddings of entities are generated.
3. The hidden states (memories) of the set of gated GRU blocks within the dynamic memory are updated over the input encoder vectors and key vectors.
4. The output layer applies a softmax over the query q and hidden states of the memory cells to generate a probability distribution over the potential answers.
5. The entire network is trained to minimize the error between the output layer candidate and answer.

The architecture of the dynamic memory cell written in TensorFlow is provided below:

```
class DynamicMemoryCell(tf.contrib.rnn.RNNCell):
    def get_gate(self, state_j, key_j, inputs):
        a = tf.reduce_sum(inputs * state_j, axis=1)
        b = tf.reduce_sum(inputs * key_j, axis=1)
        return tf.sigmoid(a + b)

    def get_candidate(self, state_j, key_j, inputs, U, V, W,
    U_bias):
        key_V = tf.matmul(key_j, V)
        state_U = tf.matmul(state_j, U) + U_bias
        inputs_W = tf.matmul(inputs, W)
        return self._activation(state_U + inputs_W + key_V)

    def __call__(self, inputs, state):
        state = tf.split(state, self._num_blocks, axis=1)
        next_states = []
        for j, state_j in enumerate(state):
            key_j = tf.expand_dims(self._keys[j], axis=0)
            gate_j = self.get_gate(state_j, key_j, inputs)
            candidate_j = self.get_candidate(state_j,
```

```
20                                                        key_j ,
21                                                        inputs ,
22                                                        U, V, W, U_bias )
23            state_j_next = state_j + tf.expand_dims(gate_j ,
        −1) * candidate_j
24            state_j_next_norm = tf.norm(tensor=state_j_next ,
25                                        ord='euclidean ',
26                                        axis=−1,
27                                        keep_dims=True )
28            state_j_next_norm = tf.where(tf.greater(
        state_j_next_norm , 0.0),
29                                        state_j_next_norm ,
30                                        tf.ones_like(
        state_j_next_norm ))
31            state_j_next = state_j_next / state_j_next_norm
32            next_states.append(state_j_next )
33        state_next = tf.concat(next_states , axis=1)
34        return state_next , state_next
```

We train an EntNet using the extended bAbI training set with 100-dim embeddings, 20 blocks, batch size of 32, and the ADAM optimizer with gradient clipping for 200 epochs. The performance on tasks QA1, QA2, and QA3 is given in Table 9.7. The performance of our implementation on bAbI tasks QA1, QA2, and QA3 ex-

Table 9.7: EntNet performance

Task	Test set accuracy
QA1	1.00
QA2	0.97
QA3	0.90

ceeds all previous architectures. Note that with proper hyperparameter tuning, the performance of EntNet and the previous architectures can be improved on the bAbI tasks.

9.4.5 Exercises for Readers and Practitioners

The readers and practitioners can consider extending the case study to the following problems in order to expand their knowledge:

1. Memory and complexity can be limited when using the same embedding matrix for both the encoder and decoder. What would need to change to address this problem?
2. Tune and increase the number of epochs for the baseline LSTM model during training. Does adding dropout help?

3. Add a second and third hop to the end-to-end memory network and see if performance improves on bAbI tasks QA2 and QA3.
4. How does restricting the size of the memory representation affect performance?
5. Is there a significant effect by using a different similarity scoring function instead
   of the softmax within the memory controller of a MemN2N network?
6. Explore the architectures in this case study on bAbI tasks 3-20. Does the simple
   baseline LSTM outperform on certain tasks?

References

[BCB14b]   Dzmitry Bahdanau, Kyunghyun Cho, and Yoshua Bengio. "Neural Ma-
           chine Translation by Jointly Learning to Align and Translate". In:
           CoRR abs/1409.0473 (2014).
[Bah+16b] Dzmitry Bahdanau et al. "End-to-end attention-based large vocabu-
           lary speech recognition". In: *2016 IEEE International Conference on
           Acoustics, Speech and Signal Processing ICASSP 2016, Shanghai,
           China, March 20–25, 2016*. 2016, pp. 4945–4949.
[Cha+16a] William Chan et al. "Listen, attend and spell: A neural network for
           large vocabulary conversational speech recognition". In: *2016 IEEE
           International Conference on Acoustics, Speech and Signal Processing
           ICASSP 2016, Shanghai, China, March 20–25, 2016*. 2016, pp. 4960–
           4964.
[Cho+15b] Jan Chorowski et al. "Attention-Based Models for Speech Recogni-
           tion". In: *Advances in Neural Information Processing Systems 28: An-
           nual Conference on Neural Information Processing Systems 2015, De-
           cember 7–12, 2015, Montreal, Quebec, Canada*. 2015, pp. 577–585.
[Dan+17]   Michal Daniluk et al. "Frustratingly Short Attention Spans in Neural
           Language Modeling". In: *CoRR* abs/1702.04521 (2017).
[DGS92]   Sreerupa Das, C. Lee Giles, and Guo-Zheng Sun. "Using Prior Knowl-
           edge in a {NNPDA} to Learn Context-Free Languages". In: *Advances
           in Neural Information Processing Systems 5, [NIPS Conference, Den-
           ver, Colorado, USA, November 30 - December 3, 1992]*. 1992, pp. 65–
           72.
[Den+12]   M. Denil et al. "Learning where to Attend with Deep Architectures for
           Image Tracking". In: *Neural Computation* (2012).
[GWD14b]   Alex Graves, Greg Wayne, and Ivo Danihelka. "Neural Turing Ma-
           chines". In: *CoRR* abs/1410.5401 (2014).
[Gra+16]   Alex Graves et al. "Hybrid computing using a neural network with
           dynamic external memory". In: *Nature* 538.7626 (Oct. 2016), pp. 471–
           476.
[Gre+15]   Edward Grefenstette et al. "Learning to Transduce with Unbounded
           Memory". In: *Advances in Neural Information Processing Systems 28:
           Annual Conference on Neural Information Processing Systems 2015,*

December 7–12, 2015, Montreal, Quebec, Canada. 2015, pp. 1828–1836.

[Hen+16]   Mikael Henaff et al. "Tracking the World State with Recurrent Entity Networks". In: *CoRR* abs/1612.03969 (2016).

[Kum+16]   Ankit Kumar et al. "Ask Me Anything: Dynamic Memory Networks for Natural Language Processing". In: *Proceedings of the 33nd International Conference on Machine Learning, ICML 2016, New York City, NY, USA, June 19–24, 2016*. 2016, pp. 1378–1387.

[LH10]   Hugo Larochelle and Geoffrey E Hinton. "Learning to combine foveal glimpses with a third-order Boltzmann machine". In: *Advances in Neural Information Processing Systems 23*. Ed. by J. D. Lafferty et al. Curran Associates, Inc., 2010, pp. 1243–1251.

[Lin+17]   Zhouhan Lin et al. "A Structured Self-attentive Sentence Embedding". In: *CoRR* abs/1703.03130 (2017).

[LPM15]   Minh-Thang Luong, Hieu Pham, and Christopher D. Manning. "Effective Approaches to Attention-based Neural Machine Translation". In: *CoRR* abs/1508.04025 (2015).

[Moz94]   Michael C. Mozer. "Neural Net Architectures for Temporal Sequence Processing". In: Addison-Wesley, 1994, pp. 243–264.

[RCW15]   Alexander M. Rush, Sumit Chopra, and Jason Weston. "A Neural Attention Model for Abstractive Sentence Summarization". In: *Proceedings of the 2015 Conference on Empirical Methods in Natural Language Processing, EMNLP 2015, Lisbon, Portugal, September 17–21, 2015*. 2015, pp. 379–389.

[SP63]   Karl Steinbuch and Uwe A. W. Piske. "Learning Matrices and Their Applications". In: *IEEE Trans. Electronic Computers* 12.6 (1963), pp. 846–862.

[Suk+15]   Sainbayar Sukhbaatar et al. "End-To-End Memory Networks". In: *Advances in Neural Information Processing Systems 28: Annual Conference on Neural Information Processing Systems 2015, December 7–12, 2015, Montreal, Quebec, Canada*. 2015, pp. 2440–2448.

[Vas+17c]   Ashish Vaswani et al. "Attention is All you Need". In: *Advances in Neural Information Processing Systems 30: Annual Conference on Neural Information Processing Systems 2017, 4–9 December 2017, Long Beach, CA, USA*. 2017, pp. 6000–6010.

[Vin+15a]   Oriol Vinyals et al. "Grammar as a Foreign Language". In: *Advances in Neural Information Processing Systems 28: Annual Conference on Neural Information Processing Systems 2015, December 7–12, 2015, Montreal, Quebec, Canada*. 2015, pp. 2773–2781.

[Wan+16b]   Yequan Wang et al. "Attention-based LSTM for Aspect-level Sentiment Classification". In: *Proceedings of the 2016 Conference on Empirical Methods in Natural Language Processing, EMNLP 2016, Austin, Texas, USA, November 1–4, 2016*. 2016, pp. 606–615.

[WCB14]   Jason Weston, Sumit Chopra, and Antoine Bordes. "Memory Networks". In: *CoRR* abs/1410.3916 (2014).

[Yan+16]   Zichao Yang et al. "Hierarchical Attention Networks for Document Classification". In: *NAACL HLT 2016, The 2016 Conference of the North American Chapter of the Association for Computational Linguistics: Human Language Technologies, San Diego California, USA, June 12–17, 2016*. 2016, pp. 1480–1489.

[Zha+18]   Yuanyuan Zhang et al. "Attention Based Fully Convolutional Network for Speech Emotion Recognition". In: *CoRR* abs/1806.01506 (2018).

Chapter 10
Transfer Learning: Scenarios, Self-Taught Learning, and Multitask Learning

10.1 Introduction

Most supervised machine learning techniques, such as classification, rely on some underlying assumptions, such as: (a) the data distributions during training and prediction time are similar; (b) the label space during training and prediction time are similar; and (c) the feature space between the training and prediction time remains the same. In many real-world scenarios, these assumptions do not hold due to the changing nature of the data.

There are many techniques in machine learning to address these problems, such as incremental learning, continuous learning, cost-sensitive learning, semi-supervised learning, and more. In this chapter, we will focus mainly on transfer learning and related techniques to address these issues.

DARPA defines transfer learning as the ability of the system to learn and apply knowledge from previous tasks to new tasks [Dar05]. This research gave rise to many successes in various domains for 7–10 years using mostly traditional machine learning algorithms with transfer learning as the focus. This research impacted various domains, such as wireless telecommunications, computer vision, text mining, and many others [Fun+06, DM06, Dai+07b, Dai+07a, TS07, Rai+07, JZ07, BBS07, Pan+08, WSZ08].

As the deep learning field is evolving rapidly, the main focus these days is on unsupervised and transfer learning. We can classify transfer learning into various sub-fields, such as self-taught learning, multitask learning, domain adaptation, zero-shot learning, one-shot learning, few-shot learning, and more. In this chapter, we will first go over the definitions and fundamental scenarios of transfer learning. We will cover the techniques involved in self-taught learning and multitask learning. In the end, we will carry out a detailed case study with multitask learning using NLP tasks to get hands-on experience on the various concepts and methods related in this chapter.

© Springer Nature Switzerland AG 2019
U. Kamath et al., *Deep Learning for NLP and Speech Recognition*,
https://doi.org/10.1007/978-3-030-14596-5_10

10.2 Transfer Learning: Definition, Scenarios, and Categorization

As shown in Fig. 10.1, in traditional machine learning, different models need to be learned for different sources (data and labels). Figure 10.1 shows that for a source (task or domain) with training data and labels the system learns models (model A and model B) that are only effective on targets (task or domain) that are similar to the source, respectively, learned by each model. In most cases, the model learned for a specific source cannot be used for predicting on a target that is different. If there is a model which requires a large number of training data, then the effort of collecting data, labeling the data, training the models, and validating the models has to be done per source. This effort becomes unwieldy with a large number of systems from a cost and resource perspective.

Figure 10.2 shows a general transfer learning system which can extract **knowledge** from the source system or the model and **transfers** it in some way so that it can be useful on a target. This model A trained for a task using training data for source A can be used to extract knowledge and transfer it to another target task.

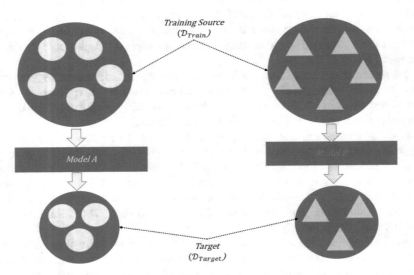

Fig. 10.1: Traditional machine learning system on two different sources and target

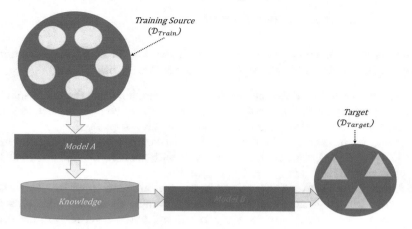

Fig. 10.2: Transfer learning system on different source and target

10.2.1 Definition

In order to define transfer learning precisely, we will first define a couple of concepts as given by Pan and Yang, i.e., domains and tasks [PY10]. A **domain** $\mathcal{D} = (\mathcal{X}, P(X))$ is defined in terms of (a) the feature space $\mathcal{X}$ and (b) the marginal probability distribution $P(X)$, where X represents the training data samples $X = x_1, x_2 \ldots x_n \in \mathcal{X}$. For example, in the task of sentiment analysis with binary classification, the $\mathcal{X}$ corresponds to a bag-of-words representation and x_i corresponds to the ith term in the corpus. Thus, when either the feature spaces or the marginal probability distribution are different for two systems, we say that the domains do not match.

A **task** $\mathcal{T} = (\mathcal{Y}, f(\cdot))$ is defined in terms of (a) a label space $\mathcal{Y}$ and (b) an objective prediction function $f(\cdot)$ that is not directly observed but learned from the input and label pairs (x_i, y_i). The label space consists of a set of all actual labels, for example, *true* and *false* for binary classification. The objective prediction function $f(\cdot)$ is used to predict the label given the data and can be interpreted in probabilistic view as $f(\cdot) \approx p(y|x)$.

Given a source domain $\mathcal{D}_S$, source task $\mathcal{T}_S$, target domain $\mathcal{D}_T$, and target task $\mathcal{T}_T$, **transfer learning** can be defined as the process of learning the target predictive function $f_T(\cdot) = P(Y_T|X_T)$ in the target domain $\mathcal{D}_T$ using the knowledge from the source domain $\mathcal{D}_S$ and the source task $\mathcal{T}_S$, such that $\mathcal{D}_S \neq \mathcal{D}_T$ or $\mathcal{T}_S \neq \mathcal{T}_T$.

10.2.2 Transfer Learning Scenarios

Based on the different components of domain and task for both source and target, there are four different transfer learning scenarios that are listed below:

1. Feature spaces are different, $\mathcal{X}_S \neq \mathcal{X}_T$. An example of this for sentiment classification would be that the features are defined for two different languages. In NLP, this term is often referred to as cross-lingual adaptation.
2. Marginal probability distributions between source and target are different, $P(X_S) \neq P(X_T)$, for example, a chat text with short forms and an email text with formal language both discussing sentiments.
3. Label spaces between source and target are different, $\mathcal{Y}_S \neq \mathcal{Y}_T$. This really means that the source and target tasks are completely different, for example, one can have labels corresponding to sentiments (positive, neutral, negative), and the other corresponding to emotions (angry, sad, happy).
4. The predictive function or conditional probability distributions are different, $P(Y_S|X_S) \neq P(Y_T|X_T)$. An example of this is how the distribution in one can be balanced, and in the other completely skewed or highly imbalanced; the source has equal cases of positive and negative sentiments, but the target has very few positives as compared to negatives.

10.2.3 Transfer Learning Categories

Based on "how to transfer" and "what to transfer" between the source and target, transfer learning can be further categorized into many different types of which many have become an independent field for research and applications. In this section, we will not cover many traditional machine learning based categorizations already defined in the survey by Pan and Yang [PY10]. Instead, we will cover only those categories that have been explored or made an impact in the deep learning area.

Based on the label availability and task similarities between the source and target, there can be various sub-categories of transfer learning, as shown in Fig. 10.3.

When the source labels are unavailable, but a large volume of source data exists and few to large numbers of target data exist, then the category of learning is known as **self-taught learning**. Many real-world applications in speech and text, where the cost or the effort of labeling poses constraints, and the large volume of data can be used to learn and transfer to specific tasks with labels, this technique has been very successful. Employing some form of unsupervised learning on the source to capture features that can help transfer knowledge to the target is the core assumption made in these learning systems.

When the goal is not only to do well on the target tasks but somehow learn jointly and do well in both source and target, where the tasks are slightly different, the form of transfer learning is called **multitask learning**. The core assumption made is that sharing information among related tasks, which should have some similarities, improves the overall generalization.

Related to multitask learning, where the tasks between source and target differ, **domain adaptation** is a form of learning where the domains (i.e., either the feature space or marginal distribution in data) are different between source and the target. The core principle is to learn a domain-invariant representation from the source that can be transferred to the target with a different domain in an effective manner.

In domain adaptation, the domains differ and small to large labeled data is available in source. Domain adaptation can be **zero-shot**, **one-shot**, and **few-shot** learning, based on the available number of labeled data $(0, 1, n)$.

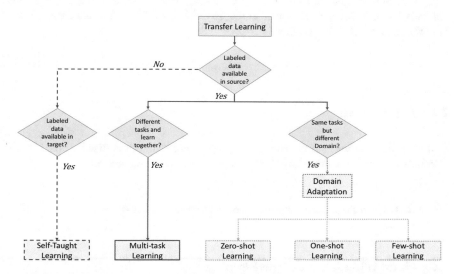

Fig. 10.3: Transfer learning categories based on labeled data, tasks, and domains for source and target

10.3 Self-Taught Learning

Self-taught learning, as shown in Fig. 10.4, consists of two distinct steps: (a) learning features in an unsupervised manner from the unlabeled source dataset and (b) tuning these learned features with a classifier on the target dataset which has labels.

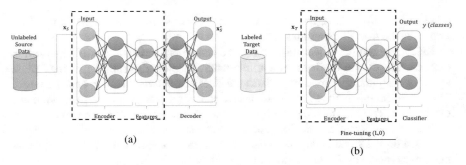

Fig. 10.4: Self-taught learning using pre-training and fine-tuning steps. (**a**) Using unlabeled source dataset to learn features. (**b**) Using the labeled target data to fine-tune the features with a classifier

10.3.1 Techniques

In this section we will summarize various approaches and then discuss specific algorithms or techniques that have been successful in NLP and speech.

10.3.1.1 Unsupervised Pre-training and Supervised Fine-Tuning

---

Algorithm 1: Unsupervised feature learning

Data: Training Dataset $\mathbf{x}_{1(S)}, \mathbf{x}_{2(S)}, \cdots, \mathbf{x}_{n(S)}$ such that $\mathbf{x}_{i(S)} \in \mathbb{R}^d$, layers= L
Result: Weight matrix $\mathbf{W}_l \in \mathbb{R}^d$ and $b_l \in \mathbb{R}$ for each layer l
begin
    appendClassifierLayer(h_L)
    for $l = k\ to\ L$ **do**
        $\mathbf{W}_l, b_l = trainUnsupervised((\mathbf{x}_{1(S)}, \mathbf{x}_{2(S)}..\mathbf{x}_{n(S)}))$
    return $\mathbf{W}_l, b_l$ for each layer l

The input to Algorithm 1 is the source unlabeled dataset of size n; the (S) in the subscript denotes the source. The first part of learning proceeds in an unsupervised way from the source, as shown in Algorithm 1. This has many similarities to feature or dimensionality reduction and manifold learning in traditional machine learning. This process generally employs linear and non-linear techniques to find a latent representation of the input that has a smaller dimension than the input. In deep learning, the *train* in the algorithm above corresponds to many unsupervised techniques such as PCA or ICA layers, restricted Boltzmann machines, autoencoders, sparse autoencoders, denoising autoencoders, contractive autoencoders, and sparse coding techniques, to name a few that can be used for feature learning. Training can be done per layer or over all layers, based on the algorithm. The function R corresponds to the general call to the underlying algorithm.

Autoencoders are the most popular technique among unsupervised learning approaches; basic encoding and decoding happens between the layers to match the input. The number of neurons or the layer size can play an important role in the autoencoder learning. When the size is smaller than the input, it is called **undercomplete representation** and can be seen as a compression mechanism to find the representation in a lower dimension. When the size is greater than input, it is called **overcomplete representation** and requires regularization techniques such as sparsity to enforce learning important features. In many practical applications, autoencoders are stacked together to create hierarchical or advanced features from the inputs.

Once these features are learned, the next step is to use the target dataset to **fine-tune** them with a classifier layer such as **softmax**. There are various choices, such as freezing the state of the learned layers at some level $k > 1$ and only using the rest of the layers for tuning or using all the layers for tuning. Algorithm 2 shows how the fine-tuning process uses the labeled target dataset of size m.

Algorithm 2: SupervisedFineTuning

Data: Training Dataset $(\mathbf{x}_{1(T)}, y_2), (\mathbf{x}_{2(T)}, y_2), ..(\mathbf{x}_{m(T)}, y_n)$ such that $\mathbf{x}_{i(T)} \in \mathbb{R}^d$ and
$\quad\quad y_i \in \{+1, -1\}$, Trained Layers $h_1, h_2, \cdots, h_L$, Training layer start k
Result: Weight matrix $\mathbf{W}_l \in \mathbb{R}^d$ and $b_l \in \mathbb{R}$ for each layer l
begin
$\quad$ appendClassifierLayer(h_{L+1})
$\quad$ **for** $l = k$ *to* L **do**
$\quad\quad$ $\mathbf{W}_l, b_l = train((\mathbf{x}_{1(T)}, y_1), (\mathbf{x}_{2(T)}, y_2), ..(\mathbf{x}_{m(T)}, y_n))$
$\quad$ **return** $\mathbf{W}_l, b_l$ for each layer l

10.3.2 Theory

In their seminal work, Erhan et al. give interesting theoretical and empirical insights around unsupervised pre-training and fine-tuning [Erh+10]. They use various architectures such as feed-forward neural networks, deep belief networks, and stacked denoising autoencoders on different datasets to empirically verify the various theoretical conclusions in a step-by-step, controlled manner.

They show that pre-training not only gives a good starting condition but captures complex dependencies among the parameters as well. The research also shows that unsupervised pre-training can be a form of regularization that guides the weights towards a better basin of attraction of minima. The regularization obtained from the pre-training process influences the starting point in supervised learning, and the effect does not disappear with more data in comparison with standard regularization techniques such as L_1/L_2. Unsurprisingly, the research concludes that in small training data settings, unsupervised pre-training has many advantages. It also shows that, in some cases, the order of training examples impacts the results, but pre-training reduces the variance even in such cases. The experiments and results that show unsupervised pre-training as a general variance reducing technique and even optimization technique for better training are insightful.

10.3.3 Applications in NLP

Using unsupervised techniques for word embeddings from a large corpus of data and employing it for various supervised tasks has been the most basic application in NLP. Since this was discussed at length in Chap. 5, we will focus more on other NLP tasks. Dai and Le show that unsupervised feature learning using sequence autoencoders or language model-based systems and then using supervised training achieves great results in text classification tasks on various datasets, such as IMDB, DBpedia, and 20 Newsgroup [DL15]. The sequence autoencoder uses an LSTM encoder–decoder to capture the dependencies in an unsupervised manner. The weights from the LSTM are used to initialize the LSTM with softmax classifier in a supervised setting. The unsupervised autoencoder training shows superior results across all datasets, and the generality of the technique gives it an edge for all sequence-to-sequence problems.

Ramachandran et al. show that the LSTM encoder pre-trained for language modeling can be used very effectively without fine-tuning in sentiment classification [RLL17]. Deing et al. show that TopicRNN, an architecture using RNN for local syntactic dependencies and topic modeling for global semantic latent representations, can be a very effective feature extractor [Die+16]. TopicRNN achieves nearly state-of-the-art results on sentiment classification task. Turian et al. show that learning features in an unsupervised manner from multiple embeddings and applying it to various supervised NLP tasks, such as chunking and NER, can give nearly state-of-the-art results [TRB10].

10.3.4 Applications in Speech

Very early, Dahl et al. showed in their research that unsupervised pre-training gives a great initialization for the weights, and using labeled fine-tuning on deep belief networks further improves results in automatic speech recognition task [Dah+12]. Hinton et al. show unsupervised pre-training for learning layer by layer in RBMs and then fine-tuning with labeled examples not only reduces overfitting but reduces time to learn on labeled examples [Hin+12]. Lee et al. show that unsupervised feature learning done on large dataset can learn phonemes that can help various audio classification tasks using deep convolutional networks [Lee+09].

10.4 Multitask Learning

Whether specifically in deep learning or generically in machine learning, the overall process is to learn a model for a task at hand given the dataset corresponding to that task. This can be seen as **single task learning**. An extension of this is **multitask learning (MTL)**, where one tries to learn jointly from multiple tasks and their corresponding datasets [Rud17]. Caruana defines the goal of multitask learning as "MTL improves generalization by leveraging the domain-specific information contained in the training signals of related tasks." Multitask learning can also be referred to as **inductive transfer** process. The inductive bias introduced in MTL is through forcing the model to prefer hypothesis which explains multiple tasks rather than a single task. Multitask learning has been generally effective when there is limited labeled data for each task, and there is an overlap between knowledge or learned features between the tasks.

10.4.1 Techniques

The two general ways of handling multitask learning in deep learning are through **hard or soft parameter sharing** as shown in Fig. 10.5. Hard parameter sharing is one of the oldest techniques in NNs with a single model, where the hidden layers share the common weights, and task-specific weights are learned at the output layers [Car93]. The most important benefit of hard parameter sharing is the prevention of overfitting by enforcing more generalization across tasks. Soft parameter sharing, on the other hand, has individual models with separate parameters per tasks, and a constraint is put to make the parameters across tasks more similar. Regularization techniques are often used in soft parameter sharing for enforcing the constraints. In the next section, we will go through selected deep learning networks that have proven useful for multitask learning.

10.4.1.1 Multilinear Relationship Network

One of the earliest deep learning networks for multitask learning was introduced by Long and Wang, and was known as the multilinear relationship network (MRN) [LW15]. MRN showed state-of-the-art performance in different tasks in image recognition. The MRN, as shown in Fig. 10.6, is a modification of the AlexNet architecture that was discussed in Chap. 6. The first few layers are convolutional, and a fully connected layer learns the transferable features, while the rest of the fully connected layers closer to the output learn task-specific features. If there are T tasks with training data $\mathcal{X}_t, \mathcal{Y}_{t\,t=1}^{T}$, where $\mathcal{X}_t = \mathbf{x}_1^t, \cdots, \mathbf{x}_N^t$ and $\mathcal{Y}_t = \mathbf{y}_1^t, \cdots, \mathbf{y}_N^t$, N_t number of training examples and labels of the tth task with D-dimensional feature space and C-cardinality label space, network parameters of t task in the lth layer are

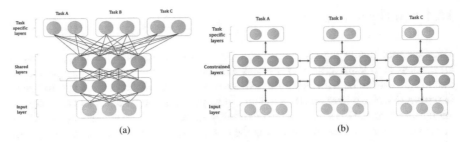

Fig. 10.5: Two generic methods of multitask learning. (**a**) Hard parameter sharing in the hidden layers. (**b**) Soft parameter sharing across the hidden layers for various tasks

given by $\mathbf{W}^{t,l} \in \mathbb{R}^{D_1^l \times D_2^l}$, where D_1^l and D_2^l are the dimensions of matrix $\mathbf{W}^{t,l}$ and parameter tensor $\mathcal{W}^l = [\mathbf{W}^{1,l}; \cdots; \mathbf{W}^{T,l}] \in \mathbb{R}^{D_1^l \times D_2^l \times T}$. The fully connected layers $(fc6 - fc8)$ learn the mappings given by $\mathbf{h}_n^{t,l} = a^l(\mathbf{W}^{t,l}\mathbf{h}_n^{t,l-1} + \mathbf{b}^{t,l})$, where $\mathbf{h}_n^{t,l}$ is the hidden representation for each data instance $\mathbf{x}_n^t$, $\mathbf{W}^{t,l}$ is the weight, $\mathbf{b}^{t,l}$ is the bias, and a^l is the activation function, such as ReLU. The classifier of tth task is given by $\mathbf{y} = f_t(\mathbf{x})$, and the empirical error is given by:

$$\min \sum_{n=1}^{N_t} J(f_t(\mathbf{x}_n^t), \mathbf{y}_n^t) \tag{10.1}$$

where $J(\cdot)$ is the cross-entropy loss function and $f_t(\mathbf{x}_n^t)$ is the conditional probability that the network assigns for the data point $\mathbf{x}_n^t$ to the label $\mathbf{y}_n^t$. MRN has tensor normal priors over the parameter tensors in the fully connected task-specific layers similar to Bayesian models that acted as regularization on task related learning.

The maximum a posteriori (MAP) estimation of network parameters $\mathcal{W} = \mathcal{W}^l : l \in \mathcal{L}$ for task-specific layers $\mathcal{L} = fc7, fc8$ given the training data is:

$$P(\mathcal{W}|\mathcal{X}, \mathcal{Y}) \propto P(\mathcal{W}) \cdot P(\mathcal{Y}|\mathcal{X}, \mathcal{W}) \tag{10.2}$$

$$P(\mathcal{W}|\mathcal{X}, \mathcal{Y}) = \prod_{l \in \mathcal{L}} P(\mathcal{W}^l) \cdot \prod_{t=1}^{T} \prod_{n=1}^{N_t} P(\mathbf{y}_n^t | \mathbf{x}_n^t, \mathcal{W}^l) \tag{10.3}$$

with assumptions made that the prior $P(\mathcal{W}^l)$ and the parameter tensors $\mathcal{W}^l$ for each layer are independent of the other layers.

The maximum likelihood estimation (MLE) part $P(\mathcal{Y}|\mathcal{X}, \mathcal{W})$ is modeled to learn the transferable features in the lower layers, and all the parameters for the layers $(conv1 - fc6)$ are shared. The task-specific layers $(fc7, fc8)$ are not shared to avoid negative transfer. The prior part $p(\mathcal{W})$ is defined as the **tensor normal distribution** and is given as:

$$p(\mathcal{W}) = \mathcal{TN}_{D_1^l \times D_2^l \times T}(\mathbf{O}, \Sigma_1^l, \Sigma_2^l, \Sigma_3^l) \tag{10.4}$$

where Σ_1^l, Σ_2^l, and Σ_3^l are the modes of covariance matrices. In the tensor prior, the row covariance matrix $\Sigma_1^l \in \mathbb{R}^{D_1^l \times D_1^l}$ learns the relationships between the features, the column covariance matrix $\Sigma_2^l \in \mathbb{R}^{D_2^l \times D_2^l}$ learns the relationship between classes, and the covariance matrix $\Sigma_3^l \in \mathbb{R}^{T \times T}$ learns the relationships between tasks in the lth layer parameters $\mathcal{W}^l = \mathbf{W}^{1,l}; \cdots ; \mathbf{W}^{T,l}$. The empirical error given in Eq. 10.1 is integrated with the prior given in Eq. 10.4 into the MAP estimation given in Eq. 10.3 and following the process of taking a negative logarithm, the equation to optimize is:

$$\min_{f_t|_{t=1}^T, \Sigma_k^l|_{k=1}^K} \sum_{t=1}^T \sum_{n=1}^{N_t} J(f_t(\mathbf{x}_n^t), \mathbf{y}_n^t)$$

$$+ \frac{1}{2} \sum_{l \in \mathcal{L}} \left(vec(\mathcal{W}^l)^T (\Sigma_{1:K}^l)^{-1} vec(\mathcal{W}^l) - \sum_{k=1}^K \frac{D^l}{D_k^l} \ln(|\Sigma_k^l|) \right) \quad (10.5)$$

where $D^l = \prod_{k=1}^K D_k^l$ and $K = 3$ is the number of modes in parameter tensor $\mathcal{W}$ or $K = 4$ for the convolutional layers and $\Sigma_{1:3}^l = \Sigma_1^l \otimes \Sigma_2^l \otimes \Sigma_3^l$ is the Kronecker product of feature, class, and task covariances. The optimization problem given in Eq. 10.5 is jointly non-convex with respect to parameter tensors and the covariance matrix, and hence one set of variables is optimized while keeping the rest of them fixed. The experiments with MRN on different computer vision multitask learning datasets show that it can achieve state-of-the-art performance.

10.4.1.2 Fully Adaptive Feature Sharing Network

Lu et al. take the approach of task-specific learning as a search, starting from a thin network and then branching out in a principled way to form wide networks during the training process [Lu+16]. The approach also introduces a new technique, simultaneous orthogonal matching pursuit (SOMP), for initializing a thin network from a wider pre-trained network for faster convergence and improved accuracy. The methodology has three phases:

1. Thin Model Initialization: Since the network (thin) is of different dimension than the pre-trained network, the weights cannot be copied. As a result, it uses SOMP for learning how to select the subset of rows d' from the original rows d for every layer l. This is a non-convex optimization problem, and hence a greedy approach is used for solving it, described in detail in the paper.
2. Adaptive Model Widening: After the initialization process, each layer starting from the top layer goes through a widening process. The widening process can be defined as creating sub-branches in the network, so that each branch does a subset of tasks performed by the network. A point where it branches is called a junction, and it is widened by having more output layers. Figure 10.7 shows the iterative widening process. If there are T tasks, the final output layer l of the thin network has a junction with T branches and each can be considered as a sub-branch. The iterative process starts with finding t branches by grouping

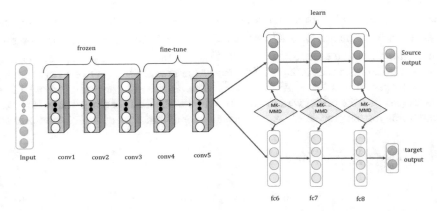

Fig. 10.6: A multilinear relationship network, in which the first few layers learn the shared features, and the final layers learn task-specific features with tensor normal priors

things such that $t \leq T$ at the layer l and then recursively move in a top-down manner to the next layer $l - 1$ and so on. The grouping of the tasks is done by associating a concept of "affinity" which is the probability of concurrently observing simple or difficult examples from the training data for the pair of tasks.

3. Final Model Training: The last step is to train the final model after the thin model initialization and the recursive widening process.

10.4.1.3 Cross-Stitch Networks

As shown in Fig. 10.8, these deep networks are modifications of AlexNet, where shared and task-specific representations are learned using linear combinations [Mis+16]. For each task, there is a deep network such as AlexNet and cross-stitch units have a connection between pooling layers as input to either convolution or fully connected ones. The cross-stitch units are linear combinations between the task outputs to learn the shared representation. They were shown to be very effective in a data-starved multitask setting.

Consider two tasks A and B and a multitask learning on the same input data. A cross-stitch unit shown in Fig. 10.9 plays the role of combining two networks into a multitask network, such that the tasks control the amount of sharing. Given

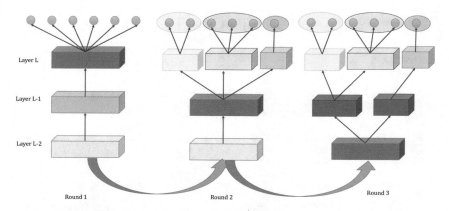

Fig. 10.7: An iterative process showing how the network is widened at a layer on a specific iteration to group the tasks

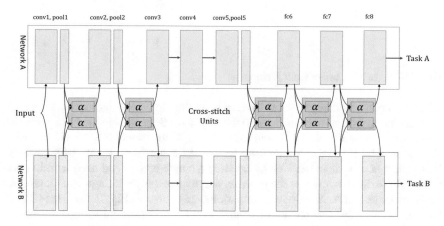

Fig. 10.8: A cross-stitch network trying to learn a latent representation that is useful for two tasks

two outputs of activations x_A, x_B from a layer l, a linear combination is learned to produce outputs $\tilde{x}_A, \tilde{x}_B$ using parameters α, which flows into the next layers and for a location (i, j) is given by:

$$\begin{bmatrix} \tilde{x}_A^{i,j} \\ \tilde{x}_B^{i,j} \end{bmatrix} = \begin{bmatrix} \alpha_{AA} & \alpha_{AB} \\ \alpha_{BA} & \alpha_{BB} \end{bmatrix} \begin{bmatrix} x_A^{i,j} \\ x_B^{i,j} \end{bmatrix} \tag{10.6}$$

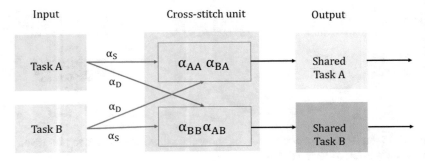

Fig. 10.9: Cross-stitch unit

10.4.1.4 A Joint Many-Task Network

NLP tasks generally can be considered to be in the pipeline of hierarchy, where one task can be useful and used as input to the next task. Søgaard and Golberg show that supervised multitasking at different layers using bidirectional RNN architecture so that the low-level tasks feed into high-level tasks can achieve great results [SG16]. Hashimoto et al. extend the idea by creating a single end-to-end deep learning network, where the network has the growing depth to accomplish linguistic hierarchies from syntactic and semantic representations, as shown in Fig. 10.10 [Has+16]. It has been shown that a single end-to-end network with this architecture can achieve state-of-the-art results in different tasks such as chunking, dependency parsing, semantic relatedness, and textual entailment.

A given sentence s of length l has w_t words. For each word, there is skip-gram word embedding and character embedding. The word representation x_i is done by concatenating both word and n-gram character embeddings which are learned using skip-gram with negative sampling for words. The character n-grams are used to give morphological features for the tasks. The first task is of POS tagging and is performed using bidirectional LSTM with embedded inputs and softmax for classifying the tags. The POS tags are learnable embeddings which is used in the next chunking layer. The label embedding for POS tagging (and many other tasks) is given by:

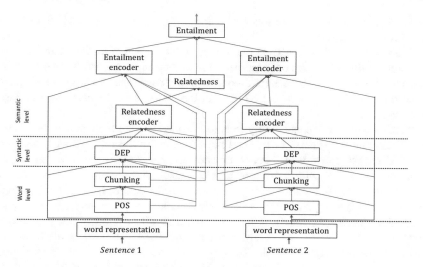

Fig. 10.10: A joint multitask network

$$y_t^{pos} = \sum_{j=1}^{C} p(y_t^1 = j|h_t^1)l(j) \tag{10.7}$$

where C is the number of POS tags, $p()$ is the probability that jth POS tag is assigned to the wth token, and l is the label embedding for jth POS tag. The second task is chunking, which uses bidirectional LSTM and takes the hidden state from POS bidirectional LSTM, the hidden state of its LSTM, embedded token and label embedding from POS tagging. The third task is dependency parsing with inputs from hidden states from chunking layer, previous hidden state from dependency parsing, embedded token and label embeddings of POS layer and the chunking layer. POS tagging layer and the chunking layer with hidden states are useful in generating low-level features that are useful for many tasks as known from traditional feature engineering in NLP. The fourth task is dependency parsing, again using bidirectional LSTM with inputs as hidden LSTM states, embedded tokens, and label embedding from POS tagging and chunking layer. The next two tasks are semantically related as compared to syntactic tasks in the previous layers. The semantic relatedness task is to compare two sentences and give a real-valued output for a measure of their relatedness. The sentence level representation is obtained via max pooling of the hidden states of the LSTM and is given by:

$$h_s^{relat} = \max(h_1^{relat}, h_2^{relat}, \cdots, h_L^{relat}) \tag{10.8}$$

The relatedness for two sentences (s, s') is given by:

$$d_1(s, s') = [|h_s^{relat} - h_{s'}^{relat}|; h_s^{relat} \odot h_{s'}^{relat}] \tag{10.9}$$

The values of $d_1(s,s')$ are given to softmax layer with maxout hidden layer to give a relatedness score.

The last task is that of textual entailment, which again takes two sentences and gives one of the categories of entailment, contradiction, or neutrality. The label embeddings from the relatedness task along with distance measure similar to Eq. 10.9 in the relatedness derived from LSTM layer feed into a softmax classifier for classification.

When a network is sequentially trained for one task and then trained on another task, it generally "forgets" or has bad performance on the first. This phenomenon is called **catastrophic interference** or **catastrophic forgetting**. The training for each layer is similar with a loss function that takes into account (a) a measure for classification loss for the layer using predictions and label, (b) L2-norm of its weight vectors, and (c) a regularization term for parameters of previous tasks if they are inputs. Joint learning gives the framework robustness from catastrophic interference according to the authors. An example is for chunking layer, with inputs from POS tagging given by weights and bias θ_{POS} and one after POS layer with current epoch given by θ'_{POS}, weights of chunking layer W_{CHK}, and probability $p(y_t^{CHK} = \alpha | h_t^{CHK})$ of assigning correct label α to w_t in the sentence is:

$$J_2(\theta_{CHK}) = -\sum_s \sum_t \log p(y_t^{CHK} = \alpha | h_t^{CHK}) + \lambda \|W_{CHK}\|^2 + \delta \|\theta_{POS} - \theta'_{POS}\|^2$$

(10.10)

10.4.1.5 Sluice Networks

Ruder et al. recently proposed a general deep learning architecture, known as sluice networks, that combines concepts from many previous types of research such as hard parameter sharing, cross-stitch networks, block-sparse regularization, and NLP linguistic hierarchical multitask learning [Rud17]. The sluice network for the main task A and an auxiliary task B consists of the shared input layer, three hidden layers per task, and two task-specific output layers as shown in Fig. 10.11. Each hidden layer for the task is an RNN divided into two subspaces, for example, task A and layer 1 has $G_{A,1,1}$ and $G_{A,1,2}$, which allows them to learn task-specific and shared representations effectively. The output of hidden layers flows through α parameters to the new layer, which carries out linear combinations of the inputs to weigh the importance of sharing and task-specific learning. By making the subspaces each have their weights and controlling how they share, sluice networks have an adaptive way of learning in multitask settings only things that are useful. The final recurrent hidden layers pass the information to β parameters which try to combine all the things the layers have learned. Ruder et al. empirically show how main tasks, such as NER and SRL, can benefit from auxiliary tasks such as POS and improve on errors by a significant value.

Ruder et al. cast the entire learning as a **matrix regularization problem**. If there are M tasks that are loosely related with M non-overlapping datasets

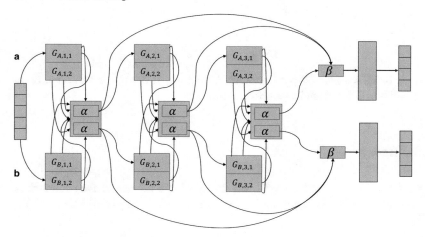

Fig. 10.11: Sluice networks for multitask learning across loosely connected tasks

$\mathcal{D}_1, \mathcal{D}_2, \cdots, \mathcal{D}_M$, K layers given by $L_1, L_2, \cdots, L_K$, and models $\theta_1, \theta_2, \cdots, \theta_M$ each with D parameters and an explicit inductive bias Ω as penalty, then the loss function to minimize is given by:

$$\lambda_1 \mathcal{L}_1(\mathbf{f}(x; \theta_1), y_1) + \cdots + \lambda_M \mathcal{L}_M(\mathbf{f}(x; \theta_M), y_M) + \Omega \qquad (10.11)$$

The loss functions $\mathcal{L}_i$ are cross-entropy loss functions, and the weights λ_i determine the importance of the task i during the training. If $G_{m,k,1}$ and $G_{m,k,2}$ are the two subspaces for each layer, the inductive bias is given by the orthogonality constraints:

$$\Omega = \sum_{m=1}^{M} \sum_{k=1}^{K} \left\| G_{m,k,1}{}^T G_{m,k,2} \right\|_F^2 \qquad (10.12)$$

The matrix regularization is carried out by updating the α parameters with similarity to Misra et al.'s cross-stitch units [Mis+16]. For the two tasks (A, B) and k layers for one subspace, the extension to cross-stitch linear combination looks like:

$$\begin{bmatrix} \tilde{h}_{A_{1,k}} \\ \vdots \\ \tilde{h}_{B_{1,k}} \end{bmatrix} = \begin{bmatrix} \alpha_{A_1 A_1} & \cdots & \alpha_{A_1 B_2} \\ \vdots & \ddots & \vdots \\ \alpha_{A_1 B_2} & \cdots & \alpha_{B_2 B_2} \end{bmatrix} \begin{bmatrix} h_{A_{1,k}} \\ \vdots \\ h_{B_{1,k}} \end{bmatrix} \qquad (10.13)$$

where $h_{A_{1,k}}$ is the output of first subspace for task A in the layer k, and $\tilde{h}_{A_{1,k}}$ is the linear combination of that first subspace and task A. The input to layer $k+1$ is the concatenation of the two, given as $h_{A,k} = [\tilde{h}_{A_{1,k}}, \tilde{h}_{A_{2,k}}]$. The hierarchical relationship between the low-level tasks and the high-level tasks is learned using the **skip-connections** between the layers with the β parameters. This acts as a mixture model and can be written as:

$$\tilde{h}_A^T = \begin{bmatrix} \beta_{A,1} \\ \cdots \\ \beta_{A,k} \end{bmatrix} \begin{bmatrix} h_{A,1}{}^T \cdots h_{A,k}{}^T \end{bmatrix} \tag{10.14}$$

where $h_{A,k}$ is the output of layer k for task A, and $\tilde{h}_{A,t}$ is the linear combination of all layer outputs that gets fed to a softmax classifier.

10.4.2 Theory

Caruana in his early research on MTL and then Ruder in his work have summarized different reasons why and when multitask learning works and is effective [Car97, Rud17].

1. **Implicit Data Augmentation**—When the constraint is limited data per task, by jointly learning different tasks which are similar, the total training data size increases. As learning theory suggests, the more the training data, the better the model quality.
2. **Attention Focusing**—When the constraint is noisy data per task, by jointly learning different tasks, focus on relevant features that are useful across tasks to get more attention. This joint learning, in general, helps as an implicit feature selection mechanism.
3. **Eavesdropping**—When the training data is limited, the features that may be needed for a particular task may not be in the data. By having multiple datasets for multiple tasks, features can eavesdrop, i.e., the features learned for a separate task can be used for the task in question and help in the generalization of that specific task.
4. **Representation Bias**—Multitask learning enforces a representation that generalizes across the tasks and thus forces better generalization.
5. **Regularization**—Multitask learning is also considered as a regularization technique through inductive bias, which theoretically and empirically is known to improve model quality.

10.4.3 Applications in NLP

In his work, Rei shows that using language modeling as an auxiliary task along with sequence labeling tasks such as POS tagging, chunking, and named entity detection for the main task can improve the results significantly over the benchmarks [Rei17]. Fang and Cohn illustrate the advantage of cross-lingual multitask joint learning for POS tagging in a low-resource language [FC17]. Yang et al. show that a deep hierarchical neural network with cross-lingual multitask learning can achieve state-of-the-art results in various sequence tagging tasks, such as NER, POS tagging, and chunking [YSC16]. Duong et al. use cross-lingual multitask learning to achieve high accuracy in a low-resource language for dependency parsing [Duo+15]. Collobert and Weston show that multitask learning using CNNs across various tasks can achieve great accuracies [CW08].

Multitask learning has been most successful in machine translation tasks, either employing them at the encoder stage, or the decoder stage, or both. Dong et al. successfully employ single source to multiple language translation using MTL at the encoder stage in sequence-to-sequence network [Don+15]. Zoph and Knight employ multi-source learning as MTL, using French and German sources to translate effectively to English using MTL at the decoder stage [ZK16]. Johnson et al. show that jointly learning the encoders and decoders enables to have a single model for multiple source and targets in a unified way [Joh+16]. Luong et al. perform a more comprehensive study of the sequence-to-sequence and multitask learning at various stages of encoding–decoding on several NLP tasks, including translation, to show benefits [Luo+15]. Niehues and Cho in their research on German–English translation explore how tasks, such as POS tagging and NER, can help machine translations, as well as improve results in these tasks [NC17].

Choi et al. use multitask learning to learn sentence selection in comprehension first and using that for question–answer model to get superior results [Cho+17]. Another exciting work uses a large corpus of data to learn and rank the passages that are likely for question–answers and then uses joint training of these passages with QA models to give state-of-the-art results in open QA tasks [Wan+18].

Jiang shows how multitask learning, when applied together with weakly supervised learning for extracting different relation or role type using a joint model, can improve results [Jia09]. Liu et al. show that joint multitask learning using a deep neural network in low-resource datasets can improve results in query classification and web search ranking [Liu+15]. Katiar and Cardie show how joint extractions of relations and mentions using attention-based recurrent networks improve on traditional deep networks [KC17]. Yang and Mitchell highlight how a single model, which can learn two tasks of semantic role labeling and predicting relations learned jointly, can improve over the state of the art [YM17].

Isonuma et al. show how summarization using a small number of summaries and document classification done together give comparable results to the state of the art [Iso+17]. In a specific domain such as legal, Luo et al. show how classification with relevant article extraction, when learned jointly, can give improved results [Luo+17]. Balikas et al. show how separate sentiment analysis tasks of learning ternary and fine-grained classification can be improved using joint multitask learning [BMA17]. Augenstein and Søgaard showcase improvements in keyphrase boundary classification when learning auxiliary tasks, such as semantic super-sense tagging and identification of multi-word expressions [AS17].

10.4.4 Applications in Speech Recognition

Watanabe et al. highlight how multiple tasks associated with speech recognition can be performed in a hybrid end-to-end deep learning framework [Wat+17]. This architecture combines two main architectures, CTC loss and attention-based sequence-to-sequence, to give results that are comparable with previous HMM-deep learning based methods. Watanabe et al. again highlight how multiple tasks such as automatic speech recognition (ASR) and language identification/classification across ten languages can be performed at the same time using end-to-end deep learning with multitask learning [WHH17]. Watanabe et al. highlight how multitask learning on ASR and speaker identification can improve the total performance significantly compared to separately trained models [Wat+18].

10.5 Case Study

In this case study, we explore how multitask learning can be applied to some common NLP tasks such as POS tagging, chunking, and named entity recognition. The overall performance depends on many choices such as sequence-to-sequence architecture, embeddings, and sharing techniques.

We will try to answer whether the low-level tasks such as POS tagging can benefit the high-level tasks such as chunking? What would be the impact of joint learning with closely related tasks and loosely related tasks? Is there an impact of connectivity and sharing on learning? Is there a negative transfer and how that impacts the learning? Do the neural architecture and embedding choices impact the multitask learning? We will use CoNLL-2003 English dataset which has annotations at token levels for each of the tasks in our experiments. CoNLL-2003 dataset already has the standard splits of train, validation, and test. We will use accuracy on the test set as our performance metric for the case study.

- Exploratory data analysis
- Multitask learning experiments and analysis

10.5.1 Software Tools and Libraries

We will describe the main open source tools and libraries we have used below for our case study:

- **PyTorch**: We use http://github.com/pytorch/pytorch as our deep learning toolkit in this case study.
- **GloVe**: We use https://nlp.stanford.edu/projects/glove/ for our pre-trained embeddings in the experiments. https://github.com/SeanNaren/nlp_multi_task_learning_pytorch/ for multitask learning experiments.

10.5.2 Exploratory Data Analysis

The raw data for training, validation, and testing have a columnar format with annotations for each token as given in Table 10.1.

Table 10.1: Raw data format

Tokens	POS	CHUNK	NER
U.N.	NNP	I-NP	I-ORG
official	NN	I-NP	O
Ekeus	NNP	I-NP	I-PER
heads	VBZ	I-VP	O
for	IN	I-PP	O
Baghdad	NNP	I-LOC	I-LOC
–	–	O	O

Basic analysis for total articles, sentences, and tokens for each dataset is given in Table 10.2. The tags follow the "inside–outside–beginning" (IOB) scheme for chunking and NER.

NER categories and number of tokens for each are given in Table 10.3.

Table 10.2: Data analysis of CoNll-2003

Dataset	Articles	Sentences	Tokens
Training	946	14,987	203,621
Validation	216	3466	51,362
Test	231	3684	46,435

Table 10.3: NER tags analysis of CoNll-2003

Dataset	LOC	MISC	ORG	PER
Training	7140	3438	6321	6600
Validation	1837	922	1341	1842
Test	1668	702	1661	1617

10.5.3 Multitask Learning Experiments and Analysis

We base our model on Søgaard and Golberg's research using bidirectional RNNs for encoder and decoder networks in "joint learning" mode. We explore "joint learning" in two different configurations: (a) shared layers between all the tasks that are connected to three different softmax layers (POS, chunk, and NER) and (b) each RNN is in different layer and the hidden layer of the lower layer flows into the next higher layer as shown in Fig. 10.12.

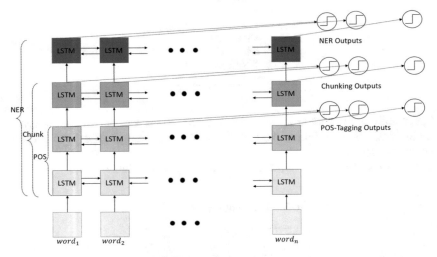

Fig. 10.12: Bidirectional LSTM configured for multitask learning with cascading layered architecture

We highlight the code below for the **JointModel** class where all the configurations of (a) individual learning, (b) joint with shared layers, and (c) joint with cascading are defined.

```
# initialization of the graph
def forward(self, input, *hidden):
        if self.train_mode == 'Joint':
            # when the number of layers is same, hidden layers
    are shared
            # and connected to different outputs
            if self.nlayers1 == self.nlayers2 == self.nlayers3
    :
                logits, shared_hidden = self.rnn(input, hidden
    [0])
                outputs_pos = self.linear1(logits)
                outputs_chunk = self.linear2(logits)
```

```
10        outputs_ner = self.linear3(logits)
11        return outputs_pos, outputs_chunk, outputs_ner
    , shared_hidden
12            # cascading architecture where low-level tasks
    flow into high level
13        else:
14            # POS tagging task
15            logits_pos, hidden_pos = self.rnn1(input,
    hidden[0])
16            self.rnn2.flatten_parameters()
17            # chunking using POS
18            logits_chunk, hidden_chunk = self.rnn2(
    logits_pos, hidden[1])
19            self.rnn3.flatten_parameters()
20            # NER using chunk
21            logits_ner, hidden_ner = self.rnn3(
    logits_chunk, hidden[2])
22            outputs_pos = self.linear1(logits_pos)
23            outputs_chunk = self.linear2(logits_chunk)
24            outputs_ner = self.linear3(logits_ner)
25            return outputs_pos, outputs_chunk, outputs_ner
    , hidden_pos, hidden_chunk, hidden_ner
26        else:
27            # individual task learning
28            logits, hidden = self.rnn(input, hidden[0])
29            outputs = self.linear(logits)
30            return outputs, hidden
```

Since we have different tasks (POS, chunking, and NER), input layer choices (pre-trained embeddings or embeddings from the data), neural architecture choices (LSTM or bidirectional LSTM), and MTL techniques (joint shared and joint separate), we perform the following experiments to gain the insights in a step-by-step manner:

1. **LSTM + POS + Chunk**: We use LSTM in our encoder–decoder, no pre-trained embeddings, and use different sharing techniques to see the impact on two tasks, POS tagging and chunking.
2. **LSTM + POS+ NER**: We use LSTM in our simple encoder–decoder, no pre-trained embeddings, and use different sharing techniques to see the impact on two tasks, POS tagging and NER.
3. **LSTM + POS + Chunk + NER**: We use LSTM in our simple encoder–decoder, no pre-trained embeddings, and use different sharing techniques to see the impact on all three tasks, POS tagging, chunking, and NER.
4. **Bidirectional LSTM + POS + Chunk**: We use bidirectional LSTM in our encoder–decoder, no pre-trained embeddings, and use different sharing techniques to see the impact on two tasks POS tagging and chunking. The impact of the neural architecture on the learning will be evident from this experiment.
5. **LSTM + GloVe + POS + Chunk**: We use LSTM in our encoder–decoder, pre-trained GloVe embeddings, and use different sharing techniques to see the impact on two tasks, POS tagging and chunking. The impact of pre-trained embeddings on the learning will be evident from this experiment.

6. **Bidirectional LSTM + GloVe + POS + Chunk**: We use bidirectional LSTM in our encoder–decoder, pre-trained GloVe embeddings, and use different sharing techniques to see the impact on two tasks, POS tagging and chunking. This experiment gives us insight into how the combination of architecture and embeddings impacts the learning for the two tasks.
7. **Bidirectional LSTM + GloVe + POS + NER**: We use bidirectional LSTM in our encoder–decoder, pre-trained GloVe embeddings, and use different sharing techniques to see the impact on two tasks, POS tagging and NER. This experiment gives us insight into how the combination of architecture and embeddings impacts the learning for the two tasks.
8. **Bidirectional LSTM + GloVe + POS + Chunk + NER**: We use bidirectional LSTM in our encoder–decoder, pre-trained GloVe embeddings, and use different sharing techniques to see the impact on all three tasks, POS tagging, chunking, and NER. This experiment gives us insight into how the combination of architecture and embeddings impacts the learning when there are multiple tasks.

We run all the experiments with parameters of input embeddings with or without pre-trained of 300 dimensions, 128 as the number of hidden units, 128 as the batch size, 300 as the number of epochs, ADAM optimizer, and cross-entropy loss.
In the tables below we have given individual experiment results, and color-coded the results which show improvement with green and where it deteriorates with red.

Table 10.4: Expt 1: LSTM + POS + Chunk

Models	POS Acc %	Chunk Acc %
POS single task	86.33	–
Chunk single task	–	84.69
MTL joint shared	83.91	85.23
MTL joint separate	86.88	85.78

Table 10.5: Expt 2: LSTM + POS + NER

Models	POS Acc %	NER Acc %
POS single task	86.33	-
NER single task	–	84.92
MTL joint shared	85.62	88.28
MTL joint separate	86.72	89.745

Table 10.6: Expt. 3: LSTM + POS + Chunk + NER

Models	POS Acc %	Chunk Acc %	NER Acc %
POS single task	87.42	–	–
Chunk single task	–	85.16	–
NER single task	–	–	90.08
MTL joint shared	85.94	85.00	88.05
MTL joint separate	87.11	86.72	88.83

Some interesting observations from the experiments are:

Table 10.7: Expt 4: Bidirectional LSTM + POS + Chunk

Models	POS Acc %	Chunk Acc %
POS single task	86.56	–
Chunk single task	–	86.88
MTL joint shared	84.53	88.20
MTL joint separate	87.34	87.11

Table 10.8: Expt 5: LSTM + GloVe + POS + Chunk

Models	POS Acc %	Chunk Acc %
POS single task	90.55	–
Chunk single task	–	88.05
MTL joint shared	89.84	88.12
MTL joint separate	90.86	87.73

Table 10.9: Expt 6: Bidirectional LSTM + GloVe + POS + Chunk

Models	POS Acc %	Chunk Acc %
POS single task	92.42	–
Chunk single task	–	89.69
MTL joint shared	91.72	89.53
MTL joint separate	92.34	89.61

Table 10.10: Expt 7: Bidirectional LSTM + GloVe + POS + NER

Models	POS Acc %	NER Acc %
POS single task	92.42	–
NER single task	–	95.08
MTL joint shared	92.89	95.70
MTL joint separate	92.19	95.0

Table 10.11: Expt. 8: Bidirectional LSTM + GloVe + POS + Chunk + NER

Models	POS Acc %	Chunk Acc %	NER Acc %
POS single task	92.662	–	–
Chunk single task	–	88.52	–
NER single task	–	–	95.78
MTL joint shared	92.89	89.53	94.92
MTL joint separate	91.95	90.00	95.31

- Tables 10.4 and 10.5 show that joint multitask learning with separate LSTM layers as compared to shared layer between both improve the performance for both combinations, i.e., POS tagging and chunking and POS tagging and NER.
- Table 10.6 shows that when all three tasks are combined joint MTL with shared as well as with separate layers, the results deteriorate except for chunking. These results are in contrast with Tables 10.4 and 10.5 and show that when there is a mix of tasks which are not all related strongly, the "negative transfer" comes into the picture.
- The experiment results in Table 10.7 use bidirectional LSTM and show similar performance as LSTM models in Table 10.4, indicating that just by adding architectural complexity by itself does not change the multitask behavior at least in this case.
- Introducing pre-trained embeddings using GloVe vectors shows a huge increase of around 4% in performance of single tasks for both POS tagging and chunking as shown in Table 10.8. The marginal improvements in MTL are similar to without GloVe.
- Experiment 6 as given in Table 10.9 shows that when both bidirectional LSTM and pre-trained GloVe vectors are used, not only the individual tasks improve but the behavior of multitask learning is different as that of the basic first experiment in Table 10.4. The shared and separate layers both show worse performance than single tasks here. Somehow, better the individual task performance is, the impact of multitask learning diminishes.
- Experiment 7, where we combine bidirectional LSTM and pre-trained GloVe for POS tagging and NER results as given in Table 10.10, shows very different results than experiment 2 as given in Table 10.5. The joint multitask learning using shared shows performance boost for both tasks which has not been seen in the previous experiments.
- Experiment 8, results given in Table 10.11, where we combine all tasks with bidirectional LSTM and GloVe, shows different performance as compared to the experiment 3 as given in Table 10.6. The POS tagging and chunking show improvements with shared but NER shows deterioration in performance. Except for chunking, all others show worse performance with separate layers as compared to experiment 3.

10.5.4 Exercises for Readers and Practitioners

Some of the extensions and extra ideas for researchers to try are given below:

1. What is the impact of using different pre-trained embeddings such as word2vec?
2. What is the impact of adding more layers to RNN for both shared and separate? Does that change the MTL behavior?
3. We tried MTL with LSTM but not with GRU or even base RNN, is there a significant difference in the performance of MTL with the choice of recurrent networks?
4. What is the impact of hyperparameters like the number of hidden units, batch size, and epochs on the MTL?
5. If we add more tasks such as language models, sentiment classification, semantic role labeling, to name a few in the mix, what would be the performance impact on MTL?
6. Use the same dataset with other research like cross-stitch networks, sluice networks, and others to get comparative analysis across the methods.

References

[AS17]   Isabelle Augenstein and Anders Søgaard. "Multi-Task Learning of Keyphrase Boundary Classification". In: *Proceedings of the 55th Annual Meeting of the Association for Computational Linguistics*. 2017, pp. 341–346.

[BMA17] Georgios Balikas, Simon Moura, and Massih-Reza Amini. "Multitask Learning for Fine-Grained Twitter Sentiment Analysis". In: *Proceedings of the 40th International ACM SIGIR Conference on Research and Development in Information Retrieval*. 2017, pp. 1005–1008.

[BBS07] Steffen Bickel, Michael Brückner, and Tobias Scheffer. "Discriminative Learning for Differing Training and Test Distributions". In: *Proceedings of the 24th International Conference on Machine Learning*. ICML '07. 2007, pp. 81–88.

[Car97] Rich Caruana. "Multitask Learning". In: *Machine Learning* 28.1 (1997), pp. 41–75.

[Car93] Richard Caruana. "Multitask Learning: A Knowledge-Based Source of Inductive Bias". In: *Proceedings of the Tenth International Conference on Machine Learning*. Morgan Kaufmann, 1993, pp. 41–48.

[Cho+17] Eunsol Choi et al. "Coarse-to-Fine Question Answering for Long Documents". In: *Proceedings of the 55th Annual Meeting of the Association for Computational Linguistics*. 2017, pp. 209–220.

[CW08]   Ronan Collobert and Jason Weston. "A Unified Architecture for Natural Language Processing: Deep Neural Networks with Multi-task Learning". In: *Proceedings of the 25th International Conference on Machine Learning*. ICML '08. 2008, pp. 160–167.

[Dah+12]   George E. Dahl et al. "Context-Dependent Pre-Trained Deep Neural Networks for Large-Vocabulary Speech Recognition". In: *IEEE Trans. Audio, Speech & Language Processing* 20.1 (2012), pp. 30–42.

[DL15]   Andrew M Dai and Quoc V Le. "Semi-supervised Sequence Learning". In: *Advances in Neural Information Processing Systems 28*. Ed. by C. Cortes et al. 2015, pp. 3079–3087.

[Dai+07a] Wenyuan Dai et al. "Boosting for Transfer Learning". In: *Proceedings of the 24th International Conference on Machine Learning*. ICML '07. 2007, pp. 193–200.

[Dai+07b] Wenyuan Dai et al. "Transferring Naive Bayes Classifiers for Text Classification". In: *Proceedings of the 22nd National Conference on Artificial Intelligence - Volume 1*. AAAI'07. 2007, pp. 540–545.

[DM06]   Hal Daumé III and Daniel Marcu. "Domain Adaptation for Statistical Classifiers". In: *J. Artif. Int. Res.* 26.1 (May 2006), pp. 101–126.

[Die+16]   Adji B. Dieng et al. "TopicRNN: A Recurrent Neural Network with Long-Range Semantic Dependency." In: *CoRR* abs/1611.01702 (2016).

[Don+15]   Daxiang Dong et al. "Multi-Task Learning for Multiple Language Translation." In: *ACL (1)*. 2015, pp. 1723–1732.

[Duo+15]   Long Duong et al. "Low Resource Dependency Parsing: Cross-lingual Parameter Sharing in a Neural Network Parser". In: *Proceedings of the 7th International Joint Conference on Natural Language Processing (Volume 2: Short Papers)*. 2015, pp. 845–850.

[Erh+10]   Dumitru Erhan et al. "Why Does Unsupervised Pre-training Help Deep Learning?" In: *J. Mach. Learn. Res.* 11 (Mar. 2010).

[FC17]   Meng Fang and Trevor Cohn. "Model Transfer for Tagging Low-resource Languages using a Bilingual Dictionary". In: *CoRR* abs/1705.00424 (2017).

[Fun+06]   Gabriel Pui Cheong Fung et al. "Text Classification Without Negative Examples Revisit". In: *IEEE Trans. on Knowl. and Data Eng.* 18.1 (Jan. 2006), pp. 6–20.

[Has+16]   Kazuma Hashimoto et al. "A Joint Many-Task Model: Growing a Neural Network for Multiple NLP Tasks". In: *CoRR* abs/1611.01587 (2016).

[Hin+12]   Geoffrey Hinton et al. "Deep Neural Networks for Acoustic Modeling in Speech Recognition". In: *Signal Processing Magazine* (2012).

[Iso+17]   Masaru Isonuma et al. "Extractive Summarization Using Multi-Task Learning with Document Classification". In: *Proceedings of the 2017 Conference on Empirical Methods in Natural Language Processing, EMNLP 2017*. 2017, pp. 2101–2110.

[Jia09]   Jing Jiang. "Multi-Task Transfer Learning for Weakly-Supervised Relation Extraction". In: *ACL 2009, Proceedings of the 4th International Joint Conference on Natural Language Processing of the AFNL*. 2009, pp. 1012–1020.

[JZ07]   Jing Jiang and Chengxiang Zhai. "Instance weighting for domain adaptation in NLP". In: *In ACL 2007*. 2007, pp. 264–271.

[Joh+16]   Melvin Johnson et al. "Google's Multilingual Neural Machine Translation System: Enabling Zero-Shot Translation". In: *CoRR* abs/1611.04558 (2016).

[KC17]   Arzoo Katiyar and Claire Cardie. "Going out on a limb: Joint Extraction of Entity Mentions and Relations without Dependency Trees". In: *Proceedings of the 55th Annual Meeting of the Association for Computational Linguistics*. 2017, pp. 917–928.

[Lee+09]   Honglak Lee et al. "Unsupervised feature learning for audio classification using convolutional deep belief networks". In: *Advances in Neural Information Processing Systems 22: 23rd Annual Conference on Neural Information Processing Systems*. 2009, pp. 1096–1104.

[Liu+15]   Xiaodong Liu et al. "Representation Learning Using Multi-Task Deep Neural Networks for Semantic Classification and Information Retrieval". In: *NAACL HLT 2015, The 2015 Conference of the North American Chapter of the Association for Computational Linguistics*.

[LW15]   Mingsheng Long and Jianmin Wang. "Learning Multiple Tasks with Deep Relationship Networks". In: *CoRR* abs/1506.02117 (2015).

[Lu+16]   Yongxi Lu et al. "Fully-adaptive Feature Sharing in Multi-Task Networks with Applications in Person Attribute Classification". In: *CoRR* abs/1611.05377 (2016).

[Luo+17]   Bingfeng Luo et al. "Learning to Predict Charges for Criminal Cases with Legal Basis". In: *Proceedings of the 2017 Conference on Empirical Methods in Natural Language Processing, EMNLP 2017*, 2017, pp. 2727–2736.

[Luo+15]   Minh-Thang Luong et al. "Multi-task Sequence to Sequence Learning". In: *CoRR* abs/1511.06114 (2015).

[Mis+16]   Ishan Misra et al. "Cross-stitch Networks for Multi-task Learning". In: *CoRR* abs/1604.03539 (2016).

[NC17]   Jan Niehues and Eunah Cho. "Exploiting Linguistic Resources for Neural Machine Translation Using Multi-task Learning". In: *Proceedings of the Second Conference on Machine Translation*. Association for Computational Linguistics, 2017, pp. 80–89.

[PY10]   Sinno Jialin Pan and Qiang Yang. "A Survey on Transfer Learning". In: *IEEE Trans. on Knowl. and Data Eng.* 22.10 (Oct. 2010), pp. 1345–1359.

[Pan+08]   Sinno Jialin Pan et al. "Transfer Learning for WiFi-based Indoor Localization". In: 2008.

[Rai+07]   Rajat Raina et al. "Self-taught Learning: Transfer Learning from Unlabeled Data". In: *Proceedings of the 24th International Conference on Machine Learning*. ICML '07. 2007, pp. 759–766.

[RLL17]   Prajit Ramachandran, Peter J. Liu, and Quoc V. Le. "Unsupervised Pretraining for Sequence to Sequence Learning". In: *Proceedings of the 2017 Conference on Empirical Methods in Natural Language Processing, EMNLP 2017, Copenhagen, Denmark, September 9–11, 2017*. 2017, pp. 383–391.

[Rei17]   Marek Rei. "Semi-supervised Multitask Learning for Sequence Labeling". In: *CoRR* abs/1704.07156 (2017).

[Rud17]   Sebastian Ruder. "An Overview of Multi-Task Learning in Deep Neural Networks". In: *CoRR* abs/1706.05098 (2017).

[SG16]   Anders Søgaard and Yoav Goldberg. "Deep multi-task learning with low level tasks supervised at lower layers". In: *Proceedings of the 54th Annual Meeting of the Association for Computational Linguistics, ACL 2016, August 7–12, 2016, Berlin, Germany, Volume 2: Short Papers*. 2016.

[TS07]   Matthew E. Taylor and Peter Stone. "Cross-domain Transfer for Reinforcement Learning". In: *Proceedings of the 24th International Conference on Machine Learning*. ICML '07. 2007, pp. 879–886.

[Dar05]   "Transfer Learning Proposer Information Pamphlet (PIP) for Broad Agency Announcement". In: Defense Advanced Research Projects Agency (DARPA), 2005.

[TRB10]   Joseph Turian, Lev Ratinov, and Yoshua Bengio. "Word Representations: A Simple and General Method for Semi-supervised Learning". In: *Proceedings of the 48th Annual Meeting of the Association for Computational Linguistics*. ACL '10. 2010.

[Wan+18]   Shuohang Wang et al. "R^3: Reinforced Ranker-Reader for Open-Domain Question Answering". In: *Proceedings of the Thirty-Second AAAI Conference on Artificial Intelligence*. 2018.

[WSZ08]   Zheng Wang, Yangqiu Song, and Changshui Zhang. "Transferred Dimensionality Reduction". In: *Proceedings of the European Conference on Machine Learning and Knowledge Discovery in Databases - Part II*. ECML PKDD '08. 2008, pp. 550–565.

[WHH17]   Shinji Watanabe, Takaaki Hori, and John R. Hershey. "Language independent end-to-end architecture for joint language identification and speech recognition". In: *ASRU*. IEEE, 2017, pp. 265–271.

[Wat+17]   Shinji Watanabe et al. "Hybrid CTC/Attention Architecture for End-to-End Speech Recognition". In: *J. Sel. Topics Signal Processing* 11.8 (2017), pp. 1240–1253.

[Wat+18]   Shinji Watanabe et al. "A Purely End-to-End System for Multi-speaker Speech Recognition". In: *ACL (1)*. Association for Computational Linguistics, 2018, pp. 2620–2630.

[YM17]   Bishan Yang and Tom M. Mitchell. "A Joint Sequential and Rela-
            tional Model for Frame-Semantic Parsing". In: *Proceedings of the 2017
            Conference on Empirical Methods in Natural Language Processing
            EMNLP 2017*. 2017, pp. 1247–1256.

[YSC16]   Zhilin Yang, Ruslan Salakhutdinov, and William W. Cohen. "Multi-
            Task Cross-Lingual Sequence Tagging from Scratch". In: *CoRR*
            abs/1603.06270 (2016).

[ZK16]   Barret Zoph and Kevin Knight. "Multi-Source Neural Translation". In:
            CoRR abs/1601.00710 (2016).

Chapter 11
Transfer Learning: Domain Adaptation

11.1 Introduction

Domain adaptation is a form of transfer learning, in which the task remains the same, but there is a domain shift or a distribution change between the source and the target. As an example, consider a model that has learned to classify reviews on electronic products for positive and negative sentiments, and is used for classifying the reviews for hotel rooms or movies. The task of sentiment analysis remains the same, but the domain (electronics and hotel rooms) has changed. The application of the model to a separate domain poses many problems because of the change between the training data and the unseen testing data, typically known as **domain shift**. For example, sentences containing phrases such as "loud and clear" will be mostly considered positive in electronics, whereas negative in hotel room reviews. Similarly, usage of keywords such as "lengthy" or "boring" which may be prevalent in domains such as book reviews might be completely absent in domains such as kitchen equipment reviews.

As discussed in the last chapter, the central idea behind domain adaptation is to learn from the source dataset (labeled and unlabeled) so that the learning can be used on a target dataset with a different domain mapping. To learn the domain shift between the source and the target, traditional techniques that are employed fall under two broad categories: instance-based and feature-based. In instance-based, the discrepancy between the source and the target domain is reduced by reweighting source samples and learning models from the reweighed ones [BM10]. In feature-based, a common shared space or a joint representation is learned between the source and the target where the distributions match [GGS13]. In recent times, deep learning architectures have been successfully implemented for domain adaptation in various applications especially in the field of computer vision [Csu17]. In this chapter, we discuss at length some of the techniques using deep learning in domain adaptation and their applications in text and speech. Next, we discuss techniques in zero-shot, one-shot, and few-shot learning that have gained popularity in the domain adapta-

© Springer Nature Switzerland AG 2019
U. Kamath et al., *Deep Learning for NLP and Speech Recognition*,
https://doi.org/10.1007/978-3-030-14596-5_11

tion field. We perform a detailed case study using many techniques discussed in the chapter to give the readers practical aspects of domain adaptation at the end.

> In this chapter, we will use the notations that are similar to the research papers they cite for easy mapping to the references.

11.1.1 Techniques

In this section, we will highlight some of the well-known techniques that can be very effective and generic enough for solving domain adaptation problem in text and speech.

11.1.1.1 Stacked Autoencoders

One of the earliest works in domain adaptation comes from Glorot et al. in the area of sentiment classification [GBB11b]. The source domain contains a large number of sentiments on Amazon reviews, while the target is completely different products with small labeled data. In this work, the researchers use stacked denoising autoencoders (SDA) on the source and target data combined to learn the features as shown in Fig. 11.1 as the first step. Then, a linear SVM is trained on the features extracted by the encoder part of the autoencoder and is used to predict unseen target data of different domains. The researchers report state-of-the-art results for sentiment classification across the domains.

Variants, such as stacked **marginalized denoising autoencoders** (mSDA), which have better optimal solutions and faster training times, have also been employed very successfully in classification tasks, such as sentiment classification [Che+12]. To explain the method, let us assume that for source S and target T, we have sample source data $D_S = \{\mathbf{x}_1, \cdots, \mathbf{x}_{n_S}\} \in \mathbb{R}^d$ and labels $L_s = \{y_1, \cdots, y_{n_S}\}$ and the target sample data $D_T = \{\mathbf{x}_{n_{s+1}}, \cdots, \mathbf{x}_n\} \in \mathbb{R}^d$ and no labels. The goal is to learn the classifier $h \in \mathbb{H}$ with labeled source training data D_S to predict on the unlabeled target data D_T.

The basic building block in this work is the one-layered denoising autoencoder. The input for this is the entire set of the source and the target data, i.e., $D = D_S \cup D_T = \{\mathbf{x}_1, \cdots, \mathbf{x}_n\}$ and it is corrupted by removal of feature with probability $p \geq 0$. For example, if the representation of the vector is a bag-of-words vector, some values can be flipped from 1 to 0. Let us consider $\tilde{\mathbf{x}}_i$ as the corrupted version of $\mathbf{x}_i$. Instead of using the two-level encoder–decoder, a single mapping $\mathbf{W} : \mathbb{R}^d \to \mathbb{R}^d$ is used that minimizes the squared reconstruction loss given by:

$$\frac{1}{2n} \sum_{i=1}^{n} \| \mathbf{x}_i - \mathbf{W}\tilde{\mathbf{x}}_i \| \tag{11.1}$$

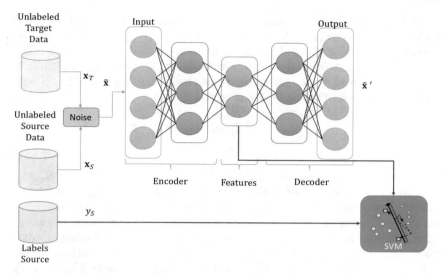

Fig. 11.1: Stacked denoising autoencoders for learning features and SVM as a classifier

If we repeat this m times, the variance gets lowered and the solution for $\mathbf{W}$ can be obtained from:

$$\mathcal{L}_{squared}(\mathbf{W}) = \frac{1}{2mn} \sum_{j=1}^{m} \sum_{i=1}^{n} \| \mathbf{x}_i - \mathbf{W}\tilde{\mathbf{x}}_{i,j} \| \tag{11.2}$$

where $\tilde{\mathbf{x}}_{i,j}$ represents the jth corrupted version of the original $\mathbf{x}_i$ input.

In matrix notation, with inputs $\mathbf{X} = [\mathbf{x}_1, \cdots, \mathbf{x}_n] \in \mathbb{R}^{d \times n}$ and its m-times repeated $\bar{\mathbf{X}}$ and corrupted being $\tilde{\mathbf{X}}$. The loss equation can be written as:

$$\mathcal{L}_{squared}(\mathbf{W}) = \frac{1}{2mn} \operatorname{tr}\left[(\bar{\mathbf{X}} - \mathbf{W}\tilde{\mathbf{X}})^\mathsf{T}(\bar{\mathbf{X}} - \mathbf{W}\tilde{\mathbf{X}}) \right] \tag{11.3}$$

The solution to this in closed-form is:

$$\mathbf{W} = \mathbf{P}\mathbf{Q}^{-1} \; with \; \mathbf{Q} = \tilde{\mathbf{X}}\tilde{\mathbf{X}}^\mathsf{T} \; and \; \mathbf{P} = \bar{\mathbf{X}}\tilde{\mathbf{X}}^\mathsf{T} \tag{11.4}$$

In the limiting case of $m \longrightarrow \inf$, $\mathbf{W}$ can be expressed in terms of the expectations of $\mathbf{P}$ and $\mathbf{Q}$.

$$\mathbf{W} = E[\mathbf{P}]E[\mathbf{Q}]^{-1} \tag{11.5}$$

Let us consider the $E[\mathbf{Q}]$, which is

$$E[\mathbf{Q}] = \sum_{i=1}^{n} E[\tilde{\mathbf{x}}_i \tilde{\mathbf{x}}_i^\mathsf{T}] \tag{11.6}$$

The off-diagonal entries in matrix $[\tilde{\mathbf{x}}_i \tilde{\mathbf{x}}_i^\mathsf{T}]$ are uncorrupted if two features α and β both survive the corruption. This has a probability of $(1-p)^2$. For the diagonal, it holds with probability $1-p$. If we define a vector $\mathbf{q} = [1-p, \cdots, 1-p, 1] \in \mathbb{R}^{d+1}$, where $\mathbf{q}_\alpha$ represents the probability of the feature α surviving the corruption, then the scatter matrix of the original uncorrupted input can be represented as $\mathbf{S} = \mathbf{X}\mathbf{X}^\mathsf{T}$ and the expectation of matrix $\mathbf{Q}$ can be written as:

$$E[\mathbf{Q}]_{\alpha,\beta} = \begin{cases} \mathbf{S}_{\alpha,\beta}\mathbf{q}_\alpha\mathbf{q}_\beta & \text{if } \alpha \neq \beta \\ \mathbf{S}_{\alpha,\beta}\mathbf{q}_\alpha & \text{if } \alpha = \beta \end{cases} \tag{11.7}$$

In a similar way, expectation of matrix $\mathbf{P}$ can be derived as $E[\mathbf{P}]_{\alpha,\beta} = \mathbf{S}_{\alpha,\beta}\mathbf{q}_\beta$.

Thus with these expectation matrices, the reconstructive mapping $\mathbf{W}$ can be computed in closed-form without corrupting a single instance $\mathbf{x}_i$ and "marginalizing" the noise. Next, instead of just single layer, the research "stacks" the layer one after another similar to the stacked autoencoders. The output of the $(t-1)$th layer feeds into the tth layer after a squashing function such as tanh to give a non-linearity and thus can be expressed as $\mathbf{h}^t = \tanh(\mathbf{W}^t\mathbf{h}^{t-1})$. The training is performed layer by layer, i.e., each layer greedily learns $\mathbf{W}^t$ (in the closed-form) and tries to reconstruct the previous output $\mathbf{h}^{t-1}$. For domain adaptation, they use the inputs and all the hidden layers concatenated as features for SVM classifier to train and predict.

Some of the advantages of mSDA as compared to others are:

1. Optimization problem is convex and guarantees optimal solution.
2. Optimization is non-iterative and closed-form.
3. One pass through the entire training data to compute the expectations $E[\mathbf{P}]$ and $E[\mathbf{Q}]$ gives a huge training speed boost.

11.1.1.2 Deep Interpolation Between Source and Target

Very similar to traditional machine learning, research by Chopra et al. uses source and target with different domains to be mixed in different proportions to learn intermediate representations [CBG13]. This work is known as deep learning for domain adaptation by interpolating between domains (DLID). The researchers use convolutional layers with pooling and predictive sparse decomposition method to learn the non-linear features in an unsupervised way. The predictive sparse decomposition method is similar to sparse coding models but with fast and smooth approximator [KRL10]. The labeled data is passed through the same transformation to get features, concatenate them, and use classifier such as logistic regression to get a joint model. In this way, the model learns useful features in an unsupervised manner from both source and target. These features can be employed for domain transfer on the target alone. Figure 11.2a, b show schematically how this process works.

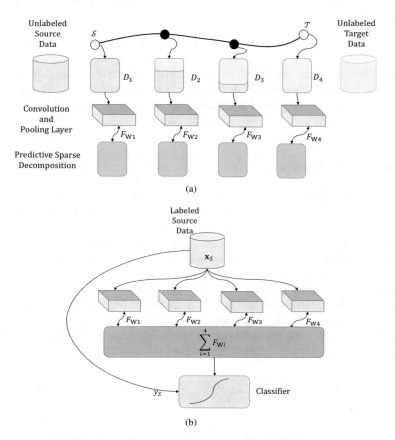

Fig. 11.2: DLID method. (**a**) Unsupervised latent representation learning. The top circles show the intermediate path between the source and the target, where the filled circles are the intermediate representations, and the empty circles are the source/target representations. (**b**) Unsupervised features and labels with classifier for learning model

Let S be the source domain with data samples D_S, T be the target domain with D_T as the data samples, and $p \in [1, 2, \cdots, P]$ be the index over P datasets. The mix between source and target is done in such a way that at $p = 1$, $D_S = D_T$ and from then onwards the number of source samples decreases and that of target increases in exact same proportion. For each dataset $p \in [2, \cdots, P-1]$, D_S, the number of samples goes down and D_T goes up incrementally for next p. Each dataset D_p as an input to a non-linear feature extractor F_{W_p} with weights W_p trained in an unsupervised manner generates output $Z_p^i = F_{W_p}(X^i)$. Once this is trained in unsupervised manner, any labeled training data goes through this DLID representation path extracting features F_{W_p} as output, and a concatenation of all the outputs forms the representation for that input as:

$$Z^i = [F_{W_1}(X^i) F_{W_2}(X^i) \cdots F_{W_p}(X^i)] = [Z_1^i Z_2^{i \cdots} Z_p^i] \tag{11.8}$$

That representation and the label Z^i, Y^i are passed to the classifier or regressor for the task and uses standard loss functions to optimize. The unseen data goes through the same path, and predictions from the classifier are used for obtaining the class and probability.

11.1.1.3 Deep Domain Confusion

The deep domain confusion (DDC) architecture, shown in Fig. 11.3, is proposed by Tzeng et al. and is one of the popular discrepancy-based domain adaptation frameworks [Tze+14]. The researchers introduce a domain adaptation layer and a confusion loss to learn a representation that is semantically meaningful and provides domain invariance. A Siamese convolutional network-based architecture is proposed, where the main goal is to learn a representation that minimizes the distribution distance between the source and the target domain. The representation can be used as features along with the source-labeled dataset to minimize the classification loss and is applied directly to unlabeled target data. In this work, the task of minimizing the distribution distance is done using maximum-mean discrepancy (MMD), which computes the distance on a representation $\phi()$ for both source and target as:

$$MMD(X_s, X_t) = \| \frac{1}{|X_s|} \sum_{x_s \in X_s} \phi(x_s) - \frac{1}{|X_t|} \sum_{x_t \in X_t} \phi(x_t) \| \tag{11.9}$$

The representation learned from this is used in the loss function as a regularizer with the regularization hyperparameter λ also acting as the amount of **confusion** between the source and the target domain:

$$L = L_C(X_L, y) + \lambda * MMD(X_s, X_t) \tag{11.10}$$

where $L_C(X_L, y)$ is the classification loss from the labeled data X_L, y is the label or the ground truth, and $MMD(X_s, X_t)$ is the maximum-mean discrepancy (MMD) between the source X_S and the target X_t. The hyperparameter λ controls the amount of confusion between the source and the target domain. The researchers use the standard AlexNet and modify it to have an additional lower -dimensional **bottleneck** layer "fc adapt." The lower-dimensional layer acts as a regularizer and prevents from overfitting on the source distribution. The MMD loss discussed above is added on top of this layer so that it learns the representation useful for both source and the target.

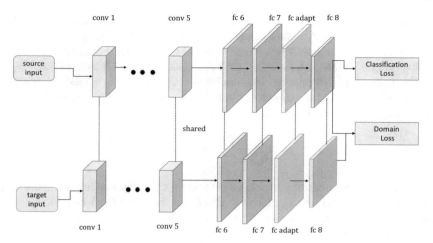

Fig. 11.3: Deep domain confusion network (DDCN) for domain adaptation

11.1.1.4 Deep Adaptation Network

Long et al. propose the deep adaptation network (DAN) as shown in Fig. 11.4, which is a modified AlexNet, where the discrepancy loss happens at the last fully connected layers [LW15]. If the null hypothesis is that the samples are drawn from the same distribution, and the alternate hypothesis is that they come from two different distributions, maximum-mean discrepancies (MMD) is one of the statistical approaches [Sej+12]. The multiple kernel variant of MMD (MK-MMD) measures the reproducing kernel Hilbert space (RKHS) distance between the mean embeddings of two distributions (source and target) with a characteristic kernel k. If $\mathcal{H}_k$ is the reproducing kernel Hilbert space endowed with a characteristic kernel k, the mean embedding of distribution p in $\mathcal{H}_k$ is a unique element $\mu_k(p)$, such that $\mathbb{E}_{\mathbf{x} \sim p} f(\mathbf{x}) = \langle f(\mathbf{x}), \mu_k(p) \rangle_{\mathcal{H}_k}$. The squared distance for any layer l, kernel k between source (S), and target (T) is given by:

$$d_k^2(\mathcal{D}_S^l, \mathcal{D}_T^l) \triangleq \|\mathbb{E}_{\mathcal{D}_S}[\phi(\mathbf{x}^S)] - \mathbb{E}_{\mathcal{D}_T}[\phi(\mathbf{x}^\mathsf{T})]\|_{\mathcal{H}_k}^2 \qquad (11.11)$$

The characteristic kernel associated with the feature map ϕ, $k(\mathbf{x}^S, \mathbf{x}^\mathsf{T}) = \langle \phi(\mathbf{x}^S), \phi(\mathbf{x}^\mathsf{T}) \rangle$ and is a combination of m positive semi-definite kernels $\{k_u\}$ with constraints on the coefficients β_u as given by:

$$\mathcal{K} \triangleq \left\{ k = \sum_{u=1}^{m} \beta_u k_u \ : \ \sum_{u=1}^{m} \beta_u = 1, \beta_u \geq 0 \right\} \qquad (11.12)$$

where the derived multi-kernel k is characteristic because of the constraints on coefficients $\{\beta_u\}$.

The modified AlexNet has three layers of convolution network ($conv1 - conv3$) as the general transferable feature layers that are frozen after training on one domain. The next two convolution layers ($conv4 - conv5$) are more specific, and hence fine-tuning is done for learning domain-specific features. The final fully connected layers ($fc6 - fc8$) are highly specific and non-transferable, so they get adapted using the MK-MMD. If all the parameters in the network are given by $\Theta = \{\mathbf{W}^l, \mathbf{b}^l\}_{l=1}^{l}$ for all the layers l, the empirical risk is given by:

$$\min_{\Theta} \frac{1}{n_a} \sum_{i=1}^{n_a} J(\Theta(\mathbf{x}_i^a), y_i^a) \tag{11.13}$$

where J is the cross-entropy loss function and $\Theta(\mathbf{x}_i^a)$ is the conditional probability of assigning the data point $\mathbf{x}_i^a$ a label y_i^a. By adding the MK-MMD-based multi-layer adaptation regularizer to the above risk, we get a loss similar to the DDC loss that can be expressed as:

$$\min_{\Theta} \frac{1}{n_a} \sum_{i=1}^{n_a} J(\Theta(\mathbf{x}_i^a), y_i^a) + \lambda \sum_{l=l_1}^{l_2} d_k^2(\mathcal{D}_S^l, \mathcal{D}_T^l) \tag{11.14}$$

where $\lambda > 0$ is a regularization constant and $l_1 = 6$ and $l_2 = 8$ for the DAN setup.

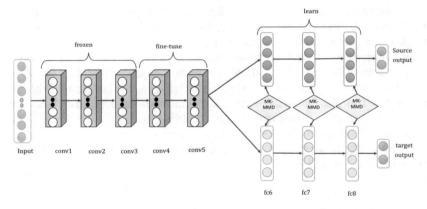

Fig. 11.4: Deep adaptation network (DAN) for domain adaptation

11.1.1.5 Domain-Invariant Representation

Many techniques employ domain-invariant representation using the source and the target data as a way to learn a common representation that can help in domain adaptation.

CORrelation ALignment (CORAL) is a technique to align the second-order statistics (covariances) of the source and target using a linear transformation. Sun and Saenko extend the framework to learn the non-linear transformation that aligns the correlations of the layers, known as Deep CORAL [SS16]. Deep CORAL extends AlexNet and has the second-order statistics loss computed at the last layer, i.e., a fully connected layer before the output. If $D_S = \{\mathbf{x}_i\}, \mathbf{x} \in \mathbb{R}^d$ are the source domain training data of size n_S, and $D_T = \{\mathbf{u}_i\}, \mathbf{u} \in \mathbb{R}^d$ are the unlabeled target data of size n_T, $D_S^{i,j}$ indicates jth dimension of the ith source data instance, $D_T^{i,j}$ indicates jth dimension of the ith target data instance, C_S is the source feature covariance matrix, and C_T is the target covariance matrix, then the CORAL loss is measured as the distance between the covariances:

$$CORAL = \frac{1}{4d^2}|C_S - C_T|_F^2 \tag{11.15}$$

where $|\cdot|_F^2$ represents the squared matrix Frobenius norm.

The covariance matrices for the source and the target are given by:

$$C_S = \frac{1}{(n_S - 1)}\left(\mathcal{D}_S^\mathsf{T}\mathcal{D}_S - \frac{1}{n_S}(\mathbf{1}^\mathsf{T}\mathcal{D}_S)(\mathbf{1}^\mathsf{T}\mathcal{D}_S)\right) \tag{11.16}$$

$$C_T = \frac{1}{(n_T - 1)}\left(\mathcal{D}_T^\mathsf{T}\mathcal{D}_T - \frac{1}{n_T}(\mathbf{1}^\mathsf{T}\mathcal{D}_T)(\mathbf{1}^\mathsf{T}\mathcal{D}_T)\right) \tag{11.17}$$

where $\mathbf{1}$ is the column vector. The joint training that reduces the classification loss l_{CLASS} and the CORAL loss is given by:

$$l = l_{CLASS} + \sum_{i=1}^{t} \lambda_i CORAL \tag{11.18}$$

where t is the number of layers, and λ is used to balance between classification and domain adaptation, aiming at learning a representation common between the source and the target (Fig. 11.5).

There are other domain-invariant representations that have been successfully employed in various works. Pan et al. use domain-invariant representation via **transfer component analysis** that uses **maximum-mean discrepancies (MMD)** and tries to reduce the distance between the two domains in the subspace [Pan+11]. Zellinger et al. propose a new distance function—**central moment discrepancy (CMD)**—to match the higher-order central moments of probability distributions [Zel+17]. They show the generality of their techniques in domain adaptation across object recognition and sentiment classification tasks (Fig. 11.5).

11.1.1.6 Domain Confusion and Invariant Representation

The research by Tzeng et al. on deep domain confusion has the disadvantage that it needs both large labeled data in the source domain and sparsely labeled data in the target domain. Tzeng et al. in their work propose domain confusion loss over both labeled and unlabeled data to learn the invariant representation across domains

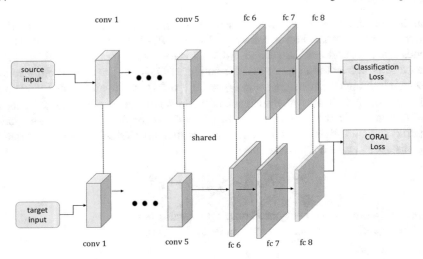

Fig. 11.5: Deep CORAL network for domain adaptation

and tasks [Tze+15]. The transfer learning between the source and the domain is achieved by a) maximizing the domain confusion by making marginal distributions between source and target as similar to each other as possible; and b) transfer of correlation between classes learned on source examples to target examples. The completely labeled source data (x_S, y_S) and the sparsely labeled target data (x_T, y_T) are used to produce a classifier θ_C that operates on feature representation $f(x; \theta_{repr})$ parameterized by representation parameters θ_{repr} and has good accuracy in classifying the target samples:

$$\mathcal{L}_C(x, y; \theta_{repr}, \theta_C) = -\sum_k \mathbf{1}[y = k] \log(p_k) \qquad (11.19)$$

where $p = \text{softmax}(\theta_C^\mathsf{T} f(x; \theta_{repr}))$.

To ensure that there is alignment between the classes in source and target instead of having "hard labels" to train on, the "soft label" is averaged over the softmax of all activations of labeled source data for a particular class. A high temperature parameter τ is used in the softmax function, so that related classes have similar effects on the probability mass during fine-tuning. The soft label loss is given by:

$$\mathcal{L}_{soft}(x_T, y_T; \theta_{repr}, \theta_C) = -\sum_i l_i^{y_T} \log(p_i) \qquad (11.20)$$

where $p_i = \text{softmax}(\theta_C^\mathsf{T} f(x_T; \theta_{repr}) / \tau)$.

A domain classifier layer with parameters θ_D is used to identify whether the data comes from the source or the target domain. The best domain classifier on the representation can be learned using the objective:

$$\mathcal{L}_D(x_S, x_T, \theta_{repr}; \theta_D) = -\sum_d \mathbf{1}[y_D = d] \log(q_d) \tag{11.21}$$

where $q = \text{softmax}(\theta_D^\mathsf{T} f(\mathbf{x}; \theta_{repr}))$.

Thus, for a particular domain classifier, θ_D, the loss that maximizes the confusion can be seen as a cross-entropy loss between the prediction of the domain and the uniform distribution over the labels and can be written as:

$$\mathcal{L}_{confusion}(x_S, x_T, \theta_D; \theta_{repr}) = -\sum_d \frac{1}{D} \log(q_d) \tag{11.22}$$

The parameters θ_D and θ_{repr} are learned iteratively by the following objectives:

$$\min_{\theta_D} \mathcal{L}_D(x_S, x_T, \theta_{repr}; \theta_D) \tag{11.23}$$

$$\min_{\theta_{repr}} \mathcal{L}_{confusion}(x_S, x_T, \theta_D; \theta_{repr}) \tag{11.24}$$

Thus, the joint loss function can be written as:

$$\begin{aligned} \mathcal{L}(x_S, y_S, x_T, y_T; \theta_{repr}, \theta_C) &= \mathcal{L}_C(x_S, y_S, x_T, y_T; \theta_C, \theta_{repr}) \\ &+ \lambda \mathcal{L}_{confusion}(x_S, x_T, \theta_D; \theta_{repr}) \\ &+ \nu \mathcal{L}_{soft}(x_T, y_T; \theta_{repr}, \theta_C) \end{aligned} \tag{11.25}$$

where λ and ν are the hyperparameters that control the domain confusion and the soft label influence during the optimization.

11.1.1.7 Domain-Adversarial Neural Network

Ganin et al. employ an interesting technique of "gradient reversal" layer for domain shift adaptation through domain-adversarial neural network (DANN) [Gan+16b]. The process is generic to all neural networks and can be easily trained using standard stochastic gradient methods. They show state-of-the-art results in different domains of computer vision and sentiment classification.

Let $S = \{(\mathbf{x}_i, y_i)\}_{i=1}^n \sim (\mathcal{D}_S)^n$; $T = \{\mathbf{x}_i\}_{i=n+1}^N \sim (\mathcal{D}_T^X)^{n'}$ be the source and target data drawn from $\mathcal{D}_S$ and $\mathcal{D}_T$ as distribution; $N = n + n'$ is the total number of samples. $\mathcal{D}_T^X$ is the marginal distribution of $\mathcal{D}_T$ over the input space X, and $Y = 0, 1, \cdots, L-1$ is the set of labels. The network has three important layers: (a) the feature generation layers which learn features from the inputs with parameters. The hidden layer $G_f : X \to \mathbb{R}^D$ parameterized by matrix–vector pair $\theta_f = (\mathbf{W}, \mathbf{b})$:

$$G_f(\mathbf{x}; \theta_f) = \sigma(\mathbf{W}\mathbf{x} + \mathbf{b}) \tag{11.26}$$

(b) the label prediction layer $G_y : \mathbb{R}^D \to [0,1]^L$ parameterized by matrix–vector pair $\theta_y = (\mathbf{V}, \mathbf{c})$:

$$G_y(G_f(\mathbf{x}); \theta_y) = \text{softmax}(\mathbf{V}\mathbf{x} + \mathbf{c}) \tag{11.27}$$

and (c) the domain classification layer $G_d : \mathbb{R}^D \to [0,1]$ is a logistic regressor parameterized by vector–scalar pair $\theta_d = (\mathbf{u}, z)$ that predicts whether the example is from the source or the target domain. Figure 11.6 shows the training across the three different layers.

The prediction loss for $(\mathbf{x}_i, y_i)$ can be written as:

$$\mathcal{L}_y^i(\theta_f, \theta_y) = \mathcal{L}_y(G_y(G_f(\mathbf{x}_i; \theta_f); \theta_y), y_i) \tag{11.28}$$

The domain loss $(\mathbf{x}_i, d_i)$, where d_i is the domain, can be written as:

$$\mathcal{L}_d^i(\theta_d, \theta_f) = \mathcal{L}_d(G_d(G_f(\mathbf{x}_i; \theta_d); \theta_f), d_i) \tag{11.29}$$

The total training loss for a single layer network can be written as:

$$\mathcal{L}_{total}(\theta_f, \theta_y, \theta_d) = \frac{1}{n} \sum_{i=1}^{n} \mathcal{L}_y^i(\theta_f, \theta_y) - \lambda \left(\frac{1}{n} \sum_{i=1}^{n} \mathcal{L}_d^i(\theta_f, \theta_d) + \frac{1}{n'} \sum_{i=n+1}^{N} \mathcal{L}_d^i(\theta_f, \theta_d) \right) \tag{11.30}$$

The hyperparameter λ controls the trade-off between the losses. The parameters are obtained by solving the equations:

$$(\hat{\theta}_f, \hat{\theta}_y) = \underset{(\theta_f, \theta_y)}{\arg\min} \, \mathcal{L}_{total}(\theta_f, \theta_y, \hat{\theta}_d) \tag{11.31}$$

$$(\hat{\theta}_d) = \underset{(\theta_d)}{\arg\max} \mathcal{L}_{total}(\hat{\theta}_f, \hat{\theta}_y, \theta_d) \tag{11.32}$$

The gradient updates are very similar to standard stochastic gradient descent with a learning rate μ except the reversal with $\lambda \frac{\partial \mathcal{L}_d^i}{\partial \theta_f}$. The gradient reversal layer has no parameter and its forward pass is the identity function, while the backward pass is the gradient from subsequent layer multiplied by -1:

$$\theta_f \longleftarrow \theta_f - \mu \left(\frac{\partial \mathcal{L}_y^i}{\partial \theta_f} - \lambda \frac{\partial \mathcal{L}_d^i}{\partial \theta_f} \right) \tag{11.33}$$

$$\theta_y \longleftarrow \theta_y - \mu \frac{\partial \mathcal{L}_y^i}{\partial \theta_y} \tag{11.34}$$

$$\theta_d \longleftarrow \theta_d - \mu \frac{\partial \mathcal{L}_d^i}{\partial \theta_d} \tag{11.35}$$

11.1.1.8 Adversarial Discriminative Domain Adaptation

Tzeng et al. propose the adversarial discriminative domain adaptation (ADDA) which uses a discriminative approach for learning the domain shifts, has no weights

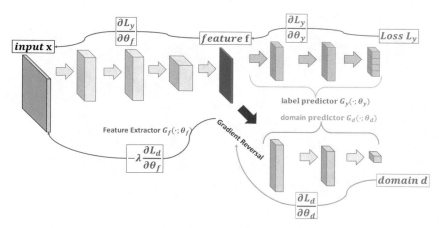

Fig. 11.6: Domain-adversarial neural network

tied between source and the target, and has a GAN loss for computing adversarial loss [Tze+17].

Let us assume we have source data $\mathbf{X}_s$ and labels Y_s drawn from source distribution $p_s(x,y)$, and target data $\mathbf{X}_t$ from a target distribution $p_t(x,y)$ with no labels. The goal is to learn a target mapping M_t and a classifier C_t that can classify K categories. In adversarial methods, the goal is to minimize the distribution distance between the source and target mapping $M_s(X_s)$ and $M_t(X_t)$, so that the source classification model C_s can be used directly on the target so that $C = C_s = C_t$. The standard supervised loss can be written as:

$$\min_{M_s,C} \mathcal{L}_{class}(\mathbf{X}_s,Y_s) = -\mathbb{E}_{(\mathbf{x}_s,y_s)\sim(\mathbf{X}_s,Y_s)} \sum_{k=1}^{K} \mathbf{1}_{[k=y_s]} \log C(M_s(\mathbf{x}_s)) \tag{11.36}$$

The domain discriminator D classifies if the data is from source or target, and D is optimized using $\mathcal{L}_{adv_D}$:

$$\min_{D} \mathcal{L}_{adv_D}(\mathbf{X}_s,\mathbf{X}_t,M_s,M_t) = -\mathbb{E}_{\mathbf{x}_s\sim\mathbf{X}_s}[\log D(M_s(\mathbf{x}_s))]$$
$$- \mathbb{E}_{\mathbf{x}_t\sim\mathbf{X}_t}[\log(1 - D(M_t(\mathbf{x}_t)))] \tag{11.37}$$

The adversarial mapping loss is given by $\mathcal{L}_{adv_M}$:

$$\min_{M_s,M_t} \mathcal{L}_{adv_M}(\mathbf{X}_s,\mathbf{X}_t,D) = -\mathbb{E}_{\mathbf{x}_s\sim\mathbf{X}_t}[\log D(M_t(\mathbf{x}_t))] \tag{11.38}$$

The training happens in phases, as shown in Fig. 11.7. The process begins with $\mathcal{L}_{adv_{class}}$ over M_s and C, using the labeled data $\mathbf{X}_s$ and labels Y_s. We can then perform adversarial adaptation by optimizing $\mathcal{L}_{adv_D}, \mathcal{L}_{adv_M}$.

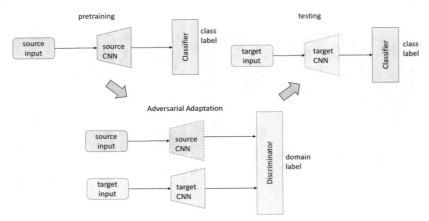

Fig. 11.7: Adversarial discriminative domain adaptation

11.1.1.9 Coupled Generative Adversarial Networks

Liu and Tuzel propose a coupled generative adversarial network (CoGAN) for learning joint distribution between two domains and show it to be very successful in computer vision [LQH16]. As discussed in Chap. 4, GANs consist of generative and discriminative models. The generative model is used to generate synthetic data resembling real data, while the discriminative model is used to distinguish between the two. Formally, a random vector $\mathbf{z}$ is input to the generative model that outputs $g(\mathbf{z})$ that has the same support as the input $\mathbf{x}$. The discriminative model outputs $f(\mathbf{x}) = 1$ if drawn from real $\mathbf{x} \sim p_X$ and $f(\mathbf{x}) = 0$ if drawn from synthetic or generated $\mathbf{x} \sim p_G$. Thus, GANs can be seen as a minimax two-player game solving through optimization:

$$\max_g \min_f V(f,g) \equiv \mathbb{E}_{\mathbf{x} \sim p_X}[-\log f(\mathbf{x})] + \mathbb{E}_{\mathbf{z} \sim p_Z}[-\log(1 - f(g(\mathbf{z})))] \qquad (11.39)$$

In CoGAN, as shown in Fig. 11.8, there are two GANs for two different domains. Generative models try to decode from higher-level features to lower-level features as opposed to discriminative models. If $\mathbf{x}_1$ and $\mathbf{x}_2$ are two inputs drawn from the marginal distribution of first ($\mathbf{x}_1 \sim p_{X_1}$) and second ($\mathbf{x}_2 \sim p_{X_2}$), respectively, then generative models GAN_1 and GAN_2 map a random vector $\mathbf{z}$ to examples having the same support as $\mathbf{x}_1$ and $\mathbf{x}_2$. The distribution of $g_1(\mathbf{z})$ and $g_2(\mathbf{z})$ is p_{G_1} and p_{G_2}. When

g_1 and g_2 are realized as MLP, then we can write:

$$g_1(\mathbf{z}) = g_1^{(m_1)}(g_1^{(m_1-1)}(\cdots g_1^{(2)}(g_1^{(1)}(\mathbf{z})))) \qquad (11.40)$$

$$g_2(\mathbf{z}) = g_2^{(m_2)}(g_2^{(m_2-1)}(\cdots g_2^{(2)}(g_2^{(1)}(\mathbf{z})))) \qquad (11.41)$$

where g_1^i and g_2^i are the layers in the corresponding GANs with layers m_1 and m_2, respectively. The structure for first few layers and the weights are identical, thus having the constraint of

$$\theta_{g_1^i} = \theta_{g_2^i} \; for \; i = 0,1,\dots k \qquad (11.42)$$

where k is the shared layers, and $\theta_{g_1^i}$ and $\theta_{g_2^i}$ are the parameters of g_1^i and g_2^i, respectively. This constraint enables the first layers that decode high-level features to decode it in the same way for both generators g_1 and g_2.

Discriminative models map the input to a probability, estimating the likelihood that the input is from the data distribution. If f_1^i and f_2^i correspond to the layers of discriminative networks for two GANs with n_1 and n_2 layers, it can be written as:

$$f_1(\mathbf{x}_1) = f_1^{(n_1)}(f_1^{(n_1-1)}(\cdots f_1^{(2)}(f_1^{(1)}(\mathbf{x}_1)))) \qquad (11.43)$$

$$f_2(\mathbf{x}_1) = f_2^{(n_2)}(f_2^{(n_2-1)}(\cdots f_2^{(2)}(f_2^{(1)}(\mathbf{x}_2)))) \qquad (11.44)$$

where f_1^i and f_2^i are the layers in the corresponding f_1 and f_2 with layers n_1 and n_2, respectively. The discriminative models work in contrast to the generative models and extract low-level features in the first layers and high-level features in the last layers. To ensure the data has the same high-level features, we share the last layers using:

$$\theta_{f_1^{(n_1-i)}} = \theta_{f_2^{(n_2-i)}} \; for \; i = 0,1,\dots (l-1) \qquad (11.45)$$

where l is the shared layers and $\theta_{f_1^i}$ and $\theta_{f_2^i}$ are the parameters of f_1^i and f_2^i, respectively. It can be shown that learning in CoGAN corresponds to a constrained minimax game given by:

$$\max_{g_1,g_2} \min_{f_1,f_2} V(g_1,g_2,f_1,f_2)$$
$$subject \; to \; \theta_{g_1^i} = \theta_{g_2^i} \; for \; i = 0,1,\dots k$$
$$\theta_{f_1^{(n_1-i)}} = \theta_{f_2^{(n_2-i)}} \; for \; i = 0,1,\dots (l-1) \qquad (11.46)$$

where the value function V is given by:

$$\max_{g_1,g_2} \min_{f_1,f_2} V(g_1,g_2,f_1,f_2) = \mathbb{E}_{\mathbf{x}_1 \sim p_{X_1}}[-\log(f_1)(\mathbf{x}_1)] + \mathbb{E}_{\mathbf{z} \sim p_Z}[-\log(1-f_1(g_1(\mathbf{z})))]$$
$$+ \mathbb{E}_{\mathbf{x}_2 \sim p_{X_2}}[-\log(f_2(\mathbf{x}_2))] + \mathbb{E}_{\mathbf{z} \sim p_Z}[-\log(1-f_2(g_2(\mathbf{z})))] \qquad (11.47)$$

The main advantage of CoGAN is that by drawing the samples separately from the marginal distributions, CoGAN can learn the joint distribution from the two domains very effectively.

11.1.1.10 Cycle Generative Adversarial Networks

Cycle-consistent adversarial networks (CycleGAN) proposed by Zhu et al. have been one of the most innovative generative adversarial networks in recent times and have wide applicability in different domains [Zhu+17]. The concept of cycle-consistency means that if we translate a sentence from language A to language B then translating it from language B to language A should give a similar sentence. The main idea is to learn to transfer from source domain X to the target domain Y when there are no examples corresponding to them available in the training data. This is done in two steps: a) learning a mapping $G : X \longrightarrow Y$, such that it is indistinguishable to know whether data came from $G(X)$ or Y using adversarial loss; and b) learning the inverse mapping $F : Y \longrightarrow X$ and introduce cycle-consistency loss so that $F(G(X)) = X$ and $G(F(Y)) = Y$ (Fig. 11.9).

The learning where $G(x)$ tries to generate data that looks similar to y while the discriminator D_Y aims at distinguishing $G(x)$ and real y can be expressed as:

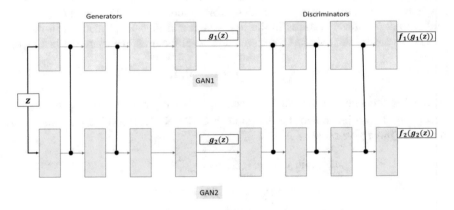

Fig. 11.8: Coupled generative adversarial networks

$$\min_{G} \max_{D_Y} \mathcal{L}_{GAN}(G, D_Y, X, Y) \tag{11.48}$$

where

$$\mathcal{L}_{GAN}(G, D_Y, X, Y) = \mathbb{E}_{y \sim p_{data}(y)}[\log D_Y(y)] + \mathbb{E}_{x \sim p_{data}(x)}[\log(1 - D_Y(G(x)))] \tag{11.49}$$

Similarly

$$\min_{F} \max_{D_X} \mathcal{L}_{GAN}(F, D_X, Y, X) \qquad (11.50)$$

$$\mathcal{L}_{GAN}(F, D_X, Y, X) = \mathbb{E}_{x \sim p_{data}(x)}[\log D_X(x)] + \mathbb{E}_{y \sim p_{data}(y)}[\log(1 - D_X(F(y)))] \qquad (11.51)$$

The cycle-consistency loss is about bringing the original data x from the translation $x \rightarrow G(x) \rightarrow F(G(x)) \approx x$ for x domain and y from the translation $y \rightarrow F(y) \rightarrow G(F(y)) \approx y$ captured as:

$$\mathcal{L}_{cyc}(G, F) = \mathbb{E}_{x \sim p_{data}(x)}[\| F(G(x)) - x \|_1] + \mathbb{E}_{y \sim p_{data}(y)}[\| G(F(y)) - y \|_1] \quad (11.52)$$

Thus the total objective thus can be written as:

$$\mathcal{L}_{total}(G, F, D_X, D_Y) = \mathcal{L}_{GAN}(G, D_Y, X, Y) + \mathcal{L}_{GAN}(F, D_X, Y, X) + \lambda \mathcal{L}_{cyc}(G, F) \qquad (11.53)$$

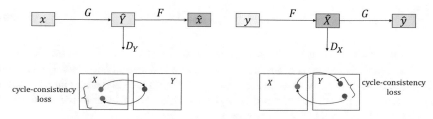

Fig. 11.9: Cycle generative adversarial networks (CycleGAN) with forward cycle-consistency and backward cycle-consistency

CycleGAN does not need the data pairs in the domains to match. It can learn the underlying relationship and help transfer between the domains.

11.1.1.11 Domain Separation Networks

Domain separation networks by Bousmalis et al. have private encoders for learning individual domains, shared encoders for learning common representations across domains, shared decoder for effective generalization using reconstruction loss, and a classifier using shared representations for robustness [Bou+16].

The source domain $\mathcal{D}_s$ has N_s labeled data $\mathbf{X}_s = \mathbf{x}_i^s, \mathbf{y}_i^s$ and the target domain $\mathcal{D}_t$ has N_t unlabeled data $\mathbf{X}_t = \mathbf{x}_i^t$. Let $E_p(\mathbf{x}; \theta_p)$ be the function that maps input $\mathbf{x}$ to a hidden representation $\mathbf{h}_p$ for a representation that is private for the domain.

Let $E_c(\mathbf{x}; \theta_c)$ be the function that maps input $\mathbf{x}$ to a hidden representation $\mathbf{h}_c$ that is common across the source and the target. Let $D(\mathbf{h}; \theta_d)$ be the decoding function that maps the hidden representation $\mathbf{h}$ to the original reconstruction $\hat{\mathbf{x}}$. Reconstruction can be given by $\hat{\mathbf{x}} = D(E_c(\mathbf{x}) + E_p(\mathbf{x}))$. Let $G(\mathbf{h}; \theta_g)$ be the classifier function that maps the hidden representation $\mathbf{h}$ to predictions $\hat{\mathbf{y}}$ given by $\hat{\mathbf{y}} = G(E_c(\mathbf{x}))$. Figure 11.10 captures the entire process of DSN.

The total loss can be written as:

$$\mathcal{L}_{total}(\theta_c, \theta_p, \theta_d, \theta_g) = \mathcal{L}_{class} + \alpha \, \mathcal{L}_{recon} + \beta \, \mathcal{L}_{difference} + \gamma \, \mathcal{L}_{similarity} \qquad (11.54)$$

where the hyperparameters α, β, γ control the weight of each loss term. The classification loss is the standard negative log-likelihood given by:

$$\mathcal{L}_{class} = -\sum_{i=0}^{N_s} \mathbf{y}_i^S \cdot \log(\hat{\mathbf{y}}_i^s) \qquad (11.55)$$

The reconstruction loss is computed using scale invariant mean-squared error:

$$\mathcal{L}_{recon} = -\sum_{i=0}^{N_s} \mathcal{L}_{si\_mse}(\mathbf{x}_i, \hat{\mathbf{x}}_i) \qquad (11.56)$$

The difference loss, as the name suggests, is applied to both domains and is meant to capture different aspects of inputs for the private and shared encoders. Let $\mathbf{H}_c^s$ and $\mathbf{H}_c^t$ be the rows of matrices which are common between the source and target hidden layers. Let $\mathbf{H}_p^s$ and $\mathbf{H}_p^t$ be the rows of matrices which are private to the source and the target hidden layers. The difference loss is given by:

$$\mathcal{L}_{difference} = \|\mathbf{H}_c^{s\,\mathsf{T}}\mathbf{H}_p^s\|_F^2 + \|\mathbf{H}_c^{t\,\mathsf{T}}\mathbf{H}_p^t\|_F^2 \qquad (11.57)$$

where $|\cdot|_F$ is the squared Frobenius norm.

The domain-adversarial similarity loss, which aims at maximizing the "confusion," is achieved via a gradient reversal layer and a domain classifier to predict the domain. If $d_i \in 0, 1$ is the ground truth of the domains for the data and $\hat{d}_i \in 0, 1$ is the predicted value of the domain, then adversarial learning can be achieved by:

$$\mathcal{L}_{similarity}^{DANN} = \sum_{i=0}^{N_s+N_t} \left\{ d_i \log \hat{d}_i + (1 - d_i) \log(1 - \hat{d}_i) \right\} \qquad (11.58)$$

The maximum-mean discrepancy (MMD) loss can also be used instead of the DANN described above.

Domain separation networks capture explicitly and jointly both the private and shared components of the domain representations making it less vulnerable to noise that is correlated with the shared distributions.

11.1.2 Theory

We will describe two topics that have been studied in the last couple of years to give a formal mapping to domain adaptation that is applicable in the deep learning area. One is the generalization of most domain adaptation networks by Tzeng et al. [Tze+17], and another is the optimization transport theory for giving a theoretical foundation to domain adaptation [RHS17].

11.1.2.1 Siamese Networks Based Domain Adaptations

Tzeng et al. present a generalized Siamese architecture which captures most implementations in domain adaptations using deep learning as shown in Fig. 11.11 [Tze+17]. The architecture has two streams, the source input which is labeled, and the target input which is unlabeled. The training is done with a combination of classification loss with either discrepancy-based loss or adversarial loss. The classification loss is computed only using the labeled source data. The discrepancy loss is computed based on the domain shift between the source and the target. The adversarial loss tries to capture latent features using the adversarial objective with respect to the domain discriminator. This study helps to put all the architectures seen as various extensions of the general architecture with changes to how classification loss, discrepancy loss, and adversarial loss are computed.

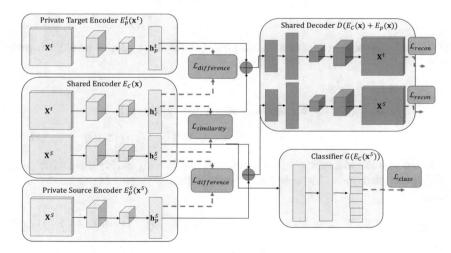

Fig. 11.10: Domain separation networks (DSN)

The setup can be generalized to drawing source-labeled samples $(\mathbf{X}_s, Y_s)$ from a distribution $p_s(x, y)$ and unlabeled target samples $\mathbf{X}_t$ from a distribution $p_t(x, y)$. The goal is to learn from source examples and a classifier C_s a representation mapping

M_s, and also have a target mapping M_t with a classifier C_t at prediction time that learns to classify unseen examples in k categories.

The goal of most adversarial methods is to minimize the distance between the distributions of $M_s(X_s)$ and $M_t(X_t)$, which implicitly means that in most cases the source and target classifiers can be same $C = C_s = C_t$. Source classification can be given in a generic loss optimization form as:

$$\min_{M_S,C} \mathcal{L}_{class}(\mathbf{X}_S, Y_S) = -\mathbb{E}_{(\mathbf{x}_s,y_s) \sim (\mathbf{X}_S, Y_S)} \sum_{k=1}^{K} \mathbb{1}_{[k=y_s]} \log C(M_S(\mathbf{x}_s)) \qquad (11.59)$$

A domain discriminator D which classifies whether the data is drawn from the source or the target can be written as:

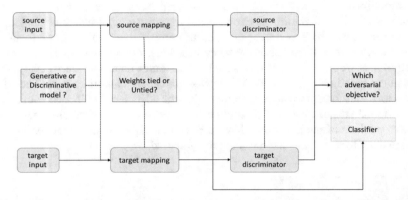

Fig. 11.11: Siamese networks for generalizing the domain adaptation implementation

$$\min_{D} \mathcal{L}_{adv_D}(\mathbf{X}_S, \mathbf{X}_T, M_S, M_T) = -\mathbb{E}_{\mathbf{x}_s \sim \mathbf{X}_S}[\log D(M_S(\mathbf{x}_s))]$$
$$- \mathbb{E}_{\mathbf{x}_t \sim \mathbf{X}_T}[\log(1 - D(M_T(\mathbf{x}_t)))] \qquad (11.60)$$

With the source and target mapping constraints given by $\psi(M_S, M_T)$, a discriminator D that can distinguish between them can be captured as an adversarial objective $\mathcal{L}_{adv_M}$:

$$\min_{D} \mathcal{L}_{adv_M}(\mathbf{X}_S, \mathbf{X}_T, M_S, M_T) \quad \min_{M_S, M_T} \mathcal{L}_{adv_M}(\mathbf{X}_S, \mathbf{X}_T, D) \quad s.t \psi(M_S, M_T) \quad (11.61)$$

Various techniques described in domain adaptation can now be understood with this general framework.

The gradient reversal process can be written in terms of optimizing the discriminator loss directly as $\mathcal{L}_{adv_M} = -\mathcal{L}_{adv_D}$.

When using GANs there are two losses: the discriminator loss and the generator loss. The discriminator loss $\mathcal{L}_{adv_D}$ remains the same, while generator loss can be written as:

$$\min_{D} \mathcal{L}_{adv_M}(\mathbf{X}_S, \mathbf{X}_T, D) = -\mathbb{E}_{\mathbf{x}_t \sim \mathbf{X}_T}[\log D(M_T(\mathbf{x}_t))] \qquad (11.62)$$

The domain confusion loss can be written as minimizing the cross-entropy loss given by:

$$\min_{D} \mathcal{L}_{adv_M}(\mathbf{X}_S, \mathbf{X}_T, D) = -\sum_{d \in s,t} \mathbb{E}_{\mathbf{x}_d \sim \mathbf{X}_D}\left[\frac{1}{2}\log D(M_d(\mathbf{x}_d)) + \frac{1}{2}\log(1 - D(M_d(\mathbf{x}_d)))\right]$$

$$(11.63)$$

11.1.2.2 Optimal Transport

In the last few years, optimal transport theory has come into prominence from various statistical, optimization, and machine learning perspectives. Optimal transport can be seen as a way of measuring the transport of data between two different distributions that are based on the geometry of the data points in the two and has a cost function related to transportation [Mon81]. This transport mechanism maps very well to domain adaptation, where the source and target domains can be seen as two different distributions, and optimal transport explains from both theory and optimization the mapping. The Wasserstein distance in optimal transport which is used to measure the distance between two distributions can also be used as a minimization objective or regularization function in the overall loss function. Optimal transport has been used to give a good generalization bound to deep domain adaptation frameworks [RHS17].

11.1.3 Applications in NLP

Glorot et al. in the very early days of deep learning showed how stacked autoencoders with sparse rectifier units could learn feature-level representations that could perform domain adaptation on sentiment analysis very effectively [GBB11c].

Nguyen and Grishman employ word embeddings along with word clustering features to show that domain adaptation in relation extraction can be very effective [NG14]. Nguyen et al. further explore the use of word embeddings and tree kernels to generate a semantic representation for relation extraction and improvements over feature based methods [NPG15]. Nguyen and Grishman show how basic CNNs with word embeddings, position embeddings, and entity type embeddings as input can learn effective representation that gives a good domain adaptation method for event detection [NG15a]. Fu et al. show the effectiveness of domain adaptation for relation extraction using domain-adversarial neural networks (DANN) [Fu+17]. They use word embeddings, position embeddings, entity type embeddings, chunking, and dependency path embeddings. They use CNNs and DANN with a gradient reversal layer to effectively learn the relationship extraction with cross-domain features.

Zhou et al. use a novel bi-transferring deep neural networks to transfer source examples into the target and vice versa for achieving close to the state-of-the-art results in sentiment classification [Zho+16]. Zhang et al. use the mapping between the keywords to the source and target and employ it in adversarial training for domain adaptation in classification [ZBJ17]. Ziser and Reichart show how pivot features (common features that are present in source and target) along with autoencoders can learn representation that is very effective in domain adaptation for sentiment classification [ZR17]. Ziser and Reichart further extend the research to a pivot-based language model in a structure-aware manner that can be employed for various classification and sequence-to-sequence-based tasks for improved results [ZR18]. Yu and Ziang combine the ideas of structural correspondence learning, pivot-based features, and joint-task learning for effective domain adaptation in sentiment classification [YJ16].

11.1.4 Applications in Speech Recognition

Falavigna et al. show how deep neural networks and automatic quality estimation (QE) can be used for domain adaptation [Fal+17]. They use a two-step process in which first manually labeled transcripts are used for evaluating WER on the data for different quality. Then adaptation is made on unseen data according to WER scores by the QE component to show significant improvements in performance.

Hosseini-Asl et al. extend the CycleGAN concepts to have multiple discriminators (MD-CycleGAN) for unsupervised non-parallel speech domain adaptation [Hos+18]. They use multiple discriminator-enabled CycleGAN to learn frequency variations in spectrograms between the domains. They use different gender speech ASR in training and testing to evaluate the domain adaptation aspect of the framework and report a good performance by using the MD-CycleGAN architecture on unseen domains.

Adapting to different speakers with different accents is one of the open research problems in speech recognition. Wang et al. in their work do a detailed analysis treating this as a domain adaptation problem with different frameworks to give important insights [Wan+18a]. They use three different speaker adaptation methods, such as linear transformation (LIN), learning hidden unit contribution (LHUC), and Kullback–Leibler divergence (KLD) on a i-vector based DNN acoustic model. They show that based on the accents using one of the methods, ASR performance can be significantly improved for not only medium-to-heavy accents but also for slight-accent speakers. Sun et al. use domain-adversarial training for solving accented speech in ASR [Sun+18a]. Employing domain adversarial training in the learning objective from unlabeled target domain with different accents to separate source and target while using labeled source domain for classification, they show a significant drop in error rates for unseen accents.

Improving ASR quality in the presence of noise by improving the robustness of the models can also be approached from the domain adaptation view based on how the noise in target domain or unseen data is different from the source domains. Serdyuk et al. use GANs for domain adaptation in unseen noisy target datasets [Ser+16]. The model has the encoder, decoder, and the recognizer with a hidden representation in between that is used to perform dual tasks of improving the recognition and minimizing domain discrimination. They show their method to be better at generalization when the target domain has more noise categories than the ones used in the source training data. Sun et al. use adversarial data augmentation using fast gradient sign method (FSGM) to show significant improvements in the robustness of acoustic models [Sun+18b]. Meng et al. use domain separation networks (DSN) for domain adaptation between source and targets for robustness on target data with different noise levels [Men+17]. The shared components learn the domain invariance between the source and the target domains. The private components are orthogonal with the shared ones and learn to increase domain invariance. They show a significant decrease in the WER over baseline with an unadapted acoustic model with their approach.

11.2 Zero-Shot, One-Shot, and Few-Shot Learning

The extremes of domain adaptation or transfer learning problem are when there are limited training examples to match the test example. The best example is the facial recognition problem from computer vision, where there is exactly 1 training example for each person, and when someone appears, the need is to match the existing or classify it as a new unseen. Based on a number of training examples corresponding to the unseen example we get at prediction time, there are different flavors such as **zero-shot learning**, **one-shot learning**, and **few-shot learning**. In the next sections, we will discuss each of them and techniques that have been popular to address them.

11.2.1 Zero-Shot Learning

Zero-shot learning is a form of transfer learning where we have absolutely no training data available for the classes we will see in the test set, or when the model is used for predictions. The idea is to learn a mapping from classes to a vector, in such a way that an unseen class in the future can be mapped to the same space, and "closeness" to the existing classes can be used to provide some information about the unseen class. An example from the NLU domain would be when the data is available about computers and knowledge bases (KB) exist for retrieving information about them, a question on "what is the cost for specific part for a function such as the display" can be formed as a query to a KB having a database of components, subcomponents, functions, and parts. Learning this mapping can be used to transfer it to another completely different domain. For example, this can be used in car-manufacturing on similar queries if the cost of parts for performing specific functions is normally used.

11.2.1.1 Techniques

We will illustrate a general method and variations that have been successful in computer vision and language/speech understanding/recognition tasks [XSA17].

The approach is to measure the similarity between source and target domains. In computer vision, for example, one way is to map the label space to a vector space based on side information, such as attributes capturing the picture. The attributes can be meta-level or image-level features, such as "presence of specific color," "size of the object," and others. The vector representation can be a one-hot vector of these attributes. The source data features are embedded in the source feature space. The next step is to find the compatibility between the source feature space, as shown in Fig. 11.12, using a **compatibility function**.

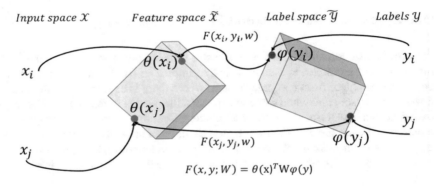

Fig. 11.12: Zero-shot learning

Formally, the source dataset $S = \{(x_n, y_n), n = 1, \cdots N\}$ with input and labels $x_n \in \mathcal{X}$, $y_n \in \mathcal{Y}$, respectively. The goal is to learn a function $f(x)$ that minimizes the loss in predicting the label y and can be written using the minimization of the empirical risk in the form:

$$\frac{1}{N} \sum_{n=1}^{N} L(y_n, f(x_n)) \tag{11.64}$$

where L is a loss measuring function. For classification, it can be 0 when matching and 1 when not matching. Let θ be the source embedding function that transforms the input data to its feature space, i.e., $\theta : \mathcal{X} \to \tilde{\mathcal{X}}$. Similarly, let $\varphi : \mathcal{Y} \to \tilde{\mathcal{Y}}$ be the label embedding function that transforms the labels into a space using the attributes.

The compatibility function $F : \mathcal{X} \times \mathcal{Y} \rightarrow$ and the function f are defined in terms of model parameters w of F, i.e., how the pair (x, y) are compatible given the parameter w:

$$f(x; w) = \arg\max_{y \in \mathcal{Y}} F(x, y; w) \tag{11.65}$$

Different forms of compatibility functions exist and are mentioned below:

1. Pairwise Ranking: A popular method which uses convex objective, pairwise ranking, and SGD updates is given by:

$$\sum_{y \in \mathcal{Y}^{train}} [\Delta(y_n, y) + F(x_n, y; W) - F(x_n, y_n; W)]_+ \tag{11.66}$$

where Δ is the $0/1$ loss, and F is the linear compatibility function.

2. Weighted Pairwise Ranking: An extension to the above which adds weights as in:

$$\sum_{y \in \mathcal{Y}^{train}} l_k [\Delta(y_n, y) + F(x_n, y; W) - F(x_n, y_n; W)]_+ \tag{11.67}$$

where $l_k = \sum_{i=1}^{k} \alpha_i$, $\alpha_i = \frac{1}{i}$, and k is the number of ranks.

3. Structured Joint Embedding (SJE): Another pairwise ranking but for multiclass scenario, where one uses the max function to find the most violating class, is given by:

$$\max_{\sum_{y \in \mathcal{Y}^{train}}} [\Delta(y_n, y) + F(x_n, y; W) - F(x_n, y_n; W)]_+ \tag{11.68}$$

4. Embarrassingly Simple Zero-Shot Learning: Extension to the above SJE method where a regularization term is added:

$$\gamma \|W\phi(y)\|^2 + \lambda \|\theta(x)^{\mathsf{T}} W\|^2 + \beta \|W\|^2 \tag{11.69}$$

where γ, λ, β are the regularization parameters.

5. Semantic Autoencoder: Another technique that uses linear autoencoder to project from $\theta(x)$ to $\varphi(y)$ space:

$$\min_W \|\theta(x) - W^{\mathsf{T}} \varphi(y)\|^2 + \lambda \|W\theta(x) - \varphi(y)\|^2 \tag{11.70}$$

6. Latent Embeddings: To overcome the limitations of the linear weights W, a piecewise-linear modification is made to the compatibility function to achieve non-linearity as given by:

$$F(x, y; W) = \theta(x)^{\mathsf{T}} W_i \varphi(y) \tag{11.71}$$

where W_i are different linear weights learned.

7. Cross Model Transfer: Performing non-linear transformation using two layered neural networks with weights W_1 and W_2 and objective function is another non-linear technique:

$$\sum_{y \in \mathcal{Y}^{train}} \sum_{x \in \mathcal{X}} \|\varphi(y) - W_1 \tanh(W_2 \theta(x)))\| \qquad (11.72)$$

8. Direct Attribute Prediction: Another technique uses attributes associated with the class to be learned directly, given by:

$$f(x) = \arg\max_c \prod_{m=1}^{M} \frac{p(a_m^c|x)}{p(a_m^c)} \qquad (11.73)$$

where M is the total number of attributes, a_m^c is the mth attribute of class c, and $p(a_m^c|x)$ is that attribute probability associated with the given data x.

11.2.2 One-Shot Learning

The general problem in one-shot learning is to learn from a dataset where there is one example for a class. The same general form is used for the similarity function in representation between the training examples, so that during the prediction the similarity function is used to find the closest available example in the training data.

11.2.2.1 Techniques

The Siamese network-based architectures with variations are generally the common way to learn similarities in these frameworks. The network parameters are learned through pairwise learning from the training dataset, as shown in Fig. 11.13. One variation is that, instead of the fully connected layers going to a softmax layer, features or encoding of the input can be used for similarity; the resulting are called **matching networks**. One way to learn the parameters of the network is during training time to minimize the difference when inputs are similar and maximize when dissimilar while during prediction to use the learned representation to compute similarity with existing training samples. If x_i and x_j are two examples from the training data, the similarity function can be the difference between the two predictions in the Siamese networks given by:

$$d(x_i, x_j) = \|f(x_i) - f(x_j)\|_2^2 \qquad (11.74)$$

Another way to learn the parameters is through the **triplet loss function** by Schroff et al. [FSP15]. The idea is to pick an **anchor** data x_A for which a positive x_P and a negative x_N sample are used to learn parameters of the network, so that the difference between anchor data and positive data is maximized, and the difference between anchor data and the negative is minimized:

$$\mathcal{L}(x_A, x_P, x_N) = \max(\|f(x_A) - f(x_P)\|_2^2 - \|f(x_A) - f(x_N)\|_2^2 + \alpha, 0) \qquad (11.75)$$

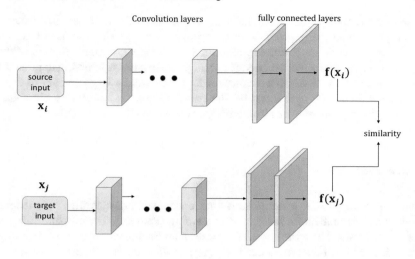

Fig. 11.13: One-shot learning

In Eq. 11.75, the parameter α is similar to the margin in SVMs. The training data is used to generate the triplets, and stochastic gradient method can be used for learning the parameters with this loss function.

11.2.3 Few-Shot Learning

Few-shot learning is a relatively easier form of learning as compared to the previous two. In general, most of the techniques mentioned in one-shot learning can be used for few-shot learning too, but we will illustrate a few additional techniques that have been successful.

11.2.3.1 Techniques

The deep learning techniques for few-shot learning can be described as either data-based approaches or model-based approaches. In the data-based approach, some form of augmenting the training data in different forms is the general process utilized to increase the number of similar samples.

In contrast, the model- or parameter-based approach enforces regularization in some form to prevent overfitting of the model from the limited training samples. Donghyun et al. use the interesting idea of correlating activations from input data to form "groups" of similar neurons or parameters per layer in the source data [Yoo+18]. The hyperparameter "number of groups" per layer is chosen using k-means clustering algorithm, and k is further learned using reinforcement techniques.

Once they are trained on source dataset, these groups of neurons are fine-tuned on the target domain using group-wise backpropagation. As the number of parameters increases, with small training data for each category, optimization algorithms such as SGD are not effective. Mengye et al. propose a **meta-learning** approach for solving this in two steps: (a) a teacher model learns from a large amount of data to capture the parameter space, and (b) then guides the actual pupil or a classifier to learn using the parameter manifold giving excellent results [Ren+18].

11.2.4 Theory

Palatucci et al. present a semantic output code mapping classifier as a theoretical base and formalization for zero-shot learning [Pal+09]. The classifier mapping helps to understand how the knowledge base and the semantic features of outputs get mapped, and how the learning can happen even when the novel classes are missing from the training data using the PAC framework.

Fei-Fei Li et al. propose a Bayesian framework for giving a theoretical base to one-shot learning in the object identification domain [FFFP06]. By modeling prior knowledge of the data as probability density function on the parameters with these models, posteriors being the categories of objects, Bayesian framework shows how models carry information even with a very few examples in training to correctly identify the categories.

Triantafillou et al. propose an information retrieval framework and implementation for modeling few-shot learning [TZU17]. This paper proposes learning a similarity metric for mapping objects into a space, where they are grouped based on their similarity relationship. The training objective optimizes relative orderings of the data points in each training batch to leverage importance in the low data regime.

11.2.5 Applications in NLP and Speech Recognition

Most of the applications of zero-shot, one-shot, and few-shot learning have been in computer vision. Only recently have there been applications in NLP and speech. Pushp and Srivastav employ zero-shot learning in classification for text categorization [PS17]. The source dataset is the news headlines crawled from the web, and the categories are the search engine. The target test data is the UCI news and the tweets categorization dataset. They employ different neural architectures based on how and what one feeds to the LSTM networks. The trained model is then applied to the dataset which has not seen the relationships before (UCI news and tweets) to get very impressive results, showing the effectiveness of zero-shot learning methods.

Levy et al. employ zero-shot learning in relation extraction by learning to answer questions from a corpus [LS17]. Yogatama et al. try to explore RNNs as generative models and empirically show the promise of generative learning in a zero-shot learning setting [Yog+17]. The relationship is learned by posing questions and having sentences in the answers that map to an entity, where the relationship is mentioned from slot-filling datasets such as WikiReading. They show that even on the unseen relationship, the zero-shot learning shows enough promise as a methodology. Mitchell et al. employ zero-shot learning using explanations about the labels or categories to learn the embedding space using constraints and show good results on email categorization [MSL18].

Dagan et al. propose a zero-shot learning framework for the event extraction problem using event ontologies and small, manually annotated labeled datasets [Dag+18]. They show transferability to even unseen types and additionally report results close to the state of the art.

Yan et al. address the difficult short text classification problem by using few-shot learning [YZC18]. They use the Siamese CNNs to learn the encoding that distinguishes complex or informal sentences. Different structures and topics are learned using the few-shot learning method and shown to generalize and have better accuracies than many traditional and deep learning methods.

Ma et al. have proposed a neural architecture for both few-shot and zero-shot learning on fine-grained named entity typing, i.e., detecting not only the entity from the sentence, but also the type (for example, "John is talking using his phone" not only identifies "John" as the entity, but can also decode that "John" is the speaker [MCG16]). They use prototypical and hierarchical information to learn label embeddings and give a huge performance boost for the classification. In their work, Yazdani and Henderson use zero-shot learning for spoken language understanding, where they assign label actions with attributes and values from the utterance output of ASR dialogs [YH15]. They build a semantic space between the words and the labels, so that it can form a representation layer that predicts unseen words and labels in very effective way.

Rojas-Barahona et al. have shown the success of deep learning and zero-shot learning in semantic decoding of spoken dialog systems. They use deep learning for learning features jointly from known and unknown categories [Roj+18]. They then use unsupervised learning to tune the weights, further using risk minimization to achieve zero-shot learning when tested on unseen data with slot pairs not known in the training set. Keren et al. use one-shot learning with Siamese networks to compute the similarity between the single exemplars from the source data to the unseen examples from the target data in the spoken term detection problem in the audio domain [Ker+18].

11.3 Case Study

We will go through a detailed case study to explore and understand different things discussed in the chapter from a practical point of view. We chose the Amazon product review dataset published in the research by Blitzer et al. [BDP07] for the **sentiment classification** task. The dataset has reviews for various product domains such as **books, DVDs, kitchen, and electronics**. All the domains have 2000 labeled examples with binary labels (positive and negative) based on the reviews. Kitchen and electronics domain has large number of unlabeled examples as well. In our experiments we have not used the unlabeled examples but treated many labeled as unlabeled when required.

We chose two different cases: (1) the source domain is kitchen and the target domain is electronics and (2) the source domain is books and the target domain is kitchen for our experiments. We divided all the datasets into training and testing with 1600 and 400 examples, respectively. The validation data is chosen from the training dataset either as a percent or a stratified sample. Though the goal is not to replicate papers or fine-tune each method to get best results, we have done some parameter tuning and kept most parameters standard or constant to see a relative impact.

11.3.1 Software Tools and Libraries

We will describe the main open source tools and libraries we have used below for our case study. There are some open source packages for specific algorithms which we have either used, adapted, or extended that are mentioned in the notebook itself:

- **Keras** (www.keras.io)
- **TensorFlow** (https://www.tensorflow.org/)
- **Pandas** (https://pandas.pydata.org/)
- **scikit-learn** (http://scikit-learn.org/)
- **Matplotlib** (https://matplotlib.org/)

11.3.2 Exploratory Data Analysis

Similar to the other case studies we will be performing some basic EDA to understand the data and some of its characteristics. The plots shown in Fig. 11.14a, b show the word distribution bar charts across entire corpus of source and target for the sentiments. It clearly shows that going from the domain kitchen to electronics may not be that different as going from books to kitchen reviews.

The plots shown in Fig. 11.15a–c illustrate the word cloud for the positive sentiment data across books, kitchen, and electronics reviews. Just visually exploring some high frequency words, the similarity between the word cloud of kitchen–electronics as also the differences between books–kitchen is very evident. The plots shown in Fig. 11.16a–c depicting the word cloud for the negative sentiment data across books, kitchen, and electronics reviews also illustrate the same characteristics.

11.3.3 Domain Adaptation Experiments

We next describe in detail all the experiments we carried out with transfer learning techniques in the form of training process, model, algorithms, and changes. Again, the goal was not to get the best tuned models for each but understand practically how each technique with its biases and processes performs on some of these complex

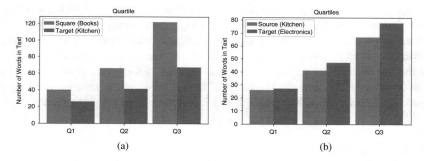

Fig. 11.14: Word distribution comparisons in quartiles of 25, 50, and 75%. (**a**) Books and kitchen comparison. (**b**) Kitchen and electronics comparison

Fig. 11.15: Word cloud for positive sentiments from (**a**) books, (**b**) kitchen, and (**c**) electronics data, respectively

Fig. 11.16: Word cloud for the negative sentiments from (**a**) books, (**b**) kitchen, and (**c**) electronics data, respectively

real-world tasks. We carry the experiments for both books–kitchen and kitchen–electronics as our source–target domains. We use classification accuracy as a metrics to see the performance as the test data was having equal number of positive and negative sentiments.

11.3.3.1 Preprocessing

We perform some basic preprocessing on the raw data for carrying out the sentiment classification tasks. The data is parsed from XML based documents, tokenized into words with basic stop words removed and some basic padding of sequences to give a constant maximum length representation for each. We create a vocabulary of words by finding all the words from source and the target sealed with maximum size of 15000 for most experiments. For some of the vector–space models we use bag-of-words representation with n-grams of size 2 and maximum features of size 10000.

11.3.3.2 Experiments

We will use Kim's CNN model shown in Fig. 11.17 as our classifier model in most experiments. We used the standard GloVe embeddings with 100 dimensions which were trained on 6 billion words. We will list the name of the experiments and the purpose behind it below:

1. **Train Source + Test Target**: The goal is to understand the transfer learning loss that happens when you train on the source data and test only on the target data due to domain change. This as we discussed can happen incrementally over time or due to completely different environment where the model is deployed. This gives a basic worst case analysis for our experiments.
2. **Train Target + Test Target**: This experiment gives us the best case analysis for the model which has not seen the source data but completely trained on the target training data and predicted on the target test data.
3. **Pre-trained Embeddings Source + Train Target**: We try to understand the impact of unsupervised pre-trained embeddings on the learning process. The embedding layer is frozen and non-trainable in this experiment. We train the model on the target domain using unsupervised embeddings and test on the target test set.
4. **Pre-trained Embeddings Source + Train Target + Fine-tune Target**: We train the model on the target domain using unsupervised embeddings but fine-tune the embeddings layer with the target train data.
5. **Pre-trained Embeddings + Train Source + Fine-tune Source and Target**: This can be the best case of pre-training and fine-tuning where you get the advantage of learning embeddings from unsupervised, train on the source, fine-tune on the target, and thus have more examples for learning useful representation across.

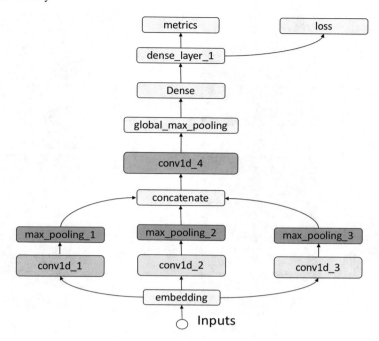

Fig. 11.17: Kim's CNN model

6. **Stacked Autoencoders and DNNs**: We use the DNNs with stacked autoencoders for the unsupervised latent feature representation learning. Train the model on the source domain in unsupervised way. Fine-tune the model with new classification layers on the target training data and test on the target data.

7. **Stacked Autoencoders and CNNs**: Goal is to understand the impact of latent representation learned from unseen source data on the target domain using CNNs for autoencoders as shown in Fig. 11.18a, b.

A sample code that shows how the autoencoder is constructed:

```
input_layer = Input(shape=(300, 300))
# encoding layers to form the bottleneck
encoded_h1 = Dense(128, activation='tanh')(input_i=
    layer)
encoded_h2 = Dense(64, activation='tanh')(encoded_h1)
encoded_h3 = Dense(32, activation='tanh')(encoded_h2)
encoded_h4 = Dense(16, activation='tanh')(encoded_h3)
encoded_h5 = Dense(8, activation='tanh')(encoded_h4)
# latent or coding layer
latent = Dense(2, activation='tanh')(encoded_h5)
# decoding layers
decoder_h1 = Dense(8, activation='tanh')(latent)
decoder_h2 = Dense(16, activation='tanh')(decoder_h1)
```

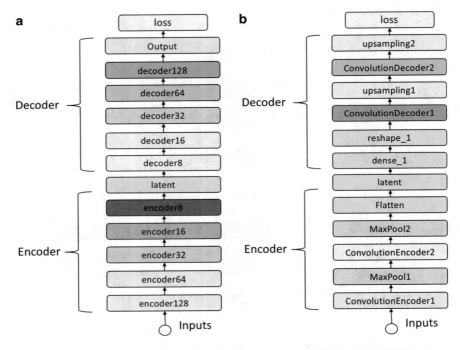

Fig. 11.18: Unsupervised training from source using (**a**) DNN and (**b**) CNN autoencoders and further trained/tested on target with classification layer in Keras

```
13   decoder_h3 = Dense(32, activation='tanh')(decoder_h2)
14   decoder_h4 = Dense(64, activation='tanh')(decoder_h3)
15   decoder_h5 = Dense(128, activation='tanh')(decoder_h4)
16   # output layer
17   output_layer = Dense(300,activation='tanh')(decoder_h5)
18   # autoencoder using deep neural networks
19   autoencoder = Model(input_layer, output_layer)
20   autoencoder.summary()
21   autoencoder.compile('adadelta', 'mse')
22
```

Using the autoencoder with encoding layers for classification:

```
1    # create a sequential model
2    classification_model = Sequential()
3    # add all the encoding layers from autoencoder
4    classification_model.add(autoencoder.layers[0])
5    classification_model.add(autoencoder.layers[1])
6    classification_model.add(autoencoder.layers[2])
7    classification_model.add(autoencoder.layers[3])
8    classification_model.add(autoencoder.layers[4])
9    classification_model.add(autoencoder.layers[5])
10   # flatten the output
11   classification_model.add(Flatten())
```

```
12    # classification layer
13    classification_model.add(Dense(2,activation='softmax'))
14    classification_model.compile(optimizer='rmsprop',
15                                 loss='
      categorical_crossentropy',
16                                 metrics=['accuracy'])
17
```

8. **Marginalized Stacked Autoencoders**: The goal of this experiment is to understand the impact of mSDA architecture in domain adaptation [Che+12]. We first learn the joint representation using the source and target data. Next, we use the last layer as the feature layer from mSDA concatenated with input layers to train SVM from labeled source data and predict on unlabeled target test data.

9. **Second-Order Statistical-based Method (Deep CORAL, CMD, and MMD)**: The goal is to see if the target data is unlabeled, can a second-order statistical-based method that can learn from source and target be useful in predicting the target under the domain shift.

10. **Domain-Adversarial Neural Network (DANN)**: The goal is to see if the target data is unlabeled, can an adversarial based method that can learn from source and target be useful in predicting the target under the domain shift (Table 11.1).

11.3.3.3 Results and Analysis

Table 11.1: Domain adaptation experiments on two different datasets for analyzing impact of source–target domain shift

Experiment	Source (books) and target (kitchen) test accuracy	Source (kitchen) and target (electronics) test accuracy
Train source + test target	69.0	78.00
Train target + test target	84.25	82.5
Pre-trained embeddings source + train target	81.5	80.25
Pre-trained embeddings source + train target + fine-tune target	85.0	84.5
Pre-trained embeddings + train source + fine-tune source and target	85.75	86.75
Stacked autoencoders and DNNs	67.75	63.75
Stacked autoencoders and CNNs	78.25	79.25
Marginalized stacked autoencoders	48.0	69.75
CORAL	63.25	69.25
CMD	63.25	69.25
MMD	63.25	69.25
DANN	75.00	80.0

Some of the observations and analysis of the results are given below:

1. Books to kitchen has higher domain transfer loss, with train accuracy (78.25) and test accuracy (69.00) it is 9.25, as compared to kitchen to electronics, with train accuracy (83.75) and test accuracy (78.00) it is 5.75. The word cloud and the data distribution confirm that reviews written for books are very different as compared to kitchen and electronics.
2. Using the pre-trained embeddings has an impact and the incremental improvement seen going from just the frozen embeddings to the embeddings trained on source and target justifies the transfer learning.
3. One of the best results is seen for both books–kitchen and kitchen–electronics is when the pre-trained embeddings are used, trained end-to-end first on source and then on the target. Thus, the advantage of learning unsupervised and fine-tuning to adapt to the domain shift is very evident.
4. Stacked autoencoders with CNNs show better results than with plain DNNs, proving the effectiveness of autoencoders in capturing the latent features and layered CNNs in capturing the signals for classification.
5. Most of the statistical techniques such as CORAL, CMD, and MMD don't show good performance
6. Adversarial methods such as DANN show a lot of promise with just the shallow networks.

11.3.4 Exercises for Readers and Practitioners

Some other interesting problems readers and practitioners can attempt on their own include:

1. What will be the impact of combining source and target training data together and test on unseen target test?
2. What will be the impact of using labeled and unlabeled data from source and target to learn embeddings and then with various techniques? Do sentiment based embeddings give better results than general embeddings?
3. What will be the impact of different embedding techniques learned in Chap. 5 on the experiments?
4. What will be the impact of different deep learning frameworks for classification that we learned in Chap. 6 on the experiments?
5. What will we see with other domain adaptation techniques such as CycleGAN or CoGAN?
6. What will be the transfer loss and improvements on other source–target such as DVD–kitchen?
7. Which of these techniques can be employed for speech recognition transfer learning problems?

References

[BDP07] John Blitzer, Mark Dredze, and Fernando Pereira. "Biographies, Bollywood, boomboxes and blenders: Domain adaptation for sentiment classification". In: *ACL*. 2007, pp. 187–205.

[Bou+16] Konstantinos Bousmalis et al. "Domain Separation Networks". In: *Advances in Neural Information Processing Systems 29*. Ed. by D. D. Lee et al. 2016, pp. 343–351.

[BM10] Lorenzo Bruzzone and Mattia Marconcini. "Domain Adaptation Problems: A DASVM Classification Technique and a Circular Validation Strategy". In: *IEEE Trans. Pattern Anal. Mach. Intell.* 32.5 (May 2010), pp. 770–787.

[Che+12] Minmin Chen et al. "Marginalized Denoising Autoencoders for Domain Adaptation". In: *Proceedings of the 29th International Conference on International Conference on Machine Learning*. ICML'12. 2012, pp. 1627–1634.

[CBG13] Sumit Chopra, Suhrid Balakrishnan, and Raghuraman Gopalan. "DLID: Deep learning for domain adaptation by interpolating between domains". In: *in ICML Workshop on Challenges in Representation Learning*. 2013.

[Csu17] Gabriela Csurka, ed. *Domain Adaptation in Computer Vision Applications*. Advances in Computer Vision and Pattern Recognition. Springer, 2017.

[Dag+18] Ido Dagan et al. "Zero-Shot Transfer Learning for Event Extraction". In: *Proceedings of the 56th Annual Meeting of the Association for Computational Linguistics, ACL 2018*. 2018, pp. 2160–2170.

[Fal+17] Daniele Falavigna et al. "DNN adaptation by automatic quality estimation of ASR hypotheses". In: *Computer Speech & Language* 46 (2017), pp. 585–604.

[FFFP06] Li Fei-Fei, Rob Fergus, and Pietro Perona. "One-Shot Learning of Object Categories". In: *IEEE Trans. Pattern Anal. Mach. Intell.* 28.4 (Apr. 2006), pp. 594–611.

[FSP15] Dmitry Kalenichenko Florian Schroff and James Philbin. "FaceNet: A unified embedding for face recognition and clustering". In: *2015 IEEE Conference on Computer Vision and Pattern Recognition, CVPR 2015*, 2015, pp. 815–823.

[Fu+17] Lisheng Fu et al. "Domain Adaptation for Relation Extraction with Domain Adversarial Neural Network". In: *Proceedings of the Eighth International Joint Conference on Natural Language Processing, IJCNLP*. 2017, pp. 425–429.

[Gan+16b] Yaroslav Ganin et al. "Domain-adversarial Training of Neural Networks". In: *J. Mach. Learn. Res.* 17.1 (Jan. 2016), pp. 2096–2030.

[GBB11b] Xavier Glorot, Antoine Bordes, and Yoshua Bengio. "Domain Adaptation for Large-Scale Sentiment Classification: A Deep Learning Approach". In: *Proceedings of the 28th International Conference on Ma-*

chine Learning, ICML 2011, Bellevue, Washington, USA, June 28 - July 2, 2011. 2011, pp. 513–520.

[GBB11c] Xavier Glorot, Antoine Bordes, and Yoshua Bengio. "Domain Adaptation for Large-Scale Sentiment Classification: A Deep Learning Approach". In: *Proceedings of the 28th International Conference on Machine Learning, ICML*. 2011, pp. 513–520.

[GGS13] Boqing Gong, Kristen Grauman, and Fei Sha. "Connecting the Dots with Landmarks: Discriminatively Learning Domain-invariant Features for Unsupervised Domain Adaptation". In: *Proceedings of the 30th International Conference on International Conference on Machine Learning - Volume 28*. ICML'13. 2013, pp. I–222–I–230.

[Hos+18] Ehsan Hosseini-Asl et al. "A Multi-Discriminator CycleGAN for Unsupervised Non-Parallel Speech Domain Adaptation". In: *CoRR* abs/1804.00522 (2018).

[KRL10] Koray Kavukcuoglu, Marc'Aurelio Ranzato, and Yann LeCun. "Fast Inference in Sparse Coding Algorithms with Applications to Object Recognition". In: *CoRR* abs/1010.3467 (2010).

[Ker+18] Gil Keren et al. "Weakly Supervised One-Shot Detection with Attention Siamese Networks". In: *CoRR* abs/1801.03329 (2018).

[LS17] Roger Levy and Lucia Specia, eds. *Proceedings of the 21st Conference on Computational Natural Language Learning (CoNLL 2017), Vancouver, Canada, August 3–4, 2017*. Association for Computational Linguistics, 2017.

[LQH16] Ming-Yu Liu and Oncel Tuzel. "Coupled Generative Adversarial Networks". In: *Advances in Neural Information Processing Systems 29*. Ed. by D. D. Lee et al. 2016, pp. 469–477.

[LW15] Mingsheng Long et al. "Learning Transferable Features with Deep Adaptation Networks". In: *Proceedings of the 32Nd International Conference on International Conference on Machine Learning - Volume 37*. ICML'15. 2015, pp. 97–105.

[MCG16] Yukun Ma, Erik Cambria, and Sa Gao. "Label Embedding for Zero-shot Fine-grained Named Entity Typing". In: *COLING 2016, 26th International Conference on Computational Linguistics*. 2016, pp. 171–180.

[Men+17] Zhong Meng et al. "Unsupervised adaptation with domain separation networks for robust speech recognition". In: *2017 IEEE Automatic Speech Recognition and Understanding Workshop*. 2017, pp. 214–221.

[MSL18] Tom M. Mitchell, Shashank Srivastava, and Igor Labutov "Zero-shot Learning of Classifiers from Natural Language Quantification". In: *Proceedings of the 56th Annual Meeting of the Association for Computational Linguistics, ACL 2018, Melbourne, Australia, July 15–20, 2018, Volume 1: Long Papers*. 2018, pp. 306–316.

[Mon81] Gaspard Monge. *Mémoire sur la théorie des déblais et des remblais*. De l'Imprimerie Royale, 1781.

[NG14]   Thien Huu Nguyen and Ralph Grishman. "Employing Word Repre-
             sentations and Regularization for Domain Adaptation of Relation Ex-
             traction". In: *Proceedings of the 52nd Annual Meeting of the Associa-
             tion for Computational Linguistics, ACL.* 2014, pp. 68–74.

[NG15a]   Thien Huu Nguyen and Ralph Grishman. "Event Detection and Do-
             main Adaptation with Convolutional Neural Networks". In: *Proceed-
             ings of the 7th International Joint Conference on Natural Language
             Processing of the Asian Federation of Natural Language Processing.*
             2015, pp. 365–371.

[NPG15]   Thien Huu Nguyen, Barbara Plank, and Ralph Grishman. "Semantic
             Representations for Domain Adaptation: A Case Study on the Tree
             Kernel-based Method for Relation Extraction". In: *Proceedings of the
             53rd Annual Meeting of the Association for Computational Linguis-
             tics and the 7th International Joint Conference on Natural Language
             Processing of the Asian Federation of Natural Language Processing,
             ACL.* 2015, pp. 635–644.

[Pal+09]   Mark Palatucci et al. "Zero-shot Learning with Semantic Output
             Codes". In: *NIPS.* Curran Associates, Inc., 2009, pp. 1410–1418.

[Pan+11]   Sinno Jialin Pan et al. "Domain Adaptation via Transfer Component
             Analysis". In: *IEEE Trans. Neural Networks* 22.2 (2011), pp. 199–
             210.

[PS17]   Pushpankar Kumar Pushp and Muktabh Mayank Srivastava. "Train
             Once, Test Anywhere: Zero-Shot Learning for Text Classification".
             In: *CoRR* abs/1712.05972 (2017).

[RHS17]   Ievgen Redko, Amaury Habrard, and Marc Sebban. "Theoretical
             Analysis of Domain Adaptation with Optimal Transport". In: *Machine
             Learning and Knowledge Discovery in Databases - European Con-
             ference, ECML PKDD 2017, Skopje, Macedonia, September 18–22,
             2017, Proceedings, Part II.* 2017, pp. 737–753.

[Ren+18]   Mengye Ren et al. "Meta-Learning for Semi-Supervised Few-Shot
             Classification". In: *CoRR* abs/1803.00676 (2018).

[Roj+18]   Lina Maria Rojas-Barahona et al. "Nearly Zero-Shot Learning
             for Semantic Decoding in Spoken Dialogue Systems". In: *CoRR*
             abs/1806.05484 (2018).

[Sej+12]   Dino Sejdinovic et al. "Equivalence of distance-based and RKHS-
             based statistics in hypothesis testing". In: *CoRR* abs/1207.6076
             (2012).

[Ser+16]   Dmitriy Serdyuk et al. "Invariant Representations for Noisy Speech
             Recognition". In: *CoRR* abs/1612.01928 (2016).

[SS16]   Baochen Sun and Kate Saenko. "Deep CORAL: Correlation Align-
             ment for Deep Domain Adaptation". In: *ECCV Workshops (3).* Vol.
             9915. Lecture Notes in Computer Science. 2016, pp. 443–450.

[Sun+18a]   Sining Sun et al. "Domain Adversarial Training for Accented Speech
             Recognition". In: *CoRR* abs/1806.02786 (2018).

[Sun+18b]   Sining Sun et al. "Training Augmentation with Adversarial Examples for Robust Speech Recognition". In: *CoRR* abs/1806.02782 (2018).

[TZU17]   Eleni Triantafillou, Richard S. Zemel, and Raquel Urtasun. "Few-Shot Learning Through an Information Retrieval Lens". In: *NIPS*. 2017, pp. 2252–2262.

[Tze+14]   Eric Tzeng et al. "Deep Domain Confusion: Maximizing for Domain Invariance". In: *CoRR* abs/1412.3474 (2014).

[Tze+15]   Eric Tzeng et al. "Simultaneous Deep Transfer Across Domains and Tasks". In: *Proceedings of the 2015 IEEE International Conference on Computer Vision (ICCV)*. ICCV '15. 2015, pp. 4068–4076.

[Tze+17]   Eric Tzeng et al. "Adversarial Discriminative Domain Adaptation". In: *2017 IEEE Conference on Computer Vision and Pattern Recognition, CVPR 2017, Honolulu, HI, USA, July 21–26, 2017*. 2017, pp. 2962–2971.

[Wan+18a]   Ke Wang et al. "Empirical Evaluation of Speaker Adaptation on DNN based Acoustic Model". In: *CoRR* abs/1803.10146 (2018).

[XSA17]   Yongqin Xian, Bernt Schiele, and Zeynep Akata. "Zero-Shot Learning - The Good, the Bad and the Ugly". In: *2017 IEEE Conference on Computer Vision and Pattern Recognition, CVPR*. 2017, pp. 3077–3086.

[YZC18]   Leiming Yan, Yuhui Zheng, and Jie Cao. "Few-shot learning for short text classification". In: *Multimedia Tools and Applications* (2018), pp. 1–12.

[YH15]   Majid Yazdani and James Henderson. "A Model of Zero-Shot Learning of Spoken Language Understanding". In: *Proceedings of the 2015 Conference on Empirical Methods in Natural Language Processing*. 2015, pp. 244–249.

[Yog+17]   Dani Yogatama et al. "Generative and Discriminative Text Classification with Recurrent Neural Networks". In: *CoRR* abs/1703.01898 (2017).

[Yoo+18]   Donghyun Yoo et al. "Efficient K-Shot Learning With Regularized Deep Networks". In: *AAAI*. AAAI Press, 2018, pp. 4382–4389.

[YJ16]   Jianfei Yu and Jing Jiang. "Learning Sentence Embeddings with Auxiliary Tasks for Cross-Domain Sentiment Classification". In: *Proceedings of the 2016 Conference on Empirical Methods in Natural Language Processing, EMNLP 2016, Austin, Texas, USA, November 1–4, 2016*. 2016, pp. 236–246.

[Zel+17]   Werner Zellinger et al. "Central Moment Discrepancy (CMD) for Domain-Invariant Representation Learning". In: *CoRR* abs/1702.08811 (2017).

[ZBJ17]   Yuan Zhang, Regina Barzilay, and Tommi S. Jaakkola. "Aspect-augmented Adversarial Networks for Domain Adaptation". In: *TACL* 5 (2017), pp. 515–528.

[Zho+16]   Guangyou Zhou et al. "Bi-Transferring Deep Neural Networks for Domain Adaptation". In: *Proceedings of the 54th Annual Meeting of the Association for Computational Linguistics, ACL*. 2016.

[Zhu+17]   Jun-Yan Zhu et al. "Unpaired Image-to-Image Translation using Cycle-Consistent Adversarial Networks". In: *Computer Vision (ICCV), 2017 IEEE International Conference on*. 2017.

[ZR17]   Yftah Ziser and Roi Reichart. "Neural Structural Correspondence Learning for Domain Adaptation". In: *Proceedings of the 21st Conference on Computational Natural Language Learning (CoNLL 2017)*. 2017, pp. 400–410.

[ZR18]   Yftah Ziser and Roi Reichart. "Pivot Based Language Modeling for Improved Neural Domain Adaptation". In: *Proceedings of the 2018 Conference of the North American Chapter of the Association for Computational Linguistics: Human Language Technologies, NAACL-HLT*. 2018, pp. 1241–1251.

Chapter 12
End-to-End Speech Recognition

12.1 Introduction

In Chap. 8, we aimed to create an ASR system by dividing the fundamental equation

$$W^* = \operatorname*{argmax}_{W \in V^*} P(W|X) \qquad (12.1)$$

into an acoustic model, lexicon model, and language model by using Bayes' theorem. This approach relies heavily on the use of the conditional independence assumption and separate optimization procedures for the different models.

Deep learning was first incorporated into the statistical framework by replacing the Gaussian mixture models to predict phonetic states based on the observations. One of the drawbacks of this approach is that the DNN/HMM hybrid models rely on training each component separately. As seen previously in other scenarios, the separate training process can lead to sub-optimal results, due to the lack of error propagation between models. In ASR, these drawbacks tend to manifest themselves as sensitivity to noise and speaker variation. Applying deep learning for end-to-end ASR allows the model to learn from the data instead of relying on heavily engineered features, allowing the models to learn from the data directly. Thus, there have been some approaches to train ASR models in an end-to-end fashion. End-to-end methods instead try to optimize the quantity $P(W|X)$ directly, rather than separating it.

With end-to-end modeling, the input–target pair need only be the speech utterance and the linguistic representation of the transcript. Many representations are possible: phones, triphones, characters, character n-grams, or words. Given that ASR focuses on producing word representations from the speech signal, words are the more obvious choice; however, there are some drawbacks. The vocabulary size requires large output layers, as well as examples of each word in training, leading to much lower accuracies than other representations. More recently, end-to-end approaches have moved towards using characters, character n-grams, and some word

© Springer Nature Switzerland AG 2019

U. Kamath et al., *Deep Learning for NLP and Speech Recognition*,
https://doi.org/10.1007/978-3-030-14596-5_12

models as well given enough data. These data pairs can be easier to produce, alleviating the requirements for linguistic knowledge when creating phonetic dictionaries. Jointly optimizing the feature extraction and sequential components together provides numerous benefits, specifically: lower complexity, faster processing, and higher quality.

The key component to accomplishing end-to-end ASR requires a method of replacing the HMM to model the temporal structure of speech. The most common methods are **CTC** and **attention**. In this chapter, the components of traditional ASR are substituted with end-to-end training and decoding techniques. We begin by introducing CTC, a method for training unaligned sequences. Next, we explore some architectures and techniques that have been used to train end-to-end models. We then review attention and how to apply it to ASR networks and some of the architectures that have been trained with these techniques. Following attention, we discuss multitask networks trained with both CTC and attention. We explore common decoding techniques for CTC and attention during inference, incorporating language models to improve prediction quality. Lastly, we discuss embedding and unsupervised techniques and then end with a case study, incorporating both a CTC and an attention network.

12.2 Connectionist Temporal Classification (CTC)

DL-HMM models rely on an alignment of the linguistic units to the audio signal to train the DNN to classify as phonemes, senones, or triphone states (plainly stated, this sequence of acoustic features should yield this phone). Manually obtaining these alignments can be prohibitively expensive for large datasets. Ideally, an alignment would not be necessary for an utterance-transcript pair. Connectionist temporal classification [Gra+06] was introduced to provide a method of training RNNs to "label unsegmented sequences directly," rather than the multistep process in the hybrid mode.

Given an acoustic input $X = [\mathbf{x}_1, \mathbf{x}_2, \ldots, \mathbf{x}_T]$ of acoustic features with the desired output sequence $Y = [y_1, y_2, \ldots, y_U]$. An accurate alignment of X to Y is unknown, and there is additional variability in the ratio of the lengths of X and Y, typically with $T \ll U$ (consider the case where there is a period of silence in the audio, yielding a shorter transcript).

How can possible alignments be constructed for a (X, Y) pair? A simple theoretical alignment, as shown in Fig. 12.1, illustrates a potential method where each input x_t has an output assigned to it, with repeated outputs combined into a single prediction.

This alignment approach has two problems: First, in speech recognition, the input may have periods of silence that do not directly align with the output assigned. Second, we can never have repeated characters (such as the two "l"s in "hello"), because they are collapsed together into a single prediction.

The CTC algorithm alleviates the issues of this naive approach by introducing a *blank* token that acts as a null delimiter. This token is removed after collapsing re-

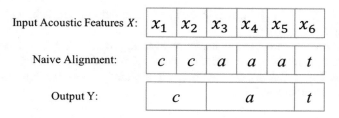

Fig. 12.1: Naive alignment for an input X and with a length of 6 and an output $Y = [c,a,t]$. Example from [Han17]

peated predictions, allowing repeated sequences and periods of "silence." Thus, the **blank** token is not included in the loss computation or decoding; however, it allows a direct alignment between the input and the output without forcing a classification to the output vocabulary. Note that the *blank* token is separate from the *space* token that is used to denote the separation of words. Figure 12.2 shows an example of the CTC alignment.

Input Acoustic Features X:	x_1	x_2	x_3	x_4	x_5	x_6	x_7	x_8	x_9	x_{10}	x_{11}	x_{12}	x_{13}	x_{14}	x_{15}	x_{16}	x_{17}
Predicted Alignment:	h	e	l	l	ε	l	o	ε	ε	_	w	o	o	ε	r	l	d
Merge Repeated Predictions:	h	e	l		ε	l	o	ε		_	w	o		ε	r	l	d
Remove *blank* token:	h	e	l			l	o			_	w	o			r	l	d
CTC predicted output Y:	h	e	l	l	o	_	w	o	r	l	d						

Fig. 12.2: CTC alignment for an input X and output $Y = [h,e,l,l,o,\_,w,o,r,l,d]$. Note that the blank token is represented by "ε" and the space character is represented by the underscore "_"

With this output representation, there is a 1:1 alignment between the lengths of the input sequence and output sequence. Furthermore, the introduction of the *blank* token implies that there can be many predicted alignments that lead to the same output. For example:

$$[h,e,l,\varepsilon,l,o,\varepsilon] \rightarrow \text{``hello''}$$
$$[h,\varepsilon,e,l,\varepsilon,l,o] \rightarrow \text{``hello''}$$

Because any token in the output can have an ε before or after, we can imagine the desired output sequence having an ε before and after each label.

$$Y = [\varepsilon, y_1, \varepsilon, y_2, \ldots, \varepsilon, y_U]$$

Multiple paths/alignments can yield a correct solution, and therefore, all correct solutions must be considered. The CTC algorithm itself is "alignment-free"; however, these "pseudo-alignments" are used to compute the probability of possible alignments.

It then produces an output distribution over all possible Ys, which can be used to infer the probability of a particular output, Y. The conditional probability, $P(Y|X)$, is computed by summing over all possible alignments between the input and the output, as shown in Fig. 12.3.

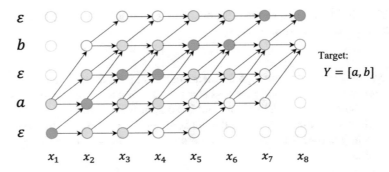

Fig. 12.3: Valid CTC paths for the target sequence, $Y = [a,b]$. Notice that the blank token, ε, is removed from the final sequence. Therefore, there are two possible initial states, ε and ja, and two possible final states, ε and b. Additionally, to achieve final output, the transition from epsilon must be to itself or the next token in the sequence, while the transition from a could be to itself, ε, or b

Mathematically we can define the conditional probability of single alignment α_t, as the product of each state in the sequence:

$$P(\alpha|X) = \prod_{t=1}^{T} P(\alpha_t|X) \tag{12.2}$$

All paths are considered mutually exclusive, so we sum the probability of all alignments, giving the conditional probability for a single utterance (X,Y):

$$P(Y|X) = \sum_{A \in A_{X,Y}} \prod_{t=1}^{T} P(\alpha_t|X) \tag{12.3}$$

where $A_{X,Y}$ is the set of valid alignments. Dynamic programming is used to improve the computation of the CTC loss function. By supplying blank tokens around each label in the sequence, the paths can be easily comparable and merged when they reach the same output at the same time step.

Combining everything gives the loss function for CTC.

$$L_{CTC}(X,Y) = -\log \sum_{a \in A_{X,Y}} \prod_{t=1}^{T} P(a_t|X) \tag{12.4}$$

The gradient for backpropagation can be computed for each time step from the probabilities at each frame.

CTC assumes conditional independence between each time step in that the output at each time step is independent of the previous time steps. Although this property allows for frame-wise gradient propagation, it limits the ability to learn sequential dependencies. Using a language model (Sect. 12.5.2) alleviates some of the issues, by providing a word or n-gram context.

12.2.1 End-to-End Phoneme Recognition

CTC was initially successful on the TIMIT [ZSG90] phoneme recognition task [GMH13]. Various architectures, trained with CTC, were explored yielding state-of-the-art performance on the task. The architecture mapped Mel filter-bank features to the phonetic sequence with a single end-to-end network. The authors explored unidirectional and bidirectional RNNs. A stacked, bidirectional LSTM architecture provided the best results. Bidirectional RNNs seemed to allow the network to leverage the context of the whole utterance, rather than the forward only context.

The authors used two regularization techniques in the training of this network: weight noise and early stopping. Weight noise adds Gaussian noise to the weights during training to reduce overfitting to specific sequences. These regularization techniques turned out to be crucial to the training of the network.

12.2.2 Deep Speech

Following the success of CTC in phoneme recognition, others attempted to use it with different output representations. The Deep Speech (DS1) architecture [Han+14a] was trained to predict a sequence of character probabilities to produce a transcript directly from the audio features (in this case the spectrogram). The Deep Speech network consisted of a DNN architecture with three fully connected layers, one bidirectional LSTM layer, which took the place of the HMM, and a fully connected output softmax layer that classifies the predictions as one of the characters in the alphabet. The input layer relied on frames from the spectrogram, a central frame with a set of 5–9 context frames on each side. An illustration of this architecture is shown in Fig. 12.4.

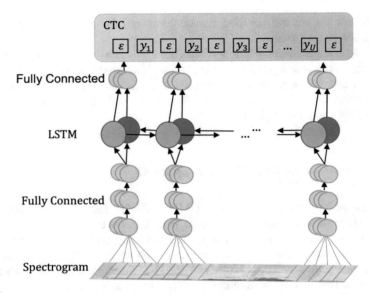

Fig. 12.4: RNN model used in the original Deep Speech paper. The architecture incorporates a single bidirectional LSTM layer after three fully connected layers that lean features on the input spectrogram

Given the complexity of the end-to-end mapping to characters, a significant component Deep Speech's success was the size of the dataset: 5000 h from 9600 speakers. Despite the increase in the size of the dataset, regularization is still essential to the generalization of the network, so the models were trained with dropout as well as data augmentation. One technique inspired by "jittering" in computer vision was leveraged, translating the audio file by 5 ms forward and backward. The output probabilities for the jittered examples are averaged before backpropagation.

One of the exciting components of the Deep Speech work is that the RNN model can learn a light character-level language model during the training procedure, producing "readable" transcripts even without a language model. The errors that appear tend to be phonetic misspellings of words, such as *bostin* instead of *boston*.

12.2.2.1 GPU Parallelism

Given the size of the dataset and computational requirements of the architecture, multiple GPUs were needed to facilitate training. The Deep Speech work was pivotal in overcoming many engineering challenges, such as how to train on large datasets. Many contributions of the paper focused on scaling the training of the architecture on multiple GPUs. Two types of parallelism were used to train the models across multiple GPUs: data parallelism and model parallelism. Data parallelism focuses on retaining a copy of the architecture on each GPU, splitting a large training batch across the separate GPUs, performing the forward and backward propagation steps on the separate data, and finally aggregating the gradient updates for all of the models. Data parallelism provides near linear scaling with the number of GPUs (it may impact the convergence rate, due to the effective batch size). The second type of parallelism is model parallelism. Model parallelism focuses on splitting the model's layers and distributing the layers across the set of available GPUs. Incorporating model parallelism can be difficult when working with recurrent neural networks, due to their sequential nature. In the Deep Speech architecture, the authors achieved model parallelism by splitting the model in half along the time dimension. These decisions allowed the authors to train on 5000 h of audio and achieve state-of-the-art results on two noisy speech benchmarks.

12.2.3 Deep Speech 2

In Deep Speech 2 (DS2) [Amo+16], a follow-on paper to Deep Speech, the authors extended the original architecture to perform character-based, end-to-end speech recognition. The authors validated the modeling techniques on both English and Mandarin Chinese transcription. The Deep Speech 2 modifications introduced many improvements to the original architecture, as well as engineering optimizations achieving 7× speedup over the original Deep Speech implementation. Figure 12.5 shows the updates architecture.

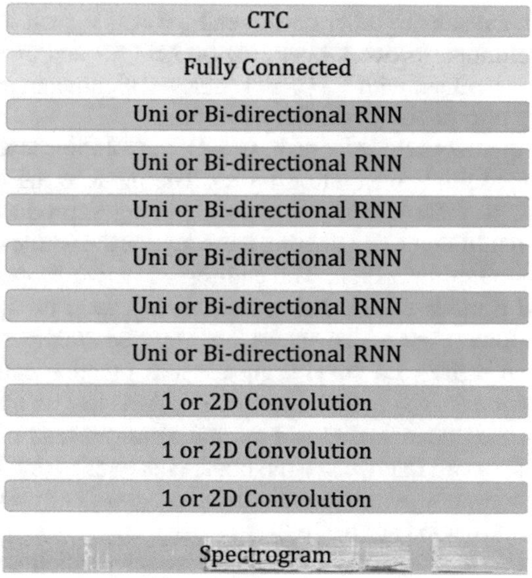

CTC
Fully Connected
Uni or Bi-directional RNN
Uni or Bi-directional RNN
Uni or Bi-directional RNN
Uni or Bi-directional RNN
Uni or Bi-directional RNN
Uni or Bi-directional RNN
1 or 2D Convolution
1 or 2D Convolution
1 or 2D Convolution
Spectrogram

Fig. 12.5: The Deep Speech 2 architecture incorporated convolutional layers that learned features from utterance spectrograms and significantly increased the depth

The main difference in the Deep Speech and Deep Speech 2 architectures is the increase in depth. In the Deep Speech 2 work, many different architectures were explored, varying the number of convolutional layers between 1 and 3 and the number of recurrent layers from 1 to 7. The optimal DS2 architecture for English transcription included 11 layers (3 convolutional, 7 bidirectional recurrent, and 1 fully connected layer). Batch normalization is incorporated after each layer (apart from the fully connected layer), and gradient clipping was also included to improve convergence. The overall architecture contained approximately 35 million parameters. With the incorporation of an n-gram language model, this leads to a 43.4% relative improvement in WER over the already competitive DS1 architecture.

Other key components in the improvements of Deep Speech 2 were training techniques and further increasing the dataset size. Training can be unstable in the early stage of CTC models. The authors use a training curriculum to improve the stability of the model when training. By first selecting the shorter utterances, the model can benefit from smaller gradient updates in the earlier part of the first epoch. Additionally, the authors increased the size of the dataset to $12,000$ h in Deep Speech 2. They note that scaling the data decreases the WER by 40% for each factor of 10 increase in the training set size.

12.2.4 Wav2Letter

Wav2Letter [CPS16] extends end-to-end models to CNN-only networks. This work showed competitive results to other end-to-end networks, such as Deep Speech 2, with a fully convolutional network operating on MFCCs and power-spectrum features. The CNN was trained with CTC, and achieved significant increases in speed, with the capability of producing real-time decoding.

After training the network, intermediate 1-D convolution layers are added between the input and initial convolution layers. The input to the network was then changed to the raw waveform, with the aim of learning to produce features similar to the MFCCs used initially. After training these layers, the whole network is trained jointly for end-to-end optimization. The end-to-end network operating on the raw waveform showed a modest degradation in accuracy, even though it operated directly on the waveform. Figure 12.6 shows the proposed architecture.

The authors also explored a novel sequence loss function called the automatic segmentation criterion (ASG) in addition to CTC. ASG has no blank label, no normalized scores on the nodes, and global normalization instead of frame-level normalization. We may recall that we use the blank character to delimit double letters. Instead, ASG incorporates an additional character specifically for repetition (e.g., "hello" could be represented as "hel2o").

The removal of RNNs from the architecture makes predictions much less computationally costly, as well as allowing for streaming transcription (the convolutions stride across the input to reveal the output at each time step). In the follow-up work

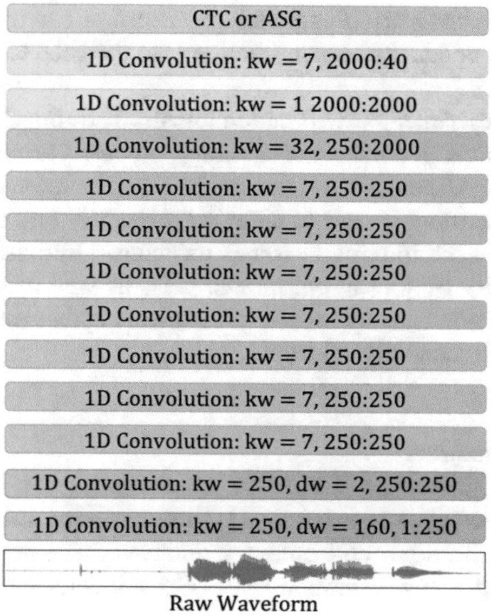

| CTC or ASG |
| 1D Convolution: kw = 7, 2000:40 |
| 1D Convolution: kw = 1 2000:2000 |
| 1D Convolution: kw = 32, 250:2000 |
| 1D Convolution: kw = 7, 250:250 |
| 1D Convolution: kw = 7, 250:250 |
| 1D Convolution: kw = 7, 250:250 |
| 1D Convolution: kw = 7, 250:250 |
| 1D Convolution: kw = 7, 250:250 |
| 1D Convolution: kw = 7, 250:250 |
| 1D Convolution: kw = 7, 250:250 |
| 1D Convolution: kw = 250, dw = 2, 250:250 |
| 1D Convolution: kw = 250, dw = 160, 1:250 |
| Raw Waveform |

Fig. 12.6: Wav2Letter architecture for recognition on a raw waveform. The first layer is not included when training on MFCCs instead of the raw waveform. The convolutional parameters are organized as $(kw, dw, \text{dim ratio})$, where kw is the kernel width, dw is the kernel stride, and dim ration is the number of input dimensions to the number of output dimensions

on Wav2Letter++ [Pra+18], the authors improved the speed of the ASR system, achieving linear scaling for training times (up to 64 GPUs).

12.2.5 Extensions of CTC

CTC provides an elegant way to compute pseudo-alignments for unaligned sequences; however, the frame independence assumption does have drawbacks. Various techniques have been introduced to relax the frame independence assumption. The most notable are Gram-CTC and the RNN transducer.

12.2.5.1 Gram-CTC

Gram-CTC [Liu+17] extended the CTC algorithm to address the fixed alphabet and fixed target decomposition. This approach focused on learning to predict the character n-grams rather than single characters, allowing the model to output multiple characters at a given time step. Using character n-grams can mildly alleviate the effects of the frame independence assumption, due to the need to learn multiple labels together.

The work also experimented with automatically learning the character n-grams (referred to as "grams") during the training process, leveraging the forward–backward algorithm. Although it is feasible for both the grams and transcription to be learned jointly, the model needs to learn the alignment and decomposition of the target in tandem, and the training becomes unstable. Multitask learning is used to combat this instability by jointly optimizing CTC as well as Gram-CTC. Overall, the incorporation of grams resulted in improvements across multiple datasets, even when the grams were manually selected.

12.2.5.2 RNN Transducer

The RNN transducer [Gra12] extends CTC by assuming a local and monotonic alignment between the input and output sequences. This approach alleviates the conditional independence assumption of CTC by incorporating two RNN layers that model the dependencies between outputs at different time steps.

$$
\begin{aligned}
P_{RNN-T}(Y|X) &= \sum_{a \in A_{X,Y}} P(a|h) \\
&= \sum_{a \in A_{X,Y}} \prod_{t=1}^{T'} P(a_t|h_t, y_{1:u_t-1})
\end{aligned}
\tag{12.5}
$$

where u_t signifies the output time step aligned to the input time step t. T' is the length of the alignment sequence including the number of blank labels predicted. Note that $y_{1:u}$ is the sequence of predictions excluding blanks up to time step u. The RNN incorporates the full history of the non-blank labels into the CTC prediction at the next time step. Training the RNN-T model requires using the forward–backward algorithm to compute the gradients (similar to the CTC computation). In online speech recognition, a unidirectional RNN can be used to model the dependencies between time steps in the forward direction.

12.3 Seq-to-Seq

The success of sequence-to-sequence models in machine translation prompted their application in speech recognition. One of the most significant benefits of seq-to-seq models in speech recognition is that they do not rely on CTC for training, natively alleviating the frame independence assumption of CTC. Typically in speech recognition, there are a large number of time steps in the input and output that make it infeasible to train basic seq-to-seq models with a single hidden state representing the full utterance.

Instead, the attention-based approach is used and can model the probability of the output sequence directly:

$$P(Y|X) = \prod_{u=1}^{U} P(y_u|y_{1:u-1}, X) \tag{12.6}$$

This quantity can be estimated by the attention-based objective function from [Bah+16c]:

$$\mathbf{h}_t = Encoder(X)$$

$$a_{ut} = \begin{cases} ContentAttention(\mathbf{q}_{u-1}, \mathbf{h}_t) \\ LocationAttention(\{a_{u-1}\}_{t=1}^{T}, \mathbf{q}_{u-1}, \mathbf{h}_t) \end{cases}$$

$$\mathbf{c}_u = \sum_{t=1}^{T} a_{ut}\mathbf{h}_t \tag{12.7}$$

$$P(y_u|y_{1:u-1}, X) = Decoder(\mathbf{c}_u, \mathbf{q}_{u-1}, y_{u-1})$$

The encoder neural network produces a hidden representation h_t of the acoustic input and decoder produces the transcript output from the encoded sequence. The attention weight, a_{ut}, is used to compute the context vector c_u for the decoder. The decoder hidden state, q_u, provides the cumulative context of the decoder's predictions into the next prediction. We consider two types of attention here: content-based and location-aware attention [Cho+15c].

12.3.0.1 Content-Based Attention

Content-based attention learns a weight vector g and two linear layers, $\mathbf{W}$ and $\mathbf{V}$ (without bias parameters), to weigh the previous prediction and the encoder hidden state $\mathbf{h}_t$. This is represented as follows:

$$e_{ut} = \mathbf{g}^\top \tanh(\mathbf{W}\mathbf{q}_{u-1} + \mathbf{V}\mathbf{h}_t) \tag{12.8}$$

$$a_{ut} = \text{Softmax}(\{e_{ut}\}_{t=1}^{T}) \tag{12.9}$$

12.3.0.2 Location-Aware Attention

Location-aware attention is an extension to support convolution. This feature accounts for the alignment at the previous step. This can be defined as:

$$\{\mathbf{f}_t\}_{t=1}^{T} = \mathbf{K} * \mathbf{a}_{u-1} \tag{12.10}$$

where $*$ represents the one-dimensional convolution operation over the time axis t with convolutional matrix $\mathbf{K}$. A linear layer U is also learned to map the output features f_t into the feature space.

$$e_{ut} = \mathbf{g}^\top \tanh(\mathbf{Wq}_{u-1} + \mathbf{Vh}_t + \mathbf{Uf}_t) \tag{12.11}$$

$$a_{ut} = \text{Softmax}(\{e_{ut}\}_{t=1}^T) \tag{12.12}$$

One of the difficulties of training attention-based networks is the simultaneous optimization of:

- the encoder weights,
- the attention mechanism for computing the correct alignment, and
- the decoder weights.

The dynamics of the network make it difficult, especially in the early stages with regularization being a key component for these models.

12.3.1 Early Seq-to-Seq ASR

Attention was successfully applied in [BCB14a] extending the work in computer vision [MHG+14] to the task of machine translation to the RNN encoder–decoder from [Cho+14].

[Bah+16c] applied seq-to-seq to speech recognition. The attention mechanism in this work focused the decoder on a range of the encoder outputs. Attention not only helped the convergence of the model but also improved the training time (Figs. 12.7, 12.8 and 12.9).

12.3.2 Listen, Attend, and Spell (LAS)

The listen, attend, and spell (LAS) network [Cha+16b] used a pyramid BiLSTM to encode the input sequence, referred to as the listener. The decoder was an attention-based RNN to predict characters.

The drawback for seq-to-seq models is that they tend to be more difficult to train (more so than CTC) and slower during inference. The decoder cannot predict until the attention mechanism has weighed all of the previous hidden states for each new time step. Some techniques have been introduced to deal with this, such as window-ing mechanisms to reduce the number of time steps considered during decoding and label smoothing, which prevent overconfidence in predictions.

One of the other difficulties for seq-to-seq models is that they cannot be used in a full online streaming fashion. The entire context must be encoded before decoding can begin.

In [VDO+16], the Wav2Text architecture used a CNN-RNN model with at-tention to predict character-based transcripts directly on the raw waveform. The encoder is a convolutional architecture combined with two bidirectional LSTMs,

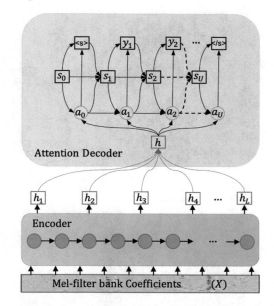

Fig. 12.7: Attention-based end-to-end ASR model from [Bah+16c]

and the decoder is a single bidirectional LSTM. The convolutional layers are used mainly to reduce the dimensionality of the input. Due to the additional complexity of attention and the utilization of the raw waveform, the network was trained via transfer learning. Initially, only the lower encoder layers predict the spectral features (MFCC and log Mel-scale spectrogram) as the target from the raw input waveform. The network is then trained with these features through the attention-based encoder–decoder with CTC to produce a transcript.

12.4 Multitask Learning

Many of the drawbacks of attention and CTC led to multitask learning approaches. Attention usually performs better in end-to-end scenarios; however, it typically has difficulties converging and tends to suffer in noisy environments. CTC, on the other hand, usually yields lower quality due to the conditional independence assumption, but is more stable. The trade-offs between CTC and attention make their combination highly valuable via multitask learning. ESPnet [KHW17, Xia+18] was trained to do just this: jointly optimizing an attention-based encoder–decoder model with CTC and attention.

The training loss for ESPnet is a multi-objective loss (MOL) defined as:

$$\mathcal{L}_{MOL} = \lambda \log P_{ctc}(C|X) + (1-\lambda) \log P_{att}^{*}(C|X) \tag{12.13}$$

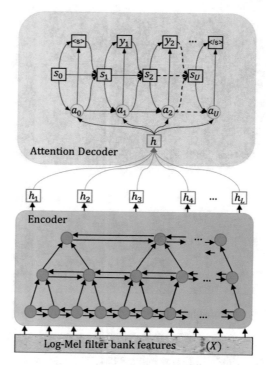

Fig. 12.8: Attention-based end-to-end ASR model from [Cha+16b], using a pyramid LSTMs in the encoder

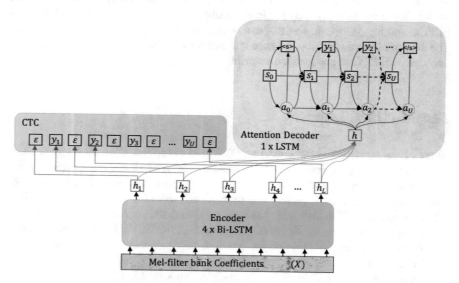

Fig. 12.9: End-to-end speech processing network from [KHW17]

where λ is the weight for each loss function and $0 \le \lambda \le 1$. P_{ctc} is the CTC objective and P_{att}^* is the attention objective.

The ESPnet architecture uses a 4-layer bidirectional LSTM encoder and a 1-layer LSTM decoder. To reduce the number of time steps of the output, the top two layers of the encoder read every second state, which reduces the length of the output h by a factor of 4.

12.5 End-to-End Decoding

CTC and attention-based models are end-to-end, producing a transcript directly from the acoustic features. Although they have the capability of learning inherent language models during training, the amount of language data seen during training is relatively small. In most circumstances, the decoding procedures can improve the predictions, and in many cases significantly improving word error rates. The desirable state is to incorporate additional information during the decoding process to improve predictions, using a beam search and language model. A beam search can incorporate a broader context into the predictions, while language models can take advantage of large text corpora that may not have utterance-transcript pairings.

In [Hor+17], two methods are introduced for decoding a combined CTC–attention model. The first approach rescores the predictions and the second method does one-pass decoding incorporating the probabilities from each of the attention and CTC predictions.

In [HCW18], the authors incorporate word and character-based RNN language models into the decoding procedure.

12.5.1 Language Models for ASR

The decoding process can be extended by providing a prior of the language in the form of a language model. These language models can be trained on large volumes of text data to accurately bias predicted transcripts to particular domains.

12.5.1.1 N-gram

In the Deep Speech 2 paper, the authors experimented with n-gram language models. Although the RNN layers included in the architecture learn an implicit language model, it tends to err on homophones and spelling of certain words. Therefore, an n-gram language model was trained using the KenLM [Hea+13] toolkit on the Common Crawl Repository,[1] using the 400,000 most frequent words from 250 million lines of text.

---

[1] http://commoncrawl.org.

The decoding step uses a beam search to optimize the quantity:

$$Q(Y) = \log(P_{CTC}(Y|X)) + \alpha \log(P_{LM}(Y)) + \beta \gamma(Y) \qquad (12.14)$$

where $\gamma(Y)$ is the number of words in Y. The weight α effects the contribution of the language model, and the weight β biases predictions to have more words. Both parameters are tuned on a development set.

The language model was incorporated into the beam search decoding and significantly improved the base WER over the no-language model baselines.

12.5.1.2 RNN Language Models

RNN language models have surfaced various times in this book. The application of RNN language models relies on utilizing the likelihood of the next word to predict the most likely sequence of words given the previous word.

These models can be incorporated as an additional score during the beam decoding in the same way as the n-gram language model or as a rescoring of the top n hypotheses.

Word-based models suffer from the OOV issue, but they have successfully beaten phoneme-based CTC models when trained on very large datasets (125 kh) [SLS16]. This limitation has prompted research on incorporating character-based prediction when encountering OOV terms [Li+17].

12.5.2 CTC Decoding

Decoding a CTC network (a deep learning network trained with CTC) refers to finding the most probable output for the classifier at inference time, similar in spirit to HMM decoding. Mathematically, the decoding process is described by the function $h(x)$:

$$h(\mathbf{x}) = \operatorname*{argmax}_{\mathbf{l} \in L^{\leq T}} P(\mathbf{l}|\mathbf{x}) \qquad (12.15)$$

In the original connectionist temporal classification publication [Gra+06], two methods were proposed: best path decoding and prefix search decoding.

Best path decoding, also known as greedy decoding, outputs the most probable output at each time step. To obtain a useful string, repeated characters are then collapsed and the blank token is removed to obtain the hypothesis, h.

$$h(x) = \mathbf{B}(\pi^*)$$
$$\pi^* = \operatorname*{argmax}_{\pi \in N^t} p(\pi|x) \qquad (12.16)$$

This decoding scheme is straight-forward. However, it is not likely to produce the best sequence because it does not consider the multiple paths to obtain the same alignment.

A beam search can be incorporated into the decoding process to improve prediction. With the beam search, the probabilities of paths leading to the same result can be summed, yielding a higher probability for that result. Algorithm 1 shows the beam search decoding process with $\varnothing$ representing the empty sequence and the set of beams, B.

Algorithm 1: CTC beam search

Input: $B \leftarrow \{\varnothing\}; P\text{-}(\varnothing, 0) \leftarrow 1$
Result: $\max_{Y \in B} P^{\frac{1}{|Y|}}(Y, T)$
begin
    for $t = 1 \ldots T$ **do**
        $\hat{B} \leftarrow$ the W most probable sequences in B
        $B \leftarrow \{\}$
        for $y \in \hat{B}$ **do**
            if $y \neq \varnothing$ **then**
                $P^+(Y, t) \leftarrow P^+(Y, t-1)P(Y^e, t|x)$
                if $\hat{y} \in \hat{B}$ **then**
                    $P^+(Y, t) \leftarrow P^+(Y, t)P(Y^e, \hat{Y}, t)$
            $P^-(Y, t) \leftarrow P^+(Y, t-1)P(-, t|x)$
            Add Y to B
            for $k = 1 \ldots K$ **do**
                $P^-(Y+k, t) \leftarrow 0$
                $P^+(Y+k, t) \leftarrow P(k, Y, t)$
                Add $(Y+k)$ to B

The beam search algorithm can be extended with an n-gram language model. A simple approach is to rescore the word sequence each time an end-of-word (*space*) token is reached. However, this relies on the model to predict full words with no misspellings.

A better approach is to use prefix search decoding, which incorporates the subword level information during the decoding process, utilizing the prefixes of the language model. Converting a word-level language model to a "label-level" or character-based model is accomplished by representing the output sequence as the concatenation of the longest completed word sequence and the remaining word prefix, denoted as w and p, respectively. The function for computing the probability of the next label given the current sequence becomes:

$$P(k|y) = \frac{\sum_{w' \in (p+k)*} P_\gamma(w'|W)}{\sum_{w' \in p*} P_\gamma(w'|W)} \tag{12.17}$$

where $P(w'|W)$ is the probability of the word history transition from W to w', $p*$ is the set of dictionary words prefixed by p, and γ is the language model weight.

During decoding, the probabilities of sequence prefixes are computed, with the option to end the current prefix or continue extending it. During the beam search, the probability of a hypothesis state is modified to also depend on the probability of a prefix, dictionary entry, or n-gram language model when determining the *extension probability*.

This method relies on the forward–backward algorithm, where the computation grows exponentially with the number of states and time steps. We can improve the efficiency of the decoding by pruning the output sequence, removing all outputs where the probability of the blank token is above a specified threshold. Because the output activations tend to be "peaky," this dramatically reduces the number of states considered and consistently outperforms best path decoding.

This algorithm can be used without a language model by setting the probabilities to 1. The prefix algorithm presented in [Han+14b] is given in Algorithm 2.

Algorithm 2: CTC prefix beam search

Input: $P_b(\varnothing;x_{1:0}) \leftarrow 1, P_{nb}(\varnothing;x_{1:0}) \leftarrow 0$
$A_{\text{prev}} \leftarrow \{\varnothing\}$
Result: most probable prefix in A_{prev}
begin

    for $t = 1 \ldots T$ **do**

        $A_{\text{next}} \leftarrow \{\}$

        for $l \in A_{prev}$ **do**

            for $c \in \Sigma$ **do**

                if $c = blank$ **then**

                    $P_b(l;x_{1:t}) \leftarrow P(\text{blank};x_t)(P_b(l;x_{1:t-1}) + P_{nb}(l;x_{1:t-1}))$

                    add l to A_{next}

                else

                    $l^+ \leftarrow$ concatenate l and c

                    if $c = l_{end}$ **then**

                        $P_{nb}(l^+;x_{1:t}) \leftarrow P(c;x_t)P_b(l;x_{1:t-1})$

                        $P_{nb}(l;x_{1:t}) \leftarrow P(c;x_t)P_b(l;x_{1:t-1})$

                    else if $c = space$ **then**

                      $P_{nb}(l^+;x_{1:t}) \leftarrow$
                        $P(W(l^+)|W(l))^\alpha P(c;x_t)(P_b(l;x_{1:t-1}) + P_{nb}(l;x_{1:t-1}))$

                    else

                      $P_{nb}(l^+;x_{1:t}) \leftarrow P(c;x_t)(P_b(l;x_{1:t-1}) + P_{nb}(l;x_{1:t-1}))$

                    if l^+ *not in* A_{prev} **then**

                      $P_b(l^+;x_{1:t}) \leftarrow P(\text{blank};x_t)(P_b(l^+;x_{1:t-1}) + P_{nb}(l^+;x_{1:t-1}))$

                      $P_{nb}(l^+;x_{1:t}) \leftarrow P(c;x_t)P_{nb}(l^+;x_{1:t-1})$

                  add l^+ to A_{next}

        $A_{\text{prev}} \leftarrow k$ most probable prefixes in A_{next}

This approach also requires length normalization, to prevent a bias towards sequences with fewer transitions.

12.5.3 Attention Decoding

Attention decoding already produces the most probable sequence given the previous predictions. Therefore, as seen previously, greedy decoding could be applied here, yielding the most probable character at each time step. However, it likely would not yield the most probable sequence $\hat{C}$.

$$\hat{C} = \underset{C \in U^*}{\operatorname{argmax}} \log P(C|X) \tag{12.18}$$

A beam search can also be applied to attention models during the decoding process. Because the previous time step is provided as an input to the next prediction, the top n most probably paths at each time step can be retained at each time step. The beam search begins by first considering the start of sentence symbol, $<s>$.

$$\alpha(h, X) = \alpha(g, X) + \log P(c|g_{l-1}, X) \tag{12.19}$$

where g is a partial hypothesis in the beam, and c is a symbol/character appended to g, yielding a new hypothesis h. An example of beam search attention decoding is shown in Fig. 12.10.

Various architectures have aimed to use this additional unpaired data in the end-to-end ASR models [Tos+18]. The term *fusion* has recently been coined, referring to the integration of these language models into the main acoustic model.

12.5.3.1 Shallow Fusion

Shallow fusion (used originally for NMT) combines the scores of the LM and ASR models during the decoding [Gul+15]. This type of language model decoding incorporates an external language model during the beam search to incorporate word or character probabilities into consideration. Shallow fusion can be used with a word or character-based language models to determine the probability of a particular sequence.

$$Y^* = \underset{Y}{\operatorname{argmax}} \log P(Y|X) + \lambda P_{LM}(Y) \tag{12.20}$$

Character language models are helpful for rescoring hypotheses before a word boundary is reached or as a rescoring mechanism for character-based languages, such as Japanese and Mandarin Chinese. Additionally, character-based language models can predict unseen character sequences, which a word-based model would not allow.

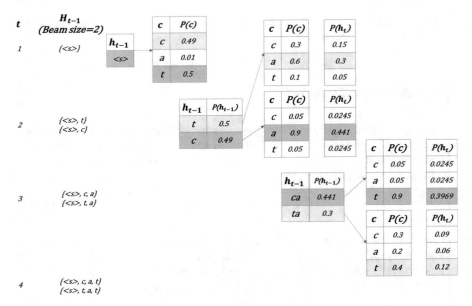

Fig. 12.10: Beam search decoding example with a beam size of 2 on a three-character alphabet $\{a, b, c\}$. With attention decoding, the previous time step is incorporated into the next character prediction. Therefore the probabilities are dependent on the path chosen. The best paths at each time step are highlighted, with the darker one being the top prediction. Note how the greedy decoding of this example would yield a sub-optimal result

Shallow fusion has been incorporated into RNN-T models, allowing the CTC training to alleviate the frame independence while also incorporating the language model bias into the prediction [He+18].

12.5.4 Combined Language Model Training

When incorporating neural language models into end-to-end ASR, it is quickly apparent that the two could be optimized together, leveraging the acoustic information as well as the linguistic information from large text corpora. The two most popular techniques for jointly training the acoustic and language models are deep fusion and cold fusion.

12.5.4.1 Deep Fusion

Deep fusion [Gul+15] on the other hand incorporates the LM into the acoustic model (specifically an encoder–decoder model), creating a combined network.

Combining the network is accomplished by "fusing" the hidden states of pre-trained AM and LM models, continuing training to learn the "fused" parameters. During this training procedure, the LM and AM parameters are fixed, reducing computation costs and converging quickly.

$$g_t = \sigma(\mathbf{v}_g^T \mathbf{s}_t^{LM} + b_g)$$
$$\mathbf{s}_t^{DF} = [\mathbf{c}_t; \mathbf{s}_t; g_t \mathbf{s}_t^{LM}] \qquad (12.21)$$
$$P(y_t | \mathbf{h}, Y_{1:(t-1)}) = \text{softmax}(\mathbf{W}_{DF} \mathbf{s}_t^{DF} + \mathbf{b}_{DF})$$

where $\mathbf{c}_t$ is the context vector, $\mathbf{h}$ is the output of the encoder, and $\mathbf{v}_g$, b_g, $\mathbf{b}_{DF}$, and $\mathbf{W}_{DF}$ are all learned during the continued training phase.

12.5.4.2 Cold Fusion

Cold fusion [Sri+17] extends the idea of deep fusion, incorporating the LM into the training procedure. However, in cold fusion the acoustic model is trained from scratch incorporating the pre-trained LM.

$$\mathbf{s}_t^{LM} = DNN(\mathbf{d}_t^{LM})$$
$$\mathbf{s}_t^{ED} = \mathbf{W}_{ED}[\mathbf{d}_t; \mathbf{c}_t] + \mathbf{b}_{ED}$$
$$\mathbf{g}_t = \sigma(\mathbf{W}_g[\mathbf{s}_t^{ED}; \mathbf{s}_t^{LM}] + \mathbf{b}_g) \qquad (12.22)$$
$$\mathbf{s}_t^{CF} = [\mathbf{s}_t^{ED}; \mathbf{g}_t \circ \mathbf{s}_t^{LM}]$$
$$\mathbf{r}_t^{CF} = DNN(\mathbf{s}_t^{CF})$$
$$P(y_t | \mathbf{h}, Y_{1:(t-1)}) = \text{softmax}(\mathbf{W}_{CF} \mathbf{r}_t^{CF} + \mathbf{b}_{CF})$$

Because cold fusion incorporates the LM into the training process from the beginning, retraining is required if there are changes in the LM. The original paper introduces a means for switching language models by using LM logits instead of the LM hidden states; however, this does increase the number of learned parameters and computation.

12.5.5 Combined CTC–Attention Decoding

Decoding with combined CTC–attention architectures relies on producing the most probable character sequence $\hat{C}$. Combining the two outputs is non-trivial. Attention produces a sequence of output labels, while CTC produces a label per frame. In [Wat+17b], the authors propose two methods for combining the CTC and attention outputs: rescoring and one-pass decoding.

12.5.5.1 Rescoring

Rescoring relies on a two-step method. The first step is to produce a set of complete hypotheses from the attention decoder. The second step is to rescore these hypotheses based on the CTC and attention probabilities (the forward algorithm is used to get the CTC probabilities).

$$\hat{C} = \underset{h \in \hat{\Omega}}{\operatorname{argmax}} \{\lambda \alpha_{CTC}(h, X) + (1 - \lambda)\alpha_{ATT}(h, X)\} \tag{12.23}$$

12.5.6 One-Pass Decoding

One-pass decoding, on the other hand, focuses on computing the probabilities of the partial hypotheses as characters are generated.

A language model can also be incorporated into the decoding process [HCW18] by adding an additional language modeling term to the decoder:

$$\hat{C} = \underset{C \in U^*}{\operatorname{argmax}} \{\lambda \log P_{CTC}(C|X) + (1 - \lambda)\log P_{ATT}(C|X) + \gamma \log P_{LM}(C)\} \tag{12.24}$$

The score in the beam search can then be described as:

$$\alpha(h) = \lambda \alpha_{CTC}(h) + (1 - \lambda)\alpha_{ATT}(h) + \gamma \alpha_{LM}(h) \tag{12.25}$$

for each incomplete hypothesis h.

Computing the attention and language model scores is straight-forward, with:

$$\begin{aligned}
\alpha_{ATT}(h) &= \alpha_{ATT}(g) + \log P_{ATT}(c|g, X) \\
\alpha_{LM}(h) &= \alpha_{LM}(g) + \log P_{LM}(c|g, X)
\end{aligned} \tag{12.26}$$

where $h = g; c$, g is a known hypothesis, and c is a character being appended to the sequence to generate h.

CTC, however, is more nuanced due to the number of sequences that could produce the character sequence. Therefore the CTC score is the sum of all sequences with h as the prefix.

$$P(h, \ldots |X) = \sum_{v \in (U \cup <\text{EOS}>)^+} P(h; v|X) \tag{12.27}$$

The CTC score becomes:

$$\alpha_{CTC}(h) = \log P(h, \ldots |X) \tag{12.28}$$

12.6 Speech Embeddings and Unsupervised Speech Recognition

The amount of unsupervised data that is available can be orders of magnitude higher than the amount of paired speech-text parallel corpora. Thus, unsupervised speech recognition and acoustic embeddings for audio processing are promising areas of research.

12.6.1 Speech Embeddings

One of the earliest works for embeddings in speech was [BH14]. In this work, the authors used a form of Siamese network to train acoustic word embeddings where similar sounding words (acoustically similar) are clustered near each other in the embedding space. In this fashion "words are nearby if they sound alike." By modeling words directly, the paradigm of speech recognition shifts away from trying to model states in the traditional HMM.

This network is trained in two parts: initially, a CNN classification model is trained to classify spoken words in a fixed segment of audio (2 s). Second, this network is fixed and incorporated in a word embedding network. The word embedding network is trained to align the embedding of the correct word with the acoustic embedding while separating wrong words. To reduce the size of the input embedding space from all words by using bag-of-letter-n-grams, only the top 50,000 letter n-grams are used to reduce the size of the input embedding space (bag-of-letter-n-grams). The architecture diagram is shown in Fig. 12.11.

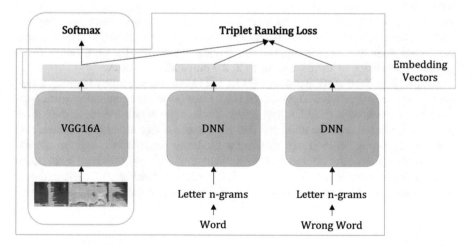

Fig. 12.11: Acoustic embedding model trained with a triplet ranking loss to align acoustic vectors and word vectors from subword units

The embeddings space yielded similarities such as (please,pleas), (plug,slug), and (heart,art).

A Siamese CNN network was also used in [KWL16] to discriminate between separate same and different word pairs given spoken instances of words. This network achieved similar results as to a strongly supervised word classification model.

12.6.2 Unspeech

In Unspeech [MB18], the authors used a Siamese network to train embeddings used with acoustic models for speaker adaptation, utterance clustering, and speaker comparison. This work relies on the assumption that similar areas of speech are likely to have the same speaker. The contexts for true and false examples of speakers are taken from neighboring contexts windows in the same utterance or from separate files. This idea is similar to the concept of negative sampling. This network, therefore, does not expect similar words to be in the same embedding space, but rather the same speaker. The architecture is shown in Fig. 12.12.

12.6.3 Audio Word2Vec

One of the drawbacks to the CNN approach is that it requires fixed-length audio segments. Audio Word2Vec [Chu+16] used a sequence-to-sequence autoencoder to learn a fixed representation for variable-length spoken words. Because the learned representation is the input itself, it can be learned in a completely unsupervised way, hence the reference to word2Vec. The resulting model is, therefore, able encode acoustic samples for use in a query-by-example system for words. Training the model does not require supervision; however, creating the word embeddings requires knowledge of word boundaries in the embedding process.

Audio Word2Vec was extended in [WLL18] to utterance level by learning a segmentation method as well. The method is an example of a segmental sequence-to-sequence autoencoder (SSAE). The SSAE learns segmentation gate to determine the word boundaries in the utterance and a sequence-to-sequence autoencoder that learns an encoding for each segment. Some guidance is needed to keep the autoencoder from splitting the utterance into too many embeddings; however, learning an appropriate estimate is not differentiable. Reinforcement learning is used to estimate this quantity, due to the non-differentiability of learning a discrete variable.

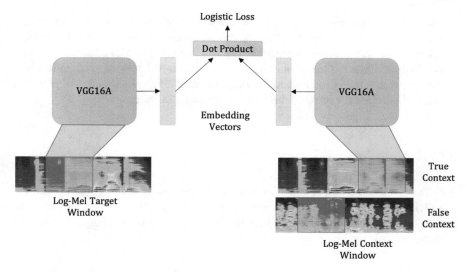

Fig. 12.12: Unspeech embeddings are trained using a Siamese CNN network (VGG16A), to compute embedding vectors. The dot product of the two vectors is computed, and a logistic loss is used to optimize a binary classification task, of whether the context window was a true or false context window of the target

12.7 Case Study

In this case study, we continue to focus on building ASR models on the Mozilla Common Voice dataset.[2] In this chapter, we focus specifically on a Deep Speech 2 model that trains an end-to-end network with CTC and a hybrid attention-CTC model.

12.7.1 Software Tools and Libraries

Since the release of the Deep Speech 2 paper, there have been multiple open sourced implementations of the architecture, with the most common difference being the deep learning framework used. The most popular are the TensorFlow implementation by Mozilla,[3] the PaddlePaddle implementation,[4] and the PyTorch version.[5] Each has a variety of benefits and drawbacks, some of which are the deep learning

---

[2] https://voice.mozilla.org/en/data.

[3] https://github.com/mozilla/DeepSpeech.

[4] https://github.com/PaddlePaddle/DeepSpeech.

[5] https://github.com/SeanNaren/deepspeech.pytorch.

framework, the amount of preprocessing required, variable-length vs. fixed-length RNNs, as well as others. We focus on the PyTorch implementation for its simplicity.

One of the most recent advancements has been the CTC+attention models, specifically ESPnet.[6] This toolkit focuses on end-to-end speech recognition and text-to-speech. It uses Chainer and PyTorch as backends for the toolkit and provides Kaldi-style recipes for some of the most modern architectures.

12.7.2 Deep Speech 2

The Deep Speech 2 implementation used is written in PyTorch. It incorporates a parallelized data loader to speed model training, an optimized CTC loss function, a CTC-decoding library with language model support, and data augmentation for acoustic model training.

12.7.2.1 Data Preparation

The data preparation requires either a directory structure or manifest file. In the first approach, a dataset directory is structured as follows (Figs. 12.13, 12.14, and 12.15).

There is no additional need for phonetic dictionaries for character-based models; the data is processed into a spectrogram and then converted to a tensor at data loading time.

In this implementation, one can also use a "manifest" file to define the datasets used. The manifest is similar to the Kaldi and Sphinx structures, containing a list of the examples in each dataset split. Manifest files can be useful for filtering longer files when using variable-length RNNs.

12.7.2.2 Acoustic Model Training

First, we train a base model given the default configuration. The resulting model has two convolutional layers and five bidirectional GRU layers, yielding approximately 41 million learnable parameters. We enable the augmentation step during training as well, which applies small changes to the tempo and gain to reduce overfitting.

---

[6] https://github.com/espnet/espnet.

```
1  / common_voice
2      / train
3          txt /
4                  train_sample000000 . txt
5                  train_sample000001 . txt
6                  . . .
7          wav /
8                  train_sample000000 . wav
9                  train_sample000001 . wav
10                 . . .
11     / val
12         txt /
13                 . . .
14         wav /
15                 . . .
16     / test
17         txt /
18                 . . .
19         wav /
20                 . . .
21
```

Fig. 12.13: Directory structure for Deep Speech 2

```
1  / path / to / train_sample000000 . wav , / path / to / train_sample000000 . txt
2  / path / to / train_sample000001 . wav , / path / to / train_sample000001 . txt
3  . . .
4
```

Fig. 12.14: Manifest structure for the training set for Deep Speech 2

```
1  python  train . py  —train—manifest  data / train_manifest . csv  —val—
      manifest  data / val_manifest . csv
2
```

Fig. 12.15: Training function for Deep Speech 2

We train all models on a GPU,[7] with early stopping based on the WER of the validation set. In our case, the model began diverging after about 15 epochs, as shown in Fig. 12.16 and achieves its best validation WER of 23.470. Once the model is trained we evaluate the best model on the test set, where we achieve an average WER of 22.611 and CER of 7.757, using greedy decoding. A few samples of the greedy decoding of the trained model are shown in Fig. 12.17.

---

[7] Although it is possible to train this model on a CPU, it is unrealistic due to the computationally intensive nature of the convolutional and recurrent layers.

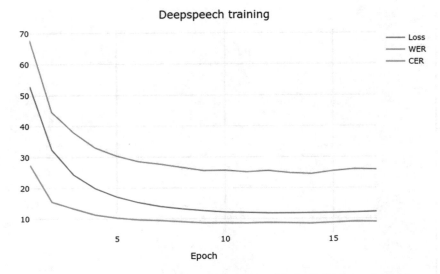

Fig. 12.16: Training curve of Deep Speech 2 with the default configuration

```
1 Ref: i understand sheep they're no longer a problem and they can
       be good friends
2 Hyp: i understand shee they're no longery problem and they can be
       good friends
3 WER: 0.214 CER: 0.027
4
5 Ref: as he looked at the stones he felt relieved for some reason
6 Hyp: ashe looked at the stones he felt relieved for som ason
7 WER: 0.333 CER: 0.051
8
```

Fig. 12.17: Output from the base Deep Speech 2 model. Note how many of the mistakes seem phonetic and create nonlogical words, such as *shee* and *ashe*

12.7.3 Language Model Training

The character-based predictions produce reasonable transcripts, without a language model. However, we can improve the greedy predictions by providing a language model during the decoding phase. We leverage the ctcdecode[8] package to apply different decoding schemes, which is integrated into the PyTorch Deep Speech 2

---

[8] https://github.com/parlance/ctcdecode.

implementation. One thing to note about this language model is that it incorporates a character FST as well. The FST acts as a spell checker, enforcing the production of words.

Decoding schemes can be applied to improve the error rates of the predictions. These results are summarized in Table 12.1.

The KenLM toolkit [Hea+13] is used to train an n-gram language model. The language model is created from transcripts of the training corpus to provide comparable results to previous case studies. In practice, language models are usually trained on very large training corpora such as the previously mentioned, Common Crawl (Fig. 12.18).[9]

```
1  kenlm/build/bin/lmplz -o 2 < training_transcripts.txt >
     cv_2gram_lm.arpa
2
3  kenlm/build/bin/build_binary cv_2gram_lm.arpa cv_2gram_lm.trie
4
```

Fig. 12.18: Train a 2-gram language model with KenLM on the training transcripts. The first command creates an ARPA language model from the transcripts, and the second command creates a binary trie-structure from the language model used in the decoding phase

We determine the best language model for the system by evaluating them on the validation set, and the best model is chosen to apply to the testing set. Table 12.1 summarizes the WER and CER for different language models.

Table 12.1: Validation results for different decoding methods. The best results are in bold

Decoding method	WER	CER
None	22.832	8.029
2-gram	12.919	7.292
3-gram	12.027	6.990
4-gram	**11.865**	**6.915**
5-gram	11.977	6.955

After applying the language model with the default beam size (beamwidth = 10), we see that our best model is the 4-gram model. Now, we can increase the size of the beam to evaluate the impact on the predictions. The results are summarized in Table 12.2.

---

[9] http://commoncrawl.org.

Table 12.2: Validation results for different beam sizes. The best results are in bold

Decoding method	WER	CER
4-gram, beam=10	11.865	6.915
4-gram, beam=64	7.742	4.458
4-gram, beam=128	6.939	3.984
4-gram, beam=256	6.288	3.616
4-gram, beam=512	**5.857**	**3.375**

```
1 Ref: i understand sheep they're no longer a problem and they can
       be good friends
2 Hyp: i understand sheep they're no longer a problem and they can
       be good friends
3 WER: 0.0 CER: 0.0
4
5 Ref: as he looked at the stones he felt relieved for some reason
6 Hyp: as he looked at the stones he felt relieved for some as
7 WER: 0.083 CER: 0.068
8
9
```

Fig. 12.19: Test output with language model decoding. Note many of the phonetic mistakes are corrected when incorporating the language model during the decoding; however, it can also cause different mistakes. In the second example, greedy decoding output *ason* instead of *reason*, but after the application of the language model, the hypothesis reduced this to *as*, reducing the WER and increasing the CER for this example

The computation time increases linearly with the beam size. In practice, it is best to choose a beam size that is a good trade-off between performance and quality. After applying our best LM (4-gram) with a beam size of 512 to the test set, we achieve a WER of **5.587** and CER of **3.232**. Some examples of the decoded output are in Fig. 12.19.

12.7.4 ESPnet

ESPnet[10] is an end-to-end speech processing toolkit that draws inspiration from Kaldi. It incorporates hybrid CTC–attention architectures, mainly the ones contained within [KHW17] and [Wat+17b]. Much of the toolkit is bash script focused,

---

[10] https://github.com/espnet/espnet.

similar to Kaldi, with Chainer and PyTorch backends. In this portion of the case study, a hybrid CTC–attention architecture is trained on the Common Voice dataset, using the ESPnet toolkit.

12.7.4.1 Data Preparation

The data preparation is very similar to Kaldi, with a reliance on Kaldi for some of the preprocessing. The main difference is the lack of phonetic lexicons and dictionaries required in Kaldi. We generate MFCC features and store them in a JSON format. This format contains the target transcript, the tokenized transcript, location of the features, and some additional information for various components of the training. An example of the formatted training data is shown in Fig. 12.20.

After extracting the features and creating the input file, the network is ready to train.

12.7.4.2 Model Training

The model training procedure also follows the Kaldi scripts to some degree; however, once the features are extracted, we run the training scripts.

The model trained is a 4-layer bidirectional LSTM encoder and a 1-layer uni-directional LSTM decoder. We train this model with Adadelta for 20 epochs on a single GPU. The full list of training arguments is shown in Fig. 12.21.

During the training procedure the losses for both CTC and attention can be monitored to ensure that there is consistency in the convergence. The overall loss for training and validation is the weighted sum of the two components. We also notice that the validation loss trends with the training data loss until the final epoch. In this example, we set a hard stop for computational reasons on the number of epochs run. To obtain our best model, we would continue training until the validation consistently diverges from the training loss, and choose the model that performs best on the validation data (Fig. 12.22).

The accuracy curves, shown in Fig. 12.23, display the network performance as training progresses. The first two epochs show significant gains in the early stages, with modest improvements as training progresses. Our best model in training achieves a WER of 12.07 on the validation data.

We can inspect the output attention weights during the decoding process by plotting the weight of each time step during the decoding. Visualizing attention, as before, shows what portion of the input is attended to during inference. This is shown in Fig. 12.24. We notice that the output generally correlates with the input audio file, yielding an aligned output that is capable of segmenting the audio, as well as dealing with the offsets in time. During the early stages, we notice some breaks in the attention alignments to the input, whereas in the latter case, the attention alignments appear seamlessly aligned to the input.

```
1  {
2      "utts": {
3          "cv-valid-dev-sample-000000": {
4              "input": [
5                  {
6                      "feat": ".valid_dev/deltafalse/feats.1.ark
   :27",
7                      "name": "input1",
8                      "shape": [
9                          502,
10                         83
11                     ]
12                 }
13             ],
14             "output": [
15                 {
16                 "name": "target1",
17                 "shape": [
18                     55,
19                     31
20                 ],
21                 "text": "BE CAREFUL WITH YOUR PROGNOSTICATIONS
   SAID THE STRANGER",
22                 "token": "B E <space> C A R E F U L <space> W I T
   H <space> Y O U R <space> P R O G N O S T I C A T I O N S <
   space> S A I D <space> T H E <space> S T R A N G E R",
23                 "tokenid": "5 8 3 6 4 21 8 9 24 15 3 26 12 23 11
   3 28 18 24 21 3 19 21 18 10 17 18 22 23 12 6 4 23 12 18 17 22
   3 22 4 12 7 3 23 11 8 3 22 23 21 4 17 10 8 21"
24                 }
25             ],
26             "utt2spk": "cv-valid-dev-sample-000000"
27         },
28         ...
29 }
30
```

Fig. 12.20: *data.json* input file format for ESPnet training

```
1  python asr_train.py —backend pytorch —outdir exp/results —dict
      data/lang_1char/train_nodev_units.txt —minibatches 0 —
      resume —train-json dump/cv_valid_train/deltafalse/data.json
      —valid-json dump/cv_valid_dev/deltafalse/data.json —etype
      blstmp —elayers 4 —eunits 320 —eprojs 320 —subsample 1
      _2_2_1_1 —dlayers 1 —dunits 300 —atype location —adim 320
      —aconv-chans 10 —aconv-filts 100 —mtlalpha 0.5 —batch-
      size 30 —maxlen-in 800 —maxlen-out 150 —sampling-
      probability 0.0 —opt adadelta —epochs 20
```

Fig. 12.21: Training command for ESPnet

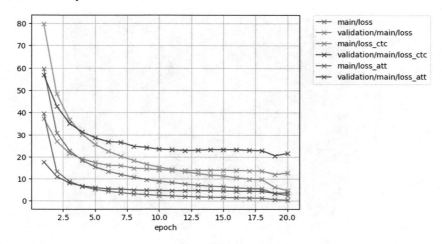

Fig. 12.22: Losses during training

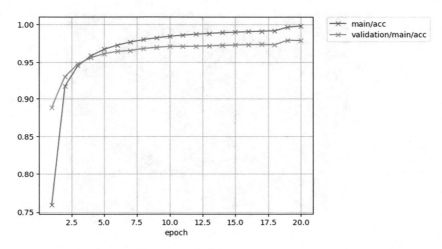

Fig. 12.23: Training and validation accuracy curves for the model

Our base model achieves a WER of 12.34 and CER of 6.25 on the testing set with greedy decoding (beam size of 1). When incorporating a beam search of 20 (ESPnet default selection) into the predictions on the testing set, we reduce the WER to 11.56 and CER to 5.80. We leave tuning the beam size and incorporating a language model as an exercise. Note the significant improvement when we added this to the Deep Speech 2 architecture.

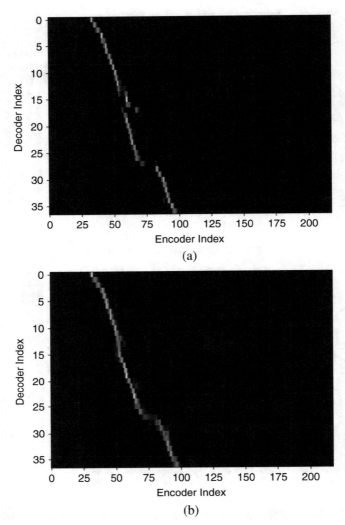

Fig. 12.24: Attention weights for a single file on the input audio after the (**a**) 1st epoch and (**b**) after the 20th epoch

12.7.5 Results

We now provide a summary of the techniques evaluated in this case study. The testing results are displayed in Table 12.3.

Table 12.3: End-to-end speech recognition performance on Common Voice test set (best result highlighted)

Approach	WER
Deep Speech 2 (no decoding)	22.83
Deep Speech 2 (4-gram LM, beam size of 512)	5.59
ESPnet (no decoding)	12.34
ESPnet (no LM, beam size of 20)	11.56
Kaldi TDNN (Chap. 8)	4.44

Overall, with a CTC–attention model, we get faster, more stable convergence and a lower WER for the base acoustic model compared to the Deep Speech 2 baseline (WER of 22.83).

Although this result is not better than the one achieved with Kaldi in Chap. 8 case study, the results between the Deep Speech 2 (with a language model) and Kaldi models are comparable, even without a lexicon model. The training procedure is more straight-forward than the training steps required for the traditional approaches to ASR, such as removing the requirement of iterative training and aligning. Additional benefits can also be gained from the inclusion of a language model during decoding to provide compelling results without significant linguistics resources.

12.7.6 Exercises for Readers and Practitioners

Some other interesting problems readers and practitioners can try on their own include:

1. What changes are required to train a Deep Speech 2 model on a new language?
2. What would be the effects of training a language model on more data?
3. Would the incorporation of the testing transcripts improve the results on the validation data? What about the testing data?
4. Does the incorporation of the testing transcripts in the language model corrupt the validity of the results?
5. How could an RNN language model be incorporated into the decoding process for Deep Speech 2? For ESPnet?
6. Perform a grid search for the beam size on the ESPnet model.

References

[Amo+16]   Dario Amodei et al. "Deep speech 2: End-to-end speech recognition in English and Mandarin". In: *International Conference on Machine Learning*. 2016, pp. 173–182.

[BCB14a]   Dzmitry Bahdanau, Kyunghyun Cho, and Yoshua Bengio. "Neural machine translation by jointly learning to align and translate". In: *arXiv preprint arXiv:1409.0473* (2014).

[Bah+16c]   Dzmitry Bahdanau et al. "End-to-end attention-based large vocabulary speech recognition". In: *Acoustics, Speech and Signal Processing (ICASSP), 2016 IEEE International Conference on*. IEEE. 2016, pp. 4945–4949.

[BH14]   Samy Bengio and Georg Heigold. "Word embeddings for speech recognition". In: *Fifteenth Annual Conference of the International Speech Communication Association*. 2014.

[Cha+16b]   William Chan et al. "Listen, attend and spell: A neural network for large vocabulary conversational speech recognition". In: *Acoustics, Speech and Signal Processing (ICASSP), 2016 IEEE International Conference on*. IEEE. 2016, pp. 4960–4964.

[Cho+14]   Kyunghyun Cho et al. "Learning phrase representations using RNN encoder-decoder for statistical machine translation". In: *arXiv preprint arXiv:1406.1078* (2014).

[Cho+15c]   Jan K Chorowski et al. "Attention-based models for speech recognition". In: *Advances in neural information processing systems*. 2015, pp. 577–585.

[Chu+16]   Y.-A. Chung et al. "Audio Word2Vec: Unsupervised Learning of Audio Segment Representations using Sequence-to-sequence Autoencoder". In: *ArXiv e-prints* (Mar 2016).

[CPS16]   Ronan Collobert, Christian Puhrsch, and Gabriel Synnaeve. "Wav2letter: an end-to-end ConvNet-based speech recognition system". In: *arXiv preprint arXiv:1609.03193* (2016).

[Gra12]   Alex Graves. "Sequence transduction with recurrent neural networks". In: *arXiv preprint arXiv:1211.3711* (2012).

[GMH13]   Alex Graves, Abdel-rahman Mohamed, and Geoffrey Hinton. "Speech recognition with deep recurrent neural networks". In: *Acoustics, speech and signal processing (ICASSP), 2013 IEEE international conference on*. IEEE. 2013, pp. 6645–6649.

[Gra+06]   Alex Graves et al. "Connectionist temporal classification: labelling unsegmented sequence data with recurrent neural networks". In: *Proceedings of the 23rd international conference on Machine learning*. ACM. 2006, pp. 369–376.

[Gul+15]   Caglar Gulcehre et al. "On using monolingual corpora in neural machine translation". In: *arXiv preprint arXiv:1503.03535* (2015).

[Han17]   Awni Hannun. "Sequence Modeling with CTC". In: *Distill*. (2017).

[Han+14a]   Awni Hannun et al. "Deep speech: Scaling up end-to-end speech recognition". In: *arXiv preprint arXiv:1412.5567* (2014).

[Han+14b]   Awni Y Hannun et al. "First-pass large vocabulary continuous speech recognition using bi-directional recurrent DNNs". In: *arXiv preprint arXiv:1408.2873* (2014).

[He+18]   Yanzhang He et al. "Streaming End-to-end Speech Recognition For Mobile Devices". In: *arXiv preprint arXiv:1811.06621* (2018).

[Hea+13]   Kenneth Heafield et al. "Scalable modified Kneser-Ney language model estimation". In: *Proceedings of the 51st Annual Meeting of the Association for Computational Linguistics (Volume 2: Short Papers)* Vol. 2. 2013, pp. 690–696.

[HCW18]   Takaaki Hori, Jaejin Cho, and Shinji Watanabe. "End-to-end Speech Recognition with Word-based RNN Language Models". In: *arXiv preprint arXiv:1808.02608* (2018).

[Hor+17]   Takaaki Hori et al. "Advances in joint CTC-attention based end-to-end speech recognition with a deep CNN encoder and RNN-LM". In: *arXiv preprint arXiv:1706.02737* (2017).

[KWL16]   Herman Kamper, Weiran Wang, and Karen Livescu. "Deep convolutional acoustic word embeddings using word-pair side information". In: *Acoustics, Speech and Signal Processing (ICASSP), 2016 IEEE International Conference on.* IEEE. 2016, pp. 4950–4954.

[KHW17]   Suyoun Kim, Takaaki Hori, and Shinji Watanabe. "Joint CTC-attention based end-to-end speech recognition using multi-task learning". In: *Acoustics, Speech and Signal Processing (ICASSP), 2017 IEEE International Conference on.* IEEE. 2017, pp. 4835–4839.

[Li+17]   J. Li et al. "Acoustic-To-Word Model Without OOV". In: *ArXiv e-prints* (Nov.2017).

[Liu+17]   Hairong Liu et al. "Gram-CTC: Automatic unit selection and target decomposition for sequence labelling". In: *arXiv preprint arXiv:1703.00096* (2017).

[MB18]   Benjamin Milde and Chris Biemann. "Unspeech: Unsupervised Speech Context Embeddings". In: *arXiv preprint arXiv:1804.06775* (2018).

[MHG+14]   Volodymyr Mnih, Nicolas Heess, Alex Graves, et al. "Recurrent models of visual attention". In: *Advances in neural information processing systems.* 2014, pp. 2204–2212.

[Pra+18]   Vineel Pratap et al. "wav2letter++: The Fastest Open-source Speech Recognition System". In: *arXiv preprint arXiv:1812.07625* (2018).

[SLS16]   Hagen Soltau, Hank Liao, and Hasim Sak. "Neural speech recognizer: Acoustic-to-word LSTM model for large vocabulary speech recognition". In: *arXiv preprint arXiv:1610.09975* (2016).

[Sri+17]   Anuroop Sriram et al. "Cold fusion: Training seq2seq models together with language models". In: *arXiv preprint arXiv:1708.06426* (2017).

[Tos+18]   Shubham Toshniwal et al. "A comparison of techniques for language model integration in encoder-decoder speech recognition". In: *arXiv preprint arXiv:1807.10857* (2018).

[VDO+16]   Aäron Van Den Oord et al. "WaveNet: A generative model for raw audio." In: *SSW.* 2016, p. 125.

[WLL18]   Yu-Hsuan Wang, Hung-yi Lee, and Lin-shan Lee "Segmental audio word2vec: Representing utterances as sequences of vectors with applications in spoken term detection". In: *2018 IEEE International Conference on Acoustics, Speech and Signal Processing (ICASSP)*. IEEE. 2018, pp. 6269–6273.

[Wat+17b]   Shinji Watanabe et al. "Hybrid CTC/attention architecture for end-to-end speech recognition". In: *IEEE Journal of Selected Topics in Signal Processing* 11.8 (2017), pp. 1240–1253.

[Xia+18]   Zhangyu Xiao et al. "Hybrid CTC-Attention based End-to-End Speech Recognition using Subword Units". In: *arXiv preprint arXiv:1807.04978* (2018).

[ZSG90]   Victor Zue, Stephanie Seneff, and James Glass. "Speech database development at MIT: TIMIT and beyond". In: *Speech communication* 9.4 (1990), pp. 351–356.

Chapter 13
Deep Reinforcement Learning for Text and Speech

13.1 Introduction

In this chapter, we investigate deep reinforcement learning for text and speech applications. Reinforcement learning is a branch of machine learning that deals with how agents learn a set of actions that can maximize expected cumulative reward. In past research, reinforcement learning has focused on game play. Recent advances in deep learning have opened up reinforcement learning to wider applications for real-world problems, and the field of deep reinforcement learning was spawned. In the first part of this chapter, we introduce the fundamental concepts of reinforcement learning and their extension through the use of deep neural networks. In the latter part of the chapter, we investigate several popular deep reinforcement learning algorithms and their application to text and speech NLP tasks.

13.2 RL Fundamentals

Reinforcement learning (RL) is one of the most active fields of research in artificial intelligence. While supervised learning requires us to provide labeled, independent and identically distributed data, reinforcement learning requires us to only specify a desired reward. Furthermore, it can learn sequential decision making tasks that involve delayed rewards, especially those that occur far distant in the future.

A reinforcement learning agent interacts with its environment in discrete time steps. At each time t, the agent in state s_t chooses an action a_t from the set of available actions and transitions to a new state s_{t+1} and receives reward r_{t+1}. The goal of the agent is to learn the best set of actions, termed a **policy**, in order to generate the highest overall cumulative reward (Fig. 13.1). The agent can (possibly randomly) choose any action available to it. Any one set of actions that an agent takes from start to finish is termed an episode. As we will see below, we can use Markov decision processes to capture the episodic dynamics of a reinforcement learning problem.

© Springer Nature Switzerland AG 2019
U. Kamath et al., *Deep Learning for NLP and Speech Recognition*,
https://doi.org/10.1007/978-3-030-14596-5_13

Due to the sequential decision-making nature of reinforcement learning, it suffers from a difficulty commonly known as the *credit assignment problem*. Since there are many actions that can lead to a delayed award, it is difficult for reinforcement learning methods to attribute the subset of actions that had greatest positive or negative effect on these rewards. This becomes an especially difficult problem for large state and action spaces.

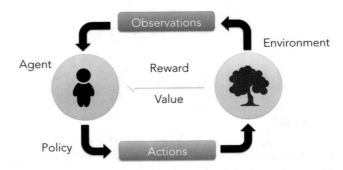

Fig. 13.1: Agent–environment interaction in reinforcement learning

13.2.1 Markov Decision Processes

A Markov decision process (MDP) is a useful mathematical framework that models situations as a discrete time stochastic control process. Mathematically, an MDP can be expressed using the tuple:

$$(s, a, p_a, r_a, \gamma) \tag{13.1}$$

where:

$$
\begin{aligned}
s &= \text{a finite set of states} \\
a &= \text{a finite set of actions} \\
p_a &= \text{the probability of each action } a \\
r_a &= \text{the reward by taking an action } a \\
\gamma &= \text{time discount factor}
\end{aligned}
$$

The process is in some state s, and at each time step, the decision maker may choose any action a that is available in state s. The process responds at the next time step by randomly moving into a new state s', and giving the decision maker a corresponding reward $R_a(s, s')$. The probability that the process moves into its new state s' from current state s is influenced by the chosen action and the reward r received. Specifically, it is defined by the state transition function $p(s'|s, a)$:

$$p(s'|s, a) = \Pr\{S_t = s' | S_{t-1} = s, A_{t-1} = a\} = \sum_{r \in \mathcal{R}} p(s', r|s, a) \tag{13.2}$$

such that:

$$\sum_{s' \in \mathcal{S}} \sum_{r \in \mathcal{R}} p(s', r | s, a) = 1, \text{for all } s \in \mathcal{S}, a \in \mathcal{A}(s) \tag{13.3}$$

Thus, the next state s' depends on the current state s and the decision maker's action a. But given s and a, it is conditionally independent of all previous states and actions; in other words, the state transitions of an MDP satisfies the Markov property.

Markov decision processes are an extension of Markov chains where the difference is the addition of a set of actions (allowing choice) and rewards (giving motivation). Conversely, if only one action exists for each state and all rewards are equal, a Markov decision process reduces to a Markov chain.

13.2.2 Value, Q, and Advantage Functions

We define r_t as the reward we receive at time t. We can define the **return** as the sum of the sequence of future rewards:

$$G_t = r_{t+1} + r_{t+2} + \dots \tag{13.4}$$

Normally, we include a time discount factor $\gamma \in (0, 1)$, and the future cumulative reward can be expressed as:

$$G_t = \sum_{k=0}^{\infty} \gamma^k r_{t+k+1} \tag{13.5}$$

With this definition, we can define the concept of a **value function** of a state s as the expected cumulative return:

$$V(s) = \mathbb{E}[G_t | s_t = s] \tag{13.6}$$

The value function for any particular state is not unique. It depends on the set of actions we take going forward in the future. We define a set of future actions known as a policy π:

$$a = \pi(s) \tag{13.7}$$

Then the value function associated with this policy is unique:

$$V_\pi(s) = \mathbb{E}_\pi[G_t | s_t = s] \tag{13.8}$$

$$= \mathbb{E}_\pi \left[\sum_{k=0}^{\infty} \gamma^k R_{t+k+1} | s_t = s \right] \tag{13.9}$$

Note that while this policy-associated value function is unique, the actual value can be stochastic under a non-deterministic policy (e.g., one where we sample from a distribution of possible actions defined by the policy):

$$\pi(a|s) = \mathbb{P}[a|s] \tag{13.10}$$

In addition to finding the value function of a particular state, we can also define a value function for a particular action given a state. This is known as the **action-value function** or **Q function**:

$$Q_\pi(s,a) = \mathbb{E}_\pi[G_t|s_t = s, a_t = a] \tag{13.11}$$

$$= \mathbb{E}_\pi\left[\sum_{k=0}^{\infty} \gamma^k r_{t+k+1}|s_t = s, a_t = a\right] \tag{13.12}$$

Like the value function, the Q function is uniquely specified for a particular policy π of actions. The expectation takes into account the randomness in future actions according to the policy, as well as the randomness of the returned state from the environment. Note that:

$$V_\pi(s) = \mathbb{E}_{a \sim \pi}[Q_\pi(s,a)] \tag{13.13}$$

The **advantage function** for a policy π measures the importance of an action by finding the difference between the state-value and state-action-value functions:

$$A_\pi(s,a) = Q_\pi(s,a) - V_\pi(s) \tag{13.14}$$

Because the value function V measures the value of state s following policy π while the Q function measures the value of following action a from state s, the advantage function measures the benefit or loss of following a particular action from state s.

13.2.3 Bellman Equations

The fundamental breakthrough of reinforcement learning is a set of propagation equations for the value and Q functions. These equations are commonly known as the Bellman equations, named after Richard Bellman, an American applied mathematician. For the state value function, the Bellman equation is given by:

$$V_\pi(s) = \mathbb{E}_{s'}\left[r + \gamma V_\pi(s')|s_t = s\right] \tag{13.15}$$

What this equation tells us is that the state value function associated with policy π is the expectation of the reward received at the next state and its discounted state value function. Similarly, the Bellman equation for the Q function is given by:

$$Q_\pi(s,a) = \mathbb{E}_{s',a'}\left[r + \gamma Q_\pi(s',a')|s_t = s, a_t = a\right] \tag{13.16}$$

The importance of the Bellman equations is that they let us express values of states as values of other states. This means that if we know the value of s_{t+1}, we can very easily calculate the value of s_t. This opens the door to iterative approaches for calculating the value for each state, since if we know the value of the next state, we can calculate the value of the current state. Sound familiar? This is similar to the notion of backpropagation.

13.2.4 Optimality

The goal of any reinforcement learning problem is to find the optimal decisions that lead to highest expected cumulative reward. Reinforcement methods fall under one of several main categories depending on how they optimize the policy π for:

1. for the expected reward:

$$\max_{\pi} \mathbb{E} \left[\sum_{k=0}^{\infty} \gamma^k r_{t+k+1} \right] \tag{13.17}$$

2. for the advantage function:

$$\max_{\pi} A_{\pi}(s,a) \tag{13.18}$$

3. for the Q function:

$$\max_{\pi} Q_{\pi}(s,a) \tag{13.19}$$

Methods such as dynamic programming or policy gradients seek to optimize expected reward, while actor-critic models and Q-learning methods focus on optimizing the advantage and Q-functions, respectively.

For any specific policy of actions, we can use the value function to determine its expected reward. There is always at least one policy that is better than or equal to all other policies. This is known as the optimal policy, called π_*, which may not be unique. All optimal policies share the same state-value function:

$$V_*(s) = \max_{\pi} V_{\pi}(s) \tag{13.20}$$

Optimal policies also share the same action-value function:

$$Q_*(s,a) = \max_{\pi} Q_{\pi}(s,a) \tag{13.21}$$

The Bellman equation can be applied to the optimal state-value function v_* to give us the *Bellman Optimality Equation* which is independent of any chosen policy:

$$V_*(s) = \max_{a} \mathbb{E}_{s'} \left[r + \gamma V_*(s') \right] \tag{13.22}$$

Similarly, the optimal action-value function is independent of chosen policy and is given by:

$$Q_*(s,a) = \mathbb{E}_{s'} \left[r + \gamma \max_{a'} Q_*(s',a') | s,a \right] \tag{13.23}$$

13.2.5 Dynamic Programming Methods

When the environment is known and completely specified, dynamic programming methods can be applied to find optimal policies. The key notion is to use value functions to search for improved policies. Commonly applied to finite Markov decision process problems, dynamic programming underpins an important class of reinforcement learning algorithms.

13.2.5.1 Policy Evaluation

Given a policy π, we can determine the state-value function v_π for this policy. Using the Bellman equation above, it is possible to start with an approximation for v_π and iteratively update an estimate v_k until it converges to v_π as $k \to \infty$:

$$V_{k+1}(s) = \mathbb{E}_\pi \left[r_{t+1} + \gamma V_k(s_{t+1}) | s_t = s \right] \tag{13.24}$$

$$= \sum_a \pi(a|s) \sum_{s',r} p(s',r|s,a) \left[r + \gamma V_k(s') \right] \tag{13.25}$$

The above is an *expected* update since it is based on the expectation over all possible next states and actions (Fig. 13.2).

13.2.5.2 Policy Improvement

Consider the next action a for state s that is not from policy π. The value of taking this action is given by the action-value function:

$$Q_\pi(s,a) = \mathbb{E} \left[r_{t+1} + \gamma V_\pi(s_{t+1}) | s_t = s, a_t = a \right] \tag{13.26}$$

$$= \sum_{s',r} p(s',r|s,a) \left[r + \gamma V_\pi(s') \right] \tag{13.27}$$

If we compare the value of taking this action to our policy π, we can decide if we should adopt a new policy that takes action a. This leads to the *policy improvement theorem*, which states that for any two deterministic policies π and π', if:

$$Q_\pi(s, \pi'(s)) \geq V_\pi(s) \tag{13.28}$$

it must be that:

$$V_{\pi'}(s) \geq V_\pi(s) \tag{13.29}$$

When we find that a new policy π' is better, we can take its value $V_{\pi'}$ and use it to find a better policy. Here, E denotes policy iteration and I denotes policy improvement. This iterative process is called **policy iteration**, where we cycle between policy evaluation $(\pi \to V_\pi)$ and policy improvement $(V_\pi \to \pi')$ until we find the optimal policy π_*:

$$\pi_0 \xrightarrow{E} V_{\pi_0} \xrightarrow{I} \pi_1 \xrightarrow{E} V_{\pi_1} \xrightarrow{I} \pi_2 \xrightarrow{E} \dots \xrightarrow{I} \pi_* \xrightarrow{E} V_* \qquad (13.30)$$

where E denotes policy evaluation and I denotes policy improvement. Because a finite MDP has a finite number of possible policies, this process will converge to π_*.

13.2.5.3 Value Iteration

There is a potentially serious drawback with policy iteration in that the evaluation of a policy π is computationally expensive as it requires iterative calculation over every state in the MDP. Instead of waiting for convergence as $k \to \infty$, we can approximate v_π by performing a single update iteration ($V_\pi \approx V_{k+1}$):

$$V_{k+1}(s) = \max_a \mathbb{E}_\pi \left[r_{t+1} + \gamma V_k(s_{t+1}) | s_t = s, a_t = a \right] \qquad (13.31)$$

$$= \max_a \sum_{s',r} p(s',r|s,a) \left[r + \gamma V_k(s') \right] \qquad (13.32)$$

This is known as **value iteration**, which is computationally efficient as it combines truncated policy evaluation with policy improvement.

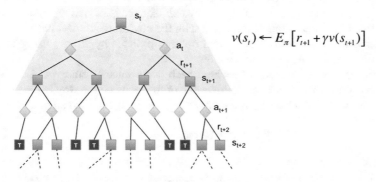

Fig. 13.2: Dynamic programming backup diagram

13.2.5.4 Bootstrapping

The concept of bootstrapping, an important concept in dynamic programming, refers to the estimation of a state or state-action values from estimates of the values of successor states. Bootstrapping is a component in other RL methods such as temporal difference learning or Q-learning and enables faster, online learning. However, since it is based on a notion of using estimates to make estimates, instability can occur, and methods that bootstrap over longer sequences of successor states will have better convergence properties.

13.2.5.5 Asynchronous DP

Dynamic programming methods operate over the entire set of states of a finite MDP. Where the set of states is large, DP is intractable, as every state must necessarily be updated before one sweep is completed. Asynchronous dynamic programming methods do not wait for all states to be updated, but instead update a subset of states during each sweep. Such methods will converge as long as all states are eventually updated. Asynchronous DP methods are very useful in that they can run in an online manner, concurrently as an agent is experiencing the states of the MDP. As such, the agent's experience can be considered in choosing the subset of states to update. This is similar to the concept of *beamsearch*.

13.2.6 Monte Carlo

Unlike dynamic programming methods that require complete knowledge of the environment, Monte Carlo (MC) methods learn from a set of agent experiences. These episodic experiences are actual or simulated sequences of actions, states, and rewards from the interaction of the agent with the environment. MC methods require no prior knowledge but can still yield optimal policies by simply using averaged sample reward for each state and action.

Consider a set of episodes E, where each occurrence of state $s \in E$ is called a visit. To estimate $v_\pi(s)$, we can follow each of the visits all the way to the end of the episode to calculate return G, and then average them to generate an update:

$$V(s_t) \leftarrow V(s_t) + \alpha \left[G_t - V(s_t) \right] \tag{13.33}$$

where α is the learning rate (Fig. 13.3). It is noteworthy that, with Monte Carlo methods, estimates for each state are independent of each other. It does not use bootstrapping. As such, Monte Carlo methods permit us to focus on a subset of relevant states to improve results.

MC methods can be used to estimate state-action values as well as state values. Instead of following visits to a state s, we can follow from an action a taken at a visit to state s, and average accordingly. Unfortunately, however, it may be that certain state-action pairs may never be visited. For deterministic policies, only one action is taken from any state and therefore only one state-action pair will be estimated. The value of all actions from each state must be estimated for policy improvement.

One method of overcoming the sufficient exploration issue of Monte Carlo is to use **exploring starts**, a method that generates episodes by starting at randomly chosen actions and states. This is called a **on-policy** method, since we seek to improve the policy that is used to generate the episodes.

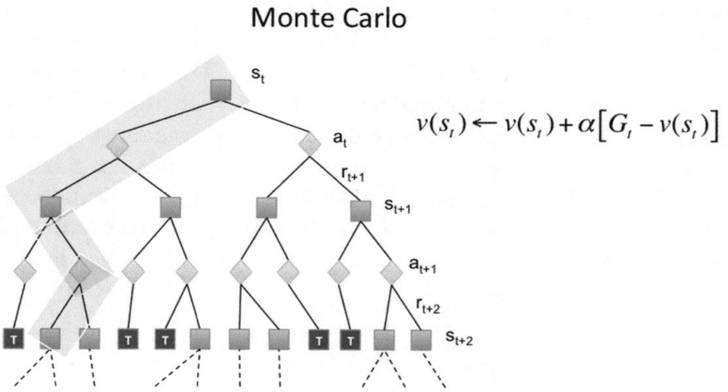

Monte Carlo

$$v(s_t) \leftarrow v(s_t) + \alpha \left[G_t - v(s_t) \right]$$

Fig. 13.3: Monte Carlo backup diagram

13.2.6.1 Importance Sampling

Off-policy methods are based on two separate policies: a target policy that will be optimized and another exploratory policy that is used to generate behavior (termed the behavior policy). Off-policy Monte Carlo methods typically use the notion of **importance sampling**, which is a technique for estimating expectations of one distribution given samples from another. The key idea is to sample values more frequently that have greater impact on the expectation by shifting the probability mass. Note that the target policy and behavior policies can be unrelated, with either or both deterministic or stochastic.

13.2.7 Temporal Difference Learning

Temporal difference (TD) learning seeks to combine the best of both worlds of dynamic programming and Monte Carlo methods. In similar fashion to dynamic programming, it uses bootstrapping to update estimates without waiting until the end of the episode. Concurrently, it can learn from experience without an explicit model of the environment like Monte Carlo methods. The simplest TD learning method is one-step TD, also known as TD(0). It is based on updating the state value function by (Fig. 13.4):

$$V(s_t) \leftarrow V(s_t) + \alpha \left[r_{t+1} + \gamma V(s_{t+1}) - V(s_t) \right] \tag{13.34}$$

This can be written as:

$$V(s_t) \leftarrow V(s_t) + \alpha \delta_t \qquad (13.35)$$

where:

$$\delta_t = r_{t+1} + \gamma V(s_{t+1}) - V(s_t) \qquad (13.36)$$

is known as the **TD error**. Whereas other methods like Monte Carlo must wait to the end of an episode (time T) to update $V(s_t)$, this method uses only estimates of the next time step to form an update. That is, one-step TD estimates the return $G_t \rightarrow r_{t+1} + \gamma V(s_{t+1})$. This is an example of bootstrapping. Like Monte Carlo, TD uses a sample return to approximate the expected return. Like dynamic programming, TD uses $V(s_{t+1})$ in place of $V_\pi(s_{t+1})$. In contrast to DP methods, TD methods do not require a model of the environment. Furthermore, TD methods update much more rapidly in an online fashion whereas Monte Carlo methods must wait until the end of a full episode to calculate returns used in the update. For very long episodes, Monte Carlo may be too slow.

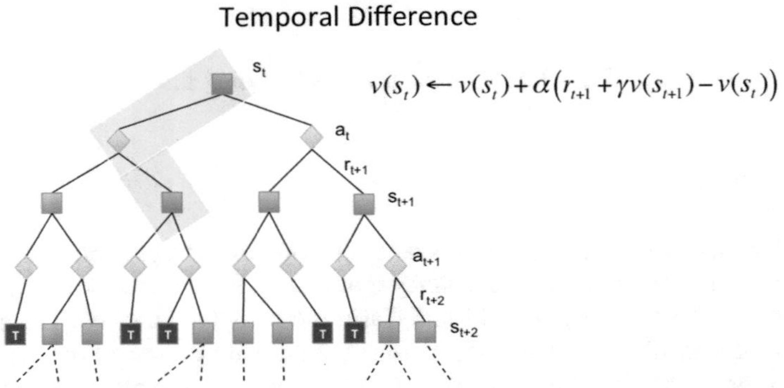

Temporal Difference

$$v(s_t) \leftarrow v(s_t) + \alpha \left(r_{t+1} + \gamma v(s_{t+1}) - v(s_t) \right)$$

Fig. 13.4: Temporal difference backup diagram

One-step TD shares some similarity with stochastic gradient descent in that it uses a one-step sample update rather than an expectation over the entire distribution of successor states. Furthermore, both can be proven to converge—one-step TD can be shown to asymptotically approach V_π. For faster convergence, TD can use batch updating where the value function is updated after computing and aggregating over a batch of experiences.

TD methods are not limited to single time steps, and n-step TD allows bootstrapping over multiple steps by using the update rule:

$$V_{t+n}(s_t) = V_{t+n}(s_t) + \alpha [G_{t:t+n} - V_{t+n-1}(s_t)] \qquad (13.37)$$

where $0 \leq t < T$ and the n-step return is given by:

$$G_{t:t+n} = r_{t+1} + \gamma r_{t+2} + \ldots + \gamma^{n-1} r_{t+n} + \gamma^n \underbrace{V_{t+n-1}(s_{t+n})}_{\text{future return estimate}} \quad (13.38)$$

This n-step return is an approximation of the full return where the last term is an estimate of the remaining returns beyond n-steps. While one-step TD can update once the successor state is computed, n-step TD must wait until after n-steps of the episode before updating. As a tradeoff, n-step TD provides better estimates for state value functions with better convergence properties than one-step TD.

Algorithm 1: One-step TD learning algorithm

---

input : the policy π
output: the value function V

initialize V randomly with $V(terminal) = 0$
for *each episode* **do**
    initialize state s
    for *each step of episode until terminal* **do**
        take action given by $\pi(a|s)$
        observe reward r, next state s'
        update $V(s) \leftarrow V(s) + \alpha[r + \gamma V(s') - V(s)]$
        update $s \leftarrow s'$

13.2.7.1 SARSA

Action-value methods are advantageous in model free formulations as they can operate on current states without access to the model of the environment. This is in contrast to state value functions which require a model since they require knowledge of future states and possible actions to be evaluated. We can apply the temporal difference method to estimate the action-value function by considering the transitions from one state-action pair to the next state-action pair:

$$Q(s_t, a_t) \leftarrow Q(s_t, a_t) + \alpha \left[r_{t+1} + \gamma Q(s_{t+1}, a_{t+1}) - Q(s_t, a_t) \right] \quad (13.39)$$

Note that this update can be applied only to transitions from non-terminal states, since $Q(s_{t+1}, a_{t+1}) = 0$ at terminal states. Because this update depends on the tuple $(s_t, a_t, r_{t+1}, s_{t+1}, a_{t+1})$, it is called SARSA. It is a fully online, on-policy method that asymptotically converges to the optimal policy and action-value function.

Algorithm 2: SARSA learning algorithm

---

input : the policy π
output: the Q function

initialize $Q(s,a)$ randomly with $Q(terminal, all) = 0$
for *each episode* **do**
    initialize state s
    choose action a from $\pi(a|s)$ derived from Q;
    for *each step of episode until terminal* **do**
        take action a, observe reward r, next state s'
        update $Q(s,a) \leftarrow Q(s,a) + \alpha[r + \gamma Q(s',a') - Q(s,a)]$
        update $s \leftarrow s', a \leftarrow a'$

13.2.8 Policy Gradient

Policy gradient methods seek to optimize the policy directly without having to learn the state or action value function. In particular, these model-free methods use a parametric representation for a stochastic policy $\pi(a|s; \theta)$ with parameters θ and seek to optimize expected return:

$$\pi(a|s; \theta) \longleftarrow \max_{\theta} \mathbb{E}_{\pi}[G_t] \tag{13.40}$$

by applying gradient ascent to update the policy parameters:

$$\theta \leftarrow \theta + \alpha \nabla_{\theta} \mathbb{E}_{\pi}[G_t] \tag{13.41}$$

Note that this formula evaluates the expectation prior to calculating the gradient, which requires us to know the transitional probability distribution of $\pi(a|s; \theta)$. For analytical tractability, we can make use of the *Policy Gradient Theorem*, given by:

$$\nabla_{\theta} \mathbb{E}_{\pi}[G_t] = \nabla_{\theta} \int_{x \sim \pi} p_{\theta}(x) G_t(\tau) dx \tag{13.42}$$

$$= \int_{x \sim \pi} p_{\theta}(x) \nabla_{\theta} \log p_{\theta}(x) G_t(x) dx \tag{13.43}$$

$$= \mathbb{E}_{\pi}[\nabla_{\theta} \log \pi(a_t|s_t; \theta) G_t] \tag{13.44}$$

which allows us to express the policy gradient update rule as:

$$\theta \leftarrow \theta + \alpha \mathbb{E}_{\pi}[\nabla_{\theta} \log \pi(a_t|s_t; \theta) G_t] \tag{13.45}$$

Thus, we can update our policy without calculating the transition probability distribution of actions and states or requiring a model.

Policy gradient methods are useful for both continuous and discrete action spaces. A popular method, known as REINFORCE, applies stochastic gradient descent such that only a single sequence is used for training at each step to estimate parameters θ. As such, it is an unbiased estimator with reduced computational burden. But because it uses a single sequence to estimate rewards, REINFORCE can suffer from high variance and take longer to converge. A way to reduce this variance is to subtract a baseline $r_b(s_t)$ reward from our expected return, which teaches the model to increase the probability of actions that generate above average expected returns:

$$\theta \leftarrow \theta + \alpha \mathbb{E}_\pi \left[\nabla_\theta \log \pi(a_t|s_t; \theta)(G_t - r_b(s_t)) \right] \tag{13.46}$$

By sampling a batch of action sequences, the average reward over this batch can be used as the baseline reward during gradient updates for each action sequence in this batch. As long as the baseline reward is not dependent on the policy parameters θ, the estimator remains unbiased.

Algorithm 3: The REINFORCE algorithm

---

input : policy $\pi(a|s; \theta)$
output: optimal policy π_*

initialize policy parameters θ
while *not converged* **do**
    generate an episode by following policy π
    for *each step in episode until terminal* **do**
        calculate return G
        update $\theta \leftarrow \theta + \alpha \gamma^t G \nabla \log \pi(a_t|s_t; \theta)$

13.2.9 Q-Learning

Q-learning is based on the notion that if the optimal Q-function is available, the optimal policy can be directly found by the relation:

$$\pi^*(s) = \arg\max_a Q^*(s,a) \tag{13.47}$$

Therefore, these methods try to learn the optimal Q-function directly by always choosing the best action from any state, without needing to consider the policy being followed. Q-learning is an off-policy TD method that updates the action-state value function by:

$$Q(s_t, a_t) \leftarrow Q(s_t, a_t) + \alpha \left[r_{t+1} + \gamma \underbrace{\max_{a'} Q(s_{t+1}, a')}_{\text{expected future reward}} - Q(s_t, a_t) \right] \tag{13.48}$$

This equation is very similar to SARSA, except it estimates future expected reward by maximizing over future actions. In effect, Q-learning uses a greedy update to iterate toward the optimal Q-function, and has been shown to converge in the limit to Q^*.

Algorithm 4: Q-learning algorithm

---

output: the Q function

initialize $Q(s,a)$ randomly with $Q(terminal, all) = 0$
for *each episode* **do**
    initialize state s
    for *each step of episode until terminal* **do**
        choose best action a from Q (ε-greedy);
        take action a, observe reward r, next state s'
        update $Q(s,a) \leftarrow Q(s,a) + \alpha[r + \gamma \max_{a'} Q(s',a') - Q(s,a)]$
        update $s \leftarrow s'$

13.2.10 Actor-Critic

Actor-critic methods, like policy gradient methods, are based on estimating a parametric policy. What makes actor-critic methods different is that they also learn a parametric function which is used to evaluate action sequences and assist in learning. The *actor* is the policy being optimized, while the *critic* is value function and can be thought of as a parametric estimate of the baseline reward in the policy gradient update equation above:

$$\theta \leftarrow \theta + \alpha \mathbb{E}_\pi \left[\nabla_\theta \log \pi(a_t|s_t; \theta) \left[\underbrace{Q_\pi(s_t, a_t)}_{\text{actor}} - \underbrace{V_\pi(s_t)}_{\text{critic}} \right] \right] \tag{13.49}$$

Note that we can replace the actor-critic by the advantage function:

$$\theta \leftarrow \theta + \alpha \mathbb{E}_\pi \left[\nabla_\theta \log \pi(a_t|s_t; \theta) A_\pi(s_t, a_t) \right] \tag{13.50}$$

where $A_\pi(s_t, a_t) = Q_\pi(s_t, a_t) - V_\pi(s_t)$. Similar to the REINFORCE algorithm, actor-critic methods can use stochastic gradient descent to sample a single sequence. In this instance, the Advantage function takes the form:

$$A_\pi(s_t, a_t) = \underbrace{r_t + \gamma V_\pi(s_{t+1})}_{\text{estimate for Q(s,a)}} - V_\pi(s_t) \tag{13.51}$$

During learning, the actor provides sample states s_t and s_{t+1} for the critic to estimate the value function. The actor then uses this estimate to calculate the advantage function used to update the policy parameters θ.

Since actor-critic methods rely on current samples to train the critic (as an on-policy model), they suffer from the fact that estimates by the actor and critic are correlated. This can be alleviated by moving to off-policy training where samples are accumulated and stored in a **memory buffer**. This buffer is then randomly batched-sampled to train the critic. This is called **experience replay**, a sample efficient technique since individual samples can be used multiple times during training. In general, batch training with Actor-critic models can yield low variance estimates, but they will be biased with a poor critic estimator. This is in contrast to policy gradient models which may have high bias but are unbiased.

Algorithm 5: Actor-Critic algorithm

input : policy $\pi(a|s;\theta)$, state-value function $v(s;w)$
output: optimal policy π_*

initialize policy parameters θ and state-value weights w
while *not converged* **do**
    initialize state s
    for *each step in episode until terminal* **do**
        take action a from $\pi(a|s;\theta)$, observe reward r, next state s'
        update $w \leftarrow w + \beta A(s,a)\nabla v(s;w)$
        $A(s,a) \leftarrow r + \gamma v(s';w) - v(s;w)$
        update $\theta \leftarrow \theta + \alpha\gamma^t A(s,a)\nabla \log \pi(a_t|s_t;\theta)$
        update $s \leftarrow s'$

13.2.10.1 Advantage Actor Critic A2C

A way to reduce variance with online training is to use multiple threads that act in parallel together as a batch to train the model. Each thread uses a single sample and calculates an update using the advantage function. When all threads have finished calculating their update, they are batched together to update the model. This is known as the *synchronous advantage actor-critic* model, or **A2C**. As an algorithm, A2C is highly efficient and does not require memory buffer. Furthermore, it can leverage modern multi-core processors very effectively to accelerate computation.

Algorithm 6: A2C algorithm

input : policy $\pi(a|s;\theta)$, state-value function $v(s;w)$
output: optimal policy π_*

initialize policy parameters θ and state-value weights w
while *not converged* **do**
    initialize state s
    for *each step in episode until terminal* **do**
        sample N actions a_i from $\pi(a|s;\theta)$, observe reward r_i, next state s'_i
        update $w_i \leftarrow w_i + \beta A(s_i,a_i)\nabla v(s_i;w_i)$
        $A(s,a) \leftarrow \frac{1}{N}\sum_i r_i + \gamma v(s'_i;w_i) - v(s_i;w_i)$
        update $\theta \leftarrow \theta + \alpha\gamma^t A(s,a)\nabla \log \pi(a_t|s_t;\theta)$
        update $s \leftarrow s'$

13.2.10.2 Asynchronous Advantage Actor Critic A3C

Rather than waiting for all threads to finish calculating an update, we can update the model asynchronously. As soon as a thread calculates an update, it can broadcast the update to other threads which immediately apply it in their calculation. This is known as *asynchronous advantage actor-critic* or **A3C** and has received tremendous attention and unprecedented success due to its light computational footprint and quick training times.

13.3 Deep Reinforcement Learning Algorithms

Deep learning methods have several important applications in reinforcement learning. Their ability to automatically learn large, distributed representations and serve as universal function approximators makes them useful for modeling parametric policies, value functions, and advantage functions. In particular, recent advances in deep learning methods for sequence-to-sequence models have led to interesting deep reinforcement learning applications for NLP.

Deep neural networks are notoriously unstable when used to approximate non-linear functions like the state value function. There are a variety of techniques to stabilize learning, including batch training, experience replay, and target networks.

13.3.1 Why RL for Seq2seq

Sequence-to-sequence (seq2seq) models, as discussed in an earlier chapter, have been widely used to solve sequential problems. The most common method for training seq2seq models is called **teacher forcing**, where ground-truth sequences are used to minimize the maximum-likelihood (ML) loss at each decoding step. However, at test time, discrete metrics like Word Error Rate (WER) are often used to evaluate a model. These discrete metrics are non-differentiable and cannot be used in an ML framework for training. It is easy to optimize for ML loss at train time only to yield suboptimal metrics at test time, a problem known as **train-test inconsistency**.

Seq2seq models suffer from another significant problem known as **exposure bias**. While teacher forcing uses a ground truth label at each step to decode the next element in the sequence, this ground truth label is not available at test time. As a result, seq2seq models can only use its predictions to decode a sequence. This means that errors will accumulate during output sequence generation. As a result, poor models may never improve during training. One way to deal with exposure bias is to use scheduled sampling during model training, where a model is first pretrained using max-likelihood and then slowly shifted to its own predictions during training [Ken+18].

Reinforcement learning offers a way to overcome these two limitations. By incorporating the discrete metric like WER as a reward function, reinforcement learning methods can avoid the train-test inconsistency. Since the state of a RL model is given at each time step by the output state of the seq2seq decoder, exposure bias can be avoided.

Attention-based models have recently been shown to significantly outperform standard seq2seq models on a variety of tasks. However, they suffer from important limitations with large output spaces. In NLP, it is common to use smaller, truncated vocabularies to reduce computational burden. Attention-based models cannot handle out-of-vocabulary words. To overcome this, pointer-generation methods have recently been proposed [SLM17]. These methods implement a switch mechanism such that when an OOV word is predicted by the model output, the input word is copied over to the output. Pointer-generation models are currently state-of-the-art for several NLP tasks.

13.3.2 Deep Policy Gradient

Deep policy gradient methods train a deep neural network to learn the optimal policy. This can be accomplished with a seq2seq model where the output state of the decoder is used to represent the state of a model. The agent is thus modeled as the deep neural network (seq2seq model), where the output layer predicts a discrete action taken by this agent (Fig. 13.5). Policy gradient methods such as REINFORCE can be applied by choosing actions according to the deep neural network during training to generate sequences. The reward is observed at the end of the sequence or when an end-of-sequence (EOS) symbol is predicted. This reward can be a performance metric evaluated on the difference between the generated sequence and ground-truth sequence.

Unfortunately, the algorithm must wait until the end of a sequence to update, causing high variance and making it slow to converge. Furthermore, at the start of training when the deep neural network is initialized randomly, early predicted actions might lead the model astray. Recent work suggest pre-training the policy gradient model using cross-entropy loss before switching over to the REINFORCE algorithm, which is a concept known as a **warm start**.

Algorithm 7: The seq2seq REINFORCE algorithm

---

input : Input sequences X, ground-truth output sequences Y
output: Optimal policy π_*

while *not converged* **do**
  select batch from X and Y
  predict sequences of actions: $[a_1, a_2, \ldots, a_N]$
  observe rewards $[r_1, r_2, \ldots, r_N]$
  calculate baseline reward r_b
  calculate gradient and update the policy network

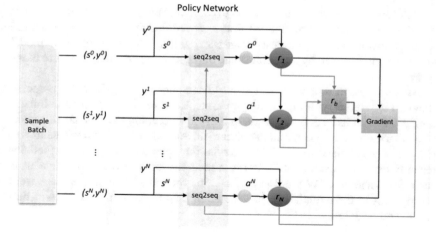

Fig. 13.5: DPG architecture

13.3.3 Deep Q-Learning

Instead of learning an estimate of the policy directly, we can use deep neural networks to approximate the action value function from which we can determine an optimal policy. These methods are commonly known as deep Q-learning, where we learn to estimate a Q-function $Q(s, a; \theta)$ with parameters θ by minimizing the loss function:

$$L(\theta) = \frac{1}{2} \mathbb{E}\left[r + \gamma \max_{a'} Q(s', a'; \theta) - Q(s, a; \theta) \right]^2 \tag{13.52}$$

Taking a gradient w.r.t. θ yields an update rule of the form:

$$\theta \leftarrow \theta + \alpha \underbrace{\left[r + \gamma \max_{a'} Q(s', a'; \theta) - Q(s, a; \theta) \right]}_{\text{temporal difference}} \nabla_\theta Q(s, a; \theta) \tag{13.53}$$

Unfortunately, the update rule has convergence issues and can be rather unstable, which limits the use of deep Q-learning models by themselves.

13.3.3.1 DQN

The deep Q-network (DQN) algorithm is a deep Q-learning model that utilizes **experience replay** and **target networks** to overcome instability (Fig. 13.6) [Mni+13]. Some have attributed the launch of the field of deep reinforcement learning to the introduction of the DQN algorithm in 2015 [HGS15]. Experience replay, as previously stated, uses a memory buffer to store transitions, which are mini-batch sampled during training. This experience buffer helps to break correlations between transitions and thereby stabilize learning.

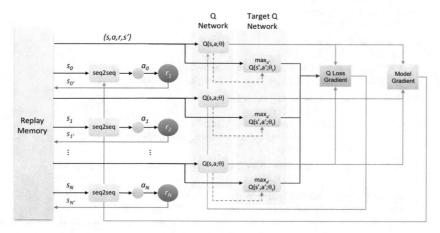

Fig. 13.6: DQN architecture

A target network is an extra copy of the deep Q-network. Its weights θ_{target} are periodically copied over from the original Q-network but remain fixed during all other times. This target network is used to compute the temporal difference during the update:

$$\theta \leftarrow \theta + \alpha \left[r + \underbrace{\gamma \max_{a'} Q(s',a';\theta_{target})}_{\text{target network}} - Q(s,a;\theta) \right] \nabla_\theta Q(s,a;\theta) \qquad (13.54)$$

Together, experience replay and a target network effectively smooth out learning and avoid parameter oscillations or divergence. Typically a finite memory buffer of length M is used for experience replay, such that only the most recent M transitions are stored and sampled. Furthermore, experiences are uniformly sampled from the buffer, regardless of significance. More recently, prioritized experience replay has been proposed [Sch+15a], where more significant transitions are sampled more frequently based on TD error and importance sampling.

Algorithm 8: Seq2Seq DQN algorithm

---

input : Input sequences X, ground-truth output sequences Y
output: Optimal Q function Q_*

Initialize seq2seq model π_θ
Initialize Q network parameters θ
Initialize target Q network parameters θ_{target}
Initialize replay memory

while *not converged* **do**
  select batch from X and Y
  sample sequences of actions from seq2seq model: $[a_1, a_2, \ldots, a_n]$
  collect experience $(s_t, a_t, r_t, s_{t'})$ and add to replay memory

  select mini-batch from replay memory
  for *each mini-batch sample* **do**
      estimate current Q value using Q network
      estimate next best action Q value using target Q network
      save estimates to buffer
  update Q network parameters θ by minimizing Q network loss with mini-batch
    estimates

  update seq2seq model π_θ with gradient based on estimated Q values
  every K steps, copy over weights to target network $\theta_{target} = \theta$

13.3.3.2 Double DQN

DQN methods suffer from the problem in that they fundamentally tend to overestimate Q-values. To see this, consider that the following relation holds:

$$\max_{a'} Q(s', a'; \theta_{target}) = Q\left(s', \arg\max_{a'} Q(s', a'; \theta_{target}); \theta_{target}\right) \qquad (13.55)$$

Using this, we can rewrite DQN loss function as:

$$L(\theta) = \frac{1}{2}\mathbb{E}\left[r + \gamma Q\left(s', \arg\max_{a'} Q(s', a'; \theta_{target}); \theta_{target}\right) - Q(s, a; \theta)\right]^2 \qquad (13.56)$$

In this expression, it can be seen that the target network is used twice; first to choose the next best action, and then to estimate the Q value of this action. As a result, there is a tendency to overestimate Q-values. **Double Deep Q-Learning** networks overcome this by using two separate target networks: one to select the next best action, and the other to estimate Q-values given the action selected.

Rather than introducing another target network, **Double Deep Q-Networks (DDQN)** uses the current Q-network to select the next best action and the target network to estimate its Q-value. The DDQN loss function can be written as:

$$L(\theta) = \frac{1}{2}\mathbb{E}\left[r + \gamma Q(s', \arg\max_{a'} Q(s', a'; \theta); \theta_{target}) - Q(s, a; \theta)\right]^2 \qquad (13.57)$$

DDQN alleviates the need for a third network as used in Double Deep Q-Learning to resolve the overestimation problem.

13.3.3.3 Dueling Networks

DQN and DDQN methods are useful when the action space is small. In NLP applications, however, the action space can be equal to the size of the vocabulary, even though only a small subset might be feasible at any one time. Estimating the Q-value of each action in such a large space can be prohibitively expensive and slow to converge. Consider the fact that in some states, the choice of action may have little to no effect, while in other states, choice of action might be life-or-death.

The **dueling network** method uses a single network to simultaneously predict both a state value function and advantage function that are aggregated to estimate the Q-function. By doing so, it avoids the need to estimate the value of each action choice. In one possible design, the dueling network is based on a Q-network architecture with CNN lower layers, followed by two separate fully connected layer streams whose outputs are summed together to estimate the Q-value.

Algorithm 9: Seq2Seq double DQN algorithm

---

input : Input sequences X, ground-truth output sequences Y
output: Optimal Q function $= Q_*$

Initialize seq2seq model π_θ
Initialize Q network parameters θ
Initialize target Q network parameters θ_{target}
Initialize replay memory

while *not converged* **do**
    select batch from X and Y
    sample sequences of actions from seq2seq model: $[a_1, a_2, \ldots, a_n]$
    collect experience $(s_t, a_t, r_t, s_{t'})$ and add to replay memory
    select mini-batch from replay memory

    for *each mini-batch sample* **do**
        estimate current Q value using Q network
        select next best action using Q-network
        estimate sample Q using target Q network
        save estimates to buffer

    update Q network parameters θ by minimizing Q network loss with mini-batch estimates

    update seq2seq model π_θ with gradient based on estimated Q values
    every K steps, copy over weights to target network $\theta_{target} = \theta$

13.3.4 Deep Advantage Actor-Critic

We have seen that the addition of a separate target network to deep Q-learning methods can help overcome high variance and overestimation. Recall in DDQN that we use the current network to select an action, and the target network to evaluate the action. In effect, the current network serves as the actor and the target network as the critic, with the caveat that the two networks are identical in architecture and the weights of the target network are periodically synchronized to the current network.

This need not be the case, as a different network can be trained to estimate the value function and act as a critic. Since deep neural networks tend to be unstable estimators of the state value function, deep actor-critic methods usually focus on estimating and maximizing the advantage function.

Instead of the advantage function defined as the difference between the state value function and Q-function, we can use the TD error:

$$\delta = r_t + \gamma V_{\pi_\theta}(s_{t+1}) - V_{\pi_\theta}(s_t) \tag{13.58}$$

since it can be proven that:

$$\mathbb{E}[\delta] = Q_{\pi_\theta}(s,a) - V_{\pi_\theta}(s_t) \tag{13.59}$$

This *value network* method is known as deep advantage actor-critic (Fig. 13.7). In this case, only a single Q network is necessary, though for stability reasons it is best trained with experience replay and a target network similar to DQN.

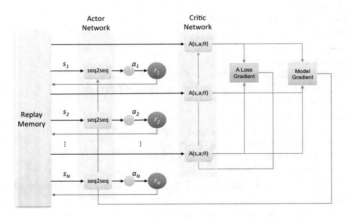

Fig. 13.7: Deep Advantage Actor-Critic architecture

Algorithm 10: Seq2Seq AC algorithm with experience replay

---

input : Input sequences X, ground-truth output sequences Y
output: Optimal policy π_*

Initialize actor (seq2seq) network, π_θ
Initialize critic network θ
Initialize replay memory

while *not converged* **do**
    select batch from X and Y
    sample sequences of actions from Actor: $[a_1, a_2, \ldots, a_n]$
    Calculate true discounted rewards: $[r_1, r_2, \ldots, r_n]$
    collect experience (a_n, v_n) and add to replay memory

    sample mini-batch from replay memory
    for *each mini-batch sample* **do**
        | compute advantage estimates from the critic network

    update critic Q network parameters θ by minimizing critic loss over mini-batch

    update actor parameters π_θ with gradient based on advantage estimates from critic

---

13.4 DRL for Text

Deep reinforcement learning methods have been recently applied to a variety of natural language processing tasks on text. In particular, they have been very successful in building conversational agents and dialogue systems. In the next sections, we provide a survey of different DRL methods for information extraction, text classification, dialogue systems, text summarization, machine translation, and natural language generation. Many of these are based on leveraging seq2seq models used to either generate embeddings or as models of the target policy. This does not say that DRL methods are restricted to use seq2seq models, as CNNs can also be successfully applied.

13.4.1 Information Extraction

Information extraction is defined as the task of automatically extracting entities, relations, and events from text. In recent years, researchers have successfully applied deep learning methods to entity extraction, including architectures that leverage CNNs and RNNs [Qi+14, GHS16]. In real domains, however, it takes very large amounts of labeled data to learn to perform high quality extraction. Furthermore, relation extraction quality depends on the results of entity extraction (and vice versa). It may also be that we care about only a subset of relations, such as in action task extraction. DRL methods have found applicability in addressing these considerations.

For large-scale domains, labeled training data is often the largest constraint to performance, as it can be prohibitively expensive to obtain accurately labeled data. Distant supervision is one method that seeks to alleviate this need by leveraging an external knowledge graph to automatically align text to extract entities or relations [Min+09b]. However, extraction generated in this manner is not directly labeled and can be incomplete. This is where reinforcement learning can be helpful.

13.4.1.1 Entity Extraction

For entity extraction tasks, external information can be used to resolve ambiguities and boost accuracy by querying similar documents and comparing extracted entities. This is a sequential task that can be addressed with a reinforcement learning agent where we model the extraction task as a Markov decision process.

Figure 13.8 is an example of the architecture proposed by K. Narasimhan et al. [NYB16] based on a DQN agent. In this model, the states are real-valued vectors that encode the matches, context, and confidence of extracted entities from the target and query documents. The actions are to accept, reject, or reconcile the entities of two documents and query the next document. The reward function is selected to maximize the final extraction accuracy:

$$R(s,a) = \sum_{entity\,j} Acc(entity_{target}(j)) - Acc(entity_{query}(j)) \tag{13.60}$$

To minimize the number of queries, a negative reward is added to each step. Since this model is based on a continuous state space, the DQN algorithm can be trained to approximate the Q-function, where the parameters of the DQN are learned using stochastic gradient descent with experience replay and a target network to reduce variance.

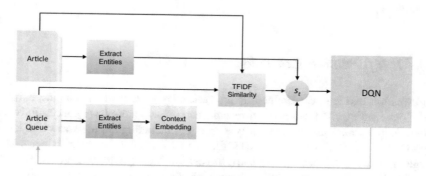

Fig. 13.8: Entity extraction with DQN

13.4.1.2 Relation Extraction

Consider a deep learning network for the task of relation extraction. This network is regarded as the DRL agent, whose role is to take as input a sequence of words in a sentence, and whose output are the extracted relations. If the sentences are regarded as states and the relations as actions, we can learn an optimal policy to perform relation extraction. The process of extracting relations from a bag of sentences becomes an episode.

Figure 13.9 depicts a deep policy gradient approach to this relation extraction task. The reward function is given by the accuracy of the predicted relations in a bag in comparison with a set of gold labels. The REINFORCE algorithm has been applied [Zen+18] to optimize the policy of this model by defining the reward function of a state s_i to be:

$$R(s_i) = \gamma^{n-i} r_n \tag{13.61}$$

where n is the number of sentences in the bag and r_n is either $+1$ or -1. The objective function for the policy gradient method is:

$$J(\theta) = \mathbb{E}_{s_1, s_2, \dots, s_n} R(s_i) \tag{13.62}$$

This leads to a gradient update of the form:

$$\theta \leftarrow \theta + \nabla J(\theta) = \sum_{i=1}^{n} \sum_{j=1}^{n_i} \nabla p(a_i | s_i; \theta)(R(s_i) - r_b) \tag{13.63}$$

where the baseline r_b is given by:

$$r_b = \frac{\sum_{i=1}^{n} \sum_{j=1}^{n_i} R(s_j)}{\sum_{i=1}^{n} n_i} \tag{13.64}$$

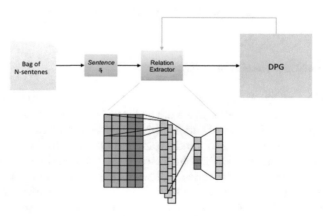

Fig. 13.9: Relation extraction with DPG

13.4.1.3 Action Extraction

The task of extracting action sequences from text is challenging in that they usually are highly affected by context. Traditional methods depend on a set of templates which do not generalize well to natural language. Sequence labeling methods do not perform well, since there is only a subset of sequences that can be considered meaningful actions. The action extractor can be modeled as a DRL agent, where the states are regarded as word sequences, and actions are the set of labels associated with the word sequence. This agent can learn an optimal labeling policy by training a DQN model. Figure 13.10 shows the architecture proposed by Feng et al. [FZK18] called EASDRL that is based on first extracting action names and then the action targets. To do so, this architecture defines two Q-functions associated with separate CNN networks for modeling the action name $Q(s, a)$ and action target $Q(\hat{s}, a)$, and is trained using a variant of experience replay that weighs positive-reward transitions higher.

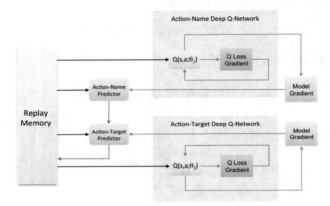

Fig. 13.10: Action extraction with DQN

13.4.1.4 Joint Entity/Relation Extraction

Usually, entity extraction occurs as a precursor to relation extraction. They can be considered interdependent tasks, since the quality of relation extraction usually depends on the quality of extracted entities. Given this sequential nature, it is possible to use reinforcement learning to jointly learn and optimize for both tasks concurrently. Figure 13.11 shows a DRL architecture [Fen+17] based on a deep Q-learning agent. In this model, the current state s is the entity extractor output from a Bi-LSTM with attention $Att(X; \theta_1)$, and the transition state s' is the relation extraction output from a Tree-LSTM $Tree(X; \theta_2)$. The actions are defined over the set (a_1, a_2, a_3, a_4) where a_1 and a_2 classify the existence of a relation mention, and a_3 and a_4 classify the type of relation mention. In other words, the DRL agent combines the tasks of entity extraction, relation mention classification, and relation classification. The DQL model is trained using stochastic gradient descent.

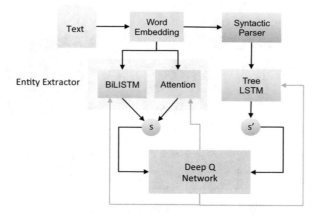

Fig. 13.11: Joint entity/relation extraction with DQL

13.4.2 Text Classification

Deep learning for text classification has mainly focused on learning representations of words and sentences that can effectively capture semantic context and structure. Current methods, however, are unable to automatically learn and optimize structure, since they are trained explicitly using supervised input or treebank annotations. In contrast, DRL can be used to build hierarchical-structure sentence representations without the need for annotations.

Figure 13.12 shows an architecture that consists of three components: a policy network, a representation model, and a classification network [ZHZ18]. The policy network is based on a stochastic policy whose states are vectors representations of both word level and phrase level structure. These vectors are the output of the representation model which consists of a two-level hierarchical LSTM that connects a sequence of words to form a phrase and a sequence of phrases to form a sentence representation. The actions of the policy network label whether a word is inside or at the end of a phrase. Whereas the policy network focuses on building sentence representations that capture structure, the classification network takes the output from the representation model and uses it to perform the classification task.

To jointly train the policy and classification networks, the hierarchical LSTM is first initialized and pre-trained using the cross-entropy loss of the classifier network, given by:

$$L = -\sum_{X \in D} \sum_{y=1}^{K} p(y, X) \log P(y|X) \tag{13.65}$$

where p and P are the target and predicted distributions, respectively. The parameters for the representation model and classifier networks are then held constant and the policy network is pre-trained using the REINFORCE algorithm. After the warm start, all three networks are jointly trained until convergence.

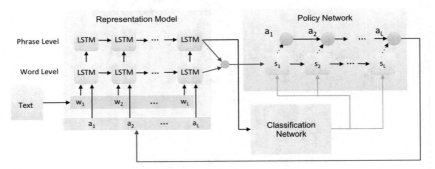

Fig. 13.12: Text classification with DPG

13.4.3 Dialogue Systems

Dialogue systems have become increasingly popular as chatbots gain widespread application across social media and customer service. Developing an intelligent dialogue system has always been a major goal of AI, dating back to the Turing Test. Dialogue agents must perform a pipeline of multiple tasks, including natural language understanding, state tracking, dialogue policy, and natural language generation. Dialogue systems have been modeled successfully as partially observable Markov decision processes.

Slot-filling dialogues are an important subclass of dialogue systems that involve filling-in a set of predefined slots in response to user dialogue and context. In these systems, the relationship between a chatbot and user is analogous to an RL agent and its environment. Conversational dialogue becomes an optimal decision making problem, where the reward function can be defined as a successful interaction between chatbot and user.

There are several fundamental problems with dialogue systems. The biggest issue is the *credit assignment* problem, where error propagation through the pipeline may make it near impossible to determine the component source of error. For instance, poor performing dialogue policy may be due to incorrect state tracking or low-quality NLU. Similarly, the reliance of downstream components on upstream tasks makes optimization particularly difficult. For instance, a tweak to the state tracker may lead to sub-optimal dialogue policy. In an ideal case, the entire pipeline is trained at once in an end-to-end manner. For these reasons, deep RL methods are finding significant use for modeling dialogue systems.

A DQN agent has been successfully applied to train a dialogue system [ZE16, GGL18] that unifies state tracking and dialogue policy and treats both as actions available to the RL agent. The architecture learns an optimal policy that generates a verbal response or updates the current dialogue state. Figure 13.13 depicts the DQN model, which uses an LSTM network to generate a dialogue state representation. The output of the LSTM serves as input to a set of policy networks in the form of multilayer perceptron networks representing each possible action. The output of these networks represents the action-state value functions for each action.

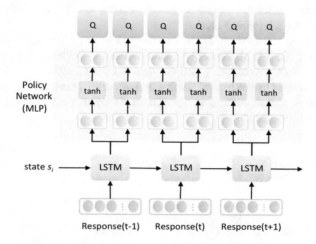

Fig. 13.13: Dialogue system with DQN

Due to the high-dimensional state and action spaces, a large number of labeled dialogues are typically required to train dialogue systems. To overcome this need for training data, a two-stage deep RL method has been proposed [Fat+16] that uses an actor-critic architecture where the policy network is first supervised-trained using a small number of high-quality dialogues via categorical cross-entropy to bootstrap learning. The value network can then be trained using the deep advantage actor-critic method.

13.4.4 Text Summarization

Text summarization is an interesting NLP task that seeks to automatically generate natural language summaries of input text in human-readable form. It has widespread use across a variety of industries and comes in two categories: extractive and abstractive summarization. In the extractive case, it seeks to eliminate superfluous text and keep only the most relevant words while maintaining natural language form.

In the abstractive case, it seeks to provide a paraphrased summary of the relevant points in the text.

Recall-Oriented Understudy for Gisting Evaluation (**ROUGE**) is the standard quality measure most often used for text summarization tasks. By definition, ROUGE-1 measures the unigrams that are shared between a predicted summarization and the ground-truth reference text. ROUGE-2 measures the bigrams that are shared, and ROUGE-L measures the longest common substring (LCS) between prediction and ground-truth. For each of these measures, precision and recall are typically quoted. The problem with ROUGE is that they provide little information about the human readability of predictions, which are usually captured by a measure like perplexity for a language model.

DQN has been successfully applied to the task of extractive text summarization [LL17, PXS17b, Çe+18b]. Figure 13.14 shows the architecture where states denote the current (partial) text summary, actions denote adding a sentence to this summary, and ROUGE is used as the reward. In this architecture, a sentence is represented as a concatenation of a document vector (DocVec), sentence vector (SentVec), and position vector (PosVec).

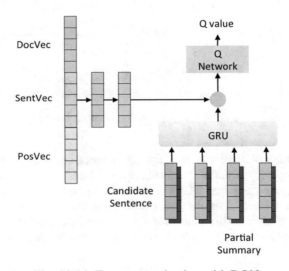

Fig. 13.14: Text summarization with DQN

Attention-based deep learning networks have found significant traction in abstractive text summarization tasks. But despite their high ROUGE scores, they often generate unnatural summaries. This has opened the door to deep RL methods that can incorporate a mixed training objective:

$$L_{mixed} = \sigma L_{rl} + (1 - \sigma)L_{ml} \tag{13.66}$$

that incorporates both the teacher-forcing maximum likelihood function:

$$L_{ml} = -\sum_{t=1}^{n} \log p(y_t|y_1, y_2, \ldots, y_{t-1}, x) \tag{13.67}$$

and a policy gradient objective:

$$L_{rl} = -[r - r_b] \sum_{t-1}^{n} \log P(y_t|y_1, y_2, \ldots, y_{t-1}, x) \tag{13.68}$$

where the reward r is a discrete objective like ROUGE.

13.4.5 Machine Translation

One of the recent breakthroughs in neural machine translation has been the use of seq2seq models. As noted above, teacher-forcing is the primary method to train these networks. These models exhibit exposure bias during prediction time. Furthermore, decoders cannot generate target sequences of interest with specific objectives. This is especially so if beam-search is employed, which tends to focus more on short-term rewards, a concept termed **myopic bias**. Machine translation is most often evaluated based on the discrete BLEU measure, which creates a train-test mismatch.

Deep RL models have been proposed to overcome some of these shortfalls. A deep PG model [Li+16] based on the REINFORCE training algorithm can address the non-differentiable nature of the BLEU metric. However, REINFORCE suffers from the inability to learn policies in large action spaces as is the case with language translation.

More recently, an actor-critic model has been proposed by using a decoding scheme that incorporates longer-term rewards through a value function estimate [Bah+16a]. In this model, the main sequence prediction model is the actor/agent and the value function acts as a critic. The current sequence prediction output is the state, and candidate tokens are actions of the agent. The critic is implemented by a separate RNN and is trained on the ground-truth output using temporal difference methods, with a target critic used to reduce variance.

13.5 DRL for Speech

Deep neural networks have significantly improved the performance of speech recognition systems nowadays. When they are used as part of a hybrid system together with GMMs or HMMs, alignment of the acoustic model is a necessity during training. This can be avoided when deep neural networks are used in end-to-end systems that learn transcriptions by directly maximizing the likelihood of the input data [YL18]. Such systems, while currently leading state-of-the-art performance, still suffer from a variety of limitations.

Drawing from the experiences with text, researchers and practitioners have begun to apply deep reinforcement learning methods to speech and audio, including tasks such as automatic speech recognition, speech enhancement, and noise suppression. In the near future, we expect to see wider adoption of deep RL techniques in other aspects of speech, including applications in speaker diarization, speaker tone detection, and stress analysis.

13.5.1 Automatic Speech Recognition

The task of automatic speech recognition (ASR) is in many ways similar to machine translation. ASR most often uses CTC maximum likelihood learning while measuring performance with a discrete measure like word-error rate (WER). As a result, train-test mismatch is a problem. Furthermore, as a sequence prediction task, ASR suffers from exposure bias since it will be trained on ground-truth labels that are not available at prediction time.

A deep RL approach using policy gradients has been shown to be effective in [ZXS17] overcoming these limitations (Fig. 13.15). In this approach, the ASR model is regarded as the agent, and training samples as the environment. The policy $\pi_\theta(y|x)$ is parameterized by θ, the actions are considered to be generated transcriptions, and the model state is the hidden data representation. The reward function is taken to be WER. The policy gradient is updated by the rule:

$$\theta \leftarrow \theta + \alpha \nabla_\theta \log P_\theta(y|x)[r - r_b] \tag{13.69}$$

13.5.2 Speech Enhancement and Noise Suppression

Machine learning speech enhancement methods have been in existence for quite a while. Enhancement techniques usually fall under four subtasks: voice-activity detection, signal-to-noise estimation, noise suppression, and signal amplification. The first two provide statistics on the target speech signal while the latter two use these statistics to extract the target signal. This can be naturally thought of as a sequential task. A deep RL method based on policy gradients has been proposed [TSN17] for the task of speech enhancement with an architecture that is based on using an LSTM network to model a filter whose parameters θ are determined by a learned policy π_θ. In this model, the filter is the agent, the state is a set of filter parameters, and actions are increases or decreases in a filter parameter. The reward function measures the mean-square error between the filter output and a ground-truth clean

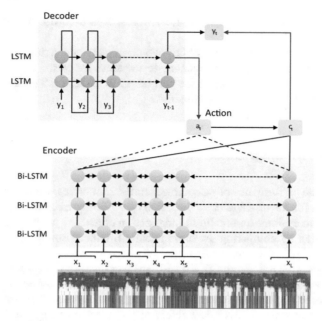

Fig. 13.15: Automatic speech recognition with DPG

signal sequence. This policy gradient model, trained using the REINFORCE algorithm, can improve signal-to-noise ratio with no algorithmic changes to the baseline
speech-enhancement process. Furthermore, by incorporating a deep reinforcement
agent, the filter can adjust to changing underlying conditions through dynamic parameter adaptation.

13.6 Case Study

In this case study, we will apply the deep reinforcement learning concepts of this
chapter to the task of text summarization. We will use the Cornell NewsRoom Summarization dataset. The goal here is to show readers how we can use deep reinforcement learning algorithms to train an agent that can learn to generate summaries of
these articles. For the case study, we will focus on deep policy gradient and double
deep Q-network agents.

13.6.1 Software Tools and Libraries

We will use the following packages in this case study:

- **TensorFlow** is an open-source software library for dataflow programming across
  a range of tasks. It is a symbolic math library, and is also used for machine

learning applications such as neural networks. It is used for both research and production at Google.

- **RLSeq2Seq** is an open-source library which implements various RL techniques for text summarization using sequence-to-sequence models.
- **pyrouge** is a python interface to the perl-based ROUGE-1.5.5 package that computes ROUGE scores of text summaries.

13.6.2 Text Summarization

To measure the performance of machine generated summaries, we will use ROUGE, which stands for *Recall-Oriented Understudy for Gisting Evaluation*. It is a set of metrics used to evaluate automatic summarization of texts as well as machine translation. It works by comparing an automatically produced summary or translation against a set of reference summaries (typically human-produced).

ROUGE-N, ROUGE-S, and ROUGE-L are measures of the granularity of texts when comparing between the system predicted summaries and reference summaries. For example, ROUGE-1 refers to overlap of unigrams between the system summary and reference summary. ROUGE-2 refers to the overlap of bigrams between the system and reference summaries. Let's take the example from above. Let us say we want to compute the ROUGE-2 precision and recall scores. For ROUGE, recall is a measure of how much of the reference summary is the captured by the system summary.

13.6.3 Exploratory Data Analysis

The Cornell Newsroom dataset consists of 1.3 million articles and summaries written by news authors and editors from 38 major publications between 1998 and 2017. The dataset is split into train, dev, and test sets of 1.1 m, 100 k, and 100 k samples. A sample of the dataset is provided below:

> Story: Coinciding with Mary Shelley's birthday week, this Scott family affair produced by Ridley for director son Luke is another runout for the old story about scientists who create new life only to see it lurch bloodily away from them. Frosty risk assessor Kate Mara's investigations into the mishandling of the eponymous hybrid intelligence (The Witch's still-eerie Anya Taylor-Joy) permits Scott Jr a good hour of existential unease: is it the placid Morgan or her intemperate human overseers (Toby Jones, Michelle Yeoh, Paul Giamatti) who pose the greater threat to this shadowy corporation's safe operation? Alas, once that question is resolved, the film turns into a passably schlocky runaround, bound for a guessable last-minute twist that has an obvious precedent in the Scott canon. The capable cast yank us through the chicanery, making welcome gestures towards a number of science-fiction ideas, but cranked-up Frankenstein isn't one of the film's smarter or more original ones.

Summary: Ridley and son Luke turn in a passable sci-fi thriller, but the horror turns to shlock as the film heads for a predictable twist ending

For our case study, we will use subsets of 10,000/1000/1000 articles and summaries from the Cornell Newsroom dataset for our training, validation, and test sets, respectively. We will tokenize and map these data sets using 100-dim embeddings generated with word2vec. For memory considerations, we limit our vocabulary to 50,000 words.

13.6.3.1 Seq2Seq Model

Our first task is to train a deep policy gradient agent that can produce summaries of the articles. Before we do so, we pre-train the seq2seq model using maximum likelihood loss, an encoder and decoder layer size of 256, batch size of 20, and adagrad with gradient clipping for 10 epochs (Fig. 13.16). After pre-training, we evaluate this model on test set to find the results shown in Table 13.1.

Table 13.1: ROUGE metrics for Seq2Seq trained on MLE

	F-score	Precision	Recall
ROUGE-1	15.6	20.6	14.5
ROUGE-2	1.3	1.6	1.3
ROUGE-L	14.3	19.0	13.3

Seq2seq: at 90-years old this tortoise has never moved better despite a horrific rat attack that caused legs
Reference: a 90-year old tortoise was given wheels after a rat attack caused her to lose her front legs

Seq2seq: a city employee in baquba the capital of diyala province vividly described his ambivalence
Reference: iraqis want nothing more than to have u.s. soldiers leave iraq but there is nothing they can less afford

Seq2seq: google reported weaker than expected results thursday for its latest quarter
Reference: the tech giant 's shares rose after it reported a smaller than expected rise in sales for its latest quarter

In comparison with the reference summaries, the generated summaries are fair but leave some room for improvement.

13.6.3.2 Policy Gradient

Let's apply a deep policy gradient algorithm to improve our summaries (Fig. 13.17). We set our reward function to ROUGE-L F1 score and switch from MLE loss to RL

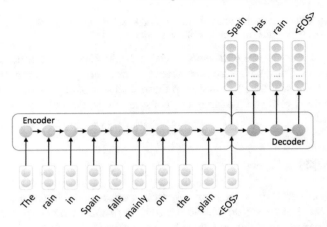

Fig. 13.16: Seq2Seq model for text summarization

loss. We continue training for 8 epochs, after which we evaluate the RL-trained model on the test set to find the results shown in Table 13.2.

Table 13.2: ROUGE metrics for DPG

	F-score	Precision	Recall
ROUGE-1	22.4	19.6	35.3
ROUGE-2	6.0	5.8	8.5
ROUGE-L	17.6	15.5	28.0

As we increase training, we expect to see the generated summaries become even closer aligned to human-generated language.

DPG: apple has disclosed the details of a streaming music service plan to recording companies sources say
Reference: apple executives have spoken to the top four recording companies about plans to offer a streaming music service free of charge to consumers multiple music industry sources told cnet

DPG: conservative pundit glenn beck says the obama administration is using churches and other faith based groups to promote its climate change agenda
Reference: glenn beck says obama uses churches on climate change green house

DPG: the zoo in georgia 's capital has reopened three months after a devastating flood that killed more than half of its 600 animals including about 20 tigers lions and jaguars
Reference: a georgia zoo that had half its animals killed during floods in june has reopened

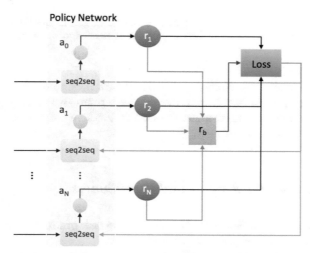

Fig. 13.17: Deep policy gradient for text summarization

13.6.3.3 DDQN

Let's see if we can improve on the results above using a double deep Q-learning agent. We start as before by pre-training the seq2seq language model using maximum likelihood loss for 10 epochs. We then train the double deep-Q network for 8 epochs using a batch size of 20, replay buffer of 5000 samples and updating the target network every 500 iterations. For better results, we will first pre-train the DDQN agent with a fixed actor for a single epoch. When we then evaluate the resulting model on the test set, we find the results in shown in Table 13.3.

Table 13.3: ROUGE metrics for DDQN

	F-score	Precision	Recall
ROUGE-1	34.6	28.8	55.5
ROUGE-2	21.4	19.0	31.1
ROUGE-L	30.4	25.7	47.7

DDQN: the commander of us forces in the middle east said that the refusal to follow orders occurred during the battle for the recently liberated town of manbij syria
Reference: a top us general said tuesday that isis fighters defied their leader 's orders to fight to the death in a recent battle instead retreating to the north

DDQN: an online discussion of the washington area rental market featuring post columnist sara gebhardt
Reference: welcome to apartment life an online discussion of the washington area rental market featuring post columnist sara gebhardt

DDQN: albania has become the largest producer of outdoor grown cannabis in europe
Reference: albania has become the largest producer of outdoor grown cannabis in europe

The DDQN agent outperforms the DPG agent for the chosen parameters. There are a myriad of possibilities to improve results further—we could use scheduled or prioritized sampling, intermediate rewards, and/or some form of attention at the encoder or decoder.

13.6.4 Exercises for Readers and Practitioners

1. How would you combine a DQN agent for the task of text classification when using a seq2seq model with soft attention?
2. Does it make sense to use two separate target networks for the double DQN agent? Why or why not?
3. What kind of deep neural networks would we use for the Q-learning model? Why would or would not CNNs be appropriate?

References

[Bah+16a] Dzmitry Bahdanau et al. "An Actor-Critic Algorithm for Sequence Prediction." In: *CoRR* abs/1607.07086 (2016).

[Fat+16]   Mehdi Fatemi et al. "Policy Networks with Two-Stage Training for Dialogue Systems." In: *CoRR* abs/1606.03152 (2016).

[FZK18]   Wenfeng Feng, Hankz Hankui Zhuo, and Subbarao Kambhampati. "Extracting Action Sequences from Texts Based on Deep Reinforcement Learning." In: *IJCAI*. ijcai.org, 2018, pp. 4064–4070.

[Fen+17]   Yuntian Feng et al. "Joint Extraction of Entities and Relations Using Reinforcement Learning and Deep Learning." In: *Comp. Int. and Neurosc.* 2017 (2017), 7643065:1–7643065:11.

[GGL18]   Jianfeng Gao, Michel Galley, and Lihong Li. "Neural Approaches to Conversational AI." In: *CoRR* abs/1809.08267 (2018).

[GHS16]   Tomas Gogar, Ondrej Hubácek, and Jan Sedivý. "Deep Neural Networks for Web Page Information Extraction." In: *AIAI*. Vol. 475. Springer, 2016, pp. 154–163.

[HGS15]   Hado van Hasselt, Arthur Guez, and David Silver "Deep Reinforcement Learning with Double Q-learning." In: *CoRR* abs/1509.06461 (2015).

[Ken+18]   Yaser Keneshloo et al. "Deep Reinforcement Learning For Se quence to Sequence Models." In: *CoRR* abs/1805.09461 (2018).

[LL17]   Gyoung Ho Lee and Kong Joo Lee. "Automatic Text Summarization
             Using Reinforcement Learning with Embedding Features." In: *IJC-
             NLP(2)*. Asian Federation of Natural Language Processing, 2017, pp.
             193–197.

[Li+16]   Jiwei Li et al. "Deep Reinforcement Learning for Dialogue Gener
             ation". In: *CoRR* abs/1606.01541 (2016).

[Min+09b]   Mike Mintz et al. "Distant supervision for relation extraction without
             labeled data." In: *ACL/IJCNLP*. The Association for Computer Linguis-
             tics, 2009, pp. 1003–1011.

[Mni+13]   Volodymyr Mnih et al. "Playing Atari with Deep Reinforcement Learn-
             ing." In: *CoRR* abs/1312.5602 (2013).

[NYB16]   Karthik Narasimhan, Adam Yala, and Regina Barzilay "Improving In-
             formation Extraction by Acquiring External Evidence with Reinforce-
             ment Learning." In: *CoRR* abs/1603.07954 (2016).

[PXS17b]   Romain Paulus, Caiming Xiong, and Richard Socher. "A Deep
             Reinforced Model for Abstractive Summarization." In: *CoRR*
             abs/1705.04304 (2017).

[Qi+14]   Yanjun Qi et al. "Deep Learning for Character-Based Information Ex-
             traction." In: *ECIR*. Vol. 8416. Springer, 2014, pp. 668–674.

[Sch+15a]   Tom Schaul et al. "Prioritized Experience Replay." In: *CoRR*
             abs/1511.05952 (2015).

[SLM17]   Abigail See, Peter J. Liu, and Christopher D. Manning. "Get To The
             Point: Summarization with Pointer-Generator Networks." In: *CoRR*
             abs/1704.04368 (2017).

[TSN17]   Andros Tjandra, Sakriani Sakti, and Satoshi Nakamura. "Sequence-to-
             Sequence ASR Optimization via Reinforcement Learning." In: *CoRR*
             abs/1710.10774 (2017).

[YL18]   Dong Yu and Jinyu Li. "Recent Progresses in Deep Learn-
             ing based Acoustic Models (Updated)." In: *CoRR* (2018).
             http://arxiv.org/abs/1804.09298

[Zen+18]   Xiangrong Zeng et al. "Large Scaled Relation Extraction With Rein-
             forcement Learning." In: *AAAI* AAAI Press, 2018.

[ZHZ18]   Tianyang Zhang, Minlie Huang, and Li Zhao. "Learning Structured
             Representation for Text Classification via Reinforcement Learning."
             In: *AAAI*. AAAI Press, 2018.

[ZE16]   Tiancheng Zhao and Maxine Eskénazi. "Towards End-to-End Learning
             for Dialog State Tracking and Management using Deep Reinforcement
             Learning." In: *SIGDIAL Conference*. The Association for Computer
             Linguistics, 2016, pp. 1–10.

[ZXS17]   Yingbo Zhou, Caiming Xiong, and Richard Socher. "Improving
             End-to-End Speech Recognition with Policy Learning." In: *CoRR*
             abs/1712.07101 (2017).

[Çe+18b]   Asli Çelikyilmaz et al. "Deep Communicating Agents for Abstractive
             Summarization." In: *NAACL-HLT*. Association for Computational Lin-
             guistics, 2018, pp. 1662–1675.

Future Outlook

Predicting the future of AI is no more possible today than it has been in years past. Furthermore, the farther into the future we project, the greater the uncertainty. In general, some things may go exactly as expected (improvements in computational speed), some expectations may have slight variability (the dominant deep learning architectures), and others are maverick innovations that are unlikely to be predicted (the intersection of big data, computational speed, and emergence of deep learning all at the same time). At the conclusion of this book, we would like to provide our predictions based on the current trajectories, trends, and usefulness of the research we've discussed. We reject all claims to be considered soothsayers or even reliable parties in these projections. We attempt to only provide considerations for the reader at the conclusion of these topics and suggest areas of awareness over upcoming years.

End-to-End Architecture Prevalence

Given the success of many end-to-end approaches in both NLP and speech, we expect that more will move towards these architectures. One of the areas where these approaches lack robustness is in the tuning to particular environments, for example, the usefulness of a lexicon model in the ASR hybrid architecture or in the adaptation of language models to new domains. This is an area that must be addressed for deep learning to make a significant impact in domains where training data is costly or unavailable.

Transition to AI-Centric

One of the simplest projections is that more companies will shift to or center around an AI-Centric strategy. Many of the leading tech companies—for example, Google,

© Springer Nature Switzerland AG 2019
U. Kamath et al., *Deep Learning for NLP and Speech Recognition*,
https://doi.org/10.1007/978-3-030-14596-5

Facebook, and Twitter—have moved in this direction, and this trend will likely continue into many other large and mid-sized companies. This shift will introduce machine learning into every level of software development and with it the need for tools and processes to ensure reliability and generality. Some have coined the term "Software 2.0[1]" in light of this shift. Transitioning to this state will require increased rigor around data, interpretability of models, an increased focus on model security, and resiliency to adversarial scenarios.

Specialized Hardware

Specialized hardware will become more common. This pattern of development is fairly common with utilization of ASIC (application-specific integrated circuit) hardware for cryptocurrency mining or image processors embedded in smartphones. The introduction of TPUs has been one of the first cases where dedicated physical hardware has been created specifically for deep learning. The introduction of the Apple A11 chip is another example of specialized hardware to support neural networks on mobile devices.

Transition Away from Supervised Learning

We expect the focus of machine learning to shift. Deep learning has seen the largest improvements with supervised data; however, the costs associated with creating large, labeled datasets are often prohibitively expensive. In many scenarios, large unlabeled sources exist that can be used by unsupervised algorithms, and we expect a greater concentration of algorithms in this area, as seen in the progression of word embeddings and language models.

Explainable AI

Though end-to-end deep learning techniques are powerful and can result in impressive performance metrics such as accuracy, they suffer from interpretability. Many applications in the financial world (such as loan applications or conduct surveillance) or in healthcare (like predicting disease) need models and predictions to be explainable. There has been a shift in the industry towards explainable AI (XAI). Many techniques such as Local Interpretable Model-agnostic Explanation (LIME), Deep Learning Important FeaTures (DeepLIFT), SHapley Additive exPlanation (SHAP) to name a few have been very promising in providing model-agnostic expla-

---

[1] https://medium.com/@karpathy/software-2-0-a64152b37c35.

nations for individual predictions as well as summarization of models. Innovations such as these and others will be necessary to overcome the hurdles of interpretability of models and trust of AI.

Model Development and Deployment Process

There is a trade-off in deep learning between the ease of experimentation during model development and deployment of these models in a high performing, low-latency production with highly optimized code. This trade-off is more prevalent in NLP and speech recognition models as they are complex dynamic graphs as compared to the preferred static graphs for optimized performance at runtime. Frameworks such as PyText, which help to tune pre-built models, perform experiments in a rapid manner, provide pre-built workflows for model designers and engineers, and support easy deployment of models to production environments with minimum intervention, will soon become the necessary part of the development process. Model testing and quality assurance is another aspect of the development and deployment process that needs to be adjusted to accommodate complex deep learning models. Google's recent research paper "The ML Test Score: A Rubric for ML Production Readiness and Technical Debt Reduction" proposes a great framework towards testing these complex deep learning based systems.

Democratization of AI

AI and deep learning are used by a still very small but rapidly growing group of researchers, educators, experts, and practitioners. To make them accessible to the masses through applications, tools, or education, there needs to be a change in attitude, policies, investment, and research, especially from top companies and universities. This phenomenon is called the "democratization of AI." Many companies such as Google, Microsoft, and Facebook, as well as many universities such as MIT, Stanford, and Oxford are contributing to software tools, libraries, datasets, courses, etc. that are freely available on the web. The positive trend in this direction will play a huge role in transforming lives through AI.

NLP Trends

Language models can be pre-trained on a large corpus of unlabeled data, giving it a considerable advantage. Language models are now considered to add enormous benefits for many NLP tasks. Language model embeddings provide features for complex tasks and have shown to provide improvements over many tasks on the state-of-the-art methods. Using adversarial methods to either understand the models, analyze fail cases, or improve the robustness of models is becoming a trend in deep learning research. Moving towards under-resourced languages and using deep learning techniques such as transfer learning is another area that many researchers are focusing especially in tasks such as machine translation.

One of the most curious areas of development is in the area of reinforcement learning. Instead of collecting data, training a model, putting it into production, and testing the result, an agent could be created to interact with the environment (real or synthetic) and learn based on its experience. Overall, we see the progression moving from supervised to unsupervised to reinforcement techniques.

Speech Trends

Many of the end-to-end deep learning techniques are able to outperform traditional hybrid HMM-based models with less tuning and linguistic expertise. These models perform very well in scenarios where training data is widely available, typically in general speech recognition tasks. However, they tend to struggle when context is crucial to prediction. Additionally, the continued pursuit of fusing speech and NLP is a direction likely to continue, with end-to-end learning taking the lead. Recent advancements are focusing on incorporating domain information into the decoding procedure via language model fusion for contextualized recognition.

Other areas where speech recognition still struggles are with acoustic environments and speaker-specific differences such as accents. Leveraging generated data from speech-to-text systems is gaining traction, providing simulated environments and speakers for improved robustness. We expect the incorporation of speech-to-text systems, similar to GAN workflows, to continue to improve, and will potentially be incorporated more fully into reinforcement workflows.

Closing Remarks

We hope that the readers have found the information in this book both informative and helpful. Deep learning has heavily impacted NLP and speech in the past few years, and the trend seems to be gaining speed. We have hopefully enabled the readers to understand both fundamental and advanced techniques that deep learning offers, while also showing how to practically apply them.

Index

A

A2C, 589
A3C, 590
activation function, 156
actor-critic, 588
Adagrad, 163
Adam, 164
advantage function, 578
Adversarial, 173
ANEW, 111
artificial intelligence, 8
ASG, 544
ASR, 369
Attention, 335
attention, 408
attention decoding, 555
autoencoder, 178
Automatic differentiation (AD), 186
automatic speech recognition, 15
average pooling, 271

B

backward propagation, 151
bag-of-words, 94
batch normalization, 170
Bellman equation, 578
bias, 145
bias-variance tradeoff, 46
Bidirectional RNNs, 329
bootstrapping, 581

C

catastrophic interference, 478
CBOW, 209
CCA, 221
CER, 388

chain rule, 150
chunking, 100
circumplex, 110
classification, 5
coherence, 105
cohesion, 105
cold fusion, 557
computational graphs, 185
computational linguistics, 87
Conditional Random Field (CRF), 75
Confusion Matrix, 47
content-based attention, 547
Convolution, 264
convolutional neural networks (CNNs), 263
coreference, 106
Cross-correlation, 265
cross-validation, 51
CTC, 538

D

data augmentation, 173
data parallelism, 543
DDQN, 594
deep advantage actor-critic, 596
Deep Belief Network (DBN), 178
deep fusion, 556
deep policy gradient, 591
deep Q-learning, 592
Deep Speech 1, 541
Deep Speech 2, 543
denoising autoencoder, 180
differentiable neural computer, 428
dilated CNN, 287
discrete cosine transform, 375
distributional semantics, 203
doc2vec, 230

© Springer Nature Switzerland AG 2019
U. Kamath et al., *Deep Learning for NLP and Speech Recognition*,
https://doi.org/10.1007/978-3-030-14596-5

Domain adaptation, 495
domain adaptation, 467
domain shift, 495
double deep Q-learning, 594
DQN, 592
Dropout, 169
dynamic memory networks, 431
dynamic programming, 580
dynamic time warping, 379

E
Energy-Based Models (EBMs), 175
entailment, 112
ew-shot learning, 517

F
fast Fourier transform, 374
feature, 264
features, 40
fine-tuning, 469
finite-state transducer, 387

G
Gated Recurrent Unit (GRU), 325
generalization, 43
Generative adversarial networks (GAN), 182
global attention, 412
GloVe, 220

H
Hamming window, 373
Hann window, 373
hard attention, 412
Hidden Markov Models (HMM), 73
Hierarchical Softmax, 161
hierarchical softmax, 212
Hinge Loss, 162
human speech, 371
hyperparameter selection, 171

I
importance sampling, 583
inductive transfer, 471
invariance, 271

K
kernel, 264
KL Divergence, 162
KL Loss, 162

L
language model, 106, 385
Laplace smoothing, 107
LAS, 548

Latent Dirichlet Allocation, 115
Latent Semantic Analysis, 114
lemmatization, 92
lexical chains, 114
local attention, 413
location-aware attention, 547
Logistic regression, 61
Long Short-Term Memory (LSTM), 324

M
Machine learning, 5
Markov decision process, 576
max pooling, 271
Mean Squared Error, 161
Mean Squared Error (MSE), 149
Mel filter bank, 375
MemN2N, 422
Memory Networks, 419
MFCC, 372
model parallelism, 543
Momentum, 163
Monte Carlo, 582
morphology, 90
multi-head attention, 416
Multilayer Perceptron (MLP), 146
multitask learning, 7, 466

N
n-grams, 94
natural language, 88
natural language processing, 11
Negative log likelihood (NLL), 162
negative sampling, 213
NER, 102
neural stack, 434
neural Turing machines, 424

O
one-shot learning, 517
OOV, 108
optimal policy, 579
overfitting, 43

P
Parameter sharing, 266
parsing, 99
part-of-speech, 97
perceptron, 143
perceptrons, 8
perplexity, 108
phones, 377
Poincaré embeddings, 237
Pointer networks, 336
policy, 575

polysemy, 102
pooling, 271
pre-emphasis, 372
precision-recall curve, 49
Principal Component Analysis (PCA), 71
Probably Approximately Correct (PAC), 44
pronunciation model, 381

Q
Q function, 578
Q-Learning, 587
quantization, 228

R
re-scoring, 558
receiver-operating characteristic, 49
Rectified Linear Unit (ReLU), 158
recurrent entity networks, 437
Recurrent Highway Networks (RHN), 329
Recurrent Neural Networks (RNNs, 315
regression, 5
Regularization, 59
REINFORCE, 586
reinforcement learning, 7
Restricted Boltzmann Machine , 176

S
SARSA, 585
semantic role labeling, 104
semi-supervised learning, 7
sense2vec, 225
sequence-to-sequence, 334
shallow fusion, 555
short time Fourier transform, 371
sigmoid, 156
skip-gram, 211
softmax, 160
sparse autoencoders, 180
Sparse Coding, 182
spectrogram, 374
stacked CNN, 286
stemming, 91

stochastic gradient descent, 63
stop words, 93
subword embeddings, 228
Supervised learning, 5
Support Vector Machine (SVM), 68
synchronic model, 89

T
tanh, 157
temporal difference learning, 583
TFIDF, 96
tokenization, 92
training error, 42
transfer learning, 7, 465
Transformer networks, 337
treebank, 101
Turing Test, 8

U
unbiased error, 43
underfitting, 43
universal approximation theorem, 153
unspeech, 560
Unsupervised learning, 6

V
value function, 577
vanishing gradients, 324
variational autoencoders, 181
VC dimension, 45
vector space model, 203

W
wav2letter, 544
weight decay, 168
weight noise, 541
WER, 387
word2vec, 208

Z
zero-shot learning, 517
Zipf's law, 93

Printed in the United States
By Bookmasters